RF EXPOSURE
and *YOU*

By Ed Hare, W1RFI
ARRL Lab Supervisor

PUBLISHED BY:

THE AMERICAN RADIO RELAY LEAGUE

225 Main Street

Newington, CT 06111

CONTENTS

FOREWORD

Some radio amateurs were understandably concerned when they first heard about the new FCC RF exposure requirements. Their reaction could be summed up as: "Oh no—not more regulations. Pretty soon you'll have to have an engineering degree before you're allowed to screw in a light bulb!"

Others were more philosophical. "After all," they reasoned, "the new rules are designed to protect my health and that of my neighbors and family. Complying with them could be a bit inconvenient, but at least it will show my neighbors and family that my station is safe."

Whichever way you may have reacted, the new RF exposure rules are now a part of the regulatory landscape and are likely to remain so. The ARRL has done its best to ease the transition. ARRL Headquarters staff and volunteers participated fully with the FCC as it determined the best advice to give amateurs on how to meet the new requirements. We were able to persuade the FCC to reconsider its rules, and to rewrite them so that amateurs would be less affected.

This book was written to communicate one simple message: For the vast majority of Amateur Radio operators, the RF exposure rules *are not difficult to understand and follow.* This book has what you need—the background information, suggestions and worksheets to help you to comply with the new RF exposure rules. With this information, you will be able to operate your station legally and safely—and you *will* be able to operate.

David Sumner, K1ZZ
Executive Vice President
January 1998

PREFACE / ACKNOWLEDGMENTS

In August 1996, the FCC announced new rules governing exposure to transmitted radiofrequency signals. The new rules set new limits on the amount of RF energy people can be exposed to. They also require that some stations be evaluated to see if they are in compliance with the rules. Almost all existing amateur operation is already in compliance with the rules. The regulations, and this book, are based on these simple concepts.

When the RF exposure rules changes were announced, a new chapter in my life began. Having been appointed as ARRL's "point man" on the subject, I had a lot to learn in a short amount of time. We wanted to ensure that things went smoothly for the Amateur Radio Service. With the help of a lot of people, the ARRL was able to work effectively with the FCC to fine tune the regulations for the Amateur Radio Service. At the same time, we wanted to be prepared to help hams do what the rules require. For the most part, ARRL's actions were successful. The rules now in place take into account the ways most hams operate their stations.

Fortunately, I did not have to do this work alone! The ARRL RF Safety Committee is a group of willing volunteers who provided ARRL with input and guidance throughout this process. We also drew on a cadre of ARRL Technical Advisors and other experts in this field for help in all areas of our interaction with the FCC and the amateur community. Assistant Technical Editor Paul Danzer, N1II, edited the manuscript and prepared it for publication. In the long run, although I was given the privilege of leading ARRL's activities, this work—and this book—represents the work of the "best of our best." That is one of the greatest strengths of any volunteer organization, and it sure worked well in this case!

When the rules were first announced, there were a number of areas that needed "fine tuning." The rules were originally scheduled to go into effect on January 1, 1997. The ARRL asked for more time to give amateurs, and the FCC, time to prepare for this rules change. The FCC agreed, extending the date to January 1, 1998.

Originally, the rules required that all amateur stations running more than 50 W PEP be evaluated for compliance with the permitted exposure limits. We asked the FCC to vary that power level to match the way the exposure limits vary with frequency. The FCC agreed, revising the 50-watt threshold upward on most amateur bands. In addition, most mobile and amateur repeater operation also was exempted from the evaluation requirement. These changes did not represent compromises with safety; rather, they ensured that the final rules better complied with the FCC's intent.

As a brand-new "expert" on the subject, I have given a number of RF-exposure presentations at ARRL conventions and local radio club meetings. In almost all cases, my audience is aware that the rules exist, but don't know much about them. As I start to speak, I usually hear concerns that these rules are going to require difficult, complicated station evaluations. An hour later, when my presentation is over, those concerns are generally gone: Most hams realize that the station evaluation they dreaded so much is not at all difficult. I feel confident you will reach the same conclusion after you've read this book.

Ed Hare, W1RFI
ARRL Laboratory Supervisor

About the
American Radio Relay League

The seed for Amateur Radio was planted in the 1890s, when Guglielmo Marconi began his experiments in wireless telegraphy. Soon he was joined by dozens, then hundreds, of others who were enthusiastic about sending and receiving messages through the air—some with a commercial interest, but others solely out of a love for this new communications medium. The United States government began licensing Amateur Radio operators in 1912.

By 1914, there were thousands of Amateur Radio operators—hams—in the United States. Hiram Percy Maxim, a leading Hartford, Connecticut, inventor and industrialist saw the need for an organization to band together this fledgling group of radio experimenters. In May 1914 he founded the American Radio Relay League (ARRL) to meet that need.

Today ARRL, with more than 170,000 members, is the largest organization of radio amateurs in the United States. The League is a not-for-profit organization that:
- promotes interest in Amateur Radio communications and experimentation
- represents US radio amateurs in legislative matters, and
- maintains fraternalism and a high standard of conduct among Amateur Radio operators.

At League headquarters in the Hartford suburb of Newington, the staff helps serve the needs of members. ARRL is also International Secretariat for the International Amateur Radio Union, which is made up of similar societies in more than 150 countries around the world.

ARRL publishes the monthly journal *QST*, as well as newsletters and many publications covering all aspects of Amateur Radio. Its Headquarters station, W1AW, transmits Morse code practice sessions and bulletins of interest to radio amateurs. The League also coordinates an extensive field organization, which provides technical and other support for radio amateurs as well as communications for public service activities. ARRL also rep-resents US amateurs with the Federal Communications Commission and other government agencies in the US and abroad.

Membership in ARRL means much more than receiving *QST* each month. In addition to the services already described, ARRL offers membership services on a personal level, such as the ARRL Volunteer Examiner Coordinator Program and a QSL bureau.

Full ARRL membership (available only to licensed radio amateurs in the US) gives you a voice in how the affairs of the organization are governed. League policy is set by a Board of Directors (one from each of 15 Divisions). Each year, half of the ARRL Board of Directors stands for election by the Full Members they represent. The day-to-day operation of ARRL HQ is managed by an Executive Vice President and a Chief Financial Officer.

No matter what aspect of Amateur Radio attracts you, ARRL membership is relevant and important. There would be no Amateur Radio as we know it today were it not for the ARRL. We would be happy to welcome you as a member! (An Amateur Radio license is not required for Associate Membership.) For more information about the ARRL and answers to any questions you may have about Amateur Radio, write or call:

ARRL
225 Main Street
Newington CT 06111-1494
860-594-0200
Prospective new amateurs call:
800-32-NEW HAM (800-326-3942)
E-mail: **newham@arrl.org**
World Wide Web: **http://www.arrl.org/**

For questions about the content of this book, contact the ARRL Laboratory Staff at 860-594-0214 or **tis@arrl.org**.

1 Introduction

Meeting the requirements of the RF-exposure rules is not difficult for radio amateurs. The chapters of this book explain the requirements in a straightforward way. Most hams will be pleasantly surprised to learn they won't have to do an evaluation on their station! Even if you have to do an evaluation, it is usually as easy as filling out the simple worksheet at the end of this chapter and looking at a few tables.

On January 1, 1998, the FCC rules on RF exposure went into effect. This is a new area for Amateur Radio, so many hams have questions about the rules and what is required to comply with them. Driving many of these questions is a concern that amateurs must perform a difficult analysis of their stations. Although the new rules do require that some amateur stations be evaluated, the evaluation is not difficult! Hams can do their own station evaluations. No paperwork need be filed with the FCC, once the station evaluation is complete. The station evaluation is usually as simple as looking at a few tables to make sure the station's antennas are located far enough away from people.

This book was written to be the tool hams need to understand the rules, and to do their station evaluations. This book covers the basics, using easy-to-understand language. It also covers a lot of ground, and includes information on the more complicated aspects such as multi-transmitter sites and amateur repeaters. This chapter starts with a little background, followed by a narrative "table of contents" on each of the chapters. It ends with a set of worksheets and instructions most hams can use to complete their station evaluations easily.

The new rules introduce a few concepts that will be new for some hams, such as electric fields, magnetic fields, near fields, far fields, antenna patterns and other electromagnetics terms. Chapter 2 presents these concepts in a fresh and educational way. While not central to complying with the rules, this chapter helps explain the fundamentals that will feed the hunger most hams feel for new knowledge and understanding.

Chapter 3 covers RF safety. It explains that the rules are not a substitute for safety, much like the rules governing electrical wiring. Written by the ARRL RF Safety Committee, the chapter discusses the RF-safety practices most amateurs have followed for years.

Once the fundamentals have been covered, Chapter 4 explains the specific requirements of the rules in a simple, straightforward way. This chapter has been reviewed by ARRL's General Counsel and the ARRL Regulatory Information Branch—it uses "plain English" to help hams understand the general and detailed requirements in the new rules. Once all the rules have been explained, it is handy to have them readily available. You will find them in Appendix A, which contains the text of the FCC rules from Parts 1 and 2 as well as the more familiar Part 97.

The next chapter is the "core" of this book—how to conduct the required routine station evaluation required of some hams. Chapter 5 is the largest chapter in the book, not because a station evaluation is necessarily difficult (it is not!), but because the FCC permits hams to use any of several different methods. Each of these methods is presented in detail, along with a discussion of the pros and cons of each. Most operators will not need the sections on multi-transmitter environments or repeater sites, but when the subject comes up, this chapter will be a valuable reference.

Chapter 6 is a condensation of the FCC's *OET Bulletin 65: Evaluating Compliance with FCC Guidelines for Human Exposure to Radiofrequency Electromagnetic Fields*. This is the information bulletin the FCC issued to all services—not just hams. It contains the background, rationale and techniques for complying with the new rules. This condensation retains all the information that applies to radio amateurs. *OET Bulletin 65 Supplement B: Additional Information for Amateur Radio Stations* is reprinted in Chapter 7. Written, as its title suggests, for radio amateurs, it complements OET 65 and provides more detail to help radio amateurs comply.

If Chapter 5 is the main course of this book, Chapter 8 is the dessert. Most sta-

(Continued on page 1.6)

Worksheet A: Instructions — Categorical Exemption for Station Evaluation

Provided as a membership service by the American Radio Relay League, Inc., 225 Main St., Newington, CT 06111.

It is easy to determine if you need to do a routine station evaluation. The requirement to do a routine station evaluation is based on Table 1.1, showing peak envelope power (PEP) input to the antenna.

A, B, C: For your records, enter the call sign of the station (A) , the name of the station licensee (B) and station location (C) onto the top of the worksheet.

D. Enter the station operating frequency band being considered for evaluation (D).

E. Enter the maximum PEP output you use on that band (E).
(This can be determined by measurement or estimated from factors such as the rated output power of your transmitter. Alternatively, you can estimate from other factors. See Chapter 5, the section titled: "How to Calculate Peak Envelope Power to the Antenna.")

F, G. Enter your feed line type (F) and length (G).

H. Enter the specification for the loss in dB per 100 feet for your cable type. Use the manufacturer's specification or use the table in Chapter 5.

I. Divide the feed line length (G) by 100, then multiply the result by the specification for your feed line type for loss in dB per 100 feet. This will give you the total feed line loss in dB (I).

J. Enter the total feed line loss in dB (I) and convert it to a percentage (J).
(See the formulas or table in Chapter 5 or, optionally, you can use 0 dB for a conservative estimate. If you use 0 dB, skip to step J and enter 0%.)

K. Multiply the maximum transmitter PEP used on this band (E) by the percentage of power lost in the feed line (J). The result is the total power lost in the feed line (K).

L. Subtract the power lost in the feed line (K) from the transmitter PEP used on this band (E). The result is the PEP input to the antenna.

Compare the PEP input to the antenna (L) to the level in Table 1.1. If the power to the antenna is greater than the level in Table 1.1 for that frequency band, it will be necessary for you to perform a routine evaluation on your station. If your PEP to the antenna does not exceed the limits in Table 1.1, the rules do not require you to do a routine station evaluation on that band.

WORKSHEET A: CATEGORICAL EXEMPTION FOR STATION EVALUATION WORKSHEET
Provided as a membership service by the American Radio Relay League, Inc., 225 Main St., Newington, CT 06111.

Use this worksheet for each band you operate to determine if you need to do a station evaluation on that band.

(A): Station Call Sign: _____ (B) Station Licensee: _____

(C) Station Location: _____

_____

(D) Frequency Band: _____

(E) Maximum Transmitter PEP used on this band: _____ W PEP

Refer to Table 1.1 — If the power on line (E) of this worksheet is less than or equal to the power limits given in the table for this band, you do not need to do an evaluation on this band. If the power exceeds the limits, continue with this worksheet.

Calculate Feed Line Loss in dB:
(F) Feed Line Type: _____ (G) Feed Line Length: _____ ft

(H) Enter Feed Line Loss in dB per 100 ft: _____ dB
(From Chapter 5 or manufacturers specification. You can use 0 dB for a conservative estimate. If you use 0 dB, skip to step J and enter 0%.)

(G) _____	/ 100	×	(H) _____ dB	=	(I) _____ dB
Feed Line Length from (G)	divide by 100	then multiply by	loss in dB per 100 feet from (H)	equals	Feed Line Loss in dB

Convert to percentage:
(I) _____ dB	=	(J) _____ %
Feed Line Loss in dB from (I)		Convert to percentage of power lost in the feed line. See Chapter 5 or use 0% as a conservative estimate.

Power to antenna:
(E) _____ W PEP	×	(J) _____ %	=	(K) _____ W PEP
Maximum transmitter PEP used on this band from (E)	times	Percentage of power lost in the feed line from (J)	equals	Power lost in the feed line

(E) _____ W PEP	–	(K)_____ W	=	(L) _____ W PEP
Maximum transmitter PEP used on this band from (E)	minus	Power lost in feed line	equals	PEP input to the antenna

Conclusion and decision:
Compare the power input to the antenna (L) to Table 1.1. If the power input to the antenna is less than or equal to this power level, you do not have to evaluate your station on this band.

Worksheet B: Instructions — Station Evaluation Worksheet

Provided as a membership service by the American Radio Relay League, Inc., 225 Main St., Newington, CT 06111

If you do have to do a station evaluation for one or more powers or modes, use this worksheet to guide you through the process. This single page worksheet and instructions will suffice for many stations. See Chapter 5 for multiple transmitter sites and repeaters.

A, B. For your records, enter the call sign of the station (A), the station licensee (B) onto the top of the worksheet.

C. Enter the frequency band being evaluated.

D. Enter the operating mode being evaluated.

E. Enter the maximum transmitter peak-envelope power being used on this band (E). (See Chapter 5, the section titled: "How to Calculate Peak Envelope Power to the Antenna.")

F. Enter the peak-envelope power input to the antenna from line L of Worksheet A (F).
(As a conservative first estimate, you can skip to steps J and K, using this power level.)

G. Enter the duty factor of the mode being evaluated (H):
(See the section in Chapter 5 titled: "Duty Factor," or use 40% for CW, 20-40% for SSB, 100% for FM or digital modes.)

H,I. Enter the maximum percentage of time the station could be on the air for controlled or uncontrolled exposure. (A good rule of thumb is to use 100% for controlled exposure, 67% for uncontrolled exposure. Also see the table in Chapter 5.)

J, K. Calculate average power.
(Multiply the PEP input to the antenna (F) by the duty factor of the mode being used (G) by the operating time percentage (H, I). The result is the average power to the antenna.

L. Refer to any of the evaluation methods described in the FCC's *OET Bulletin 65* of Chapter 5. Determine that the antenna is located far enough away from areas where people are present or that the field strength is below the maximum permissible exposure (MPE) limits in areas where people are present. Describe briefly the method used to perform this evaluation.

M. Record the results of your station evaluation. Your station evaluation for this band and mode is now complete. Although it is not required by FCC rules, it is recommended that you retain a copy of your station evaluation in your station records.

If the station is not in compliance under all circumstances of its expected operation, attach a separate sheet describing any limitations of methods that the station operator will use to ensure compliance if people are present in areas that could be out of compliance.

WORKSHEET B: STATION EVALUATION WORKSHEET

Provided as a membership service by the American Radio Relay League, Inc., 225 Main St., Newington, CT 06111.

Use this worksheet for each band, mode and antenna combination you use to determine if your station complies with the FCC regulations for RF exposure.

(A): Station Call Sign: _____   (B) Station Licensee: _____

(C) Frequency Band: _____   (D) Operating mode being evaluated: _____

(E) Maximum Transmitter PEP used on this band: _____ W PEP

(F) PEP input to the antenna on this band (from line (L) on Worksheet A): _____W PEP
For a conservative estimate, you could use your maximum transmitter PEP and skip to step (L) and use this power for your evaluation. If you "pass," you do not need to do the other steps.

Mode and duty factor:

(D) Operating mode being evaluated: _____   (G) Duty Factor for this mode: _____%
(See Chapter 5 or use 40% for CW, 20% for SSB with no speech processing, 40% for SSB with heavy speech processing, 100% for FM or digital modes)

Maximum time the station could be transmitting in:
(H) 6-min period (controlled):   ____ / 6 = _____ %

(I) 30-min period (uncontrolled): ____ / 30 = _____ %

Calculate average power — Controlled exposure:

(F) _____ W PEP × (G) _____ % × (H) _____ %   =   (J) _____ W avg
PEP input to the   times Duty Factor   times Controlled   equals   Controlled average
antenna from (F)   from (G)   operating time   power input to the
                                                   percentage   antenna

Calculate average power — Uncontrolled exposure:

(F) _____ W PEP × (G) _____ % × (I) _____ %   =   (K) _____ W avg
PEP input to the   times Duty Factor   times Uncontrolled   equals   Uncontrolled average
antenna from (F)   from (G)   operating time   power input to the
                                                   percentage   antenna

(L) Refer to any of the evaluation methods in FCC's *OET Bulletin 65* or Chapter 5. Determine if the antenna is located far enough away from areas where people are present or that the field strength is below the maximum permissible exposure (MPE) limits, based on the frequency, mode, average power and antenna type being used.

(M) Describe the method used to do the evaluation: _____

Using this method, did your station exceed the FCC RF exposure limits? (Y/N)

Controlled exposure: _____(Y/N) Uncontrolled exposure: _____ (Y/N)

If the station is not in compliance under all circumstances of its expected operation, attach a separate sheet describing any limitations of methods that the station operator will use to ensure compliance if people are present in areas that could be out of compliance.

tion evaluations can be done by finding the table that best represents how a particular station operates. Then, by looking up the average power level used, you can determine if the station's antenna is located far enough away from people. Chapter 8 consists of about 200 tables, prepared using the same methods the FCC used for the sample tables published in their RF exposure information bulletins. There is one major difference between the FCC's tables and those in Chapter 8: Because there are far more of them, the tables in Chapter 8 provide a much greater level of detail and precision.

This book also contains a number of appendices. As noted above, the paragraphs of Part 1, Part 2 and of course Part 97 of the FCC Rules that apply to the Amateur Radio Service are printed in Appendix A. To provide a full understanding of where we are, and how we got here, a condensation of three FCC documents in ET Docket 93-62 is included as Appendix B: *FCC 96-326 Report and Order*, August 1, 1996; *FCC 96-487 First Memorandum Opinion And Order*, December 23, 1996, and *FCC 97-303 Second Memorandum Opinion And Order And Notice Of Proposed Rulemaking*, August 25, 1997.

The new FCC Form 610, with the now *mandatory* radiation safety statement that all applicants must sign, is in Appendix C, along with Forms 610-A and 610-B. Appendix D is a list of FCC information sources on radiation safety and a set of *FAQ*s (frequently asked questions) and their answers from the FCC files. Of course, an extensive Resources section (Appendix E) is included—it gives names, addresses, telephone numbers, e-mail addresses and Web page addresses for all companies and organizations mentioned in this book.

To put this book together, a balance had to be struck between ease of use and completeness. The pages printed here represent this

TABLE 1.1

Wavelength Band	*Evaluation Required if Power* (watts) Exceeds:*
MF	
160 m	500
HF	
80 m	500
75 m	500
40 m	500
30 m	425
20 m	225
17 m	125
15 m	100
12 m	75
10 m	50
VHF (all bands)	50
UHF	
70 cm	70
33 cm	150
23 cm	200
13 cm	250
SHF (all bands)	250
EHF (all bands)	250
Repeater stations (all bands)	*non-building-mounted antennas*: height above ground level to lowest point of antenna < 10 m *and* power > 500 W ERP *building-mounted antennas*: power > 500 W ERP

*Power=PEP input to the antenna except, for repeater stations only, power exclusion is based on ERP (effective radiated power).

balance. Those hams who need only a simple answer will find this answer presented clearly in this book. Those who need to learn about the complete picture, or more detail about any part, will find it here, too.

All radio amateurs must decide if they have to perform an evaluation, or if their station—for the bands and modes used—is categorically exempt. To show you how simple it is for most hams, Worksheet A in this chapter lets you make this determination in few easy-to-follow steps. If the resulting answer is "yes," and evaluation must be made, Worksheet B tells you how to go about it.

2

Basic Electromagnetic Theory

This chapter explains the theory behind electromagnetic fields, antennas and the regulations. This foundation will help you understand the requirements of the rules.

This chapter was written by Kai Siwiak, KE4PT, a professional engineer working in the field of electromagnetics.

INTRODUCTION

As you approach the topic of RF-protection guidelines, standards and regulations, you will need to understand a few basic properties of electromagnetic (EM) fields and waves. In this chapter we will first develop the concepts of electric and magnetic fields. Then we'll relate them to their sources, which are electric currents and electric charges.

Although the behavior of electromagnetic fields can be described very precisely with just a few very compact, but complex equations, we will not go into the really heavy math here. For those of you who do want to dive into the governing equations, a good reference textbook on antennas will be a valuable asset.[1,2]

BASIC RADIO-WAVE AND ANTENNA TERMS

First, we will define some terms commonly encountered in the study of antennas and radio-wave propagation. These definitions are consistent with industry standards and with common engineering usage.[3,4] The definitions relate antenna and transmission-line currents and charges to electromagnetic fields.

Definitions for Impedance

Impedance with regard to transmission lines and electromagnetic fields is defined in terms of where it is applied. On a transmission line, such as coaxial cable, twin lead or open-wire line:

- *Characteristic Impedance* (Z_0) is defined as the ratio of voltage to current on a transmission line. It is a property related to the physical construction and dimensions of the transmission line.
- *Intrinsic Impedance* is defined as the ratio of the complex amplitudes of the *electric* and *magnetic* fields for a plane wave in an unbounded medium. The ratio is $\eta_0 = 376.730313\ \Omega$ in a vacuum (and essentially the same value in air). Intrinsic impedance is a property of the medium, not of the fields.
- *Wave Impedance* is defined as the ratio of the electric-field component to the magnetic-field component at the same point of the same wave. For a plane wave in unbounded space (no boundaries or conductors), the wave and intrinsic impedances are the same.

Some examples should help you understand these various impedances a bit better. Coaxial transmission lines are constructed so that there is a certain parallel capacitance per unit length, as well as a certain series inductance per unit length of the line. The characteristic impedance is the square root of the ratio of the inductance to capacitance, or $Z_0 = \sqrt{L/C}$. For coaxial lines used in most amateur applications, the characteristic impedance is between 50 to 75 Ω, with 50-Ω cable the most common. The characteristic impedance is determined by the physical dimensions chosen by the manufacturer of a particular cable. This includes the ratio of inner-to-outer conductor diameters and the dielectric constant of the insulating material between the center conductor and the outer shield.

When we measure the voltage between the inner and outer conductors of a 50-Ω coax and divide by the current flowing through the inner (or outer) conductor, the answer will be 50 Ω. This is only true when the cable is match-terminated in a resistive load equal to the characteristic impedance. If the transmission line were not match terminated, the ratio of the voltage to the current will not equal the characteristic impedance—we will have standing waves on the line. For example, on a mismatched 50-Ω line, the voltage divided by the current can be greater than or less than 50 Ω, but the characteristic impedance of the line itself always remains 50 Ω.

The same concept is true of electromagnetic waves traveling in space. We define the intrinsic impedance of free space as the product of free space permeability $\mu_0 = 4\pi \times 10^{-7}$ henry/meter (H/m) and the free space velocity of light c = 299,792,458 meters/second (m/s). These are *exact* physical constants,[5] and the exact answer is 376.730313 Ω. We usually round the number off to 376.7 or even 377 Ω.

If we were to use 300,000,000 m/s for the speed of light, we would get 120 π as an approximation for the intrinsic impedance of free space. A wave traveling in unbounded free space, with no boundaries or reflections, will have an electric-field to magnetic-field ratio of 376.7 Ω, by definition.

But here's the kicker—"unbounded free space" hardly ever exists—except in textbooks, because we are always near one *boundary* or another, such as ground, human bodies, houses, cars, and trees, even the pet cat and dog. Here's another kicker:

Those boundaries—cars, trucks, dogs, and so on—often move a lot.

Boundaries are where dielectric constants change, or permeability changes, or conductivity changes. Air-to-earth is a boundary; air-to-a-conducting-metal-sheet is a boundary. They behave just like mismatches on a transmission line. Waves reflect from these boundaries in a complicated way and form standing waves in space. The electric fields and the magnetic fields reflect differently at boundaries, just like voltages and currents reflect differently from mismatches on a transmission line.

For example, when a wave travels in air to a metal-wall boundary, it reflects and sets up a standing wave in air. The electric field component standing wave has a deep *null* at the boundary and nulls every half-wave length away from the boundary. The magnetic-field component, on the other hand, has a *maximum* value at the boundary and minimum values one quarter of a wavelength from the boundary, as well as every half wavelength away from that minimum.

This picture should sound familiar, since it is just like the voltages and currents on a short-circuited transmission line. The electric fields and magnetic fields have peak and nulls that are a quarter wave length out of step with each other. At any point in that kind of a field the wave impedance (remember, that's the electric-field magnitude divided by the magnetic-field magnitude at a point) can range from nearly zero at an electric-field null to an extremely high value at a magnetic-field null. This is the same behavior of standing waves on a transmission line.

This is the wave situation everywhere in our environment. And it constantly changes because many of our boundaries are in motion. Furthermore, there are always other conductors nearby, perhaps power, telephone lines, or other antennas. All these interact with the operation of the antenna. Because of all these boundaries and parasitic conductors, all real-world factors, an exact, deterministic assessment of electromagnetic fields everywhere is impossible, or at best, pointless.

The more boundaries there are, the more reflections there are, and the more complicated the standing-wave EM fields picture becomes. We've all experienced one common manifestation of this messy field picture in the form of "picket fencing" when we are operating mobile. This is the rapid signal-strength variation—sounding like someone dragging a stick along a picket fence—that we often find in weak-signal areas of VHF and UHF repeaters as the vehicle moves.

The wave impedances that amateurs encounter are rarely equal to 376.7 Ω, because we are always near one boundary or another. The most obvious boundary is the ground itself. The wave-impedance variation of fields near boundaries complicates our discussions of RF-exposure compliance later on. It is also one reason why the FCC standards treat exposure to electric and magnetic fields separately.

Definitions for Near and Far Regions of Antennas

Now let's turn our attention to some antenna terminology. For the moment, it will be most convenient to talk about antennas that are in "unbounded free space." Unbounded free space is a region that has neither boundaries (such as ground) nor other conductors. Later on, we'll examine in more detail the boundaries and conductors that were discussed in the previous section.

There are many ways of defining the *far field* of an antenna, depending on what concept we trying to portray. Let's see if we can "sneak up" on an understanding of the far field. Imagine a film or slide projector whose focus knob has been set to "infinity." The lens of the projector is analogous to an antenna. If you move the screen to a large distance from the projector, simulating an infinite distance, the image on the screen will be in perfect focus.

Now, move the screen closer to the projector lens. There is a range between infinity and some distance away from the projector where the image on the screen will remain pretty much in focus. This is the far-field region. How far is the far field? Well, how much image defocusing can you tolerate and still maintain that it is essentially in focus? This is a subjective definition.

Now move the screen even closer—right up to the projector lens. There is a smooth transition between the far-field image and the near-field image. The near-field image does not much resemble the far-field image. The light distribution across the lens surface is exactly equivalent to the currents and voltages on an antenna.

• The *far-field region* is defined as that region of the field of an antenna where the angular field distribution is essentially independent of the distance from the antenna.

To continue the projector analogy, we are in the far field when the image is acceptably in focus at some distance. Even if we move further away, the image on the projector screen will not change appreciably. This is admittedly a vague, subjective

sort of definition, but so is the distinction between near and far.

There are measurable things that happen when we enter the far field. For one, the wave impedances of the antenna pattern closely approach 376.7 Ω, *as long as no other boundary, like the ground, is present.* Remember, wave impedance is the ratio of the electric and magnetic fields at the same point.

Another subjective observation about the far field is that when the antenna is viewed from the far field, it appears to be small. In more scientific terms, an antenna viewed from the far field subtends a small angular extent and the distance between any point on the antenna so far as the observer is concerned is essentially the same. When viewed from the far field, the details of the antenna are not apparent to the observer.

Textbooks and standards also define other regions like *induction zone*, *reactive near-field region*, *radiating near-field region*, *Fraunhöfer region*, and *Fresnel region*, depending on the antenna concept that they need to explain. Of these, the definition of reactive near-field region or simply the near-field region is of interest to amateurs,

• *Near-field region* is defined as that region of the field immediately surrounding the antenna, wherein the reactive field dominates and where the angular field distribution is dependent upon distance from the antenna.

In the near field, the physical details of the antenna dominate the fields picture. We will touch on the near field again when we explain antenna *Q*. For now, imagine the antenna to electrically resemble a parallel tank circuit—an inductor, a capacitor and a resistor in parallel. At the resonant frequency, energy swaps between the magnetic field of the inductor and the electric field of the capacitor once every half RF cycle, with a portion of the energy dissipated in the resistor.

The energy dissipated in the parallel resistor analogy corresponds to the energy actually radiated into the antenna's far field. The fields swapping between electric and magnetic energy around the antenna are confined to the immediate vicinity of the antenna—just like the electric and magnetic fields in the tank-circuit example. Indeed, for this reason unshielded tank circuits are potential sources of RF exposure!

Just as there are measurable characteristics for the far field, there are also measurable characteristics for near fields. Like the voltages and currents in the inductor and capacitor of a tank circuit, the fields immediately near an antenna are reactive. That is, the electric and magnetic fields are 90° out of phase with each other (in *phase quadrature* or simply a quarter of a cycle out of step with each other).

People and objects within the *reactive near-field region* of an antenna will interact with an antenna, as mentioned previously. Among other things, they can influence the antenna's feed-point impedance. How far does the near field extend? The answer is again subjective—hence the many textbook definitions of the regions, depending on the particular concept that needs explaining. In the practical sense, the reactive near field diminishes to a strength below that of the radiating field components (which supply energy to the far field) within a third of a wavelength or so of the physical extent of the antenna.

ELECTRIC AND MAGNETIC FIELDS

As the name implies, there are two "genders" of electromagnetic (EM) fields—electric and magnetic. Let's look at some of the properties of EM fields by first considering *electric charges* and their properties. Electric fields are lines of force connecting charges of opposite polarity, like the lines labeled E in Figure 2.1A. The electric force is defined in units of volts per meter (V/m). A single-polarity static charge, such as that shown in Figure 2.1B, has electric lines of force E extending outward to infinity. The lines are all radially directed away from the charge. There is a definite mathematical relationship, called Gauss's Law (one of Maxwell's equations), that allows us to compute the electric field strength if we know the amount of charge. For now, however,

that's a detail.

Magnetic fields are lines of force that encircle *moving charges*. For example, look at the field lines labeled H encircling the moving charges represented by the current I in Figure 2.1C. Again, there is a precise mathematical relationship, called the Ampere-Maxwell Law, between the amplitude of the current and the amplitude of the magnetic field.

The magnetic field forms paths surrounding currents and has units of amperes per meter (A/m). We call moving charges, of course, *electric currents*. When the current is a constant direct current, the magnetic field is constant. In this discussion, we'll ignore permanent magnets and dc systems, since they don't radiate EM fields. The relationships between charges and electric fields, and between currents and magnetic fields, are described by two of Maxwell's famous four equations.

Current Flow and Charge Accumulations

The real fun begins when the charges are made to move in a regular fashion, with the resulting currents varying in amplitude as a function of time. Specifically, let's assume the current varies in direction and in amplitude according to the mathematical *sine* or *cosine* function. We'll consider an ac current supplied to a resistor. The resistor here represents the *radiation resistance* of the antenna.

Figure 2.2 shows two cycles of the sine function, just like you could see using an oscilloscope. The trace shows the voltage created by the current flowing through the resistor and is called a *sinusoidally alternating* voltage. The rapidity with which

the current-flow changes from one direction to the opposite direction and back again is called the *frequency* (in cycles per second or Hz). Note that Figure 2.2 shows two peaks A and C to the sine amplitude that are the square root of two higher in amplitude than the RMS value. In our discussions, we will always refer to the RMS value of sinusoidal currents, voltages and field strengths.

The Resonant Half-Wave Dipole

Varying currents—charges that are moving—produce electric and magnetic fields that vary with time. Let's look at a simple resonant half-wave long dipole (and we'll define what that means later), fed in the center with a sinusoidal current. See Figure 2.3. The particular instant in time shown in Figure 2.3 corresponds to the point that the current at the terminals of the antenna has risen to its maximum value, point A in Figure 2.2.

The current flows out one feed-point terminal and into the other terminal. Each terminal is *balanced* with respect to ground. The current is always zero at the dipole ends, so it builds up first positively as shown in Figure 2.3, then a half cycle later negatively along the dipole length. At the peak of one such build-up, we can "freeze" the picture in time and measure the current amplitude along the wire—as it should be, the amplitude is shaped like a cosine function. It is zero at each tip and maximum in the middle.

The corresponding magnetic field lines encircling the dipole are shown in Figure 2.4A, where the size of the magnetic-field circles correspond with the current amplitude on the dipole. What do the two fields

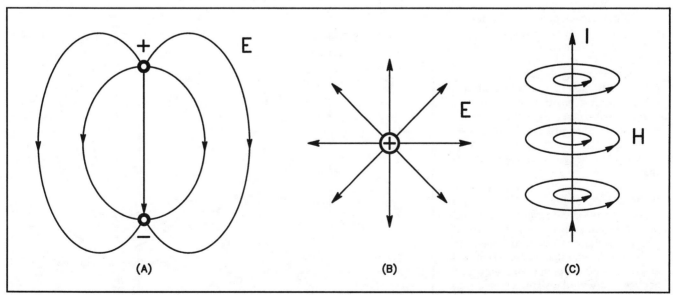

(A)　　　　　　　　(B)　　　　　　　　(C)

Figure 2.1—At A, E field of two charges; at B, E field of one charge; at C, magnetic field H of a dc current I.

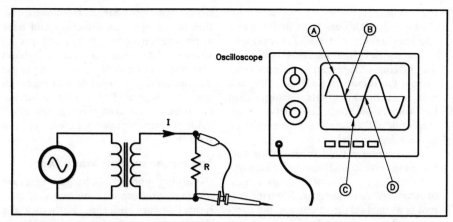

Figure 2.2—A sinusoidal alternating current, flowing through a resistor to create a voltage that can be measured by an oscilloscope. The resistor represents the antenna radiation resistance.

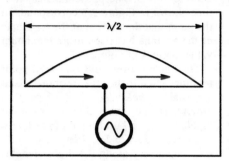

Figure 2.3—Currents on a half-wavelength resonant dipole antenna fed by a current source.

look like at that snapshot in time? Two of Maxwell's equations tell us that at the dipole wire, the magnetic field wraps around the wire and the field intensity right at the wire surface equals exactly the current density. The magnetic field right at the wire surface is the current there divided by the wire circumference. We've just found the *exact* value of the near magnetic field of the dipole in terms of the current right at the dipole surface! Finding the electric field, however, is not so easy, and will not be attempted here.

A quarter of a cycle later, as shown in Figure 2.4B, the charges have peaked at the dipole tips: positive on one tip and negative on the other tip. At the feed point, the current is zero, just before reversing in direction. At this instant of time, the magnetic field at the feed point on the dipole surface is zero.

One quarter of a cycle later, corresponding to point C of Figure 2.2, the current is at a negative maximum as the field picture of Figure 2.4B implies, but with the arrow directions reversed. And one quarter of a cycle after that, at point D of Figure 2.2D, the field picture of Figure 2.2B applies, but with the signs reversed. The whole process repeats for each and every full cycle current flow.

We also realize from the above description that the current peaks and the charge peaks are out of step in time by a quarter cycle in time. This means that the peak magnetic and peak electric fields right at the dipole are also a quarter cycle out of step. That is another distinguishing characteristic of near fields, as we discovered in the previous section. In the near field, the magnetic and electric fields are 90° out of phase, in phase quadrature.

We see here that the places on the dipole where current peaks occur (and also magnetic field peaks) are different from the places on the dipole where the charges accumulate (peak electric fields occur). Now look back to the previous section at the definition of *wave impedance*. The wave impedance varies drastically over the length of the dipole wire. The tip of a dipole is clearly a high wave-impedance point, while the feed point exhibits a relatively low wave impedance, more commonly known as the feed-point impedance.

Velocity of Moving Charges on an Antenna

Let's now look at the velocity of the moving charges on a half-wave antenna. For our purposes here, it is accurate enough to say that the charges travel along our dipole wires at the velocity of light, where $c = 299,792,458$ meters/second (approximately one foot every nanosecond). Look back at the oscilloscope picture of a sinusoidal current wave form in Figure 2.2. The *time T* taken for one complete cycle to occur is $T = 1/f$. In English, this means that the cycle duration is the inverse of the frequency. The *distance d* that a charge travels in a full cycle's worth of time T is $d = cT$. This is defined as the *wavelength* λ (Greek letter lambda), so $\lambda = c/f$.

On our half-wavelength dipole, the time it takes for the charges to move from the feed point to the tip is one quarter of an RF cycle. Thus the length of each half of the resonant dipole must be $L = \lambda/4$ and the total length of the resonant half-wave dipole is $\lambda/2$.

So far we have seen that oscillating charges produce electric fields between opposite sign charges, and magnetic fields wrapping around currents. Now we need to introduce another two physical concepts: Faraday's law of induction, and Ampere's law. Stated in plain English, a time-varying magnetic field causes an electric field to try to wrap around the magnetic field lines, and a time-varying electric field causes a magnetic field to try

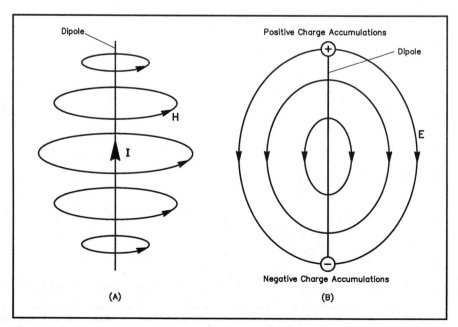

Figure 2.4—Currents and charges on a dipole antenna.

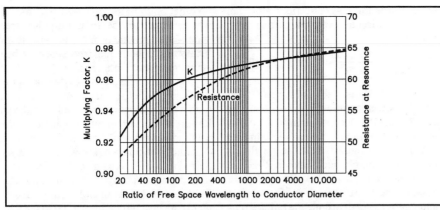

Figure 2.5–The solid curve shows the factor, K, by which the length of a half wave in free space should be multiplied to obtain the physical length of a resonant half-wave antenna versus the wavelength-to-diameter ratio. This curve does not take end effect or element tapering into account. The broken curve shows how the radiation resistance of a half-wave antenna varies with the wavelength-to-diameter ratio.

to wrap around the electric field lines. Thus once launched, a moving magnetic field will generate a moving electric field and vice versa in an ever-expanding dance that we call *wave propagation*.

Stored and Radiated Electromagnetic Fields

These near fields close to an antenna are a quarter of a time-cycle out of step with each other. The electric and magnetic fields swap between stored electric and stored magnetic energy in the immediate volume around the dipole, very much like the electric and magnetic fields in a simple *LCR* tuned circuit described earlier.

Now we can see that, once launched, those *reactive fields* will give rise to an expanding picture of *radiating fields* that sap away some fraction of the energy stored in the *reactive near field* and propagate it away from the dipole in an ever expanding sphere of *radiating energy*.

The Concept of Q

How much energy is stored in the vicinity of the dipole? Remember the parameter Q? Its precise mathematical definition is:

$$Q = 2\pi \frac{\text{Total stored energy}}{\text{Energy dissipated in one cycle}}$$

$$(Eq 1)$$

Q is a useful figure of merit when we speak of inductors and capacitors, and the same exact concept holds here for antennas. We can see now that if the Q of the antenna is high, a rather substantial fraction of energy is being swapped between the electric and magnetic fields (like in an LCR tank circuit) in the vicinity of the antenna, compared to the amount of en-

ergy radiated. Of course, the energy supplied to the antenna feed-point terminals replaces the energy dissipated. This radiated energy is associated with a load resistance (corresponding to the R in the tank circuit) that we call the radiation resistance of an antenna.

For a resonant, infinitesimally thin half-wave dipole having a sinusoidal current distribution, that radiation resistance is $73.08\ \Omega$. This is the resistance that appears as a load resistor at the dipole's feed point. The radiation resistance depends on the actual wire thickness and this also affects the resonance length for the dipole, as shown in Figure 2.5.

The wire thickness also affects the dipole reactance as shown in Figure 2.6. We can even find the Q of these antennas from an equivalent definition of Q based on the 3-dB bandwidth of a tuned circuit. Let's refer to curve A of Figure 2-6 for this

example. The resonant frequency f_R corresponds to the antenna length where the reactance is zero, $f_R = c/(2\times0.497)$. The frequency where the dipole resistance equals the negative reactance is $f_N = c/(2\times0.474)$ and the frequency where the dipole resistance equals the reactance is $f_P = c/(2\times0.51)$, and c is the velocity of light.

The Q is then:

$$Q = \frac{f_R}{f_N - f_P} \qquad (Eq 2)$$

or nearly 14. From our previous definition of Q, there is $Q/2\pi$ or about twice as much stored energy compared with radiated energy for the particular dipole represented by curve A. The near field of antennas, closer than a third of a wavelength or so, does need special consideration when dealing with RF compliance issues.

WHAT ABOUT THE RADIATING FIELDS OF ANTENNAS?

We will look into the radiation picture of antennas in this section. We will concentrate on dipoles because other antennas, like Yagis, are no more than collections of dipoles with slightly different lengths and certain physical spacings. The nature of a *spherical wave* must first be explored. Remember from the above discussion that a fraction of the energy stored around a dipole radiates away? This is replaced by the energy that we supply to the antenna with the transmitter. How is this radiated energy distributed in space?

Let's look at a sphere that is big enough to enclose our antenna. For the moment imagine that the energy radiates equally in all directions from our antenna, somewhat like a bare light bulb. The energy per unit time, that is, the *power*, flowing out of that

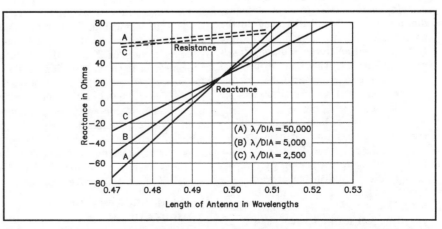

Figure 2-6—Resistance and reactance at the input terminals of a center-fed antenna as a function of its length. As shown by curves A, B and C, the reactance is affected more by the λ-to-dia ratio of the conductor than is the radiation resistance.

sphere exactly equals the power supplied to the antenna feed point. Imagine now a second sphere at twice the radius of the first surface. We see that because of the conservation of energy, the same total power also exits the second sphere. The second sphere, at twice the radius of the first surface, has four times the surface area. Therefore when we doubled the distance from the antenna (first sphere to second sphere), the area through which our constant power is flowing quadruples because the area of a sphere is proportional to the square of radius ($A = \pi R^2$). Thus the *power density*, measured in watts per square meter (W/m²), decreases with the square of distance for the radiating energy. Power density P_d (W/m²) is thus proportional to the square of field strength (V/m and A/m). When the intrinsic impedance is 376.7 Ω, and we are in unbounded free space, the wave impedance is also 376.7 Ω.

$$P_d = \frac{\text{Power supplied}}{4\pi R^2} = \frac{E^2}{376.7} = H^2 \times 376.7$$

(Eq 3)

The radiated power density decreases inversely with the square of distance from the antenna. We can take the square root of each term, so

$$E = \frac{\sqrt{P \dfrac{376.7}{4\pi}}}{R}$$

(Eq 4)

where *P* is the power supplied to the antenna, and

$$H = \frac{\sqrt{\dfrac{P}{4\pi \times 376.7}}}{R}$$

(Eq 5)

and we find that the radiating fields (those carrying energy away from the antenna) diminish inversely with distance from the antenna. To be sure, there are reactive fields in addition to the *radiating fields* near the antenna. These make the total fields picture very close to the antenna more complicated, as we discussed in the previous sections.

Dipoles don't radiate equally in all directions; they exhibit directionality. We can find the magnetic field in the peak radiation direction by applying principles already discovered in the previous sections. The first principle is that the magnetic field at the feed point of the dipole is exactly equal to the current density divided by the dipole wire's circumference. The second is that the field diminishes inversely with distance R. The exact ex-

pression for the magnetic field along an axis in the direction of peak gain is:

$$H = \frac{I}{2\pi R} = \frac{\sqrt{P/73.08}}{2\pi R}$$

(Eq 6)

where I is the dipole feed point current and P is the power supplied to the dipole. The 73.08 term is the feed-point impedance for an infinitely thin half-wave dipole in free space. The formula is suitable for determining the compliance distance R meters for a radiated power P watts and field compliance limit H in A/m. Our dipole antenna does not distribute the radiated energy omnidirectionally in space, but rather focuses it, primarily in the directions perpendicular to the dipole length. The focusing pattern, although complex in the immediate vicinity of the antenna, eventually becomes constant with distance (see the definition of far-field region again).

We call the peak value of the focusing factor the *antenna gain*. For the resonant half-wave dipole we can compare the peak field given by the right-hand side of Eq 6 with the omnidirectionally radiated field given by the right-hand side of Eq 5. The result is the dipole gain:

$$G = 20 \log \left[\frac{\sqrt{4\pi \times 376.7 / 73.08}}{2\pi} \right] = 2.15 \text{ dBi}$$

(Eq 7)

Eq 7 states that the far fields of a dipole in the maximum radiation direction, 0° and 180° in Figure 2.7A are 2.15 dB stronger than would occur if the same energy were radiated equally, like our bare light bulb, in all directions. The bare light bulb is analogous to an *isotropic radiator*, one that radiates equally well in all directions with no directionality. The gain of a dipole referenced to an isotropic radiator is 2.15 *dBi* (dB referenced to isotropic).

The exact expressions for the electric fields, both near and far-field, from a sinusoidally excited, resonant dipole are somewhat more complex than that for the magnetic field in Eq 6. See Section 11.4.5 in Reference 6, for example.[6] The expressions for a dipole of arbitrary length and arbitrary thickness are exceptionally complex, however.[7] Exact expressions for loop antennas are also available, and are similarly complex.[8] Radio amateurs rely on numerical solutions using computer codes like *MININEC*, *NEC* and similar programs.

FIELD MEASUREMENTS AND COMPUTATIONS

Up until now, the discussion about fields and radiation from a dipole has been very straightforward because there were

no other boundaries involved. The dipole was assumed to be in unbounded free space—but practical antennas are never located in unbounded free space. When actual measurements are attempted, other objects can couple parasitically to alter the readings. Even the very instruments (not to mention the operator!) used to make an EM-field measurement can get into the act to affect the readings. Great care and a detailed, specialized knowledge are required to configure and calibrate an accurate measurement of electromagnetic fields. For this reason EM-field measurements are best left to the specialists. Radio amateurs are better served by computational methods than by measurements.

The electromagnetic fields around antennas can be very accurately calculated using readily available computer software. Computer antenna modeling programs such as *MININEC* and other codes derived from *NEC* (the Numerical Electromagnetics Code) are very suitable for estimating magnetic and electric fields around amateur antenna systems. You must be sure to include the effects of ground, and also to recognize that waves reflect from *all* surfaces, including walls and vehicles. All surfaces and conductors should be explicitly included in the computer models.

We have seen in Figures 2.1 and 2.4 that fields have an orientation in space. Computer programs such as *NEC* generally are based on a Cartesian, or (x, y, z), coordinate system so geometries and fields are expressed in x, y and z components. For the purpose of RF compliance when using computer codes, be sure to calculate the *total* fields at a particular point. From the exposure standard point of view, we are interested in:

$$E_{\text{total}} = \sqrt{|E_x|^2 + |E_y|^2 + |E_z|^2}$$

(Eq 8)

and

$$H_{\text{total}} = \sqrt{|H_x|^2 + |H_y|^2 + |H_z|^2}$$

(Eq 9)

because that is how the fields interact with biological tissues. The symbol |E| means "the absolute value of the magnitude of E." Eq 8 and 9 are applicable both to near-field and far-field regions, as well as near reflectors, such as the ground.

HOW THIS ALL RELATES TO EXPOSURE STANDARDS

In transmitting applications, the measure of the rate at which energy is absorbed by the human body is called the *specific*

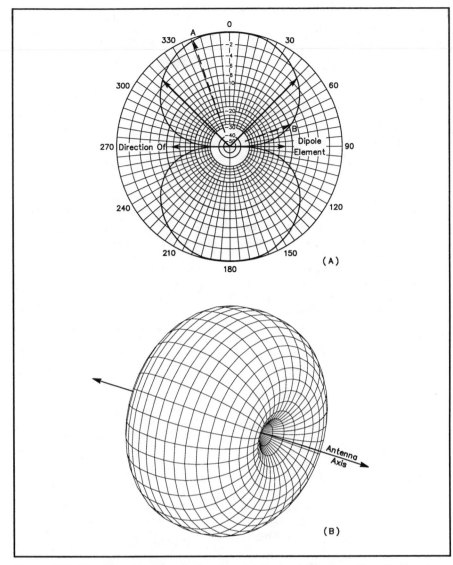

Figure 2.7—Directive diagram of a free-space dipole. At A, the pattern in the plane containing the wire axis. The length of each dashed-line arrow represents the relative field strength in that direction, referenced to the direction of maximum radiation, which is at right angles to the wire's axis. The arrows at approximately 45° and 315° are the half-power or –3 dB points. At B, a wire-grid representation of the "solid pattern" for the same antenna. These same patterns apply to any center-fed dipole antenna less than a half wavelength long.

absorption rate (SAR). This is defined as the time (t) derivative of incremental energy (dW) absorbed by an incremental mass (dm) contained in a volume element (dV) of a given density (ρ). In terms of an impressed RMS electric field E in V/m, on a dielectric material of conductivity σ in siemens/meter (S/m), and ρ mass per volume in kilograms per cubic meter (kg/m³), the SAR is:

$$SAR = \frac{\sigma E^2}{\rho} \qquad (Eq\ 10)$$

expressed in watts/kilogram (W/kg). You should note that frequency doesn't enter into the SAR. The basic premise of modern standards is that the severity of an ef-

fect is directly related to the rate of RF energy absorbed, hence the introduction of the concept of SAR. Fields external to the body are not easily related to fields like E of Eq 10 inside the body. The determination of SAR is thus complex and often relies on precise measurements, the details of which are beyond the scope of this description. Since we can not easily determine SAR, we fall back to the second line of defense, and rely on exposure guidelines that have safety factors included with respect to the SAR levels. It is the compliance to those exposure standards, codified by the FCC regulations, that is of interest to us as radio amateurs.

Biological tissues subjected to RF en-

ergy will absorb energy and convert it to heat, governed by the SAR Eq 10 above. External fields couple most efficiently to the body when the electric field is aligned with the body length in the whole-body half-wave resonance range. The upright human body acts effectively like a lossy half-wave dipole element.

For adult humans this occurs between 35 MHz for a grounded person and about 70 MHz for a person isolated from the ground. For small infants the resonant range extends upwards in frequency, so special attention is paid to RF exposure in the entire human whole-body resonant-frequency region between 30 to 300 MHz. Additionally, body parts may exhibit resonant behavior. The adult head, for example is resonant around 400 MHz, while a baby's smaller head resonates near 700 MHz.

Body size thus determines the frequency at which RF energy is absorbed most efficiently. As the frequency is increased above resonance, less RF heating generally occurs. Because RF skin depth decreases with increasing frequency, heating is increasingly confined to surface tissue. All these factors have led to RF exposure guidelines whose limiting levels of power-density exposure vary with frequency, and in some cases, different exposure limits for electric and magnetic fields.

In short, tissue heating is the primary effect of concern in the RF electromagnetic fields standards, SAR is the relevant mechanism, and the fields external to the tissue that give rise to the SAR are what we attempt to control to meet the relevant RF-exposure standards. The rest of this book is devoted to those topics of compliance.

Glossary

Controlled environment—An RF-exposure environment in which the people being exposed to an RF field are aware of the potential for exposure. Members of the household of the Amateur Radio station are considered to be in a controlled environment.

Duty cycle—The ratio between the actual RMS value of an RF signal and the RMS value of a continuous signal having the same PEP value, expressed as a percentage. A duty cycle of 100% corresponds to a continuous-wave (CW) signal.

Duty factor—The ratio of pulse duration to the pulse period of a pulse train. A duty factor of 1.0 corresponds to continuous-wave (CW) operation. [IEEE C95.1 1991]

Electric Field (E Field) Strength—This is the electromagnetic field resulting from the charge distributions present on a radiating element. It is the field vector quantity that represents the force (F) on a positive test charge (q) at a point, divided by the charge: $E = F/q$. Electric field strength is expressed in volts/meter. [IEEE C95.1-1991]

Far field—Is that region of the field of an antenna where the angular field distribution is essentially independent of the distance from a specified point in the antenna region [IEEE Std 145-1993]. In this region (also called the free space region), the field has a predominantly plane-wave character. That is, locally uniform distributions of electric field strength and magnetic field strength are in plane transverse to the direction of propagation.

Magnetic Field (H Field) Strength—This is the electromagnetic field resulting from the current distribution on a radiating element. It is the field vector quantity that results in a force (F) that acts on a charge (q) moving with velocity v, multiplied by the permeability (μ) of the medium and the vector cross product of the velocity v at which an infinitesimal unit test charge q is moving: $v \times H\mu = F/q$. It is expressed in amperes/meter. [IEEE C95.1 1991]

Maximum Permissible Exposure (MPE)—The RMS and peak electric and magnetic field strengths, their squares, or the plane-wave equivalent power densities associated with these fields and the induced and contact currents to which a person may be exposed without harmful effect and with an acceptable safety factor. [IEEE C95.1 1991]

Near field—This is that part of space between the antenna and the far-field region [IEEE Std 145-1993]. It is a region generally in proximity to an antenna or other radiating structure, in which the electric and magnetic fields do not have a substantial plane-wave character, but vary considerably from point to point. The near-field region is further divided into the reactive near-field region, which is closest to the radiating structure and that contains most or nearly all of the stored energy, and the radiating near-field region where the radiation field predominates over the reactive field, but lacks substantial plane-wave character and is complicated in structure. [IEEE C95.1 1991]

Plane wave—A wave in which the only spatial dependence of the field vectors is through a common exponential factor whose exponent is a linear function of position. [IEEE Std 100-1984]

Power density (S)—This is a measure of the power flow through per unit area normal to the direction of propagation. It is usually expressed in W/m^2. It is valid everywhere, but quantifiable most readily only in the far field of the antenna. In the near field of the antenna, a *far-field equivalent power density* is defined in terms of the near E field or near H field and the free space intrinsic impedance of 377 Ω: $S = E^2/377 = 377H^2$.

Specific Absorption Rate (SAR)—The time derivative of the incremental energy (dW) absorbed by (dissipated in) an incremental mass (dm) contained in a volume element (dV) of a given density (ρ):   SAR = (d/dt)(dW/dm) = (d/dt)(dW/ρdV). SAR is expressed in units of watts per kilogram. [IEEE C95.1 1991]

Uncontrolled environment—An RF-exposure environment in which the people being exposed to an RF field would not normally be aware that they are being exposed.

Footnotes and References

[1]E. C. Jordan and K. G. Balmain, *Electromagnetic Waves and Radiating Systems,* 2nd ed. (Englewood Cliffs, NJ: Prentice-Hall, 1968).

[2]J. D. Kraus, *Electromagnetics*, 3rd ed. (New York: McGraw-Hill Book Company, 1984).

[3]*IEEE Standard Definitions of Terms for Antennas*, IEEE Std 145-1993, SH16279, 18 March 1993.

[4]*IEEE Standard Definitions of Terms for Radio Wave Propagation*, IEEE Std 211-1990, SH13904, 1990.

[5]E. R. Cohen and B. N. Taylor, "The 1986 CODATA Recommended Values of the Fundamental Physical Constants," *Journal of Research of the National Bureau of Standards*, Vol 92, No 2, March-April 1987.

[6]K. Siwiak, *Radiowave Propagation and Antennas for Personal Communications*, (Norwood, MA: Artech House, 1995).

[7]Q. Balzano, O. Garay, K. Siwiak, "The Near Field of Dipole Antennas, Part I: Theory," *IEEE Transactions on Vehicular Technology*, Vol VT-30 No 4, pp 161-174, Nov 1981. "Part II: Experimental Results," *IEEE Transactions on Vehicular Technology*, Vol. VT-30 No 4, pp 175-181, Nov 1981.

[8]Q. Balzano, K. Siwiak, "The Near Field of Annular Antennas," *IEEE Transactions on Vehicular Technology*, Vol. VT-36, No 4, pp 173-183, Nov 1987.

RF Radiation And Electromagnetic Field Safety

3

Compliance with the FCC RF-exposure rules and RF safety may be related, but they are *not* the same thing. The new FCC rules are not a substitute for safety and common sense. This chapter fills the gap by discussing the safety aspects of working with RF energy.

The differences between regulation and safety can be understood by looking at the area of electrical safety. Various building codes cover the requirement of safe wiring and installation, but there are still many areas of safety, such as not touching live wires, that are not directly addressed by the law. This chapter discusses the safety aspects of working with RF energy. It also builds on the foundation of Chapter 2, introducing more terms and definitions relating to both safety and the rules.

Amateur Radio is basically a safe activity. In recent years, however, there has been considerable discussion and concern about the possible hazards of electromagnetic radiation (EMR), including both RF energy and power-frequency (50-60 Hz) electromagnetic (EM) fields. FCC regulations set limits on the maximum permissible exposure (MPE) allowed from the operation of radio transmitters. These regulations do not take the place of RF-safety practices, however. This section deals with the topic of RF safety.

This section was prepared by members of the ARRL RF Safety Committee and coordinated by Dr. Robert E. Gold, WBØKIZ. It summarizes what is now known and offers safety precautions based on the research to date.

All life on Earth has adapted to survive in an environment of weak, natural, low-frequency electromagnetic fields (in addition to the Earth's static geomagnetic field). Natural low-frequency EM fields come from two main sources: the sun, and thunderstorm activity. But in the last 100 years, man-made fields at much higher intensities and with a very different spectral distribution have altered this natural EM background in ways that are not yet fully understood. Researchers continue to look at the effects of RF exposure over a wide range of frequencies and levels.

Both RF and 60-Hz fields are classified as *nonionizing radiation,* because the frequency is too low for there to be enough photon energy to ionize atoms. (*Ionizing radiation,* such as X-rays, gamma rays and even some ultraviolet radiation has enough energy to knock electrons loose from their atoms. When this happens, positive and negative ions are formed.) Still, at sufficiently high power densities, EMR poses certain health hazards. It has been known since the early days of radio that RF energy can cause injuries by heating body tissue. (Anyone who has ever touched an improperly grounded radio chassis or energized antenna and received an *RF burn* will agree that this type of injury can be quite painful.) In extreme cases, RF-induced heating in the eye can result in cataract formation, and can even cause blindness. Excessive RF heating of the reproductive organs can cause sterility. Other health problems also can result from RF heating. These heat-related health hazards are called *thermal effects.* A microwave oven is a positive application of this thermal effect.

There also have been observations of changes in physiological function in the presence of RF energy levels that are too low to cause heating. These functions return to normal when the field is removed. Although research is ongoing, no harmful health consequences have been linked to these changes.

In addition to the ongoing research, much else has been done to address this issue. For example, FCC regulations set limits on exposure from radio transmitters. The Institute of Electrical and Electronics Engineers, the American National Standards Institute and the National Council for Radiation Protection and Measurement, among others, have recommended voluntary guidelines to limit human exposure to RF energy. The ARRL has established the RF Safety Committee, consisting of concerned medical doctors and scientists, serving voluntarily to monitor scientific research in the fields and to recommend safe practices for radio amateurs.

Thermal Effects of RF Energy

Body tissues that are subjected to *very high* levels of RF energy may suffer serious heat damage. These effects depend upon the frequency of the energy, the power density of the RF field that strikes the body and factors such as the polarization of the wave.

At frequencies near the body's natural resonant frequency, RF energy is absorbed more efficiently, and an increase in heating occurs. In adults, this frequency usually is about 35 MHz if the person is grounded, and about 70 MHz if insulated from the ground. Individual body parts may be resonant at different frequencies. The adult head, for example, is resonant

RF Radiation And Electromagnetic Field Safety 3.1

around 400 MHz, while a baby's smaller head resonates near 700 MHz. Body size thus determines the frequency at which most RF energy is absorbed. As the frequency is moved farther from resonance, less RF heating generally occurs. *Specific absorption rate (SAR)* is a term that describes the rate at which RF energy is absorbed in tissue.

Maximum permissible exposure (MPE) limits are based on whole-body SAR values, with additional safety factors included as part of the standards and regulations. This helps explain why these safe exposure limits vary with frequency. The MPE limits define the maximum electric and magnetic field strengths or the plane-wave equivalent power densities associated with these fields, that a person may be exposed to without harmful effect—and with an acceptable safety factor. The regulations assume that a person exposed to a specified (safe) MPE level also will experience a safe SAR.

Nevertheless, thermal effects of RF energy should not be a major concern for most radio amateurs, because of the power levels we normally use and the intermittent nature of most amateur transmissions. Amateurs spend more time listening than transmitting, and many amateur transmissions such as CW and SSB use low-duty-cycle modes. (With FM or RTTY, though, the RF is present continuously at its maximum level during each transmission.) In any event, it is rare for radio amateurs to be subjected to RF fields strong enough to produce thermal effects, unless they are close to an energized antenna or unshielded power amplifier. Specific suggestions for avoiding excessive exposure are offered later in this chapter.

Athermal Effects of EMR

Athermal effects of EMR involve lower-level energy fields that are insufficient to cause either ionization or heating effects. Research about possible health effects resulting from exposure to the lower level energy fields, the athermal effects, has been of two basic types: epidemiological research and laboratory research.

Scientists conduct laboratory research into biological mechanisms by which EMR may affect animals including humans. Epidemiologists look at the health patterns of large groups of people using statistical methods. These epidemiological studies have been inconclusive. By their basic design, these studies do not demonstrate cause and effect, nor do they postulate mechanisms of disease. Instead, epidemiologists look for associations between an environmental factor and an observed pattern of illness. For example,

in the earliest research on malaria, epidemiologists observed the association between populations with high prevalence of the disease and the proximity of mosquito infested swamplands. It was left to the biological and medical scientists to isolate the organism causing malaria in the blood of those with the disease, and identify the same organisms in the mosquito population.

In the case of athermal effects, some studies have identified a weak association between exposure to EMF at home or at work and various malignant conditions including leukemia and brain cancer. A larger number of equally well designed and performed studies, however, have found no association. A risk ratio of between 1.5 and 2.0 has been observed in positive studies (the number of observed cases of malignancy being 1.5 to 2.0 times the "expected" number in the population). Epidemiologists generally regard a risk ratio of 4.0 or greater to be indicative of a strong association between the cause and effect under study. For example, men who smoke one pack of cigarettes per day increase their risk for lung cancer tenfold compared to nonsmokers, and two packs per day increases the risk to more than 25 times the nonsmokers' risk.

Epidemiological research by itself is rarely conclusive, however. Epidemiology only identifies health patterns in groups—it does not ordinarily determine their cause. And there are often confounding factors: Most of us are exposed to many different environmental hazards that may affect our health in various ways. Moreover, not all studies of persons likely to be exposed to high levels of EMR have yielded the same results.

There also has been considerable laboratory research about the biological effects of EMR in recent years. For example, some separate studies have indicated that even fairly low levels of EMR might alter the human body's circadian rhythms, affect the manner in which T lymphocytes function in the immune system and alter the nature of the electrical and chemical signals communicated through the cell membrane and between cells, among other things. Although these studies are intriguing, they do not demonstrate any effect of these low-level fields on the overall organism.

Much of this research has focused on low-frequency magnetic fields, or on RF fields that are keyed, pulsed or modulated at a low audio frequency (often below 100 Hz). Several studies suggested that humans and animals can adapt to the presence of a steady RF carrier more readily than to an intermittent, keyed or modulated energy source.

The results of studies in this area, plus speculations concerning the effect of various types of modulation, were and have remained somewhat controversial. None of the research to date has demonstrated that low-level EMR causes adverse health effects.

Given the fact that there is a great deal of ongoing research to examine the health consequences of exposure to EMF, the American Physical Society (a national group of highly respected scientists) issued a statement in May 1995 based on its review of available data pertaining to the possible connections of cancer to 60-Hz EMF exposure. This report is exhaustive and should be reviewed by anyone with a serious interest in the field. Among its general conclusions were the following:

1. The scientific literature and the reports of reviews by other panels show no consistent, significant link between cancer and power line fields.

2. No plausible biophysical mechanisms for the systematic initiation or promotion of cancer by these extremely weak 60-Hz fields has been identified.

3. While it is impossible to prove that no deleterious health effects occur from exposure to any environmental factor, it is necessary to demonstrate a consistent, significant, and causal relationship before one can conclude that such effects do occur.

In a report dated October 31, 1996, a committee of the National Research Council of the National Academy of Sciences has concluded that no clear, convincing evidence exists to show that residential exposures to electric and magnetic fields (EMFs) are a threat to human health.

A National Cancer Institute epidemiological study of residential exposure to magnetic fields and acute lymphoblastic leukemia in children was published in the *New England Journal of Medicine* in July 1997. The exhaustive, seven-year study concludes that if there is any link at all, it is far too weak to be concerned about.

Readers may want to follow this topic as further studies are reported. Amateurs should be aware that exposure to RF and ELF (60 Hz) electromagnetic fields at all power levels and frequencies has not been fully studied under all circumstances. "Prudent avoidance" of any avoidable EMR is always a good idea. Prudent avoidance doesn't mean that amateurs should be fearful of using their equipment. Most amateur operations are well within the MPE limits. If any risk does exist, it will almost surely fall well down on the list of causes that may be harmful to your health (on the other end of the list from your automobile). It does mean, however, that hams should be aware of the potential for exposure from their stations, and take whatever reasonable steps they can take to minimize their own exposure and the exposure of those around them.

Safe Exposure Levels

How much EM energy is safe? Scientists and regulators have devoted a great deal of effort to deciding upon safe RF-exposure limits. This is a very complex problem, involving difficult public health and economic considerations. The recommended safe levels have been revised downward several times over the years — and not all scientific bodies agree on this question even today. An Institute of Electrical and Electronics Engineers (IEEE) standard for recommended EM exposure limits was published in 1991 (see Bibliography). It replaced a 1982 American National Standards Institute (ANSI) standard. In the new standard, most of the permitted exposure levels were revised downward (made more stringent), to better reflect the current research. The new IEEE standard was adopted by ANSI in 1992.

The IEEE standard recommends frequency-dependent and time-dependent maximum permissible exposure levels. Unlike earlier versions of the standard, the 1991 standard recommends different RF exposure limits in *controlled environments* (that is, where energy levels can be accurately determined and everyone on the premises is aware of the presence of EM fields) and in *uncontrolled environments* (where energy levels are not known or where people may not be aware of the presence of EM fields). FCC regulations also include controlled/occupational and uncontrolled/general population exposure environments.

The graph in Figure 3.1 depicts the 1991 IEEE standard. It is necessarily a complex graph, because the standards differ not only for controlled and uncontrolled environments but also for electric (E) fields and magnetic (H) fields. Basically, the lowest E-field exposure limits occur at frequencies between 30 and 300 MHz. The lowest H-field exposure levels occur at 100-300 MHz. The ANSI standard sets the maximum E-field limits between 30 and 300 MHz at a power density of 1 mW/cm^2 (61.4 V/m) in controlled environments—but at one-fifth that level (0.2 mW/cm^2 or 27.5 V/m) in uncontrolled environments. The H-field limit drops to 1 mW/cm^2 (0.163 A/m) at 100-300 MHz in controlled environments and 0.2 mW/cm^2 (0.0728 A/m) in uncontrolled environments. Higher power densities are permitted at frequencies below 30 MHz (below 100 MHz for H fields) and above 300 MHz, based on the concept that the body will not be resonant at those frequencies and will therefore absorb less energy.

In general, the 1991 IEEE standard requires averaging the power level over time periods ranging from 6 to 30 minutes for power-density calculations, depending on the frequency and other variables. The ANSI exposure limits for uncontrolled environments are lower than those for controlled environments, but to compen-sate for that the standard allows exposure levels in those environments to be averaged over much longer time periods (generally 30 minutes). This long averaging time means that an intermittently operating RF source (such as an Amateur Radio transmitter) will show a much lower power density than a continuous-duty station—for a given power level and antenna configuration.

Time averaging is based on the concept that the human body can withstand a greater rate of body heating (and thus, a higher level of RF energy) for a short time than for a longer period. Time averaging may not be appropriate, however, when considering nonthermal effects of RF energy.

The IEEE standard excludes any transmitter with an output below 7 W because such low-power transmitters would not be able to produce significant whole-body heating. (Recent studies show that hand-held transceivers often produce power densities in excess of the IEEE standard within the head.)

There is disagreement within the scientific community about these RF exposure guidelines. The IEEE standard is still intended primarily to deal with thermal effects, not exposure to energy at lower levels. A small but significant number of researchers now believe athermal effects also should be taken into consideration. Several European countries and localities in the United States have adopted stricter standards than the recently updated IEEE standard.

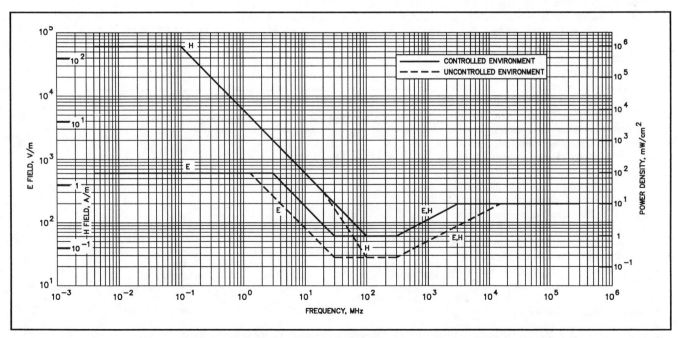

Fig 3.1—1991 RF protection guidelines for body exposure of humans. It is known officially as the "IEEE Standard for Safety Levels with Respect to Human Exposure to Radio Frequency Electromagnetic Fields, 3 kHz to 300 GHz."

Another national body in the United States, the National Council for Radiation Protection and Measurement (NCRP), also has adopted recommended exposure guidelines. NCRP urges a limit of 0.2 mW/cm^2 for nonoccupational exposure in the 30-300 MHz range. The NCRP guideline differs from IEEE in two notable ways: It takes into account the effects of modulation on an RF carrier, and it does not exempt transmitters with outputs below 7 W.

The FCC MPE regulations are based on parts of the 1992 IEEE/ANSI standard and recommendations of the National Council for Radiation Protection and Measurement (NCRP). The MPE limits under the regulations are slightly different that the IEEE/ANSI limits. Note that the MPE levels apply to the FCC rules put into effect for radio amateurs on January 1, 1998. These MPE requirements do not reflect and include all the assumptions and exclusions of the IEEE/ANSI standard.

Cardiac Pacemakers and RF Safety

It is a widely held belief that cardiac pacemakers may be adversely affected in their function by exposure to electromagnetic fields. Amateurs with pacemakers may ask whether their operating might endanger themselves or visitors to their shacks who have a pacemaker. Because of this, and similar concerns regarding other sources of electromagnetic fields, pacemaker manufacturers apply design methods that for the most part shield the pacemaker circuitry from even relatively high EM field strengths.

It is recommended that any amateur who has a pacemaker, or is being considered for one, discuss this matter with his or her physician. The physician will probably put the amateur into contact with the technical representative of the pacemaker manufacturer. These representatives are generally excellent resources, and may have data from laboratory or "in the field" studies with specific model pacemakers.

One study examined the function of a modern (dual chamber) pacemaker in and around an Amateur Radio station. The pacemaker generator has circuits that receive and process electrical signals produced by the heart, and also generate electrical signals that stimulate (pace) the heart. In one series of experiments, the pacemaker was connected to a heart simulator. The system was placed on top of the cabinet of a 1-kW HF linear amplifier during SSB and CW operation. In another test, the system was placed in close proximity to several 1 to 5-W 2-meter handheld transceivers. The test pacemaker was connected to the heart simulator in a

third test, and then placed on the ground 9 meters below and 5 meters in front of a three-element Yagi HF antenna. No interference with pacemaker function was observed in these experiments.

Although the possibility of interference cannot be entirely ruled out by these few observations, these tests represent more severe exposure to EM fields than would ordinarily be encountered by an amateur—with an average amount of common sense. Of course prudence dictates that amateurs with pacemakers, who use hand-held VHF transceivers, keep the antenna as far as possible from the site of the implanted pacemaker generator. They also should use the lowest transmitter output required for adequate communication. For high power HF transmission, the antenna should be as far as possible from the operating position, and all equipment should be properly grounded.

Low-Frequency Fields

Although the FCC doesn't regulate 60-Hz fields, some recent concern about EMR has focused on low-frequency energy rather than RF. Amateur Radio equipment can be a significant source of low-frequency magnetic fields, although there are many other sources of this kind of energy in the typical home. Magnetic fields can be measured relatively accurately with inexpensive 60-Hz meters that are made by several manufacturers.

Table 3.1 shows typical magnetic field intensities of Amateur Radio equipment and various household items. Because these fields dissipate rapidly with distance, "prudent avoidance" would mean staying perhaps 12 to 18 inches away from most Amateur Radio equipment (and 24 inches from

power supplies with 1-kW RF amplifiers).

Determining RF Power Density

Unfortunately, determining the power density of the RF fields generated by an amateur station is not as simple as measuring low-frequency magnetic fields. Although sophisticated instruments can be used to measure RF power densities quite accurately, they are costly and require frequent recalibration. Most amateurs don't have access to such equipment, and the inexpensive field-strength meters that we do have are not suitable for measuring RF power density. Chapter 5 of this book discusses this topic in detail.

Table 3.2 shows a sampling of measurements made at Amateur Radio stations by the Federal Communications Commission and the Environmental Protection Agency in 1990. As this table indicates, a good antenna well removed from inhabited areas poses no hazard under any of the IEEE/ANSI guidelines. However, the FCC/EPA survey also indicates that amateurs must be careful about using indoor or attic-mounted antennas, mobile antennas, low directional arrays or any other antenna that is close to inhabited areas, especially when moderate to high power is used.

Ideally, before using any antenna that is in close proximity to an inhabited area, you should measure the RF power density. If that is not feasible, the next best option is make the installation as safe as possible by observing the safety suggestions listed in Table 3.3.

It also is possible, of course, to calculate the probable power density near an antenna using simple equations. Such calculations have many pitfalls. For one, most of the situations where the power density would be high enough to be of

---

Table 3.1

Typical 60-Hz Magnetic Fields Near Amateur Radio Equipment and AC-Powered Household Appliances

Values are in milligauss.

Item	Field	Distance
Electric blanket	30-90	Surface
Microwave oven	10-100	Surface
	1-10	12"
IBM personal	5-10	Atop monitor
computer	0-1	15" from screen
Electric drill	500-2000	At handle
Hair dryer	200-2000	At handle
HF transceiver	10-100	Atop cabinet
	1-5	15" from front
1-kW RF amplifier	80-1000	Atop cabinet
	1-25	15" from front

(Source: measurements made by members of the ARRL RF Safety Committee)

---

concern are in the near field. In the near field, ground interactions and other variables produce power densities that cannot be determined by simple arithmetic. In the far field, conditions become easier to predict with simple calculations.

The boundary between the near field and the far field depends on the wavelength of the transmitted signal and the physical size and configuration of the antenna. The boundary between the near field and the far field of an antenna can be as much as several wavelengths from the antenna. This is discussed in Chapter 2.

Computer antenna-modeling programs are another approach you can use. *MININEC* or other codes derived from *NEC* (Numerical Electromagnetics Code) are suitable for estimating RF magnetic and electric fields around amateur antenna systems.

These models have limitations. Ground interactions must be considered in estimating near-field power densities, and the "correct ground" must be modeled. Computer modeling is generally not sophisticated enough to predict "hot spots" in the near field—places where the field intensity may be far higher than would be expected, due to reflections from nearby objects. In addition, "nearby objects" often change or vary with weather or the season, so the model so laboriously crafted may not be representative of the actual situation, by the time it is running on the computer.

Intensely elevated but localized fields often can be detected by professional measuring instruments. These "hot spots" are often found near wiring in the shack, and metal objects such as antenna masts or equipment cabinets. But even with the best instrumentation, these measurements also may be misleading in the near field.

One need not make precise measurements or model the exact antenna system, however, to develop some idea of the relative fields around an antenna. Computer modeling using close approximations of the geometry and power input of the antenna will generally suffice. Those who are familiar with *MININEC* can estimate their power densities by computer modeling, and those who have access to professional power-density meters can make useful measurements.

While our primary concern is ordinarily the intensity of the signal radiated by an antenna, we also should remember that there are other potential energy sources to be considered. You also can be exposed to RF radiation directly from a power amplifier if it is operated without proper shielding. Transmission lines also may radiate a significant amount of energy under some conditions. Poor microwave waveguide joints or improperly assembled connectors are another source of incidental radiation.

Further RF Exposure Suggestions

Potential exposure situations should be taken seriously. Based on the FCC/EPA measurements and other data, the "RF awareness" guidelines of Table 3.3 were developed by the ARRL RF Safety Committee. A longer version of these guidelines, along with a complete list of references, appeared in a *QST* article by Ivan Shulman, MD, WC2S ("Is Amateur Radio Hazardous to Our Health?" *QST*, Oct 1989, pp 31-34). For more information or background, see the list of RF Safety References in the next section.

In addition, ARRL maintains an RF-

Table 3.2
Typical RF Field Strengths Near Amateur Radio Antennas

A sampling of values as measured by the Federal Communications Commission and Environmental Protection Agency, 1990

Antenna Type	Freq (MHz)	Power (W)	E Field (V/m)	Location
Dipole in attic	14.15	100	7-100	In home
Discone in attic	146.5	250	10-27	In home
Half sloper	21.5	1000	50	1 m from base
Dipole at 7-13 ft	7.14	120	8-150	1-2 m from earth
Vertical	3.8	800	180	0.5 m from base
5-element Yagi at 60 ft	21.2	1000	10-20	In shack
			14	12 m from base
3-element Yagi at 25 ft	28.5	425	8-12	12 m from base
Inverted V at 22-46 ft	7.23	1400	5-27	Below antenna
Vertical on roof	14.11	140	6-9	In house
			35-100	At antenna tuner
Whip on auto roof	146.5	100	22-75	2 m antenna
			15-30	In vehicle
			90	Rear seat
5-element Yagi at 20 ft	50.1	500	37-50	10 m antenna

Table 3.3
RF Awareness Guidelines

These guidelines were developed by the ARRL RF Safety Committee, based on the FCC/EPA measurements of Table 3.2 and other data.

• Although antennas on towers (well away from people) pose no exposure problem, make certain that the RF radiation is confined to the antennas' radiating elements themselves. Provide a single, good station ground (earth), and eliminate radiation from transmission lines. Use good coaxial cable, not open-wire lines or end-fed antennas that come directly into the transmitter area.

• No person should ever be near any transmitting antenna while it is in use. This is especially true for mobile or ground-mounted vertical antennas. Avoid transmitting with more than 25 W in a VHF mobile installation unless it is possible to first measure the RF fields inside the vehicle. At the 1-kW level, both HF and VHF directional antennas should be at least 35 ft above inhabited areas. Avoid using indoor and attic-mounted antennas if at all possible.

• Don't operate high-power amplifiers with the covers removed, especially at VHF/UHF.

• In the UHF/SHF region, never look into the open end of an activated length of waveguide or microwave feed-horn antenna or point it toward anyone. (If you do, you may be exposing your eyes to more than the maximum permissible exposure level of RF radiation.) Never point a high-gain, narrow-bandwidth antenna (a paraboloid, for instance) toward people. Use caution in aiming an EME (moonbounce) array toward the horizon; EME arrays may deliver an effective radiated power of 250,000 W or more.

• With hand-held transceivers, keep the antenna away from your head and use the lowest power possible to maintain communications. Use a separate microphone and hold the rig as far away from you as possible. This will reduce your exposure to the RF energy.

• Don't work on antennas that have RF power applied.

• Don't stand or sit close to a power supply or linear amplifier when the ac power is turned on. Stay at least 24 inches away from power transformers, electrical fans and other sources of high-level 60-Hz magnetic fields.

exposure news page on its Web site. This site contains reprints of selected *QST* articles on RF exposure and links to the FCC and other useful sites.

RF Safety References

IEEE Standard for Safety Levels with Respect to Human Exposure to Radio Frequency Electromagnetic Fields, 3 kHz to 300 GHz, IEEE Standard C95.1-1991, Institute of Electrical and Electronics Engineers, New York, 1992.

For an assessment of ELF hazards, read the series in *Science*, Vol. 249 beginning 9/7/90 (p 1096), continuing 9/21/90 (p 1378), and ending 10/5/90 (p 23). Also see *Science*, Vol. 258, p 1724 (1992). You can find *Science* in any large library.

An excellent and timely document is available on the Internet at http://www.mcw.edu/gcrc/cop/powerlines-cancer-FAQ/toc.html

The Environmental Protection Agency publishes a free consumer-level booklet entitled, "EMF in Your Environment," document 402-R-92-008, dated December 1992. Look for the nearest office of the EPA in your phone book.

W. R. Adey, "Tissue Interactions with Nonionizing Electromagnetic Fields," *Physiology Review*, 1981; 61:435-514.

W. R. Adey, "Cell Membranes: The Electromagnetic Environment and Cancer Promotion, *Neurochemical Research*, 1988; 13:671-677.

W. R. Adey, Electromagnetic Fields, Cell Membrane Amplification, and Cancer Promotion, in B. W. Wilson, R. G. Stevens, and L. E. Anderson, *Extremely Low Frequency Electromagnetic Fields: The Question of Cancer* (Columbus, OH: Batelle Press, 1989), pp 211-249.

W. R. Adey, Electromagnetic Fields and the Essence of Living Systems, Plenary Lecture, 23rd General Assembly, International Union of Radio Sciences (URSI), Prague, 1990; in J. Bach Andersen, Ed., *Modern Radio Science* (Oxford: Oxford Univ. Press), pp 1-36.

Q. Balzano, O. Garay and K. Siwiak, "The Near Field of Dipole Antennas, Part I: Theory," *IEEE Transactions on Vehicular Technology (VT)* 30, p 161, Nov 1981. Also "Part II; Experimental Results," same issue, p 175.

R. F. Cleveland and T. W. Athey, "Specific Absorption Rate (SAR) in Models of the Human Head Exposed to Hand-Held UHF Portable Radios," *Bio-electromagnetics*, 1989; 10:173-186.

R. F. Cleveland, E. D. Mantiply and T. L. West, "Measurements of Environmental Electromagnetic Fields Created by Amateur Radio Stations," presented at the 13th annual meeting of the Bioelectromagnetics Society, Salt Lake City, Utah, Jun 1991.

R. L. Davis and S. Milham, "Altered Immune Status in Aluminum Reduction Plant Workers," *American J Industrial Medicine*, 1990; 131:763-769.

F. C. Garland, et al, "Incidence of Leukemia in Occupations with Potential Electromagnetic Field Exposure in United States Navy Personnel," *American J Epidemiology*, 1990; 132:293-303.

A. W. Guy and C. K. Chou, "Thermograph Determination of SAR in Human Models Exposed to UHF Mobile Antenna Fields," Paper F-6, Third Annual Conference, Bioelectromagnetics Society, Washington, DC, Aug 9-12, 1981.

E. Hare, "The FCC's New RF-Exposure Regulations," *QST*, Jan 1997, pp 47-49.

E. Hare, "What's New About the FCC's New RF-Exposure Regulations?" *QST*, Oct 1997, pp 51-52.

E. Hare, "FCC RF-Exposure Evaluations—the Station Evaluation," *QST*, Jan 1998, pp 50-55.

R.T. Hitchcock and R.M. Patterson, *Radio-Frequency and ELF Electromagnetic Energies, a Handbook for Health Professionals*, Van Nostrand Reinhold, New York (1995)

C. C. Johnson and M. R. Spitz, "Childhood Nervous System Tumors: An Assessment of Risk Associated with Paternal Occupations Involving Use, Repair or Manufacture of Electrical and Electronic Equipment," *International J Epidemiology*, 1989; 18:756-762.

D. L. Lambdin, "An Investigation of Energy Densities in the Vicinity of Vehicles with Mobile Communications Equipment and Near a Hand-Held Walkie Talkie," EPA Report ORP/EAD 79-2, Mar 1979.

D. B. Lyle, P. Schechter, W. R. Adey and R. L. Lundak, "Suppression of T-Lymphocyte Cytotoxicity Following Exposure to Sinusoidally Amplitude Modulated Fields," *Bioelectromagnetics*, 1983; 4:281-292.

G. M. Matanoski et al, "Cancer Incidence in New York Telephone Workers," *Proc Annual Review, Research on Biological Effects of 50/60 Hz Fields*, U.S. Dept of Energy, Office of Energy Storage and Distribution, Portland, OR, 1989.

D. I. McRee, *A Technical Review of the Biological Effects of Non-Ionizing Radiation*, Office of Science and Technology Policy, Washington, DC, 1978.

G. E. Myers, "ELF Hazard Facts" *Amateur Radio News Service Bulletin*, Alliance, OH, Apr 1994.

S. Milham, "Mortality from Leukemia in Workers Exposed to Electromagnetic Fields," *New England J Medicine*, 1982; 307:249.

S. Milham, "Increased Mortality in Amateur Radio Operators due to Lymphatic and Hematopoietic Malignancies," *American J Epidemiology*, 1988; 127:50-54.

W. W. Mumford, "Heat Stress Due to RF Radiation," *Proc IEEE*, 57, 1969, pp 171-178.

W. Overbeck, "Electromagnetic Fields and Your Health," *QST*, Apr 1994, pp 56-59.

S. Preston-Martin et al, "Risk Factors for Gliomas and Meningiomas in Males in Los Angeles County," *Cancer Research*, 1989; 49:6137-6143.

D. A. Savitz et al, "Case-Control Study of Childhood Cancer and Exposure to 60-Hz Magnetic Fields," *American J Epidemiology*, 1988; 128:21-38.

D. A. Savitz et al, "Magnetic Field Exposure from Electric Appliances and Childhood Cancer," *American J Epidemiology*, 1990; 131:763-773.

I. Shulman, "Is Amateur Radio Hazardous to Our Health?" *QST*, Oct 1989, pp 31-34.

R. J. Spiegel, "The Thermal Response of a Human in the Near-Zone of a Resonant Thin-Wire Antenna," *IEEE Transactions on Microwave Theory and Technology (MTT)*, 30(2), pp 177-185, Feb 1982.

B. Springfield and R. Ely, "The Tower Shield," *QST*, Sep 1976, p 26.

T. L. Thomas et al, "Brain Tumor Mortality Risk among Men with Electrical and Electronic Jobs: A Case-Controlled Study," *J National Cancer Inst*, 1987; 79:223-237.

N. Wertheimer and E. Leeper, "Electrical Wiring Configurations and Childhood Cancer," *American J Epidemiology*, 1979; 109:273-284.

N. Wertheimer and E. Leeper, "Adult Cancer Related to Electrical Wires Near the Home," *Internat'l J Epidemiology*, 1982; 11:345-355.

"Safety Levels with Respect to Human Exposure to Radio Frequency Electromagnetic Fields (300 kHz to 100 GHz)," ANSI C95.1-1991 (New York: IEEE-American National Standards Institute).

"Biological Effects and Exposure Criteria for Radiofrequency Electromagnetic Fields," NCRP Report No. 86 (Bethesda, MD: National Council on Radiation Protection and Measurements, 1986).

"Electric and Magnetic Fields—Background Paper," OTA-BP-E53 (Washington, DC: US Government Printing Office), 1989.

Handbook of Biological Effects of Electromagnetic Fields, CRC Press, Boca Raton (1986)

"EMF in the Web" site http://safeemf.iroe.fi.cnr.it/safeemf/emfref.htm

Why Do We Need New Rules To Control RF Bioeffects?

By Gregory D. Lapin, N9GL

It's a good question. Surely there are many more things that we encounter that could do us more harm than radio waves. Getting into collisions in our cars, chopping off limbs with power tools, slipping and breaking bones on a wet kitchen floor; house catching fire, tree branch falling on you, getting hit by lightning, the list goes on and on. These are all potentially hazardous things that we deal with every day of our lives. We have come to accept them and have learned to live with them. How can a little RF energy be mentioned in the same breath as these terrible things?

In one sense, RF energy requires more thought than all of these. Not because it can do more damage. Rather, because it is so hard to detect that you are being hurt until it is too late. With most of the hazards in our lives, the source of danger is very obvious. Because we know what the hazard looks like, we can take steps to avoid it. The main difference with RF energy is that, except for extremely high exposures, it is capable of damaging tissue without our even realizing that it is happening. Not only that, but RF energy is capable of heating our internal tissue without heating the tissues at the surface of the body.

Our bodies are designed to protect us from the elements. The tissue that comes in contact with the external world, our skin, is capable of surviving large swings in temperature and is equipped with nerve endings to allow us to feel conditions that may be damaging. If we put our hands near a lit stove, we feel the heat and draw back before getting too close and damaging our tissue. The tissues in the interior of our bodies differ from this in two fundamental ways. There is far less enervation inside our bodies—we cannot feel most things in there. Of course, it is very uncommon for anything that we would need to feel to be in there.

The tissues inside our bodies are very sensitive to changes in temperature. We have all experienced just how carefully our bodies regulate their own internal (core) temperature. Our normal body temperature is 98.6°F (37°C). When this temperature rises as little as one degree, we feel pretty lousy and call it a low grade fever. If our core temperatures rise to about 102°F (39°C) we are pretty sick and usually stay in bed. By the time our core temperatures rise above 105°F (40.5°C) our lives are in danger. The control system in our brains that keeps the core temperature so finely tuned detects these changes and does whatever it can to help decrease that temperature. Some of the tools at its disposal are the ability to change blood flow patterns, allowing the blood to carry the heat to the outer parts of our bodies where it can be radiated into the air; increasing respiration so our breath can blow off more heat; and increasing sweating, so the heat leaves our bodies as the sweat evaporates. When we get very sick with high fevers, it is usually because this control system has been disabled by disease.

RF heating of interior tissues in our bodies can be highly localized. If tissues are heated to dangerous levels but the heat does not reach the temperature control center in the brain, the control center is unaware of the danger and does not act to remove additional heat. The cells that are being heated are still in danger of dying with no relief in sight. If these cells are in vital organs, the efficacy of those organs is reduced; if too much so, the organ dies. When we are talking about organs such as heart, liver, kidney or brain, damage to even a few cells can be catastrophic.

To understand how cells are damaged by increased temperature, it is necessary to realize that the cell relies on a very delicate chemical balance. Pumps in the cell membrane make sure that the correct concentrations of various chemicals are present. The cell uses oxygen and glucose to live and generates waste products, such as carbon dioxide and water. The cell also contains complex proteins: enzymes that facilitate necessary chemical reactions and amino acid complexes that function as the cell's controlling elements to produce necessary chemicals and make cellular reproduction possible. The proteins are not only complex chemicals but also have specific shapes that determine how they function. Increasing the heat around these initially changes the molecular shape (denatures them) and they lose their efficacy. If the temperature increases to a high enough level, the pieces of the protein can start to come apart and they change into different types of chemicals. These molecules, which are necessary to keep the cell alive, lose their ability to do so and the cell eventually dies.

Cells do not die immediately when heated. Their proteins denature slowly and not all at once. The changes occur more quickly at higher temperatures. A combination of temperature and exposure time determines if a cell will live or die. The threshold of cell death is proportional to the product of temperature and time. At relatively small rises in temperature, it takes hours, or even days, for a cell to die. However, cell death can occur in minutes or seconds with much larger increases in temperature. If we are in an RF field that causes some of our cells to increase in temperature by about 5°F, we may have to be in that exposure condition for many hours before we lose any cells. In a stronger field that causes some cells to increase in temperature by 20°F, we can lose cells in minutes.

Biological damage due to heating is not cumulative on the cellular level. This means that if we heat up some of our cells by several degrees for an hour and then stop, the cells will return to normal. The next day if we have the same exposure, the cells will take just as long to die as if they had not been exposed the day before.

RF energy can be used safely. In most Amateur Radio applications there is little to no danger of heating cells to dangerous levels. Our experience tells us that hams get no sicker than anyone else. To insure that this remains so, the FCC RF Bioeffects regulations, which are based on IEEE/ANSI and NCRP RF Safety standards (which in turn are based on the probability of RF heating and are designed to provide a large margin of safety) have been developed to make us cognizant of our exposure levels while operating. At the same time we must be aware, as users of RF energy, that internal heating can occur without our feeling it. We must always make sure that we follow the safety guidelines to keep our exposure at safe levels.

The RF Exposure Rules

The FCC regulations about RF exposure are not difficult to understand. They require all hams to meet certain exposure limits. The rules also require that some stations be evaluated to determine that they are in compliance with the rules. This chapter helps you understand what the rules require you to do.

INTRODUCTION

Now that the fundamentals have been explained in the earlier chapters, it is time to learn more about the actual requirements of the rules. This chapter explains the sometimes complex requirements set for all radio services regulated by the FCC. The actual texts of the FCC regulations are in Appendix A. The RF Exposure rules can help ensure that operation in the Amateur Radio Service continues to be safe. These rules are not difficult for amateurs to follow.

In summary, FCC regulations control the amount of RF exposure that can result from your station's operation (Sections 97.13, 97.503, 1.1307(b)(c)(d), 1.1310, 1.1312(a) and 2.1093). The regulations set limits on the maximum permissible exposure (MPE) allowed from operation of transmitters in all radio services. They also require certain types of stations be evaluated to determine if they are in compliance with the MPEs specified in the rules. The FCC also has required that five questions on RF environmental safety practices be added to Novice, Technician and General license examinations.

Amateur Radio is included in these FCC rules to help ensure that amateurs are aware of the RF-exposure potential from their stations. In general, following these rules will not be difficult for most hams and will help ensure that their stations are operated within recognized exposure standards.

WHERE DID THE RULES COME FROM?

The Rules are Not New

Hams have been calling these regulations "new," but they are actually not new at all. They are the result of a long process started by medical researchers and industry, working through various research processes, national and international standards bodies and culminating in the US regulations administered by the FCC. See the sidebar, "How the IEEE/ANSI C95.1 Standard was Developed."

History of the Development of the Standards and Rules

The first US standards for RF exposure actually date to the late 1960s. The IEEE (Institute of Electrical and Electronic Engineers) has formed Standards Coordinating Committee 28 (SCC-28) to develop standards related to RF exposure. In 1982, ANSI adopted the IEEE C95.1 standard on RF exposure as IEEE/ANSI C95.1-1982. This standard described appropriate limits for human exposure to RF energy. The SCC-28 committee is comprised of a balance of medical researchers, engineers and representation from industry. It is judged by most to represent the most complete consensus of the appropriate levels for safe exposure to RF energy. (A representative from ARRL HQ serves on this committee.)

Shortly after the introduction of the C95.1 standard, the FCC wrote a set of regulations that required radio services to comply with the exposure limits in the standard. While the FCC was developing those early regulations, the ARRL commented that it was unlikely that amateur operation would exceed the proposed limits. Therefore the Amateur Radio Service should be categorically exempt from any *specific* requirements under the regulations. The ARRL further urged the FCC to rely upon the demonstrated technical competence of amateur operators and self-education as sufficient tools to ensure continued Amateur Radio safety. The FCC agreed, and Amateur Radio was *categorically* exempt from any specific requirement to perform a station evaluation under the old RF-exposure regulations.

Amateur Radio had no specific requirements under the old rules, so most hams were not concerned with them. It was quite unlikely that any amateur station would exceed the exposure limits.

Proposed Changes—1993

On April 8, 1993, the FCC released a Notice of Proposed Rulemaking (ET Docket 93-62), announcing that it intended to develop a new set of regulations for all services. The rules were to be based on the new IEEE/ANSI C95.1-1992 Standard. ARRL filed comments, asking that the Amateur Radio Service exemption continue, relying on the continued technical expertise and self-education of amateurs. The Amateur Radio Health Group filed comments requesting that Amateur Radio not be exempt under the new regulations, citing some instances where amateur installations could exceed the exposure levels in the standard. They noted that not all hams have read the educational material available on the topic. The FCC took no further action until the US Congress added a mandate to the Telecommunications Act of 1996 for FCC to complete its work on revisions to the RF-exposure regulations.

The Changes Emerge—1996

Things proceeded slowly toward the Congressionally mandated time limit of August 1, 1996. On August 1, 1996, just in time to meet the mandated date, the FCC announced the new regulations in the ET Docket 93-62, FCC 96-326 Report and Order, *Guidelines for Evaluating the Environmental Effects of Radio-Frequency Radiation.*

Fine Tuning and Change

As first announced, these regulations posed some problems for the Amateur Radio Service. Over the intervening time, these regulations have been subject to a number of important changes. (As I snipped in one of my *QST* articles, "Every time I got to where it's at, they moved it!"—*Ed Hare*) Not surprisingly, many of the changes were in response to petitions for reconsideration filed by the ARRL, as they sought to fine tune these regu-

How the IEEE/ANSI C95.1-1992 Standard was Developed

Virtually all standards for human exposure to nonionizing electromagnetic fields have derived from the collective thinking of groups of individuals – generally those who play active roles in this specialized technical area. As examples, the IEEE, NCRP and ICNIRP standards-setting committees all function through the contributions of volunteer technical experts, who are specialists in a variety of disciplines directly related to assessment of the biological effects and potential hazards of exposures to these fields.

In the IEEE, standards documents are developed within the technical committees of the various IEEE Societies and the Standards Coordinating Committees of the IEEE Standards Board. Members of these committees, often non-IEEE members, serve voluntarily and without compensation. The standards developed through this process represent a consensus of the broad expertise represented on individual committees; this is one of the strengths of the IEEE process in the development of safety levels with respect to human exposure to radio frequency fields.

I recently attended a one-day seminar that was conducted by the leadership of IEEE Standards Coordinating Committee 28, Non-Ionizing Radiation Hazards (SCC 28). SCC 28 is the sponsor, or oversight committee, with five subcommittees. These subcommittees generate IEEE standards, recommended practices and guides. SCC 28 currently has about 80 active members, including a voting member representing ARRL. One of the primary functions of SCC 28 is to ensure that the five imperative principles driving the standards process are met.

These principles are due process, openness, consensus, balance and the right of appeal. In addition to this oversight of subcommittee activities, the members of SCC 28 put the final stamp of approval on all subcommittee products through a formal voting process prescribed by the IEEE Standards Board. The purpose of the seminar I attended was to educate engineers about the IEEE/ANSI C95.1-1992 *Standard for Safety Levels with Respect to Human Exposure to Radio Frequency Electromagnetic Fields, 3 kHz to 300 GHz* and how it was developed.

Under SCC 28, Subcommittee 4 (SC-4) is charged with developing standards over the frequency range from 3 kHz to 300 GHz. SC-4 had 125 working members when the C95.1 standard was finalized in 1991. About 70% of the membership consisted of researchers working in university, nonprofit, military and government laboratories. The remainder represented industry, including consultants (12.8%), governmental administration (4%), and the general public, including independent consultants (11.2%). This membership represented a wide range of technical expertise, including medicine, biology, engineering and physical sciences. The C95.1 document describes in detail how this expertise was used to evaluate the biological database, dosimetry, statistical treatments and exposure risk, in addition to the drafting and refinement of the text.

Like its predecessor, ANSI C95.1-1982, the new C95.1 standard recognized that the rate of absorption of electromagnetic energy by the human body is frequency dependent. Whole body averaged rates of energy absorption or SAR in W/kg (Watts/kilogram) approach maximal values when the long axis of the body is parallel to the E-field vector and is 4/10 of a wavelength of the incident field. This situation is called resonant exposure. An envelope of resonant frequencies, accounting for all sizes (babies to basketball players) and positions (eg, squatting to standing with arms raised) of humans, describes the broad resonance range for which the recommended exposure guidelines are reduced in the standard. Again, like its predecessor, the new C95.1 standard incorporated dosimetry and adopted the unit-mass, time-averaged rate of electromagnetic energy absorption, or SAR. This can be applied to any element of mass of a biological body. Depending on the exposure situation, either whole-body SAR or spatial peak SAR can be used (in the broad resonance range) to determine compliance with the standard.

As noted in the C95.1 standard, there are many thousands of published papers and reports on all aspects of the subject of exposure to radio frequency fields. SC-4 made an initial selection of 321 papers from peer reviewed journals for evaluation by large Working Groups on Engineering Validation and Biological Validation. These papers were winnowed down, by means of strict acceptance criteria, to a final database of 120 critical papers that were further evaluated by a Risk Assessment Working Group. Only those papers with accurately measured fields and adequate dosimetry were considered acceptable in the final database. With regard to biological validation, SC-4 emphasized papers containing reliable evidence for debilitation or morbidity during the exposure of whole organisms.

lations to better fit the Amateur Radio Service.

Overall, the ARRL believes that the MPE limits are generally appropriate for the operation of Amateur Radio stations. These limits have been affirmed by recognized standards bodies and have been supported by the ARRL RF Safety Committee. The safety aspects of these regulations are based on the best work of the researchers and other experts on SCC-28, and the work of the National Council for Radiation Protection and Measurement (NCRP).

Although many people have quipped that there are a lot of old hams, this does not demonstrate that RF energy is unconditionally safe. Anyone who has ever received a nasty RF burn from an antenna or observed the actions of their microwave oven can see that there *can* be some danger associated with RF energy. In reality, hams have not suffered ill effects from the RF *because* most existing amateur operation is well within the exposure guidelines. The ARRL Bioeffects Committee and ARRL RF Safety Committee, comprised of some rather well-known experts in the field of biomedical research, have all supported the safety levels developed by the standards bodies.

What ARRL Asked For

Although the actual exposure levels are generally recognized as appropriate, there were still a number of problems with the rules as they were first announced in 1996. The first was the implementation date: The *original* date was set to January 1, 1997. The amateur community, having been exempt from specific requirements under the old rules, was simply not ready to meet that short of a timetable. The FCC had not

---

Because very few measurements have been made of the responses of human beings to radio frequency fields, SC-4 had to rely once again on data collected on subhuman species such as rodents and nonhuman primates. They found that most reports of biological effects involved acute exposures at relatively few frequencies. The extensive literature review showed once again that the most sensitive measures of potentially harmful biological effects were based on the interference with complex behavior in animal subjects that accompanied exposure to a radio frequency field at a whole-body SAR of about 4 W/kg. For example, a monkey trained to press a button six times to get a banana reward, decided, when exposed to a 4 W/kg field, that he didn't want the reward; when the field was removed, he soon decided that he was hungry after all and resumed pressing the button. The disruption of such behavior has been demonstrated in rodents and two monkey species, and happens despite significant differences in the characteristics of the field (such as frequency, near and far-field, multipath and planewave, CW and pulse-modulated). Because such changes in behavior are usually accompanied by an increase in the animals' body temperature, they are deemed to be thermal effects. Human volunteers exposed to such fields often ask the question, "Who turned on the sun?" because they feel warm.

The 4 W/kg whole-body SAR associated with disruption of animal behavior was adopted by the SC-4 Risk Assessment Working Group as an appropriate basis for setting exposure guidance for human beings. This decision was accepted by the SC-4 membership, which then agreed that safety factors should be applied across the broad resonance range of frequencies from 0.1 MHz to 6.0 GHz. For human exposure in a controlled (or occupational) environment, they applied a safety factor of 10, yielding an SAR of 0.4 W/kg as the basis for the MPEs. An additional factor of 5 (SAR = 0.08 W/kg) was added for exposure in an uncontrolled environment.

Outside of the broad human resonance range, other considerations apply. Below 0.1 MHz, the standard is designed to limit induced currents in the ankles during free-field exposure, and to lower the probability of inducing large body currents when conducting objects are touched. At frequencies above 6.0 GHz, the exposure is quasi-optical, penetration of energy is very superficial, and thermal time constants drop to sec-

onds as the infrared range is approached.

The safety levels published in IEEE/ANSI C95.1-1992 are believed by many to be conservative. Unlike some safety standards, the C95.1 standard is not based on ideas like "acceptable levels of risk." The exposure level for people unaware of their exposure (uncontrolled environment) is set at 2% of the level that made a monkey not want to work for a banana reward. If this level were difficult for Amateur Radio operators to meet, it would certainly appear as a problem for our radio service. But, in being able to meet the standards that are set to be conservatively safe, we have the best of both worlds. As questions about this safety standard come up with neighbors or the local zoning board, it is comforting to note that the standard is based on very conservative assumptions. These include 1) "worst-case" exposure (far-field, E-polarization), 2) an assumed but not defined hazard of behavioral disruption in animals, 3) a single contour for human resonance, and 4) direct extrapolation from animal to man that ignores the superior thermoregulation of humans.

ARRL HQ has received many inquiries from the neighbors of ham radio operators. One neighbor, after hearing about how the standard was developed (and about the banana reward), decided to attend an upcoming zoning board meeting concerning the siting of an Amateur Radio tower. She told other attendees that she didn't think the RF exposure questions should be a concern. (She told me that she disliked the tower for other reasons, but she now understood that there were some good reasons for having the amateur antenna located high in the air.)

Some have suggested that the reasons for concern about RF exposure are unfounded; that there are no adverse effects of RF energy at the levels normally encountered in the environment. Several ARRL committees, together with other technical experts, advise us that the IEEE/ANSI C95.1-1992 standard is realistic and we should support it. I serve on two US standards bodies and have participated in others. I know how difficult it is to find common ground for agreement in a large group. Given that a consensus of 120 members of SC-4 agreed upon this standard, and SCC-28 voted to approve it, it is almost certainly based on sound scientific principles.–*Ed Hare, W1RFI, ARRL Laboratory Supervisor*

yet prepared any revisions to its Office of Engineering and Technology bulletins, notably *OET Bulletin 65*, to help hams understand the rules and comply with them. The rules further required the changes to Amateur Radio examinations to go into effect immediately. In addition, the FCC had set a 50-W threshold to require the need for hams to do an evaluation of their station.

The ARRL immediately filed a number of petitions for reconsideration. In addition to pointing out some procedural flaws that had been part of the process of developing the final rules, the ARRL asked that the implementation date be extended to **January 1, 1998**. This was done to give amateurs time to understand the rules, to conduct the required station evaluation and to make any changes necessary to be in compliance.

The ARRL filed two petitions over the question pools—the first an "emergency" petition, pointing out that the rules as first written required the examination changes to take place immediately, without the appropriate number of questions available in the question pool. The ARRL also asked that amateurs be permitted to make the changes to the question pools as they were updated in their normal cycles, with the Novice and Technician changes to apply to the question pools that would be used after July 1, 1997, and the General pool to be used after July 1, 1998.

The rules also initially set a 50-W PEP threshold above which amateur stations needed to be evaluated. The ARRL asked that the 50-W limit be scaled by frequency, to match the way the permitted exposure levels varied by frequency. The ARRL was successful in all these areas, and the results are explained in various chapters of this book. The League is still seeking federal preemption of any local or state regulations on RF exposure, although they have not been successful in that area to date.

As if all that weren't enough to keep the ARRL busy, the staff continued to work closely with the FCC as FCC information on the subject was prepared. The ARRL's work with the FCC helped forge a mutual understanding on how to handle mobile installations, repeater exemptions and multitransmitter sites.

WHAT THE REGULATIONS REQUIRE AMATEURS TO DO

Okay, now that the ARRL has bragged a little, let's take a look at just what the rules expect of operators in the Amateur Radio Service. Full details of the requirements are contained in the language of the regulations, the Report and Order, the two Memorandum Opinion and Orders, three errata, and throughout this book. This chapter section summarizes the most important parts of the rules that concern the Amateur Radio operator.

EFFECTIVE DATE—A MOVING TARGET

The rules changes went into effect immediately when they were announced in August 1996, with an *original* transition period of January 1, 1997, before amateurs were required to come into compliance. After receiving a number of petitions for reconsideration, the FCC decided upon a transition period of September 1, 1997, for all services except for the Amateur Radio Service. January 1, 1998, marked the end of the transition period for amateurs. During this transition period, those stations and services in transition were governed by the older RF-exposure rules.

Starting on January 1, 1998, new amateur applications, or those for renewal or station modification requiring an FCC Form 610 application, will be required to be in compliance as of the time of application. As part of their application, hams will certify that they have read and understand the RF-exposure rules and that they will comply with them. Starting on **January 1, 1998**, the FCC will only accept FCC Forms 610 dated September 1997 or later, which include the RF-exposure certification.

The FCC has added an additional transition period for existing installations. Existing installations that are not renewed or have a modification to their station license (such as a change of address) have until **September 1, 2000**, to be in complete compliance with the RF-exposure rules. The FCC expects that hams will make good faith efforts to evaluate their stations and bring them into compliance if necessary, but those stations that are not substantially changed during the transition period do not require the submission to the FCC of a Form 610. They have until September 1, 2000, as a "date certain" to be in compliance.

However, Amateur Radio is covered by a paragraph in the rules in §1.1312(a). This paragraph discusses services that do not require FCC equipment authorization, before placing a modification to the station on the air. This section of the rules requires that if you make a change to your station that could affect compliance with these rules, such as adding a new antenna or band or increasing your transmitter power, you may have to perform the necessary evaluation on that change *before* you start using those changes on the air.

What the Rules are Not!

As discussed, the rules are *not* new. Earlier RF-exposure rules *always* applied to the Amateur Radio Service. Under the old higher MPE limits, it was unlikely that amateur stations would exceed the MPEs, so they had no specific requirements. The rules are *not just* for the Amateur Radio Service; they apply to all radio services regulated by the FCC, including high-powered broadcast stations and low-powered cellular telephones. The cellular services are *not* exempt from these rules; in fact, cellular devices are specifically mentioned as *not* being included in the evaluation exemptions for most mobile and portable transmitters. The new limits do *not* change the amateur power levels and will *not* require that most hams change the way they operate. Most important, *it is not difficult* for the Amateur Radio Service to comply with the regulations!

The Major Points

The two most important aspects of these rules are that there are limits on exposure and that some hams must do a routine evaluation on their stations.

In general, the rules have these major provisions:

- Set guideline limits on RF exposure to people that result from the operation of FCC-regulated transmitters.
- The limits vary with frequency, to match the way humans absorb RF energy.
- Stations regulated by the FCC are required to meet the guideline limits or to file a complicated Environmental Assessment (EA) with the FCC. (Note: It is unlikely that *any* amateur station would find it easier to file an EA than to comply with the MPE limits!)
- Some stations, based on power, frequency and use, need to be evaluated for compliance with the RF exposure guidelines in the rules.
- Amateurs perform their own station evaluations—if the station evaluation demonstrates to the amateur that the station is operating within the guidelines, it is not necessary to file extensive paperwork with the FCC.
- Some stations are categorically exempt from the evaluation requirement because their power, frequency, operating duty cycles and antenna separations are such that they are presumed to be in compliance.
- There are additional exemptions for most amateur mobile, portable hand-held and repeater operation.
- The FCC mandated that five questions on "radiofrequency environmental safety practices at an amateur station"

be added to the examinations for Novice, Technician and General licenses.

Exposure "Environments"

The regulations define two primary RF-exposure environments: *controlled/occupational* and *uncontrolled/general public*. The permitted exposure levels, as shown in Table 1 under the MPE section of this chapter, are lower for the uncontrolled exposure environment. Different exposure averaging times apply to each environment.

The regulations require amateurs to evaluate their stations for both controlled and uncontrolled exposure areas. Some hams may choose to apply the more stringent uncontrolled limits under all circumstances, such as to their own stations and property.

Controlled/Occupational

A *controlled* environment is one in which the people who are being exposed are aware of that exposure and can take steps to minimize that exposure, if appropriate. (In an *uncontrolled* environment, the people being exposed are *not* normally aware of the exposure.) The MPE levels are higher for a controlled environment than they are for an uncontrolled environment. In a controlled environment, exposure is averaged over a 6-minute period.

Although the permitted exposure levels in a controlled environment are safe, higher exposure levels are permitted in controlled exposure areas than are permitted in uncontrolled areas. Although they are primarily occupational environments, the FCC includes amateurs in this category. You can apply this exposure environment to your immediate families and guests if you provide them with education and training about RF exposure.

In most cases, controlled-environment limits can be applied to your home and property to which you can control physical access. You can apply a controlled environment to those general-public areas where exposure would be only transitory,

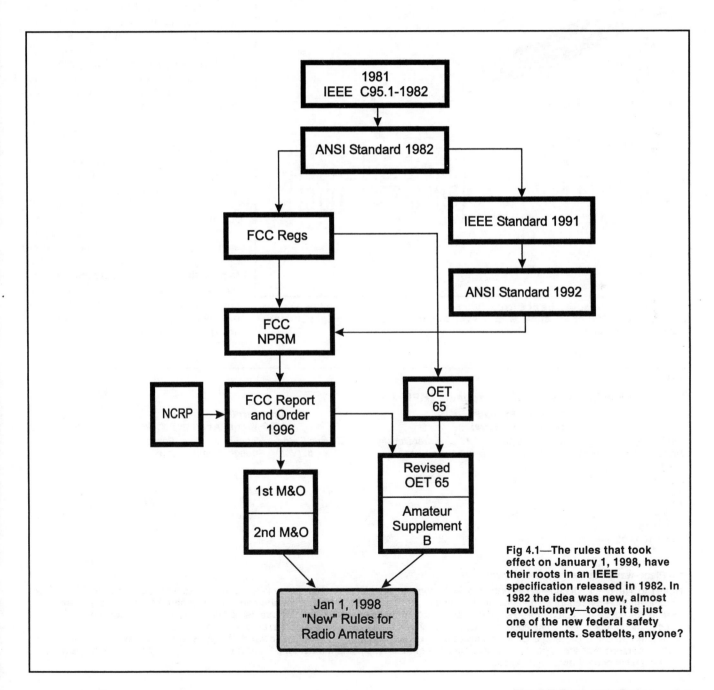

Fig 4.1—The rules that took effect on January 1, 1998, have their roots in an IEEE specification released in 1982. In 1982 the idea was new, almost revolutionary—today it is just one of the new federal safety requirements. Seatbelts, anyone?

To Be or Not to Be?

It is not always easy to know whether to consider an area as controlled or uncontrolled. As an example, a rooftop tower installation would be controlled for the service technicians. If someone were on the roof to paint the tower, the exposure probably would have be evaluated for an uncontrolled environment because painters generally have insufficient knowledge and training in RF exposure. Therefore this would not be considered a controlled environment. This means that even occupational environments may not always be controlled: At W1AW and ARRL HQ, we have what can reasonably be considered a controlled environment. We also have non-ham visitors on the premises and non-technical, non-ham employees—not all are knowledgeable about RF exposure. Therefore the ARRL HQ building area and W1AW have been evaluated as an uncontrolled exposure environment.

100% MPE

5% MPE

Fig 4.2—A bystander can be at a fixed distance from your antenna and be exposed to 100% of the MPE or 5% of the MPE—depending on the antenna height and the bystander's altitude.

such as for motorists driving past your home. If you do this, however, you should take steps to *ensure* that any exposure is transitory; if the motorist parks on the side of the road, exposure may be exceeded.

Uncontrolled/General Public

The *uncontrolled* environment is intended for areas that are accessible by the general public, normally your neighbors' properties and the public areas around your home. The MPE levels are lower for an uncontrolled environment than they are for a controlled environment. In an uncontrolled environment, exposure is averaged over a 30-minute period.

The uncontrolled environment limits are more stringent than the controlled environment limits. In an uncontrolled environment, the people being exposed are *not normally aware of the exposure*. This applies to all property near your station where you don't control access: sidewalks, neighboring homes and other areas that might have some degree of public access.

Some judgment may be necessary to determine where and when to apply the two exposure environments. Service personnel climbing a tower to make antenna repairs or working on a rooftop location making transmitter repairs can probably be assumed to have the necessary understanding of RF exposure—therefore they can be considered as being in a controlled environment. However, if a crew shows up to repair the roof, or someone is hired to paint the tower, it is likely that they are not aware of RF exposure. Under those circumstances, an uncontrolled environment would probably apply, unless they are given proper training.

Table 4.1

Limits for Maximum Permissible Exposure (MPE) Limits

Frequency Range (MHz)	Controlled Exposure (6-Minute Average)			Uncontrolled Exposure (30-Minute Average)		
	Electric Field Strength (V/m)	Magnetic Field Strength (A/m)	Power Density (mW/cm²)	Electric Field Strength (V/m)	Magnetic Field Strength (A/m)	Power Density (mW/cm²)
0.3-3.0	614	1.63	(100)*			
3.0-30	1842/f	4.89/f	(900/f²)*			
0.3-1.34				614	1.63	(100)*
1.34-30				824/f	2.19/f	(180/f²)*
30-300	61.4	0.163	1.0	27.5	0.073	0.2
300-1500	—	—	f/300	—	—	f/1500
1,500-100,000	—	—	5	—	—	1.0

f = frequency in MHz
* = Plane-wave equivalent power density

Note 1 to Table 4.1: Occupational/controlled limits apply in situations in which persons are exposed as a consequence of their employment provided those persons are fully aware of the potential for exposure and can exercise control over their exposure. Limits for occupational/controlled exposure also apply in situations when an individual is transient through a location where occupational/controlled limits apply provided he or she is made aware of the potential for exposure.

Note 2 to Table 4.1: General population/uncontrolled exposures apply in situations in which the general public may be exposed, or in which persons that are exposed as a consequence of their employment may not be fully aware of the potential for exposure or can not exercise control over their exposure.

Table 4.2

Maximum Permissible Exposure for Frequencies in the Amateur Radio Service Controlled RF Environment

Notes 1, 2 and 3 apply to all table entries.

Frequency Band	E field V/m	H field A/m	Power density mW/cm²	Notes
2.0 MHz	614.0	1.63	(100)	4, 5
4.0 MHz	460.5	1.23	(56.25)	5, 6
7.3 MHz	252.4	0.670	(16.89)	5, 6
10.15 MHz	181.5	0.482	(8.74)	5, 6
14.35 MHz	128.4	0.341	(4.38)	5, 6
18.168 MHz	101.4	0.270	(2.73)	5, 6
21.45 MHz	85.9	0.228	(1.96)	5, 6
24.99 MHz	73.8	0.196	(1.45)	5, 6
29.7 MHz	62.1	0.165	(1.03)	5, 6
50.0 MHz	61.4	0.163	1.00	4
144.0 MHz	61.4	0.163	1.00	4
219.0 MHz	61.4	0.163	1.00	4
222.0 MHz	61.4	0.163	1.00	4
420.0 MHz	—	—	1.40	7, 8
902.0 MHz	—	—	3.01	7, 8
1.24 GHz	—	—	4.14	7, 8
2.3 GHz	—	—	5.00	4, 7
3.3 GHz	—	—	5.00	4, 7
5.65 GHz	—	—	5.00	4, 7
10.1 GHz	—	—	5.00	4, 7
24.0 GHz	—	—	5.00	4, 7
47.0 GHz	—	—	5.00	4, 7
75.5 GHz	—	—	5.00	4, 7
119.98 GHz	—	—	—	9
142.0 GHz	—	—	—	9
241.0 GHz	—	—	—	9
Above 300 GHz	—	—	—	9

Notes:

Note 1: The FCC has determined that in most cases, amateurs and their immediate families may be considered as being in a controlled RF environment. See other sections of this book and FCC material for more information.

Note 2: The levels in Table 4.2 and Table 4.3 represent the level of RF fields or power density that are at the MPE. These levels assume continuous exposure. The evaluation chapter has a section on determining average exposure that offers more information. .

Note 3: The values in this table represent values averaged over a 6-minute time period. They represent the worst-case value for the listed amateur band.

Note 4: The MPE limit for this band is uniform across the entire band.

Note 5: The power density on this band is expressed as a plane-wave equivalent power density. See text.

Note 6: The MPE limit for this band varies with frequency. The MPE limit for the E field is determined by the formula $E = 1842/f$, with E in V/m and f in MHz. The MPE limit for the H field is determined by $H = 4.89/f$ with H in A/m and f in MHz. The Power density is determined by $S = 900/f^2$ with S in mW/cm² and f in MHz.

Note 7: On this band, the MPEs are specified only in terms of power density.

Note 8: The MPE limits on this band vary with frequency. The MPE limit for power density is determined by the formula $S = f/300$ with S in mW/cm² and f in MHz.

Note 9: The regulations do not specify an MPE on this band.

Maximum Permissible Exposure (MPE)

The rules and guidelines set limits to the maximum permissible exposure for humans who are near radio transmitters. The regulations control *exposure* to RF fields, not the *strength* of RF fields. There is no limit to how strong a field can be as long as no one is being exposed to it, although FCC regulations require that amateurs use the minimum necessary power at all times (§97.313(a)). If the operation of your station resulted in exposure over the limits in areas where there are no people *at the time you are operating*, the station is still in compliance.

Some amateurs have misinterpreted some of the categorical exemptions for evaluation (more about that later). All radio stations must comply with the requirements for MPEs, even QRP stations running only a few watts or less! The MPEs vary with frequency, as shown in Table 4.1. They have been summarized for each amateur band in Tables 4.2 and 4.3. The numbers in Tables 4.2 and 4.3 have been rounded up to the next highest decimal tenth. All tables discussed here are printed at the end of this chapter.

MPEs are derived from the Specific Absorption Rate (SAR)—the rate at which tissue absorbs RF energy, usually expressed in watts per kilogram (W/kg). The FCC MPEs are not based strictly on IEEE/ANSI C95.1-1992, but rather on a hybrid between that standard and the report written by the NCRP (the National Council on Radiation Protection and Measurements). NCRP is a body commissioned to develop recommendations for federal agencies. In terms of exposure levels, the NCRP report recommends lower MPE levels over some frequency ranges than are found in the IEEE/ANSI standard. The most stringent requirements are from 30 to 300 MHz, because various human-body resonances fall in that frequency range.

MPE limits are specified in maximum electric and magnetic fields for frequencies below 30 MHz, in power density for frequencies above 300 MHz and all three ways for frequencies from 30 to 300 MHz. The fundamentals chapter (Chapter 2) explains just what E and H fields are, and how they relate to power density. For compliance purposes, all these limits must be considered separately—if any one is exceeded, the station is not in compliance.

For example, if a 144-MHz amateur station had an H-field exposure level of 0.08 A/m (Amperes/meter) and an E-field exposure of 22 V/m (Volts/meter), the station is not in compliance for exposure to the general public because the H field limit has been exceeded, even though the E field is below the limits.

MPEs Vary With Frequency

The MPE limits vary with frequency, as shown in Tables 4.1, 4.2 and 4.3. The human body is roughly resonant (for different body sizes and under differing conditions) at frequencies between 30 and 300 MHz. The MPEs are the most stringent over this frequency range. In Table 4.1, there are separate limits for the electric field (E field), the magnetic field (H field) and power density. The electric field is specified in volts per meter (V/m), the magnetic field is specified in amperes per meter (A/m) and the power density is specified in milliwatts per square centimeter. In addition, for some frequencies the term "plane-wave equivalent power density" is used.

Some hams will use these MPE levels to help determine their station's compliance. Many others, however, will use the simple charts and tables discussed in this chapter and in the evaluation chapter.

Table 4.3

Maximum Permissible Exposure for Frequencies in the Amateur Radio Service Uncontrolled RF Environment

Notes 1, 2 and 3 apply to all table entries.

Frequency Band	E field V/m	H field A/m	Power density mW/cm²	Notes
2.0 MHz	412.0	1.095	(45)	4, 5
4.0 MHz	206.0	0.548	(11.25)	5, 6
7.3 MHz	112.9	0.300	(3.38)	5, 6
10.15 MHz	81.2	0.216	(1.75)	5, 6
14.35 MHz	57.5	0.153	(0.88)	5, 6
18.168 MHz	45.4	0.121	(0.55)	5, 6
21.45 MHz	38.5	0.103	(0.40)	5, 6
24.99 MHz	33.0	0.088	(0.29)	5, 6
29.7 MHz	27.8	0.074	(0.21)	5, 6
50.0 MHz	27.5	0.073	0.20	4
144.0 MHz	27.5	0.073	0.20	4
219.0 MHz	27.5	0.073	0.20	4
222.0 MHz	27.5	0.073	0.20	4
420.0 MHz	—	—	0.28	7, 8
902.0 MHz	—	—	0.61	7, 8
1.24 GHz	—	—	0.83	7, 8
2.3 GHz	—	—	1.00	4, 7
3.3 GHz	—	—	1.00	4, 7
5.65 GHz	—	—	1.00	4, 7
10.1 GHz	—	—	1.00	4, 7
24.0 GHz	—	—	1.00	4, 7
47.0 GHz	—	—	1.00	4, 7
75.5 GHz	—	—	1.00	4, 7
119.98 GHz	—	—	—	9
142.0 GHz	—	—	—	9
241.0 GHz	—	—	—	9
Above 300 GHz	—	—	—	9

Notes:

Note 1: The uncontrolled RF environment applies in general to all areas where the general public could be reasonably expected to be exposed. This may include some Amateur Radio stations such as a club station located in an area accessible to the public or certain Field Day sites, as examples. See other parts of this book and FCC material for more information.

Note 2: The levels in Table 4.2 and Table 4.3 represent the level of RF fields or power density that are at the MPE. These levels assume continuous exposure. The evaluation chapter has a section on determining average exposure that offers more information. .

Note 3: The values in this table represent values averaged over a 30-minute time period. They represent the worst-case value for the listed amateur band.

Note 4: The MPE for this band is uniform across the entire band.

Note 5: The power density on this band is expressed as a plane-wave equivalent power density. See text.

Note 6: The MPE for this band varies with frequency. The MPE for the E field is determined by the formula $E = 824/f$, with E in V/m and f in MHz. The H field is determined by $H = 2.19/f$ with H in A/m and f in MHz. The Power density is determined by $S = 180/f^2$ with S in mW/cm² and f in MHz.

Note 7: On this band, the MPEs are specified only in terms of power density.

Note 8: The MPE limits on this band vary with frequency. The MPE limit for power density is determined by the formula $S = f/1500$ with S in mW/cm² and f in MHz.

Note 9: The regulations do not specify an MPE on this band.

Average Exposure

Table 4.1 shows the MPE limits for various power levels and frequency ranges. MPEs assume continuous-duty and operation at the average rate. The levels shown assume that the exposure will be continuous over the exposure period. The regulations, however, average the total exposure over 6 minutes for controlled environments and 30 minutes for uncontrolled environments. This average includes both the duty factor of the operating mode and the actual on and off times over the worst-case averaging period. In most cases, the average power of an amateur station is considerably less than its peak power, and the ratio of transmit time to nontransmit time is less than 100%.

Thus it is permissible to exceed the MPE limits for periods of time, as long as that is offset by a corresponding reduction in the exposure limits for other periods of time within the averaging period. For example, in an uncontrolled environment, if one is in a field that is 10 times the limit for 3 minutes, this is acceptable as long as one has no exposure for the preceding 27 min-

utes *and* the following 27-minute periods. The 30-minute window is not based on arbitrary half-hour segments by the clock, but is a "sliding" window, such that in *any* 30-minute period, the total exposure must be below the limits. It would not be acceptable to have no exposure for 15 minutes, twice the exposure for 15 minutes, then be exposed at the limit for the next 15 minutes because the total exposure in the worst-case window (the last 30 minutes) exceeds the average MPE levels. There are a number of ways to calculate the time-averaged exposure; these are discussed at length in the evaluation chapter.

Who Must Comply?

All Amateur Radio stations must comply with the MPE limits, regardless of power, operating mode or station configuration. (Even W1RFI's 10-milliwatt station must comply.—*Ed.*) It is unlikely that low-power stations would exceed the limits, but rules apply equally to all.

Routine Environmental Evaluations

The cores of the requirements under these regulations are the MPE levels. However, the core of the specific actions that need to be taken by Amateur Radio operators is the requirement for some amateurs to perform a "routine environmental evaluation" for RF exposure. This will establish that the station is being operated in compliance with the FCC RF-Exposure guidelines.

A routine evaluation is not nearly as onerous as it sounds! This subject is covered in detail in the evaluation chapter, but it is summarized here, as part of the discussion about the rules. Doing an evaluation will help ensure a safe operating environment for amateurs, their families and neighbors.

The FCC is relying on the demonstrated technical skill of Amateur Radio operators to *evaluate their own stations* (although it is perfectly okay for an amateur to rely on another amateur or skilled professional to perform the evaluation).

Most evaluations will not involve measurements, but will be done with comparisons against typical charts developed by the FCC, relatively straightforward calculations or computer modeling of near-field signal strength. The FCC encourages flexibility in the analysis, and will accept any technically valid approach.

It is not difficult to do the necessary station evaluation. The FCC guidance is contained in *OET Bulletin No. 65: Evaluating Compliance With FCC-Specified Guidelines for Human Exposure to Radio Frequency Radiation* and an Amateur Radio

Table 4.4

Power Thresholds for Routine Evaluation of Amateur Radio Stations

Wavelength Band	Evaluation Required if Power* (watts) Exceeds:
MF	
160 m	500
HF	
80 m	500
75 m	500
40 m	500
30 m	425
20 m	225
17 m	125
15 m	100
12 m	75
10 m	50
VHF (all bands)	50
UHF	
70 cm	70
33 cm	150
23 cm	200
13 cm	250
SHF (all bands)	250
EHF (all bands)	250
Repeater stations (all bands)	*non-building-mounted antennas*: height above ground level to lowest point of antenna < 10 m *and* power > 500 W ERP *building-mounted antennas*: power > 500 W ERP

*Transmitter power = Peak-envelope power input to antenna. For repeater stations **only**, power exclusion based on ERP (effective radiated power).

supplement to that bulletin, *Supplement B to OET Bulletin 65: Additional Information for Amateur Radio Stations*. This book and the FCC material contain the basic information hams need to evaluate their stations, including a number of tables showing compliance distances for typical amateur power levels and antennas. Generally, hams will use these tables to evaluate their stations.

In most cases, hams will be able to use a table that best describes their station's operation to determine the minimum compliance distance for their specific operation. These tables show the compliance distances for uncontrolled environments for a particular type of antenna at a particular height. (The power levels shown in the tables are average power levels, adjusted for the duty cycle of the operating mode being used, and operating on and off time, averaged over 6 minutes for controlled environments or 30 minutes for uncontrolled environments.) The FCC tables, plus a number of similar tables developed by the ARRL, are found in the reprint of *Bulletin 65* in Chapter 6 of this book, the *Amateur Supplement B* in Chapter 7 of this book and in the tables in Chapter 8 of this book.

Categorical Exemptions

Some types of amateur stations do not need to be evaluated (but these stations must still comply with the MPE limits!) The FCC has exempted these stations from the evaluation requirement because their output power, operating mode, use, frequency or antenna location are such that they are presumed to be in compliance with the rules. These stations are *not* exempt from the rules, but are presumed to be in compliance without the need for an evaluation.

• Stations using the peak-envelope power levels or less to the antenna as shown in Table 4.4.
• Amateur repeaters using 500-W ERP or less.
• Amateur repeaters with antennas not mounted on buildings if the antenna is located more than 10 meters (32.8 feet) high above ground.
• Amateur mobile and portable hand-held stations using push-to-talk or equivalent operation.

Note that Table 4.4 cites power to the antenna. This is not the same as your transmitter output power, although you can conservatively use your transmitter output power to decide if you need to do an evaluation, if you wish. As an example, if you are running 90 W PEP and have a feed line loss of 3 dB, you are losing approximately 50% of your power in the feed line, so you have approximately 45 W PEP to the antenna.

This part of your operation would not have to be evaluated on any band. Note, too, that unlike the exposure, the levels in this table are *not* average-power levels, but are peak-envelope powers (PEP). If you transmit only one short dit per 30-minute period, and that dit is transmitted at levels above those in the chart, you will still have to do an evaluation. When you did the evaluation, however, you could use average power. Admittedly, it sounds a bit complex, but it is explained in detail in Chapter 5.

Stations that use more power than the power levels to the antenna shown in Table 4.4 must be evaluated. For the majority of amateurs, this change has virtually eliminated the need to perform station evaluations. Most HF transceivers are rated at 100-W PEP output; on 15 meters and below, stations using this power level need not be evaluated. Most VHF transceivers are rated at 50-W PEP output or less; stations using this power level on VHF need not be evaluated. (Statistically, most HF operators use "barefoot" rigs, typically 100-W PEP output.) While this change doesn't cover all barefoot HF operation, operators who wish to use 12 and 10 meters could either perform an evaluation for those two bands, or they could reduce power to the levels in Table 4.4 and forego the evaluation altogether.

News for Repeater Operators

The repeater exemption was added with an Erratum to the rules issued by the FCC in October 1997. All amateur repeaters operating at a power of 500 W ERP or less are generally categorically exempt from evaluation. All amateur repeaters whose antennas are not mounted on buildings and that have all parts of their radiating antenna located at least 10 meters (32.8 feet) above ground also are exempt. Amateur repeaters with antennas located on buildings (presumably buildings where people could be located) must be evaluated if they use more than 500 W ERP. There is more information about calculating ERP in Chapter 5, but to summarize, ERP is derived by multiplying the power to the antenna by the numerical gain of the antenna over a dipole (6 dBd, for example, represents a numerical equivalent of 3.98). This categorical exemption from evaluation will probably cover many repeater stations.

Mobile and Portable Hand-Held Operators

They are not specifically mentioned in Table 4.4, but §1.1307(b)(2) of the FCC rules and the Report and Order cover portable and mobile devices. As described in

FCC Office of Engineering and Technology (OET) *Bulletin 65*

To help hams perform the routine evaluation, the FCC has prepared a bulletin, *OET Bulletin No. 65: Evaluating Compliance With FCC-Specified Guidelines for Human Exposure to Radio Frequency Radiation* and an Amateur Radio supplement to that bulletin, *Supplement B to OET Bulletin 65: Additional Information for Amateur Radio Stations.* The FCC bulletins are available from their Web site; see Appendix E, Resources. When this book refers to *Bulletin 65*, it is generally referring to both the main *Bulletin* and *Supplement B* together. *Supplement B* makes many references to *Bulletin 65*, so most hams will want to read both. The main *Bulletin* contains a section specifically addressing Amateur Radio. The applicable parts of the *Bulletin* and *Supplement* is printed in Chapters 6 and 7 of this book.

Not Cast in Stone

Although the regulations are firm requirements, the FCC intends that *Bulletin 65* is advisory in nature. To quote directly from the bulletin:

This revised OET Bulletin 65 has been prepared to provide assistance in determining whether proposed or existing transmitting facilities, operations or devices comply with limits for human exposure to radio-frequency (RF) fields adopted by the Federal Communications Commission (FCC). The bulletin offers guidelines and suggestions for evaluating compliance. *However, it is not intended to establish mandatory procedures, and other methods and procedures may be acceptable if based on sound engineering practice.*

This flexibility applies especially to the Amateur Radio Service; the FCC is relying on the technical ability of hams to select an appropriate method of analysis for their station evaluations. While *Bulletin 65* outlines several acceptable ways for amateurs to satisfy the FCC regulations, it is not intended to define the only methods amateurs can use. Amateurs are permitted to use any method that has technical validity. This could include accurate field-strength measurements, calculation from valid field-strength formulas and principles or computer modeling using programs based on accepted algorithms such as *NEC* or *MININEC* code.

What's in the Bulletin?

In general, *Bulletin 65* outlines how hams can use formulas, tables and graphs, computer software or measurements to complete their evaluations. All these methods are discussed in detail in Chapter 5 of this book. Most hams will probably choose to use the simple tables.

The "core" *Bulletin 65* was written primarily for commercial radio stations, although the information can be used by any radio service. It begins with a brief historical introduction, followed by definitions of key terms. Next comes formulas about MPEs and some descriptions of parts of the rules. While hams can use this existing bulletin to complete their station evaluations, they need to be careful. It is easy to get lost in the complex formulas and explanations intended to be most helpful to other radio services. Most hams will find the amateur supplement a lot easier to use.

Formulas

Bulletin 65 describes how to use far-field formulas to obtain estimates of field strengths in the near field. *NEC4* modeling done by the ARRL shows that the formula applies conservatively to antennas like dipoles and Yagis. The ARRL Laboratory staff found, however, that it does *not* apply well to some antenna types such as small loops, so these formulas should be used with some caution.

The formulas apply to the field-strength levels *in the main beam of the antenna.* For this reason, they may result in an overly conservative estimate for many actual installations. The ARRL has supplied the FCC with data tables illustrating antennas modeled over real grounds to offer realistic compliance distances. *Supplement B* includes some of these tables in the amateur supplement to *Bulletin 65*, along with some simple tables based on the worst-case formulas.

Easy Way Out

You don't need to resort to complicated formulas to do the worst-case analysis. If you have access to the World Wide Web, check the University of Texas Amateur Radio Club site at **http://www.cs.utexas.edu/users/kharker/ rfsafety/**. You'll find a "form" that allows you to enter transmitter power, antenna gain and distance. After you enter the information, it calculates the field strength and tells you if you are in compliance. (If you're *not* in compliance, it tells you at what distance you would be in compliance.)

These simple calculations can be a good tool because if you pass "worst case," you pass. If you use peak-envelope power in these estimates, this is truly a worst case; the regulations are specified in terms of *average* exposure, averaged over 30 minutes for uncontrolled exposure environments, 6 minutes for controlled environments. You also should use the ground-reflection options that are part of the formulas or programs on the page, if you want to ensure that you have a realistic estimate.

Amateur Supplement

Not surprisingly, *Supplement B* offers guidance for amateurs. In addition to reviewing the basics, it contains a large section with tables for different amateur bands, antenna gains, power and antenna type to show how far people need to be from an amateur antenna to be below the MPE levels. Most hams will probably use one or more of these tables to do their station evaluation. The ARRL supplied many of these tables to the FCC, as did the W5YI Group and Wayne Overbeck, N6NB. The FCC could not print them all, so Chapter 8 of this book picks up where the FCC left off—using the same method used by the FCC for their tables.

Both documents are available in their full form for download from the FCC. The URL is **http://www.fcc.gov/ oet/info/documents/bulletins/#65**, but it is just as easy to start with the ARRL Web page, **http://www.arrl.org/ news/rfsafety.**—*Ed Hare, W1RFI, ARRL Laboratory Supervisor*

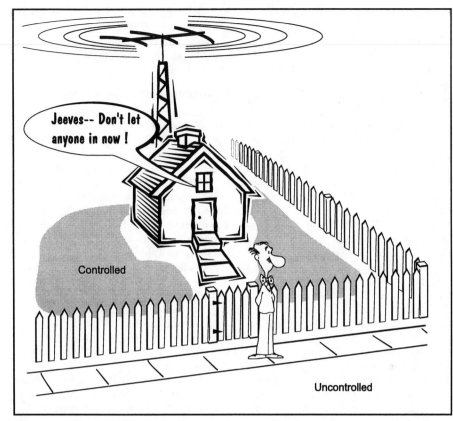

Fig 4.3—If you cannot control access to the area, the average for *uncontrolled* areas apply.

§§1.1307(b)(1), 1.1307(b)(2), 2.1091(c) and 2.1093(c) of the FCC regulations, there is no specific requirement that mobile and portable devices used under Part 97 (Amateur Radio) be evaluated. The 1996 Report and Order announcing the rules further amplified that mobile and portable devices specifically using push-to-talk operation, as used by police, taxi-cab *and Amateur Radio*, for examples, generally need not be evaluated. This is because of the low power, low operating duty cycles generally employed and the expected shielding of the vehicle occupants by the vehicle body. Most Amateur Radio mobile or portable stations that meet these general criteria do not need to be evaluated.

It must be added that the terms *mobile* and *portable* mean different things to the FCC than they might mean to hams. Both terms cover transmitters not used in a fixed location, but a portable device is one that is customarily operated with the antenna located within 20 centimeters of the body.

Exceptions to the Evaluation Exemptions

There is an exception to every rule, and this old adage could apply to stations that are categorically exempt from the requirement to evaluate. That exemption is not absolute. No station is exempt from the requirement not to exceed the MPE levels. There are some station configurations that could result in exceeding the limits, even for stations that are normally exempt. When the FCC wrote the table for categorical exemptions from the evaluation requirements, they had to balance a number of different factors. On one hand, most amateur operation is already within the guidelines, so they wanted to exclude most relatively low-power operation. On the other hand, some station configurations, even for low-power stations, could present problems. Of course, the FCC wants to ensure that the administrative burden of these regulations in minimal. They didn't want an overly complex set of requirements that could cover every possible combination of what could and could not exceed the MPE limits. The overall goal of generally increasing the awareness by Amateur Radio operators of the RF-exposure potential of their stations was included in the mix, too. The result is seen in the evaluation requirements of Table 4.4.

Many classes of amateur stations are categorically exempt from the need to do a station evaluation. This is because the exempt stations are usually operated such that the station is in compliance with the MPEs. Under some circumstances, such as an an-

tenna that is located unusually near people or in some mobile installations, it is possible to exceed the MPE levels.

Sections §1.1307(c)(d) of the FCC's rules stipulate that the *Commission may require that a station that is normally categorically exempt from the requirement to perform a routine evaluation perform such an evaluation if the FCC determines that there is reason to believe that the station may be exceeding the MPEs allowed.*

The FCC will generally handle these exceptions on a case by case basis. In addition, the FCC will also rely on amateurs to voluntarily consider whether any operating parameter of their stations might also indicate that it is prudent to do a station evaluation—even in cases where the category of that station would otherwise make it exempt. If an antenna is located unusually close to people, such as an indoor antenna in a living space or a balcony mounted antenna a foot or so away from a neighbor's balcony, the FCC could require a station evaluation or take other action.

Mobile stations should also be closely considered before an amateur automatically applies the categorical exemption. As an example, a 500 W, 10-meter mobile installation with a vehicle-mounted antenna would certainly merit a closer look. On VHF, the use of a high-power amplifier could also present problems in some cases. In general, it is recommended that in these higher power installations, the antenna be located such that the vehicle occupants will be shielded from the antenna during normal use. One good location is in the center of an all-metal roof. Locations to be avoided for high-power operation include a trunk-mounted antenna located near a rear window or a mobile installation in a vehicle with a fiberglass roof. In general, mobile installations, even higher-power ones, will not exceed the MPEs if sound installation guidelines are followed. *The ARRL Handbook* and *The ARRL Antenna Book*, available from the ARRL, have additional material on mobile installations and antennas.

Even if the regulations do not require you to do an evaluation, there could be a number of reasons to do one anyway. As a minimum, it will be good practice for the time that you make a station change that might require evaluation. The results of your evaluation will certainly demonstrate to yourself and possibly your neighbors that your station's operation is well within the guidelines, and is no cause for concern. In the case of some of the unusual circumstances just described, the regulations could require an evaluation of a station otherwise categorically exempt. In all cases, regardless of categorical exemption, the regulations require that the MPE limits not be exceeded. In most cases, the FCC will rely on amateurs to determine for themselves how the evaluation requirements apply to their stations. Remember, under the rules, the FCC can ask an evaluation be performed on any transmitter regulated by the FCC.

Multitransmitter Sites

The rules are intended to ensure that operation of transmitters regulated by the FCC doesn't result in exposure in excess of MPE limits. It is fairly easy to make this determination for single transmitters, when there are no other sources of RF to complicate things. However, many transmitters operate in proximity to other transmitters, and it is entirely possible for two or more transmitters to all be below their own limit, but the total exposure from them all operating together to be greater than the permitted MPEs.

The FCC regulations, the Report and Order and the two Memorandum Opinion and Orders all cover the likely situation of multiple transmitters. The bottom line is that, in most cases, all the *significant* RF transmitters operating at multitransmitter sites generally must be considered when determining if the site's total exposure is in compliance. In addition, all significant emitters are jointly responsible for overall site compliance.

The rules stipulate that in a multitransmitter environment, a single transmitter operator is jointly responsible with other operators at the site for all areas at the site where the exposure from that transmitter is greater than 5% *of what is permitted for that transmitter.* (This is 5% of the permitted power density or 5% of the square of the E or H-field value.) Note that this is *not* the same as 5% of the total exposure, which could sometimes be unknown.

In many cases involving Amateur Radio transmitters, only a relatively small area would be encompassed by that 5% exposure threshold, so joint responsibility might only exist in the immediate vicinity of the amateur antenna. A repeater trustee, for example, might have that 5% level extend only to those areas to 10 feet above and below the antenna up the tower, and thus be responsible for overall site compliance only to that area on the tower. In this case, the responsibility may be only to radio service personnel climbing the tower (generally a controlled exposure environment would apply) or tower maintenance people (who may or may not be trained about RF exposure, so an uncontrolled environment may be more appropriate).

However, some types of stations, such as amateur repeaters using 500 W ERP or less, do not need to be evaluated. *Bulletin 65* clarifies that these stations are presumed to be in compliance with their own individual MPE limits and *generally* do not need to be included when calculating overall site compliance. They are presumed not to be jointly responsible for site compliance. These are *not* iron-clad assumptions. The FCC rules, in §1.1312(a), allow the FCC to require any station to file an Environmental Assessment (EA) or conduct a routine environmental evaluation to demonstrate compliance —even those covered by specific categorical exemptions.

The FCC will make these determinations on a case by case basis. In cases where a station is categorically exempt from evaluation, or a station is creating exposure that is less that 5% of what is permitted to it, the FCC could determine that the particular station needs to share responsibility for site compliance.

Clearly, if an amateur station shares space with a high-power broadcast station, the "5% rule" is pretty straightforward.

However, if a number of low-power transmitters share a site, even minor emitters might have to make changes to their station if the overall site compliance is more than the MPE limits allow. It is quite possible for some sites to have literally hundreds of transmitters, most operating below the 5% level, even though the overall site's RF exposure is greater than the MPE limits. The best approach is to err on the side of caution, and cooperate with other operators on the site if there is a compliance problem. There is, of course, no substitute for your own good judgment. Use it as it appears to be appropriate in "gray" areas. This may prevent the FCC from having to make a "federal" case out of your station.

Often, you may not know much about the other transmitters on your site. In that case, you should make the best assumptions you can about the other stations' power, antenna gains and operating duty cycles, and conduct your assessment of site compliance accordingly.

The methods used to evaluate stations in a multitransmitter environment are generally straightforward, but not quite as simple as the compliance distance tables discussed in Chapter 5.

Paperwork and Proof of Evaluation

Once an Amateur Radio operator determines that a station complies, station operation may proceed. There's no need for FCC approval before operating. Other than a short certification on Form 610 station applications, the regulations do not normally require hams to file proof of evaluation with the FCC. The Commission recommends that each amateur keep a record of the station evaluation procedure and its results, in case questions arise.

Environmental Assessment

Once an amateur completes the necessary evaluation and determines that his or her station does not exceed the MPE limits, the station may be put into operation. It is not necessary to file any paperwork with the FCC. The regulations discuss EAs (environmental assessments). EAs are *not* normally required for amateur stations. An EA is required for any station that will continue to operate even though it exceeds the limits in Table 4.1.

Actually, much of what hams are considering as "regulations" are more appropriately called "guidelines." The actual regulations are simple: stations must either not exceed the limits in the published guidelines or, if they do, the operators must file an EA with the FCC. In practice, however, it is not likely that any amateur station operator would find it easier to file an EA than to make station changes to comply with the guidelines. Even in the commercial world, EAs are not often used as a means of complying with the regulations.

Examinations

The regulations add the requirement to include five questions on the topic of RF environmental safety practices to each Amateur Radio examination for Novice, Technician and General licenses. Thus the VEC Question Pool Committee (QPC) had to add 55 new questions to each of these pools. In response to a request from the ARRL, the question pools are being updated in their normal cycle. The QPC has completed the questions for the Novice and Technician license examinations in the latest revision of the pool, released December 1, 1996, for examinations beginning July 1, 1997. Completing the major revisions for the Novice and Technician license pools in such a short period of time speaks highly for the dedication of this hard working committee. ARRL is proud to serve as a participant on the QPC. The General license pool was updated in late 1997, for examinations beginning July 1, 1998.

FIXING PROBLEMS

Chapter 5, How to Evaluate an Amateur Station, discusses how to correct problems. In summary, the FCC and ARRL have estimated that most amateur stations are already in compliance with the MPE levels. Some amateurs, especially those using indoor antennas or high-power, high-duty-cycle modes such as RTTY bulletin stations and moonbounce stations, may need to make adjustments to their station or operation to be in compliance. *Bulletin 65* offers guidance and flexibility on what the FCC considers acceptable. Hams can adjust their power, mode, frequency, antenna location, antenna pointing or operating on-and-off times to bring their operation into compliance. For example, if you discovered that you were not in compliance after 25 minutes of operation with your antenna pointed in a particular direction, you could either not point your antenna in that direction, or take a five-minute break for a period after 25 minutes of operation.

IN SUMMARY

In general, these rules are not hard to understand. They are based on the sound science that went into developing the standards. Ed Hare, W1RFI, of the ARRL Laboratory has spent a good part of the last year studying all the complex issues surrounding these rules. He offers an observation, "Following these rules is important for the Amateur Radio Service, not only because we should uphold our reputation for following the rules, but because they help us to demonstrate to ourselves, our families, our neighbors and anyone else with questions that the operation of Amateur Radio Stations is safe."

How to Evaluate an Amateur Station

5

This chapter describes a number of techniques that can be used to evaluate single-transmitter installations, multiple-transmitter installations and repeaters. Most hams will elect to use the simple tables that show compliance distance at a glance.

THE ESSENCE OF THE RULES

There is an old saying: If a tree falls in the forest and there is no one there to hear it, does it make a sound? In radio, with regard to MPE limits, the answer is "No."

The FCC regulations cover *exposure* of people to RF energy, not the strength of RF energy where people are not being exposed. This principle applies to most aspects of a routine station evaluation. For example, if you find that exposure to the corner of a neighboring property is over the limit, it is only over the limit if someone remains in that area for an extended period. As another example, if you find that an area is at twice the limit, but you *know* that it is only occupied for 1 minute out of every hour, the exposure is below the limit.

The crux of the requirements of a station evaluation is found in *OET Bulletin 65*:

Before causing or allowing an Amateur Radio station to transmit from any location where the operation of the station could cause human exposure to RF electromagnetic energy in excess of the FCC RF-exposure regulations, amateur licensees are required to take certain actions. A routine RF-radiation evaluation is required unless the station is categorically excluded from the requirement to perform a station evaluation.

This chapter deals exclusively with the actual evaluation, based on compliance with the MPE (Maximum Permissible Exposure) limits. Refer to the earlier chapters in this book for information on how

these limits related to exposure, safety and specific absorption rates (SAR).

The Amateur Radio service is a lot more diverse than many radio services regulated by the FCC. If the FCC had to spell out the specific requirements of doing a station evaluation for every possible configuration in the rules, the rules would be larger than this book. Amateur Radio operators are licensed to use a wide range of frequencies and operating modes. Amateur Radio operation ranges from low-power (QRP) operation of a few milliwatts to 1500 watts PEP. Each operating mode has its own particular duty cycle and pattern of operation. Amateurs also use a wide range of antennas, from simple wires to tower-mounted gain antennas, to name just two. The diversity of Amateur Radio operation is one of its strengths, enabling amateurs to perform a wide range of technical investigations and operations under adverse conditions. The diversity, however, may require that amateurs choose from a number of methods to perform the station analysis and evaluation required by FCC regulations.

Certain Amateur Radio installations were made subject to a requirement that the station operator perform a routine analysis to establish that the station is being operated in compliance the FCC RF-Exposure regulations. The determination of just which stations need to be evaluated is based on power levels, frequency and the type of station.

The FCC is relying on the demonstrated technical skill of Amateur Radio operators to evaluate their own stations (al-

though it is perfectly okay for an amateur to rely on another amateur or skilled professional to perform the evaluation). The FCC regulations do *not* require that an amateur perform field-strength measurements. In many cases, the evaluation can be accomplished by some relatively straightforward calculations or compari-

Figure 5.1—Some stations can be rather complex. There are a lot of possible power, frequency, mode and antenna combinations that could be associated with this commercial installation. *(photo courtesy Robert Cleveland, FCC Office of Engineering and Technology)*

sons between station operation and typical graphs developed by the FCC. Once an Amateur Radio operator has performed the required routine station evaluation, and determined that the station does not exceed the permitted MPEs, the Amateur Radio station may be placed into immediate operation. It is not necessary to secure FCC approval before operating.

WHAT IS A "ROUTINE RF ENVIRONMENTAL EVALUATION"?

The core of the requirements under these regulations is the MPE levels. However, the specific actions that need to be taken by Amateur Radio operators is to perform a "routine RF environmental evaluation" to establish that the station is being operated in compliance with the FCC RF-Exposure guidelines. This generally consists of a series of calculations to determine compliance with the MPE levels—including those derived from power-density formulas and those obtained with *NEC*- or *MININEC*-based antenna-modeling programs. A routine evaluation will generally need to be done for both controlled and uncontrolled exposure environments. However, if a ham determines that his or her operation meets the requirements for uncontrolled exposure in his or her own station, home and property, it will not be necessary to evaluate the same areas for controlled exposure.

A routine environmental evaluation is not nearly as onerous as it sounds! It is generally not difficult to do the necessary station evaluation. In general terms, the FCC requires operators of radio transmitters be *aware* of the RF exposure potential from their stations. In doing the evaluation, amateurs will be considering the ways that people could be exposed to RF fields from the operation of their station. This can be done by either calculating or measuring the fields, or by using tables derived from those calculations

The following general factors can all play a part in doing a routine evaluation:
- Transmitter frequency
- Transmit power
- Operating mode
- Transmitter duty cycle
- Antenna location
- Antenna gain
- Antenna pattern
- General station configuration
- The amount of time people are exposed

Most evaluations will not involve measurements, but will be done with comparisons against typical tables that have been developed by the FCC, individual amateurs and the ARRL. In many cases, the evaluation can be as quick and easy as looking at a table that represents your op-

eration and determining that your antenna is far enough away from areas where people are located.

In most cases, hams will be able to use the table that best describes their station's operation to determine the minimum compliance distance for their specific operation. OET *Supplement B* (Chapter 7 of this book) contains a number of these tables (with the compliance distances converted to feet); additional tables are in Chapter 8 of this book, prepared using the same methods as were used for the *Supplement B* tables. The term *compliance distance* refers to the minimum distance one must be from an antenna to have the estimated fields be below the MPE limits.

Alternatively, hams could do relatively straightforward calculations of worst-case scenarios or computer modeling of near-field signal strength. The FCC encourages flexibility in the analysis, and will accept any technically valid approach.

WHO IS RESPONSIBLE FOR THE EVALUATION?

The rules generally require that the station licensee be responsible for ensuring that the evaluation is complete. If someone other than the licensee were acting as control operator, he or she also would also be responsible for the proper operation of the station under all FCC rules, including the rules on RF exposure.

WHERE CAN HAMS LEARN ABOUT DOING AN EVALUATION?

Hams could rely on their own personal technical expertise to know just what needs to be considered when doing an evaluation. However, for many hams, the whole topic is a "learning opportunity," because hams have never had *specific* requirements about RF exposure evaluations under the old RF-exposure rules and guidelines. Although most hams are enthusiastic about learning something new, they need some instruction and guidance.

The FCC didn't leave us out in the cold!

Drawing on the resources of both their staff and the amateur community, the FCC has prepared two documents, *OET Bulletin 65: Evaluating Compliance With FCC-Specified Guidelines for Human Exposure to Radio Frequency Radiation* and *OET Bulletin 65 Supplement B: Additional Information for Amateur Radio Stations*. In this chapter, *Bulletin 65* generally refers to *both* documents together. These FCC materials explain a number of different ways that hams can complete the required evaluations.

WHO NEEDS TO DO AN EVALUATION?

The good news is that most amateur stations *do not* need to be evaluated. The following classes of amateur stations are exempt *from the evaluation requirement* because their power levels, operating duty cycles or station configuration are such that they are *presumed* to be in compliance with the MPE limits:

- Stations using the peak-envelope power (PEP) input or less to the antenna shown in Table 5.1
- Amateur repeaters using 500 W or less effective radiated power (ERP)
- Amateur repeaters with antennas not mounted on buildings if the antenna is located more than 10 meters above ground
- Amateur mobile and portable hand-held stations using push-to-talk or equivalent operation

Unlike the rules for maximum amateur power, which are expressed in PEP output from the transmitter, the rules for determining which stations need to be evaluated are expressed in PEP *input to the antenna*. Table 5.1 shows peak-envelope power to the *antenna* as the deciding factor. Factors such as feed line losses and losses in accessories such as wattmeters and antenna tuners can reduce the power from your transmitter to be some fraction of its original value at the antenna.

Bulletin 65 is Not Mandatory

Although the regulations are firm requirements, *Bulletin 65* is advisory in nature. To quote directly from the bulletin:

"This revised OET *Bulletin 65* has been prepared to provide assistance in determining whether proposed or existing transmitting facilities, operations or devices comply with limits for human exposure to radio frequency (RF) fields adopted by the Federal Communications Commission (FCC). The bulletin offers guidelines and suggestions for evaluating compliance. *However, it is not intended to establish mandatory procedures, and other methods and procedures may be acceptable if based on sound engineering practice.*"

The flexibility offered by this language especially applies to the Amateur Radio Service; the FCC is relying on the demonstrated technical ability of hams to select an appropriate method of analysis for the evaluation that may be required for their station.

Table 5.1

Wavelength Band	Evaluation Required if Power* (watts) Exceeds:
MF	
160 m	500
HF	
80 m	500
75 m	500
40 m	500
30 m	425
20 m	225
17 m	125
15 m	100
12 m	75
10 m	50
VHF (all bands)	50
UHF	
70 cm	70
33 cm	150
23 cm	200
13 cm	250
SHF (all bands)	250
EHF (all bands)	250
Repeater stations (all bands)	non-building-mounted antennas: height above ground level to lowest point of antenna < 10 m and power > 500 W ERP building-mounted antennas: power > 500 W ERP

* Power = PEP input to antenna except, for repeater stations only, power exclusion is based on ERP (effective radiated power).

Figure 5.2—This repeater antenna is not mounted on a building and is located more than 10 meters above ground, so the operator of the repeater is not required to do a routine station evaluation.

power levels, but are peak-envelope powers (PEP), specified as power input to the antenna. If you transmit only one word per 30-minute period, and that word is transmitted at levels above those in the chart, you will still have to do an evaluation. When you do the evaluation, however, you can use average power. Admittedly, it sounds a bit complex, but it will be much more clear after you have read this chapter.

For the majority of amateurs, the power levels in Table 5.1 have virtually eliminated the need to perform station evaluations! Most HF transceivers are rated at 100-W PEP output; on 15 meters and below, stations using this power level need not be evaluated. Most VHF transceivers are rated at 50-W PEP or less; stations using this power level on VHF need not be evaluated. Statistically, most HF operators use "barefoot" rigs, typically 100-W PEP. Operators who wish to use 12 and 10 meters could either perform an evaluation for those two bands, or they could reduce power to the levels in Table 5.1 and forgo the evaluation altogether.

CATEGORICAL EXEMPTIONS

No station is exempt from the rules and the MPE levels, but many amateur stations are categorically exempt *from the requirement to perform a station evaluation.* Stations using the power levels in Table 5.1, or less, do not need to be evaluated. Mobile and portable (hand-held) stations using PTT operation do not need to be evaluated. Amateur repeaters using 500 W ERP or less also are categorically exempt from the requirement to evaluate.

News for Repeater Operators

The evaluation exemption for amateur repeater operation is determined by the *effective radiated power* (ERP) of the repeater. ERP is referenced to the gain of a half-wave dipole in free space (unlike equivalent isotropically radiated power, EIRP, which is referenced to an isotropic source). *Bulletin 65* describes how to calculate feed line losses and determine ERP for an amateur repeater.

All amateur repeaters using 500 W ERP or less generally do not need to be evaluated. This repeater exemption was added with an Erratum to the rules issued by the FCC in October 1997. Those that operate with more than 500 W ERP need to be evaluated if they have an antenna mounted on a building, or if any part of a non-building-mounted antenna is less than 10 meters (32.8 feet) above ground.

There is more information about calculating ERP later in this chapter, but to summarize, ERP is derived by multiplying the power to the antenna by the

The Rules chapter (Chapter 4) discusses the "letter of the law" about who needs to do a station evaluation. Many hams may find that they don't need to evaluate their station at all, because their power is low enough and their antennas are located far enough away from areas of exposure that they are not required to evaluate their stations. They are presumed to be in compliance with the MPE (maximum permissible exposure) levels. Those hams whose transmitter power is not more than the limits shown in Table 5.1 can stop right now; you do not need to do an evaluation, except perhaps in some rather unusual circumstances.

Note, too, that unlike the MPE limits, the levels in Table 5.1 are *not* average-

numerical gain of the antenna over a dipole (6 dBd, for example, represents a numerical equivalent of 3.98). This categorical exemption from evaluation covers many repeater stations.

Mobile and Portable (Hand-Held) Stations

According to *Supplement B*, all amateur mobile and portable hand-held operation is categorically exempt from the requirement to evaluate, although it is often a good idea to do so anyway. To clarify right up front, "portable" means something different to the FCC than it usually does to hams. To the FCC, a portable device is defined in the FCC rules as a non-fixed station customarily operated with its antenna within 20 cm of the body. Under the rules, mobile devices are evaluated to the MPE limits, while portable devices are generally evaluated to SAR limits. (See Chapters 1 and 2.)

As described in the FCC rules, there is no specific requirement that mobile and portable devices used under Part 97 (Amateur Radio) be evaluated. *Bulletin 65* explained that this applies particularly to amateur mobile operation using push-to-talk operation. Most Amateur Radio mobile or portable stations that meet these general criteria do not need to be evaluated.

They are not specifically mentioned in Table 5.1, but Section 1.1307(b)(2) of the FCC rules and the 1996 Report and Order cover portable and mobile devices. As described in Sections 1.1307 (b)(1), 1.1307 (b)(2), 2.1091 (c) and 2.1093 (c) of the FCC regulations, there is no specific requirement that mobile and portable devices used under Part 97 (Amateur Radio) be evaluated. The 1996 Report and Order announcing the rules further amplified that mobile and portable devices specifically using push-to-talk operation, as used by police, taxicab and *Amateur Radio*, generally need not be evaluated. This is because of the low power, low operating duty cycles generally employed and the expected shielding of the vehicle occupants by the vehicle body.

This is explained in *Bulletin 65* and *Supplement B. Bulletin 65* emphasizes that although this applies to all mobile and portable hand-held operation in the Amateur Radio Service, it is intended that this general categorical exemption apply to mobile or portable operation using push-to-talk (PTT) operation. In general, most mobile operation would be considered as being a controlled environment, as long as the operator and passengers were aware of the RF exposure.

The FCC has prepared another supplement to *Bulletin 65* that discusses evaluation of mobile and portable devices. While intended for evaluation of devices such as cellular telephones, this supplement may be of some passing interest to amateurs. It is known as *Supplement C to OET Bulletin 65*. It is available from the FCC or can be downloaded from the FCC web site.

If You Don't Need to Do an Evaluation

There is an exception to every rule, and this old adage could apply to stations that are categorically exempt from the requirement to evaluate. That exemption is not absolute. No station is exempt from the requirement not to exceed the MPE levels. There are some station configurations that could result in exceeding the limits, even for stations that are normally exempt.

If the regulations do not specifically require you to perform an evaluation, there could be a number of reasons to do one anyway. If nothing else, doing an evaluation now would be good practice for the day when you upgrade your station (by adding an amplifier or antenna, for instance) in such a way that makes an evaluation necessary. More importantly, the results of your evaluation will certainly

Figure 5-3—This station would technically be classified as a mobile station, although it is not likely that hams will duplicate it exactly.

Figure 5.4—Should this mobile installation be evaluated? The regulations are not a substitute for the RF-safety concerns that have been addressed in ARRL publications for years.

demonstrate to yourself, and possibly your neighbors, that your station is operating well within FCC guidelines and is no cause for concern. Finally, if you have an antenna that is located very close to people, you may be operating in excess of the MPEs. It's a good idea to evaluate and be on the safe side, just in case.

Many classes of amateur stations are categorically exempt from the need to do a station evaluation. This is because the circumstances under which exempt stations are usually operated are such that the station is presumed to be in compliance with the MPEs. Under some circumstances, such as an antenna that is located unusually near people or in some mobile installations, it is possible to exceed the MPE levels.

Sections 1.1307(c) and (d) of the FCC's rules stipulate that the Commission may require that a station that is normally categorically exempt from the requirement to perform a routine evaluation, perform such an evaluation—if the FCC determines that there is reason to believe that the station may be exceeding the MPEs allowed.

The FCC will generally handle these exceptions on a case by case basis. In addition, the FCC also will rely on amateurs to voluntarily consider whether any operating parameter of their stations also make it prudent to do a station evaluation—even in cases where the category of that station would otherwise make it exempt. If an antenna is located unusually close to people, such as an indoor antenna in a living space, or a balcony-mounted antenna a foot or so away from a neighbor's balcony, the FCC *could* require a station evaluation or take other action.

Mobile stations also should be closely considered before an amateur automatically applies the categorical exemption. As an example, a 500-watt, 10-meter mobile installation with a vehicle-mounted antenna would certainly merit a closer look. On VHF, the use of a high-power amplifier also could present problems in some cases. In general, it is recommended that in these higher power installations, the antenna be located such that the vehicle occupants will be shielded from the antenna during normal use. One good location is in the center of an all-metal roof. Locations to be avoided for high-power operation would be a trunk-mounted antenna, or installation in a vehicle with a fiberglass roof. In general, mobile installations will not exceed the MPEs if sound installation guidelines are followed. *The ARRL Handbook for Radio Amateurs*, *Your Mobile Companion*, *Your Ham Antenna Companion* and *The ARRL Antenna Book*, available from the ARRL, have additional material on mobile installations and antennas.

How to Calculate Peak Envelope Power to the Antenna

A number of hams are a bit confused about peak-envelope power. PEP is defined as the *average* power of a single cycle of RF at the modulation peak when the transmitter is being operated normally. See Figure 5.5 and the sidebar "What's Power?". A very good explanation of power is found in the *Lab Notes* column of the May 1995 issue of *QST*, page 88 (*Watt's It All About*, by Mike Gruber, W1DG).

Table 5.1 uses PEP input to the antenna as the threshold to trigger the need to do a station evaluation. This can easily be calculated. Because the PEP input to the antenna can't be more than the PEP output from the transmitter, the simplest way to calculate power to your antenna is not to bother with any calculations—you can assume that your transmitter power output and the power reaching the antenna are the same. This is, of course, a conservative estimate, but you are allowed (and perhaps even encouraged) to be conservative in doing your evaluation. If you assume that *all* the power from your transmitter is reaching your antenna, you can safely use that as the power that will determine if you need to do an evaluation. If you "pass," there would be no need to calculate other factors, such as feed line losses, etc. Most hams will easily pass their evaluation, so some of these steps may not be necessary.

Supplement B contains information and a worksheet about how to calculate power to the antenna. The worksheet makes use of a convenient tool: the decibel (dB). The convenient thing about doing this calculation using dB is that one can easily add and subtract to ultimately obtain a power level. See the worksheet in Chapter 1 of this book.

Doing the Calculation

To calculate PEP to the antenna, start with your transmitter's PEP output, or the PEP of an external amplifier, if you are using one. Many commercially manufactured transmitters and amplifiers have a power meter built in. These meters can provide a measurement of PEP with reasonable accuracy for this purpose. Also, commercially manufactured external PEP reading power meters are available for stations that use common coaxial cables as feed lines. If there isn't any capability to measure the PEP output, the maximum PEP capability specified by the manufacturer may be used. Another approach would be to use a reasonable estimate, based on factors such as measured power input, the maximum capability of the final amplifier devices or the power supply. If the PEP output of your transmitter is at the levels in Table 5.1 or less, you can stop right here: You don't need to do an evaluation. If your power is greater than the levels in Table

What's Power?

The peak envelope of an SSB or AM signal occurs at the highest crest of the modulation envelope. (The point at which PEP occurs has been labeled in Figure 5.5.) The easiest way to appreciate the meaning of PEP is to calculate it. Let's assume a 50-Ω load and a peak voltage at the modulation crest of 110 V.

$$PEP = \frac{(V_{peak} * 0.707)^2}{R} = \frac{(110 * 0.707)^2}{50} = 121W \; PEP$$

The peak envelope power calculation uses the peak voltage during the maximum RF cycle, and converts it to an RMS value by multiplying by 0.707. The instantaneous peak voltage during the maximum modulation crest is treated as if it were a complete cycle of a sine wave. This is why the terms "average" and "peak" are not mutually exclusive in this case. Although PEP is the peak power, it is averaged over one complete RF cycle as if it were a sine wave.

Wattmeters and PEP

To determine your power, you could, of course, measure that power with an accurate wattmeter. (Virtually any wattmeter with its scale in watts is accurate enough for this job.) If you do use a wattmeter, to determine power at the transmitter or at the antenna, *ensure that the wattmeter is capable of measuring PEP*, if you are measuring modes such as single-sideband or full-carrier, double sideband AM. If you are measuring CW or FM, the PEP is the same as the average power that will be measured by non-PEP-reading wattmeters. Remember, too, that most wattmeters are only accurate if they are measuring power in a 50-ohm resistive system. (If your SWR is 1.5:1 or better, you can safely assume that the wattmeter is reasonably accurate. If not, consult the owner's manual for your meter or consult with the meter's manufacturer.)

If you do accurately measure the power at the antenna, you can compare the result with the values in Table 5.1. If your power is at those levels or less, you do not need to do a station evaluation for that band at that power level.

5.1, you will need to calculate or determine the power input to your antenna.

Feed Line System Losses

The power at the transmitter will be reduced by any losses between the transmitter and antenna. This usually includes losses in the feed line and any external accessories such as power meters or antenna tuners. Most of the time, these losses are expressed in decibels (dB), either dB/100 feet for feed lines, or in dB for each accessory. In most cases, the published loss for feed lines is fairly accurate and it can be used directly in making your calculations.

To obtain an estimate of your feed line losses, refer to the graph of Figure 5.6. This graph provides estimates of feed line losses for common types of feed lines. It is not meant to represent the actual attenuation performance of any particular product made by any particular manufacturer. The actual attenuation of any particular sample of a feed line type may vary somewhat from other samples of the same type because of differences in materials or manufacturing. If the feed line manufacturer's specification is available, use that instead of the values listed in this table.

Feed line loses also vary with SWR. The higher the SWR, the higher the losses over

and above the attenuation loss discussed above. For further information see *The ARRL Handbook for Radio Amateurs*, *Your Ham Antenna Companion* or *The ARRL Antenna Book*. You can ignore the additional losses caused by SWR for a conservative evaluation.

The graph gives the feed line loss in

dB/100 feet. If your feed line is exactly 100 feet long, you already know your feed line losses. This is, however, unlikely, so you are going to have to multiply the loss in dB/100 feet by the ratio between your actual feed line length and 100 feet.

Other Losses

You can factor in other losses between the transmitter and the antenna, if you know them. Although the feed-line loss specification is reasonably realistic, the specifications for accessory items is often a "maximum" specification. The actual losses can be less. An antenna tuner might have a specification of 3 dB insertion loss, or loss of 50% of the available power, but this would be worst case—on most bands, the losses would be less. A conservative estimate on HF might be to assume that these components are lossless. On VHF and above, it would be reasonable to add 0.1 dB to the total losses for each accessory item that is connected between the output and the feed line going to the antenna. Do not include accessories that are between an exciter and the final amplifier. It would be conservative to assume that *connectors* have 0 dB loss.

Using Arithmetic

Decibels can only be added or subtracted with decibels. To obtain the power at the antenna, you will either have to convert your power to a form that is expressed in decibels or you will have to convert the decibel value to a number.

If you know the loss in dB, you can convert that to the percentage of loss using the following formula:

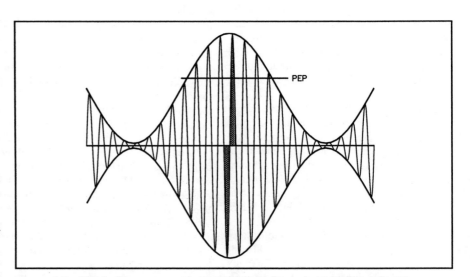

Figure 5.5—PEP is the average power of the single cycle highlighted in this graph. If the peak of the RF waveform is 100 volts and the resistance is presumed to be 50 ohms, the RMS voltage of the cycle is 70.7 and the power is 100 watts, using the classic formula, P = E²/R.

Step by Step

Let's look at a hypothetical example of an amateur station and run through the evaluation steps. Assume that Al, N9AT, has the following station configuration:

• 80 meters, 100 W and 1000 W CW and SSB with a half wavelength dipole antenna 10 feet above ground. (This is a terrible height for an 80-meter dipole, but it serves as a worst case!)

• 40 meters 100 W and 1000 W CW and SSB with a half wavelength dipole antenna 10 feet above ground. (Ditto the height comments above!)

• 10 meters, 100 W and 1500 W CW and SSB with a 3-element beam 30 feet above ground, 8.5 dBi gain.

• 2 meters, 35 W FM, 100 W CW and SSB with a 4-element Yagi, 8 dBi gain 60 feet above ground.

Al first looks at Table 5.1 to see which operation requires a station evaluation. In this case, his 100-W 80- and 40-meter operation and his 35-W 2-meter FM operation do not need to be evaluated. (Al intends to evaluate them anyway, just to learn more about the subject.)

He could calculate his average power for the remaining operation, but this may not be necessary. Al first tries his evaluation with PEP, using Table 5.7 in this chapter in conjunction with Table 5.5. Rounding up to 3 dBi for the antenna gain, Table 5.5 estimates that on 80 meters at 1000 W his antenna needs to be located 2.8 feet from areas of controlled exposure and 6.2 feet from areas of uncontrolled exposure. The antenna is located about 10 feet from the property line and is attached to the house with 5-feet of rope, so this band would be in compliance for operation at a 1000-W continuous carrier level.

On 40 meters at 1000 W, Al first rounds his dipole gain up to 3 dBi. Table 5.5 shows 5.1 feet for controlled exposure and 11.4 feet for uncontrolled exposure. On this band the end of his antenna is located 5 feet from the property line and tied to the house with a 4-foot rope. It doesn't quite pass with full power. Al has a few choices. He can relocate the antenna, reduce power, or calculate his average power and try again or use the antenna-specific table at the same height. In this case, he calculates his average power and determines that he

is using 133 W average power on SSB and 266 W average power on CW. Rounding up, he selects 500 W in Table 5-9 and determines that his antenna needs to be 3.6 feet from controlled exposure and 8.0 feet from uncontrolled exposure. He meets the requirements for controlled exposure, but the antenna would be located 6.4 feet from a person standing on the property line, so the station may still not be in compliance. Al decides to move the antenna 10 feet from the property line sometime next week. In the meantime, he will reduce his power on 40 meters.

On 10 meters, he is using a 3-element Yagi 30 feet in the air. Rounding his gain up to 9 dBi, using Table 5.5 he determines that his antenna needs to be 50.6 feet from controlled exposure and 113.2 feet from uncontrolled exposure. The tower is located 40 feet from the house, and solving for the hypotenuse of the distance between his residence and the tower (his one-floor house has the top of the first floor 12 feet above ground), he calculates that the antenna is located 43.9 feet from areas of controlled exposure. Thus there is a problem for full power, but not when he calculates his average power. The tower is 50 feet from the property line, for a total distance of 55.5 feet from ground level exposure *on the property line*. This does not pass for uncontrolled exposure. Al doesn't give up, though, he goes to Table 5.9 and determines that at ground level, the *NEC* model shows that the compliance distance needs to be 57.1 feet from the center of the antenna at 1500 W average power. He clearly cannot do 30 minutes of tune-up if his neighbor is on the property line. At 500 W average power, however, Al notes that his antenna could be built on the property line and ground-level exposure would be below the limits. He has met the requirements and does not need to make any changes to his station except to limit his tune-up time.

On 2 meters, his antenna has 8 dBi of gain. Rounding up to 9 dBi, he determines that at 100 W his antenna needs to be 13.2 feet from controlled exposure and 29.5 feet from uncontrolled. This antenna is at the top of his 45 foot tower, so he can run continuous power on 2 meters. Al gathers all the papers containing these calculations (along with his notes) and files them with his station records. Within 20 minutes he has completed his station evaluation!

$$Loss\% = 100 - \frac{100}{10^{\frac{dB}{10}}} \qquad \text{Eq 5.1}$$

Most electronic calculators have exponent functions (10^x) that can do this calculation handily. For those who don't want to do the mathematics, Table 5.2 handles the conversion in convenient steps. To be conservative, round the calculated feed line losses *down* to the next lowest step in this table. As you can see, if the losses are greater than a few dB, a *lot* of power is getting lost in your feed line. On the other hand, if your loss were 12 dB, about 94% of your power is lost as heat.

If you use the calculated feed line system loss in the above formula or table, multiply the power at the transmitter by the result of the above calculation percentage. This will give you the amount of power being lost in your feed line system. Subtract this power from the output of your transmitter and you will have calculated the amount of power being delivered to the antenna.

Using dBW

You also can convert your power into a decibel unit. This is the method generally used in the radio engineering field. This method is outlined in the FCC worksheet in *Bulletin 65*. The power unit dBW expresses

---

Table 5.2
dB to Decimal Number Loss Table

dB	Loss%	dB	Loss%	dB	Loss%	dB	Loss%
0.0	0.00	1.5	29.21	7.0	80.05	15.0	96.84
0.1	2.28	2.0	36.90	7.5	82.22	16.0	97.49
0.2	4.50	2.5	43.77	8.0	84.15	17.0	98.00
0.3	6.67	3.0	49.88	8.5	85.87	18.0	98.42
0.4	8.80	3.5	55.33	9.0	87.41	19.0	98.74
0.5	10.88	4.0	60.19	9.5	88.78	20.0	99.00
0.6	12.90	4.5	64.52	10.0	90.00	22.0	99.37
0.7	14.89	5.0	68.38	11.0	92.05	25.0	99.69
0.8	16.82	5.5	71.82	12.0	93.69	30.0	99.90
0.9	18.72	6.0	74.88	13.0	94.99	35.0	99.97
1.0	20.57	6.5	77.61	14.0	96.02	40.0	99.99

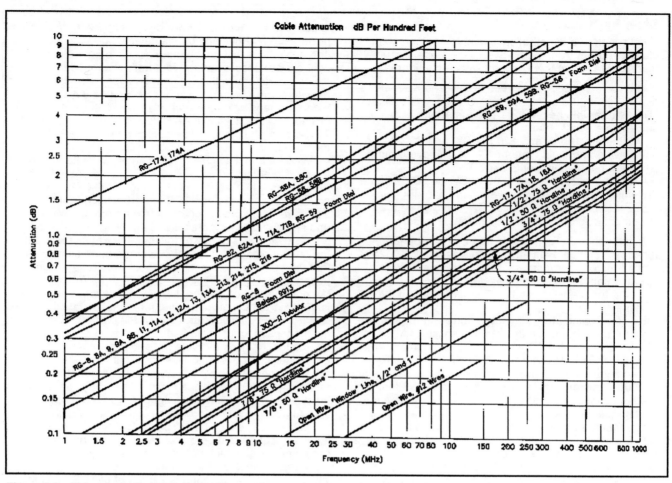

Figure 5.6—This graph shows the actual losses for many common feed lines.

the ratio of the power in question to 1 watt, in decibels. To obtain power in dBW, use the following formula:

$$power_{dBW} = 10\log_{10}\left(power_{Watts}\right) \qquad Eq\ 5.2$$

Table 5.3 gives the power in dBW for a number of power levels that will be useful to do this calculation. If you use this table, you will have to round up your actual transmitter power to the nearest value in the table. The power levels in this table were selected to correspond with various power levels that are part of the FCC RF-exposure rules, or that result from average power calculations of 1500 watt transmitters using various modes. This table can be used to convert dBW to watts, or watts to dBW. Ensure that any rounding up or down that you do with this table is in the "conservative" direction.

Working with the Decibel

Now that you have the power in dBW, you can easily subtract the feed line and other losses directly from the power in dBW, giving you power at the antenna in dBW. You can convert this back to power in watts, either using Table 5.3 (rounding up the dBW as required) or the formula:

$$power_{watts} = 10^{\frac{dBW}{10}} \qquad Eq\ 5.3$$

Practice converting power in watts to power in dBW, then from dBW back to watts. If you are doing the math correctly, you will end up with the same power you started with.

Table 5.3

Conversion of Power In Watts to dBW

Watts	dBW	Watts	dBW
1	0.00	125	20.97
2	3.01	150	21.76
3	4.77	200	23.01
5	6.99	225	23.53
10	10.00	250	23.98
15	11.76	300	24.77
20	13.01	400	26.02
25	13.98	425	26.28
30	14.77	500	26.99
40	16.02	600	27.78
50	16.99	750	28.75
70	18.45	1000	30.00
75	18.75	1200	30.79
100	20.00	1500	31.76

Once you have calculated power at the antenna using one of these methods, the power at the antenna can be compared to the power levels in Table 5.1 to see if you need to do a station evaluation. If you do, the peak-envelope power at the antenna will be used later in the evaluation to calculate average power and average exposure that will be used in doing your station evaluation. (It is a lot easier than it sounds!)

As an example, if you are running 100 watts PEP and have a feed line loss of 3 dB, you would convert 100 watts to 20 dBW, then subtract 3 dB. This would leave you with 17 dBW, which by using the table gives you 50 watts to the antenna. You could also look to Table 5.2 and determine that you are losing 50% of your power in the feed line. Follow the instructions on using Table 5.2 and you will calculate that you have 50 watts to the antenna. According to Table 5.1, this part of your operation would not have to be evaluated on any band.

PERFORMING AN EVALUATION FOR CONTROLLED AND UNCONTROLLED ENVIRONMENTS

In general terms, controlled exposure ap-

Multiple Evaluations

To comply with the requirements, an evaluation must be made for each transmitter, duty cycle and antenna. Different modes usually correspond to different duty cycles. In addition, where applicable, each combination has to be made of both controlled and uncontrolled areas. For the example in the text, each of the following modes and antennas will have to be evaluated twice—for controlled and uncontrolled spaces.

Band	Mode	Power	Antenna
146 MHz	FM	>50 W	Groundplane
146 MHz	FM	>50 W	5-element Yagi
144 MHz	SSB	>50 W	Groundplane
144 MHz	SSB	>50 W	5-element Yagi
222 MHz	FM	>50 W	Vertical collinear
222 MHz	SSB	>50 W	10-element Yagi

- Antenna modeling software (*NEC*, *MININEC*, etc)
- Power-density and field-strength formulas
- Graphs made from power-density formulas
- Software developed from field-strength formulas
- Calibrated field-strength measurements

The First Step—Decide On a Method

Most amateurs will probably select one or more calculation methods to perform their station evaluations. The selection of method is based on the needed accuracy, the specific factors that must be considered and the available software, hardware or information "tools."

The first step in doing an evaluation is to determine in advance what method you will use. The list above shows some examples of the ways most hams will use for their evaluations. Once you have selected a method, you can either apply that method directly to your transmitter's output power as a shortcut, or you can determine the actual average exposure.

Average Exposure

FCC rules define maximum permitted amateur power in PEP output from the transmitter. They also define the threshold that triggers the need to do a station evaluation in PEP input to the antenna. The MPE limits, however, are based on *average exposure*, not peak exposure, using an average of the power density, or an average of the square of the electric or magnetic fields.

The concept of averaging RF exposure means that the total exposure for the averaging period must be below the limits. For example, someone could be at twice the MPE limit for half of the averaging period. As long as there was no exposure for that same amount of time before and after the exposure that was double the limit, you would meet the MPE requirements.

Another way of factoring in average exposure could be to determine the average transmitter power, and use that power in all your following calculations. Those who use the power-density formulas to calculate the power density to areas of exposure will probably find this method to be the most useful way of determining average exposure.

The easiest way to calculate average power is not to do the calculation. First use your transmitter's PEP output, or PEP to the antenna, and assume continuous exposure. You may meet the requirements. In that case, you don't need to calculate average exposure or average power at all!

plies to you, your immediate household and property areas that you control. Uncontrolled exposure is a "general public" exposure, generally applied to neighboring properties and public areas.

A routine evaluation will generally need to be done for both controlled and uncontrolled exposure environments. However, if a ham determines that his or her operation meets the requirements for uncontrolled exposure in his or her own station, home and property, it will not be necessary to evaluate the same areas for controlled exposure. The definitions and scope of these terms are discussed in the Rules chapter.

Evaluation Must Be Done by Mode, Power, Antenna and Band

Amateur stations must be evaluated for each frequency, mode and station configuration used. Separate evaluations will probably need to be made for both controlled and uncontrolled environments, if it is possible that fields in these areas could exceed the MPEs. For example, if an amateur operates more than 50 W FM and/or SSB on 144 and 222 MHz, using one of two different antennas on 144 MHz and one antenna for each mode on 222 MHz, the evaluations shown in the "Multiple Evaluations" sidebar would have to be performed, including both controlled and uncontrolled environments:

Each mode has a specific duty cycle and each antenna has a specific gain and/or distance from areas of exposure, so each combination must be tested. In most cases, if an amateur uses two different transmitters with the same power for a single band and mode, the evaluation made for one will apply to the other. (This may not *always* be true, however. See the section on Duty Factor later in this chapter.)

One would find different average field strengths and resultant compliance distances for each mode, so it may be necessary to evaluate each mode separately.

There are a few shortcuts, however. If a station meets the MPE requirements with a mode like FM with a 100% duty factor, it also will pass using a mode like SSB or CW with a smaller duty factor. In general, the compliance distance with a low-gain antenna such as the ground plane will be less than it will for the Yagi. Thus, if the station complies at a certain distance with the Yagi, the compliance distance with the ground-plane antenna will almost always be less.

How to Do an Evaluation

Most amateurs will probably select one or more of several calculation methods to perform their station evaluations. If appropriate, different methods may be applied to different station configurations. The selection of method is based on the needed accuracy, the specific factors that must be used to determine improvements from "worst-case," and the available tools.

General Methods Overview

Bulletin 65 outlines several ways that hams can evaluate their stations. However, hams may use any other technically appropriate methods. Many hams envision complicated measurements when they think about evaluating their stations. While precise measurements could be used, most hams will probably meet the requirements using one of the easier methods. The FCC notes, however, that some of these formula-based calculations and tables can give results that are *much* higher than would be actually encountered. In some cases, a more specific analysis, perhaps using computer modeling or the tables in Chapter 8 derived from computer modeling may help a ham prove compliance.

In general, you can estimate compliance by using:
- Tables developed from the field-strength formulas
- Tables derived from antenna modeling

Ground Reflections

A precise calculation in the near field is not very straight forward!

The presence of boundaries such as earth ground alters the wave impedance, so that electric and magnetic fields must be considered separately, even in the far field of the antenna. This is illustrated by considering the case of a horizontal dipole 15 m above the earth, operating at 29 MHz with 1,500 W supplied power. The electric and magnetic fields each obey the boundary conditions at the air-earth interface, and the magnetic field is enhanced, while the electric field is diminished. When normalized to the MPE of the 1996 FCC standard, the total magnetic field in decibels relative to the standard is shown in Figure A.

The total electric field contours similarly normalized are picture in Figure B. Ignoring the exposure averaging time in the standards, permissible general population exposure levels are the regions outside the "0 dB" contours. Significantly, the magnetic field contours of Figure A are substantially different from the electric field contours shown in Figure B. Magnetic fields peak at ground level while electric fields peak a quarter wavelength above ground. This is a consequence the ground reflection, and has nothing to do with whether the fields are near or far with respect to the dipole. The wave impedance evaluated on the total fields is simply not equal to the intrinsic impedance associated with the medium.

The exposure standard is written around the maximum of the either the electric or magnetic field limit. That quantity is pictured in Figure C. The "0 dB" contours represent the limits where either the electric or magnetic fields exceed the MPE level of the standard. If the power transmitted by the dipole were reduced by 5 dB, then the MPE limit contour would be represented by the "5 dB" contour in Figure C.

The figure illustrates that the determination field levels relative to MPE levels is complex, even for the very simple case of a dipole antenna in the presence of a single boundary—the ground.

Figures A - C show the fields near the ground. Those complicated contours make it awkward to specify a single distance as the compliance distance for this antenna and power combination. First, the electric or magnetic field alone produces different compliance contours, Figures A and B. We must comply with the worst case of both figures, which is represented by Figure C.

Even then, near ground level, the compliance distance along the ground is 7 m, as shown by point "A," whereas at a height ground of 7 m the compliance distance, point "B," is almost 11 m. This helps illustrate why the compliance distances in the ARRL compliance distance tables sometimes might appear to be unusual.—*Kai Siwiak, KE4PT*

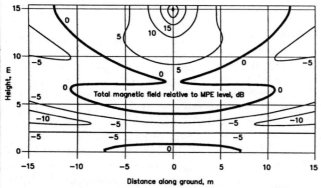

Figure A—Magnetic fields relative to MPE limits. The contours "0 dB" and greater are regions where the magnetic fields are not in compliance.

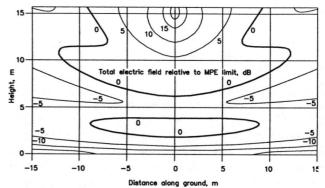

Figure B—Electric fields relative to MPE limits. The contours "0 dB" and greater are regions where the electric fields are not in compliance.

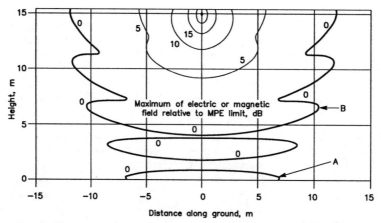

Figure C—Greater of the magnetic or electric fields relative to MPE limits. Any point outside the "0 dB" contours is in compliance with the FCC standards.

Table 5.4

Operating Duty Factor of Modes Commonly Used by Amateurs

Mode	Duty Cycle	Notes
Conversational SSB	20%	1
Conversational SSB	40%	2
SSB AFSK	100%	
SSB SSTV	100%	
Voice AM, 50% modulation	50%	3
Voice AM, 100% modulation	25%	
Voice AM, no modulation	100%	
Voice FM	100%	
Digital FM	100%	
ATV, video portion, image	60%	
ATV, video portion, black screen	80%	
Conversational CW	40%	
Carrier	100%	4

Note 1: Includes voice characteristics and syllabic duty factor. No speech processing.
Note 2: Includes voice characteristics and syllabic duty factor. Heavy speech processor employed.
Note 3: Full-carrier, double-sideband modulation, referenced to PEP. Typical for voice speech. Can range from 25% to 100%, depending on modulation.
Note 4: A full carrier is commonly used for tune-up purposes

Duty Factor

Duty factor is an expression between the peak-envelope power of a transmitter and its average power during the time it is on the air. It is usually expressed as a percentage, although it is not uncommon for it to be expressed as a decimal. It is sometimes called "duty cycle."

If all else is equal, some emission modes will result in less RF electromagnetic energy exposure than others. For example, modes like RTTY or FM voice transmit full power during the entire transmission (100% duty factor). On CW, you transmit at full power during dots and dashes and at zero power during the space between these elements. A single-sideband (SSB) phone signal generally produces the lowest exposure because the transmitter is not at full power all the time during a single transmission. The **duty factor** of an emission takes into account the amount of time a transmitter is operating at full power. Duty factor can either consider the time of a single transmission, or the time of a series of transmissions over a specific time period. The duty-factor tables and text in this section assume 100% transmission time. An emission mode with a lower duty factor produces less exposure for the same PEP output.

Lower duty factors, then, result in lower RF exposures. That also means the antenna can be closer to people without exceeding their MPE limits. Compared to a 100% duty-factor mode, people can be closer to your antenna if you are using a 40% duty-factor mode.

Duty factor is used as part of your calculation of average power. If you do want to determine your average power, you will need to know about how different modes have different average powers. *The MPE limits are based on exposures averaged over 6 minutes for controlled exposure or 30 minutes for uncontrolled exposure.* To obtain this average, we need to consider the mode being used, its duty factor and the total operating time.

Using a duty-factor correction for some modes, SSB, for example, would give an accurate MPE for conversational SSB. However, if the same transmitter were used for extended tune-up purposes on the air using a carrier, the MPE could be exceeded. If you apply duty factor to two different transmitters using the same mode, consider whether the speech processing, or CW keying characteristics might be different. This could result in a different duty factor and average power than would be obvious from the mode and power used.

Table 5.4 shows the duty factors of a number of modes in common use by amateurs. The actual PEP to the antenna can be multiplied by these values to yield a power level that has been corrected by the duty factor of the mode being used. The resultant average power can then be used in the various calculation methods described elsewhere in this bulletin. If so used, they are based on 100% operating "on" time for the mode described.

Determining Average Power

The concept of power averaging includes both on and off times and the "duty

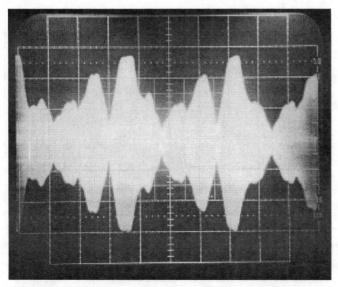
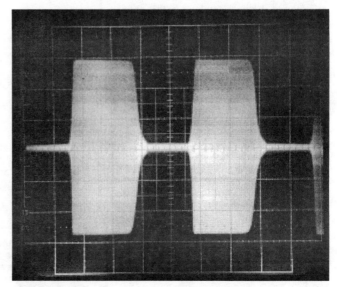

Figure 5.7—These two signals have different average power, but the same PEP.

factor" of the transmitting mode being used. Each mode of operation has its own duty factor that is representative of the ratio between average and peak power. Table 5.4 shows the duty factors for several modes commonly in use by amateur operators. To obtain an easy estimate of average power, multiply the transmitter peak envelope power by the duty factor. Then multiply that result by the worst-case percentage of time the station would be on the air in a 6-minute period for controlled exposure, or a 30-minute period for uncontrolled exposure.

For example, if a 1500-watt PEP amateur single-sideband station operates 10 minutes on, 10 minutes off, then 10 minutes on, this would be:

1500 W * 20% * (20 out of 30 minutes) = 200 watts for uncontrolled exposure
1500 W * 20% * (6 out of 6 minutes) = 300 watts for controlled exposure
A 500-watt CW station that is used in a DX pileup, transmitting 15 seconds every two minutes would be:
500 W * 40% * (15 out of 120 seconds) = 25 watts for controlled or uncontrolled exposure
A 250-watt FM base station used to talk for 5 minutes on, 5 minutes off, 5 minutes on, would be:
250 W * 100% * (5 out of 6 minutes) = 208 watts for controlled exposure
250 W * 100% * (15 out of 30 minutes) = 125 watts for uncontrolled exposure

The percentages (%) shown are taken from Table 5.4 for the mode used.

If the station might transmit for more than 6 minutes, one can assume continuous exposure in a controlled environment, so the average power for controlled exposure is 300 watts. Additional examples are shown elsewhere in this chapter under the "Step by Step" section. If an amateur does consider on and off operating time in determining average power, it is recommended that this generally not be applied to evaluation for controlled environments. It is very likely that in the long run, any one mode would be in continuous use for at least 6 minutes, resulting in the maximum exposure for controlled environments.

If an amateur corrects the duty factor for time for an uncontrolled environment, the worst-case 30-minute period must be considered. For example, in an HF contest operation, it is likely that the on time/off time could be 4:1. Thus the station is on the air 80% of the time for a long period. At first glance, an amateur might assume that if the station is operated for half the time, the duty factor correction is 0.5, but that is not always the case. For example, if a

station were operated for 10 minutes on, 10 minutes off, then 10 minutes on, over the worst-case 30-minute period, the station would be on the air 67% of the time, resulting in a duty factor correction of 0.67.

Compliance Distance Tables

Most amateurs will use the tables in *Bulletin 65* to estimate their compliance with the MPE levels. The *Bulletin 65* tables do have advantages: they generally offer conservative estimates and they are easy to use. The tables in *Bulletin 65* are all formatted with distances in meters. These tables, plus a larger number created using the same methods as the FCC tables, are featured in Chapter 8, formatted in feet. These tables show the compliance distance—the minimum distance one must be from the antenna to be in compliance with the FCC rules for the frequency, antenna gain and average power involved. You can use PEP for the power levels shown in all the tables for a conservative estimate, or calculate average power for a more precise estimate.

Bulletin 65 contains three major sets of tables. The first features a list of antenna gains, frequencies and power levels, with

the necessary compliance distance for each. The concept for this table was submitted to the FCC by the W5YI Group. The W5YI Group and the ARRL then worked together to expand the number of listings. Additional entries have been made to the version of this table featured in Chapter 8. The distances in these tables were derived using the far-field, power-density formula shown in Eq 5.7 later in this chapter. The tables assume that the exposure is taking place in the main beam, at the height of the antenna as a conservative estimate. This equation includes the "EPA" ground-reflection factor.

The second set of tables features specific antennas and transmitter powers, by frequency. These tables were supplied to the FCC by Wayne Overbeck, N6NB, Kai Siwiak, KE4PT, and the FCC staff. The tables assume that the exposure is taking place in the main beam, at the height of the antenna as a conservative estimate.

The third set of tables features specific antennas and transmitter powers, by frequency, modeled using *NEC4* at various heights above average ground. In these tables, the *horizontal* compliance distance was calculated from the center of radiation for various antenna heights, at heights

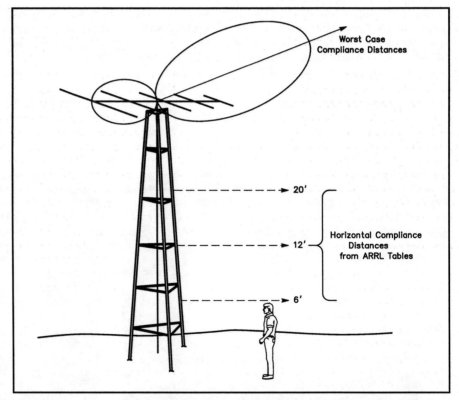

Figure 5.8—The power-density and field-strength formulas give the compliance distance in the main beam of the antenna, at any angle, as the uppermost line shown on this drawing. If this same distance is applied to ground-level exposure, the estimate is generally conservative. The tables based on antenna modeling have calculated the horizontal compliance distances at ground level, and at first and second story exposure levels.

where exposure occurs of 6 feet, 12 feet, 20 feet and at the height of the antenna. The 6-foot height estimates ground-level or first-story-level exposure. The 12-foot height represents the ceiling of a typical first-story exposure, or the floor of a second-story exposure. The 20-foot height represents the ceiling of a second story or the floor of a third story. These heights were chosen to accommodate different building structures. This is shown in Figure 5.8.

The tables calculate *actual* exposure at the various points being evaluated. The modeling process automatically includes the specific gain of the antenna and the actual ground conditions. These tables demonstrate that the exposure below an antenna is often much less than the exposure in the main beam. Figure 5.9 shows how these various tables and methods relate to the areas being evaluated.

Tables Developed from Far-Field, Power-Density Formulas

The easiest-to-use of these tables were developed from the far-field, power-density formula. They have been calculated with a "ground-reflection factor." This includes the "ground gain" of an antenna over typical ground. This allows hams to use manufacturer's antenna gain figures in dBi with confidence that the result represents a conservative real-world estimate. (Many antenna gains are expressed in decibels relative to a dipole. Add 2.15 dB to the gain in dBd to obtain dBi.) This model, although simplified, has been verified by the ARRL Laboratory staff using *NEC* antenna-modeling software against a number of dipole, ground plane and Yagi antennas modeled over ground. These tables do not necessarily apply to all antenna types. *NEC* models of small HF loops, for example, give fields near the antenna that are much higher than the far-field formula predicts. The table for the small loop was calculated using different, more accurate, techniques.

In most cases, however, the power-density-formula derived tables give results that are conservative. Examples of the easiest-to-use of these tables are shown in Table 5.5 and Table 5.6, followed by a number of tables based on specific antenna types.

The first step for an amateur is to select the simple tables that best applies to his or her station and determine the estimated compliance distance per band. *Bulletin 65* contains a number of these tables. If the compliance distance is less than the actual distance to the exposure, the station "passes" and the evaluation is complete. It can be that simple. Remember that these

distances are for the absolute distance from the antenna *at any angle*. Figure 5.9 shows an example of how to determine the distance between an antenna and any point being evaluated.

This distance can be used with the tables derived from the power-density formula. The ARRL tables of modeled antennas use distance b or b' in Figure 5.9.

One shortcut is to use the highest power you use on each band. First, use your transmitter's PEP output to see if you are in compliance. Next select the table entry of antenna that represents your station configuration. Finally, look up your frequency and power and determine if areas where people might be exposed are farther away than the compliance distance in the table.

Tables Based on Antenna Gain

Tables 5.5 and 5.6 are derived from the method used in the tables in the FCC *Bulletin 65* submitted by the W5YI Group. They show the distances required to meet the power-density limits for different amateur bands, power and antenna gain, for occupational/controlled exposures (*con*), or for general population/uncontrolled exposures (*unc*). (All FCC tables give all the distances in meters; the tables in this article have been converted to feet.)

Tables 5.5 and 5.6 probably represent the *easiest* approach to doing a station

evaluation. They can be conservatively applied to most antenna types. The frequency represents the "worst-case" for each band; the antenna gains are in dBi. (Some antenna gains are expressed in decibels relative to a dipole. Add 2.15 dB to the gain in dBd to obtain dBi.) Hams can use PEP or average power to obtain either a conservative or more precise estimate of compliance distances. Select the appropriate band and "round up" antenna gain and power to match the table. The distances are the minimum separation that must be maintained between the antenna and any area where people will be exposed. See Figures 5.8 and 5.9 for examples of how this distance applies.

To obtain a conservative estimate using Tables 5.5 and 5.6, hams should follow the following steps.

- Select the table entry for the frequency band being evaluated.
- Determine the estimated free-space antenna gain in dBi from the antenna manufacturer or from Table 5.7.
- First, assume full PEP and 100% operation, then look up the compliance distance on the chart. If the antenna is located at least this far from areas of exposure, either horizontally, vertically, or diagonally, the station "passes" on that antenna/band combination.
- If necessary, calculate average power,

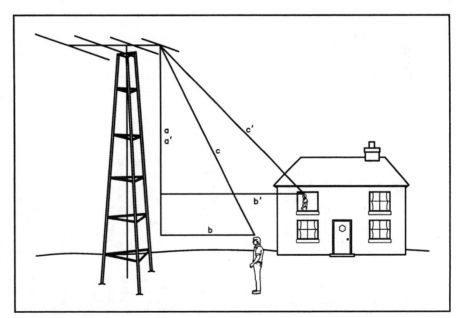

Figure 5.9— In calculating the actual worst-case horizontal compliance distances between the antenna and areas being evaluated, you must consider the antenna height, the height of the exposure and the horizontal distance between the antenna and the exposure point. This drawing illustrates exposures at ground and second-story levels. (Use the a´ and b´ for the second-story exposure.) From there, you can use the formula:

$$c = \sqrt{a^2 + b^2}$$

Table 5.5

Estimated distances from transmitting antennas necessary to meet FCC power-density limits for Maximum Permissible Exposure (MPE) for either occupational/controlled exposures ("Con") or general-population/uncontrolled exposures ("Unc"). The estimates are based on typical amateur antennas and assuming a 100% duty cycle and typical ground reflection. (The figures shown in this table generally represent worst-case values, primarily in the main beam of the antenna.) The compliance distances apply to average exposure and average power, but can be used with PEP for a conservative estimate. An expanded version of this table appears in Chapter 8.

Frequency (MHz)	Gain (dBi)	100 W Con	100 W Unc	500 W Con	500 W Unc	1,000 W Con	1,000 W Unc	1,500 W Con	1,500 W Unc
2	0	0.5	0.7	1.0	1.6	1.5	2.2	1.8	2.7
	3	0.7	1.0	1.5	2.2	2.1	3.1	2.6	3.8
4	0	0.6	1.4	1.4	3.1	2.0	4.4	2.4	5.4
	3	0.9	2.0	2.0	4.4	2.8	6.2	3.4	7.6
7.3	0	1.1	2.5	2.5	5.7	3.6	8.1	4.4	9.9
	3	1.6	3.6	3.6	8.0	5.1	11.4	6.2	13.9
	6	2.3	5.1	5.1	11.4	7.2	16.1	8.8	19.7
10.15	0	1.6	3.5	3.5	7.9	5.0	11.2	6.1	13.7
	3	2.2	5.0	5.0	11.2	7.1	15.8	8.7	19.4
	6	3.2	7.1	7.1	15.8	10.0	22.4	12.2	27.4
14.35	0	2.2	5.0	5.0	11.2	7.1	15.8	8.7	19.4
	3	3.2	7.1	7.1	15.8	10.0	22.4	12.3	27.4
	6	4.5	10.0	10.0	22.3	14.1	31.6	17.3	38.7
	9	6.3	14.1	14.1	31.6	20.0	44.6	24.4	54.7
18.168	0	2.8	6.3	6.3	14.2	9.0	20.1	11.0	24.6
	3	4.0	9.0	9.0	20.0	12.7	28.3	15.5	34.7
	6	5.7	12.7	12.7	28.3	17.9	40.0	21.9	49.0
	9	8.0	17.9	17.9	40.0	25.3	56.5	31.0	69.2
21.45	0	3.3	7.5	7.5	16.7	10.6	23.7	13.0	29.0
	3	4.7	10.6	10.6	23.6	15.0	33.4	18.3	41.0
	6	6.7	14.9	14.9	33.4	21.1	47.2	25.9	57.9
	9	9.4	21.1	21.1	47.2	29.8	66.7	36.5	81.7
24.99	0	3.9	8.7	8.7	19.5	12.3	27.6	15.1	33.8
	3	5.5	12.3	12.3	27.5	17.4	39.0	21.3	47.7
	6	7.8	17.4	17.4	38.9	24.6	55.0	30.1	67.4
	9	11.0	24.6	24.6	55.0	34.8	77.7	42.6	95.2
29.7	0	4.6	10.4	10.4	23.2	14.7	32.8	18.0	40.1
	3	6.5	14.6	14.6	32.7	20.7	46.3	25.4	56.7
	6	9.2	20.7	20.7	46.2	29.3	65.4	35.8	80.1
	9	13.1	29.2	29.2	65.3	41.3	92.4	50.6	113.2

Distance from antenna (feet)

Table 5.6

Frequency	Gain	50 W Con	50 W Unc	100 W Con	100 W Unc	500 W Con	500 W Unc	1,000 W Con	1,000 W Unc
50, 144, 222	0	3.3	7.4	4.7	10.5	10.5	23.4	14.8	33.1
	3	4.7	10.5	6.6	14.8	14.8	33.1	20.9	46.8
	6	6.6	14.8	9.3	20.9	20.9	46.7	29.5	66.1
	9	9.3	20.9	13.2	29.5	29.5	66.0	41.7	93.3
	12	13.2	29.5	18.6	41.7	41.7	93.2	59.0	131.8
	15	18.6	41.6	26.3	58.9	58.9	131.7	83.3	186.2
	20	33.1	74.0	46.8	104.7	104.7	234.1	148.1	331.1
420	0	2.8	6.3	4.0	8.8	8.8	19.8	12.5	28.0
	3	4.0	8.8	5.6	12.5	12.5	28.0	17.7	39.5
	6	5.6	12.5	7.9	17.7	17.7	39.5	25.0	55.8
	9	7.9	17.6	11.2	24.9	24.9	55.8	35.3	78.9
	12	11.1	24.9	15.8	35.2	35.2	78.8	49.8	111.4
	15	15.7	35.2	22.3	49.8	49.8	111.3	70.4	157.4
1240	0	1.6	3.6	2.3	5.2	5.2	11.5	7.3	16.3
	3	2.3	5.1	3.3	7.3	7.3	16.3	10.3	23.0
	6	3.2	7.3	4.6	10.3	10.3	23.0	14.5	32.5
	9	4.6	10.3	6.5	14.5	14.5	32.5	20.5	45.9
	12	6.5	14.5	9.2	20.5	20.5	45.8	29.0	64.8
	15	9.2	20.5	13.0	29.0	29.0	64.8	41.0	91.6

based on duty cycle and on/off times. See the Power Averaging section of this chapter or, as a rough rule of thumb, for CW or SSB you can use 40% of your output power as a conservative estimate of average power.

- In the unlikely event that your station still doesn't pass, you should refer to the more precise tables of antennas over ground in Chapter 8, use some of the other methods for estimating compliance or follow some of the steps described in this chapter under Correcting Problems.

Tables for Specific Antenna Types

Bulletin 65 also contains tables for specific antenna types. Table 5.8 is an example of those supplied *for Bulletin 65* by Wayne Overbeck, N6NB. These tables have been reproduced, with distances in feet, in Chapter 8. It shows the estimated compliance distance *in the main beam* of a typical specific three-element Yagi HF antenna. These tables also are based on the far-field, power-density equations, with the frequency identifying the amateur band, the antenna gains in dBi. Hams can use PEP to obtain a conservative estimate of compliance distance or use average power to obtain a more precise estimate. Select the appropriate band and "round up" antenna gain and power to match the table. The distances are the minimum separation that must be maintained between the antenna and any area where people will be exposed.

To obtain a conservative estimate using these tables, hams should follow the following steps

- Select the correct table entry for the frequency band and antenna being evaluated
- First, assume full PEP and 100% operation, then look up the compliance distance on the chart. If the antenna is located at least this far from areas of exposure in any direction, the station meets the requirements on that antenna/band combination. Figure 5.9 shows how to determine the actual distance to the antenna.
- If necessary, calculate average power, based on duty cycle and on/off times. See the Power Averaging section of this chapter or, as a rough rule of thumb, for CW or SSB you can use 40% of your output PEP as a conservative estimate of average power.
- In the unlikely event that your station still doesn't pass, you should refer to the tables of antennas over ground in Chapter 8, use some of the other methods for estimating compliance or follow some of the steps described in

Table 5.7

Typical Antenna Gains in Free Space

	Gain in dBi	Gain in dBd
Quarter-wave ground plane or vertical	1.0	−1.1
Half-wavelength dipole	2.15	0.0
2-element Yagi array	6.0	3.9
3-element Yagi array	7.2	5.1
5-element Yagi array	9.4	7.3
8-element Yagi array	13.2	11.1
10-element Yagi array	14.8	12.7
17-element Yagi array	16.8	14.7

Note: Use the number of active elements on each band.

---

this chapter under Correcting Problems.

These simple tables give conservative estimates of compliance. They estimate the required distance one needs to be from the antenna *in the main beam of the antenna* (see Figures 5.8 and 5.9).

Like many tables, the ones shown in this article and *Bulletin 65* paint with a broad brush. They provide conservative answers to generalized conditions. If you want to bolster your confidence by using more precise evaluation methods, those are certainly available to you as well.

Tables Derived from NEC Modeling

The tables just described are all fairly easy to use. In many cases, however, exposure *near* an antenna in some areas can be much less than that indicated by the far-field tables. If a station "passes" using the simple tables, this could be a moot point. Even so, some hams may find it useful to use other methods to demonstrate that the exposure from their station is much less than what the rules allow.

A number of antenna-modeling programs (see the sidebar, "Available Software") will give much more accurate estimates of field strength in the near field of an antenna. However, many hams do not have the necessary experience to use them.

The ARRL Laboratory staff came up with a solution, but it involved consider-

able work on their part. To provide tables for specific antennas modeled at various heights over real ground, they selected the *NEC4* software package. Using *NEC4* they modeled a number of antennas, heights and power levels and calculated the compliance distances at ground level, first story and second story exposure points. (My personal 75-MHz Pentium PC had to chew on some of these calculations for as long as four hours!—*Ed.*) The antennas were modeled over "average" ground, with a conductivity of 5 milliseimens and a dielectric constant of 13, considered as being average ground by most antenna experts. Although the regulations permit whole-body exposure averaging, these tables are generally more conservative, calculating the field strength only at specific points.

The results were distilled into tables like Table 5.9, showing the 10-meter Yagi from Table 5.8, modeled 30 feet over average ground. Figures 5.8 and 5.9 show how these tables relate to the areas being evaluated. In many cases, a station that does not pass "worst-case" can easily be demonstrated to be in compliance using these tables.

Tables such as Table 5.9 provide a more accurate estimate of actual exposure than tables such as Table 5.8, derived from the far-field power-density formula. However, the antenna and its height must match the table to be applicable. (If the antenna is located higher than the heights in these tables, the exposure should be less than the predicted values.) The ARRL offered a number of these tables to the FCC for inclusion in *Bulletin 65*. *Supplement B* features a number of these antennas at heights of both 30 feet *and* 60 feet, helping to demonstrate that "higher is better"! In addition to the tables originally printed in *Supplement B*, Chapter 8 of this book contains a number of tables prepared using the same method as the tables in *Bulletin 65*.

To obtain a conservative estimate using these tables, hams should follow the following steps:

• Select the correct table for the frequency

band, antenna and antenna height being evaluated

• First, assume full PEP and 100% operation, then look up the compliance distance on the chart. If the antenna is located at least this far from areas of exposure, either horizontally, vertically or diagonally, the station "passes" on that antenna/band combination. Figure 5.10 shows how to determine the actual distance to the antenna.

• If necessary, calculate average power, based on duty cycle and on/off times. See the Power Averaging section of this chapter or, as a rough rule of thumb, for CW or SSB you can use 40% of your output power as a conservative estimate of average power.

• In the unlikely event that your station still doesn't meet the more precise requirements, you should refer to the tables of antennas over ground in Chapter 8 , use some of the other methods for estimating compliance or follow some of the steps described in this chapter under Correcting Problems.

You will have to use these tables to look up the compliance distance for ground level, first story and second story exposures, if applicable. The distance shown is the *horizontal* distance at the exposure height, from the center of the antenna. This is shown in Figures 5.8 and 5.9. It was calculated using *NEC4*, in the direction of the main beam of the antenna. If you are calculating worst-case exposure in the main beam, you can assume that this distance is from the tower to the exposure point. If you are calculating exposure in areas other than where the antenna is pointing, a conservative approach is to assume that these distances are from any part of the antenna.

Let's Compare

Tables similar to Table 5.8 can be used for a conservative estimate of compliance; tables like Table 5.9 show compliance under specific "real-world" conditions. Let's look at the differences between these tables.

In both tables, the maximum distances are similar. The 1500-watt distance for the 10-meter Yagi in Table 5.8 corresponds closely with the 1500-watt distance at the height of the antenna in Table 5.9. This is to be expected; Table 5.8 calculates the estimated distance in the main beam of the antenna and the *NEC4* calculation at 30 feet is in the main beam of the antenna. It can be seen in Table 5.9 that the exposure at 20 feet above ground also is in the same ballpark.

Table 5.9, however, represents a model

Table 5.8

Estimated distances (in feet) to meet RF power density guidelines in the main beam of a typical three-element "triband" (20-15-10 meter) Yagi antenna assuming surface (ground) reflection. Distances are shown for controlled (con) and uncontrolled (unc) environments.

	14 MHz, 6.5 dBi		21 MHz, 7 dBi		28 MHz, 8 dBi	
	con	unc	con	unc	con	unc
100	4.7	10.4	7.4	16.5	11.0	24.6
500	10.4	23.1	16.5	36.8	24.6	54.9
1000	14.7	32.7	23.3	51.9	34.8	77.7
1500	17.9	40.1	28.5	63.6	42.6	95.1

Available Software and Freeware

The calculations used to create the far-field tables have been written in BASIC by Wayne Overbeck, N6NB, and made available for download from the Web at **ftp://members.aol.com/cqvhf/97issues/rfsafety.bas**. This software also has been written into a Web-page calculator by Ken Harker, KM5FA. It can be accessed at **http://www.utexas.edu/students/utarc**.

Brian Beezley, K6STI, has made a scaled-down version of his *Antenna Optimizer* software available. Download *NF.ZIP* from the Web at **http://oak.oakland.edu:8080/pub/hamradio/arrl/bbs/programs/**. These programs are based on *MININEC* and will generally give the same results as you can obtain from using the tables derived from *NEC4* modeling. Contact Brian Beezley, K6STI, 3532 Linda Vista Drive, San Marcos, CA 92069; Telephone 760-599-4962, e-mail **k6sti@n2.net**.

Roy Lewallen, W7EL, sells *ELNEC* and *EZNEC* antenna-modeling software. *ELNEC* is based on *MININEC*, but does not have near-field capability. *EZNEC* is based on *NEC2* and can be used to predict the near-field strength. This software is available from W7EL Software, PO Box 6658, Beaverton, OR 97007; Telephone 503-646-2885; fax 503-671-9046; e-mail **w7el@teleport.com; ftp://ftp.teleport.com/vendors/w7el/**.

NEC2 and documentation is available from the "NEC Home—Unofficial" at **http://www.dec.tis.net/~richesop/nec/index.html**. Beware, however, that "native" *NEC* is *not* a user-friendly program. These are used best in the hands of experienced antenna modelers.

---

of a real antenna. In real-world conditions, the fields under an antenna do not vary smoothly. In many cases, the field directly under an antenna is *not* the maximum field to be expected! That maximum often occurs some distance away from the antenna. As the power is lowered, the level of the maximum also lowers in proportion. When the maximum field at a particular height drops below the MPE level, the compliance distance will suddenly go to 0.0 feet! This can be seen in several of the entries in Table 5.9. In comparing a number of the entries in both tables, it can be seen that

Table 5.8 indicates that one must be more distant from the antenna under some circumstances than what is shown in Table 5.9.

Note that the requirements for this real model shown in Table 5.9 are in many cases much less difficult to meet than the worst-case requirements shown in Table 5.8. As you can see, things are difficult to predict in the near field. In several cases, the table takes some pretty wild jumps, as noted between 600 watts and 750 watts at the 6-foot compliance point level. This is due to the distribution of fields under the antenna; the field is actually less

right under the antenna than it is some distance away. Chapter 2 has additional information about what effects can be found in the near field of an antenna.

Antenna Modeling

In *Bulletin 65*, the FCC suggests that *NEC*, *MININEC* and other computer modeling can be used to satisfy the requirements of the regulations. The software used to create the tables in Chapter 8 can model virtually any antenna system. Hams sometimes use some exotic antennas and it is not practical to create a table for each one. Some hams may want to evaluate the effect of multiple antennas or other conductors in proximity to their antennas to have a more accurate answer than can be derived from any other calculation method. In these cases, many hams will elect to use antenna-modeling software.

To use antenna-modeling program calculations, the amateur must first accurately model the antenna systems associated with his or her station. This generally requires that the location of the antenna conductors be entered into the computer program as rectangular coordinates (the horizontal and vertical positions of the end of each conductor). It is generally agreed that computer modeling using *NEC* or *MININEC* code yields accurate results under most conditions *if the model entered is accurate*. The latter point is important because this usually requires that the antenna and *all nearby conductors* be entered into the model. This would include the antenna, tower, guy wires and conductors such as electrical and telephone wiring.

A specific evaluation of RF fields in the near field of an antenna is not a simple issue. The relationship between the E and H fields is not constant in the near field, being determined mainly by the characteristics of the radiating element. Some antennas exhibit more E field than H field close to the antenna; others radiate more H field and less E field. (As these fields propagate away from the antenna, the ratio of the E to H fields converges toward the far-field value of 377 ohms.) There are a number of factors that affect the specific value of the E or H field in the near field.

These factors do not follow the classic "inverse square" law that applies to the far field of a spherical wave. Both the near field and far field additionally may contain components due to direct fields and to those that are scattered and reflected from objects and surfaces near the observer. The presence of these scatterers (both conducting and non-conducting) will affect both the near- and far-field calculations or measurements. All field values can be perturbed by nearby scatterers and sur-

---

Table 5.9

10-meter band horizontal, 3-element Yagi, Frequency = 29.7 MHz, Antenna height = 30 feet

Horizontal distance (feet) from any part of the antenna for compliance with occupational/controlled or general population/uncontrolled exposure limits

Height above ground (feet) where exposure occurs

Average Power	6 feet		12 feet		20 feet		30 feet	
(watts)	con	unc	con	unc	con	unc	con	unc
10	0	0	0	0	0	0	8	9
25	0	0	0	0	0	0	8.5	11
50	0	0	0	0	0	0	9	13.5
100	0	0	0	0	0	0	10.5	18.5
200	0	0	0	0	0	21.5	12.5	25
250	0	0	0	0	0	25	13.5	27.5
300	0	0	0	0	0	28.5	14.5	30
400	0	0	0	39	0	35	16.5	34
500	0	0	0	47	0	48	18.5	37.5
600	0	0	0	52.5	0	59.5	20	40.5
750	0	36	0	59	16.5	70.5	22	45.5
1000	0	46.5	0	67	21.5	82.5	25	61.5
1250	0	53	0	73.5	25	91.5	27.5	95.5
1500	0	58.5	0	79	28.5	99	30	108

faces, such as guy wires, power and telephone wiring inside the home of the operator or his or her neighbors.

These points are made because no *simple* calculation can yield an *exact* answer in the near field. Specific near-field calculations often require a lot of work. This is where antenna modeling comes in! The sophisticated software used in most antenna-modeling programs considers all these factors, often using computer methods just past what could reasonably be done with a human and a calculator.

Modeling programs do require some amount of user skills, although they should not be too difficult for the average ham. A list of software vendors is found in the "Software" sidebar. The ARRL Web page also maintains a list of software vendors who sell antenna modeling software. See **http://www.arrl.org/news/rfsafety/**.

General Considerations

Once you have selected an appropriate antenna-modeling program, you can consult the users manual and/or the vendor for specific applications information. In general terms, using antenna-modeling software is relatively easy. First enter the parameters for your antenna. This will include the location of all conductors in your antenna, element diameter, feed point, loading coils and traps, etc. As discussed in the section "Real World Considerations," you may want to include nearby conductors (tower, guy wires, telephone

and electrical wiring, etc) in the model to have the most accurate possible estimate. You should be able to use the program to verify that the model is accurate. If you see an antenna pattern and antenna gain and feed point SWR or impedance that is reasonable for the antenna type, you have probably done it right. Most of the programs come complete with example models for common antenna types. Of course, this will not help with some of the unusual antennas hams are known to use, although they will serve as good examples of how to model antennas in general.

When you have the model right, use the program's "near-field" capability to calculate the electric (E field) and magnetic (H field) in those areas you want to evaluate. Input the *average* power of your station in a 6-minute period for controlled exposure, and in a 30-minute period for uncontrolled exposure. (See the discussion under "Average Exposure" earlier in this chapter.) In most programs, this is done by specifying a line and calculating the field along that line in the increments you specify. For a Yagi antenna, calculating the near field, starting at a point directly below the antenna in a horizontal direction in the main beam would probably be most useful. This can be done at various heights above ground, to determine ground level exposure and the exposure to nearby buildings.

The near-field analysis capability of most of these programs shows the field

value for each of the points and increments you have specified. You can then compare these results with the MPE limit. It is safe for people to remain indefinitely in all areas that are below the MPE limit for the operating mode, power and on/off times you used to determine your average power. The ARRL tables in *Supplement B* show the *farthest* compliant distance. For some antenna configurations, however, it is possible that some areas *closer* to the antenna might be in compliance. An example of this is shown in Figure 5.10. The only way to know exactly what areas are above or below the limit is to use the near-field model.

Measurements

While amateurs are not required to specifically measure the field strength from their station operation, the FCC would consider accurate measurement to be a valid method of complying with the regulations. However, most amateurs will not need to make measurements to perform a routine station evaluation.

Some hams, however, might *choose* to make actual measurements of the electric and magnetic field strengths around their antenna while they are transmitting a signal. If you happen to have a calibrated field-strength meter with a calibrated field-strength sensor, you can make accurate measurements. Unfortunately, such calibrated meters are expensive and not normally found in an amateur's tool box.

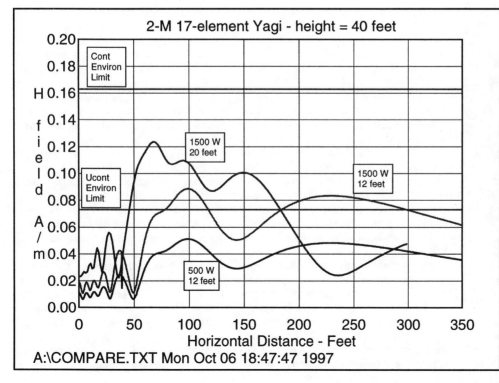

Figure 5.10—This plot shows the way the H field varies under an antenna. The X axis represents the horizontal distance from the center of the antenna in the main beam. Note that the field reaches a peak some distance in front of the antenna.

The relative field-strength meters many amateurs use are not accurate enough to make this type of measurement.

Making field-strength measurements, especially in the near field of an antenna, can be tricky. Measurements require accurate calibrated equipment, calibrated E and H-field probes and a sound understanding of the proper use, and limitations, of the equipment involved. Fortunately, the FCC regulations do not require actual field-strength measurements.

Measurements are one way to perform an analysis, but they're very tricky. With calibrated equipment and skilled measuring techniques, ±2 dB error is pretty good. In untrained hands, errors exceeding 10 dB are likely. A ham who elects to make measurements will need calibrated equipment (including probes) and knowledge of its use. Many factors can confound measurements in the near field. In most cases, various calculation methods, especially computer antenna modeling, can give results that are more accurate—if the model is right.

Usually you need to use a calibrated field-strength meter to make accurate measurements. These come in two varieties—tuned and wideband. Most of the instruments available are broadband devices. A broad-bandwidth instrument used to measure RF fields is calibrated over a wide frequency range, and responds instantly to any signal within that range. The nice thing about a wide-bandwidth instrument is that it requires no tuning over its

entire operating range. Broad-band instruments offer some significant advantages—one can enter an RF environment and not have to carefully adjust the instrument for a peak response. They can be tricky to use in other ways, though, because the response of the probes used often varies with frequency, so one would have to have some knowledge of the signals present. In multiple-transmitter environments, it may not be possible to obtain an accurate measurement with some broad-band instruments. Other, more sophisticated instruments have compensation networks built in, tailored to match the frequency variation of any particular standard or regulation. With these instruments, you get a reading in a multiple-transmitter site that does not need to consider the frequencies involved. The instrument automatically compensates, and expresses the reading as a total of the permitted MPE level.

A narrow-bandwidth instrument, on the other hand, may be able to cover a wide frequency range, but would have a bandwidth of perhaps only a few kilohertz at any instant. You have to tune the instrument to the particular frequency of interest before making your measurements. Narrowband, tunable instruments can overcome some of the problems inherent with simple broadband instruments, although they are often a bit more complex to use. In essence, these are calibrated receivers. If the characteristics of the probe are known, the field-strength level can be determined directly for each frequency being measured.

All these instruments are used in conjunction with calibrated E-field, H-field or power-density probes. E-field probes generally consist of multiple short dipoles, mounted at right angles to each other to read E fields of any polarity. An H field-probe similarly consists of multiple small loops, mounted at right angles to read H fields of any polarity. A well-designed E- or H-field probe will have the response of the "wrong" field that is at least 20 dB less than the desired response. Power-density probes are usually thermocouple devices. One significant disadvantage of thermocouples is that they can be damaged by fields that are significantly higher than what they are designed to measure. They can sometimes be damaged even if the measurement instrument is not turned on, so they are generally used only in those areas where the test engineer has some knowledge of the strength of the RF energy.

To use most of these instruments, one needs to consider the overall accuracy and frequency response of the instrument, the accuracy and frequency response and ori-

entation of the probes, and the interaction of the fields with nearby objects, the test equipment or the test engineer. Some test engineers have cited accuracy and repeatability of 6 dB as being typical. Others have noted that with "heroic" precautions taken, it is possible to obtain an accuracy of 1 dB. But this often consisted of taking and averaging multiple readings, setting the instrumentation on a small table and having the operator walk away and look at the reading through a pair of binoculars!

Even if you do have access to a laboratory-grade calibrated field-strength meter, you must be aware of factors that can upset your readings. Reflections from ground and nearby conductors (power lines, other antennas, house wiring, etc) can easily confuse field-strength readings. For example, if the measuring probe and the person making the measurement are located in the near-field zone, they can both interact with the antenna fields. In addition, you must know the frequency response of the test equipment and probes, and use them only within the appropriate range. Even the orientation of the test probe with respect to the test antenna polarization is important.

Why should we be concerned with the separation between the source antenna and the field-strength meter, which has its own receiving antenna? One important reason is that if you place a receiving antenna very close to an antenna when you measure the field strength, *mutual coupling* between the two antennas may actually alter the radiation pattern from the antenna you are trying to measure.

Actual measurements are best left to the professionals. In untrained hands, the errors can mount up fast. Some instruments just do not have the needed accuracy and consistent frequency response. If a ham, or the neighbor of a ham, uses these "instruments" to do field-strength measurements, the results are apt to be so far off as to cause undue alarm, of give a false sense of security.

It should be mentioned that many of the field-strength meters, especially the inexpensive ones, give only a relative field-strength measurement. Many of them have probes with a response that varies with frequency and is non-linear with power level. Most of these inexpensive instruments measure either the relative E field or H field. Although they may be calibrated in power-density units, they are really reporting the approximation of power density represented by equivalent plane-wave power density, usually for just one field component. For purposes of complying with these regulations, uncalibrated field-strength meters should be avoided.

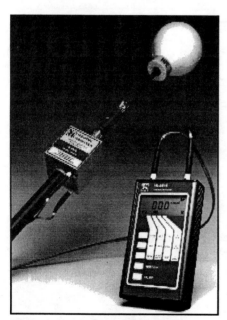

Figure 5.11—This calibrated field-strength meter and probes can be used to make measurements of the fields near a radio transmitter. (Photo courtesy of Holaday Industries, Inc.)

Formulas

Most of the methods that hams will use to complete their station evaluations involve some form of calculation. The results of these calculations can be compared with the MPE limits. The tables published in Chapter 8, *Bulletin 65* and *Supplement B*—were derived from various calculational methods. Even the tables derived from computer modeling involved calculations, except in that case, the calculations were done by the computer. Fortunately, for those hams who want to "homebrew" their own evaluation, the equations involved are all quite straightforward. A knowledge of square roots and simple algebra is all that is required.

While most hams will probably prefer to use one of the table or software methods to estimate compliance, the power-density equations contained in *Bulletin 65* may be useful in some cases. The "basic" power density equation in *Bulletin 65* is shown in Eq 5.4:

$$S = \frac{PG}{4\pi R^2} \qquad \text{Eq 5.4}$$

where S = the power density, G = the numerical gain of the antenna in dBi expressed as a decimal number, R = the distance from the center of radiation and P = power input to the antenna. This is the equation for power density *in free space*. It will give the power density for areas located "R" distance away from the center of the antenna, in the main beam of the antenna. It assumes that all areas being considered are in the far-field region (see Chapter 2), a reasonable *approximation* for estimating compliance for most antenna types.

S, P and R must be expressed in the same units. S is the power density per square unit. If S is in milliwatts per square centimeter, then P must be in milliwatts and R must be in centimeters. G is the gain of the antenna, expressed as a decimal, not in dB. To convert the gain in dBi to a decimal number, use Eq 5.5 or consult Table 5.10.

$$G = 10^{dB/10} \qquad \text{Eq 5.5}$$

where G is the numerical gain of the antenna whose gain is expressed in dBi.

Most antennas are not in free space; they are located above ground. Placing an antenna above ground modifies the pattern such that the main beam of the antenna contains more energy than it would in free space. This is known as *ground gain*.

If an antenna is placed over a perfect ground, Eq 5.6 can be used to calculate power density. This formula assumes 100% reflection of the E and H fields from an infinite, perfect ground plane under the antenna.

$$S = \frac{PG}{\pi R^2} \qquad \text{Eq 5.6}$$

where S = the power density in mW/cm², G = the numerical gain of the antenna in dBi expressed as a decimal number, R = the distance from the center of radiation in centimeters and P is the power input to the antenna in milliwatts.

In reality, however, actual surface reflections are never 100% efficient. Various factors and losses reduce the actual reflection. The Environmental Protection Agency has made a recommendation that Eq 5.7 be used to estimate actual ground reflections under real-world conditions:

$$S = \frac{0.64\,PG}{\pi R^2} \qquad \text{Eq 5.7}$$

where S = the power density in mW/cm², G = the numerical gain of the antenna in dBi expressed as a decimal number, R = the distance from the center of radiation in centimeters and P = power input to the antenna in milliwatts.

If you know the power to your antenna, the gain of your antenna and the distance to any area for which you want to know the power density, these formulas can give a reasonable estimate. They tend to be conservative in the near field of an antenna, where one might be close to only part of the antenna.

As an example of the use of these formulas, assume a 1000 watt transmitter is operating into an antenna system with 3 dBi of gain. (To keep it simple, assume the feed line is lossless.) If you want to know the exposure at a point that is 20 feet from the center of the antenna, expressed in mW/cm², with the EPA ground-reflection factor, use Eq 5.7. First convert the 1000 watts to 1,000,000 milliwatts, convert 20 feet to 609.6 centimeters and convert 3 dBi gain to 2.0. The solution then is:

$$S = \frac{0.64 * 1,000,000 * 2.0}{3.14 * 609.6^2} = 1.097\,mW/cm^2$$

Feet!

All this converting from feet to meters can get tedious. Here are a few variations on the equations, expressing P in watts, R in feet, using the ground-reflection factor of Eq 5.7. In the equations that follow, all the conversion, square root and π factors have been considered in simplifying the formula.

$$S = \frac{0.219\,PG}{R^2} \qquad \text{Eq 5.8}$$

Eq 5.8 will give the power density in mW/cm² if P is in watts, R is in feet and G is the antenna gain expressed as a decimal number.

Perhaps the most useful derivation of these equations is one that tells you how far away from a particular antenna and power people must be for a given power density S.

$$R = \sqrt{\frac{0.219\,PG}{S}} \qquad \text{Eq 5.9}$$

where S = the power density in mW/cm², G = the numerical gain of the antenna in dBi expressed as a decimal number, R = the distance from the center of radiation in feet and P = power input in watts.

Another variation on this theme is shown in Eq 5.10. This formula lets you input your antenna gain (G), the distance to the antenna (R) and the power-density limit and determine the maximum allowed *average* transmitter power.

$$P = \frac{SR^2}{0.219\,G} \qquad \text{Eq 5.10}$$

where S = the power density in mW/cm², G = the numerical gain of the antenna in dBi expressed as a decimal number, R = the distance from the center of radiation in feet and P = power input in watts.

Eq 5.8, 5.9 and 5.10 can be helpful in a number of ways. If you run 500 watts average power and have a 10-meter dipole (2.15 dBi, G = 1.64) located 20 feet from

---

Table 5.10

dB to Decimal Number Gain Conversion Table

dB	Gain	dB	Gain	dB	Gain
0.0	1.00	5.5	3.55	12.0	15.84
0.5	1.12	6.0	3.98	13.0	19.95
1.0	1.26	6.5	4.47	14.0	25.12
1.5	1.41	7.0	5.01	15.0	31.62
2.0	1.58	7.5	5.62	16.0	39.81
2.5	1.78	8.0	6.31	17.0	50.12
3.0	2.00	8.5	7.08	18.0	63.10
3.5	2.24	9.0	7.94	19.0	79.43
4.0	2.51	9.5	8.91	20.0	100.00
4.5	2.82	10.0	10.00	22.0	158.49
5.0	3.16	11.0	12.59	25.0	316.23

an upstairs bedroom in your neighbor's home, you can use Eq 5.8 to calculate that the power density is 0.49 mW/cm². Unfortunately, the uncontrolled MPE limit on 10 meters is 0.2 mW/cm², so this is not in compliance for 500 watts of average power. You can then use Eq 5.9 to calculate that you would be in compliance if you move your antenna 30 feet away. You also could use Eq 5.10 to calculate that if you reduce your average power to 222.7 watts, you are in compliance.

Last but not least, because the MPE power-density level is frequency dependent, equations can be derived that include the frequency. For MF/HF *only*, the following formula can be used to calculate the required compliance distance in feet:

$$R = 0.03049 \sqrt{PG} \qquad \text{Eq 5.11}$$

Where R = the required minimum distance from the antenna in feet, P = power input to the antenna in watts and G = the gain of the antenna in dBi expressed as a decimal number. This formula has been simplified to remove all the feet-to-centimeter conversions, the watts to milliwatts conversions, π and square roots of fixed numbers.

All these formulas generally give conservative results. They are assuming that the distances involved are in the main beam of the antenna. In the examples given, the actual exposure could well have been in areas below the antenna, which generally give less exposure than areas at or slightly above the antenna. Although these examples showed the proper and easy use of the formulas, a better alternative might have been to use the antenna-over-ground tables in Chapter 8. FCC *Bulletin 65* features a number of variations on some of these formulae. Figures 1 and 2 in *Bulletin 65* show these formulas graphically. The formulas and graphs are reprinted in Chapter 6.

E to H to Power Density Formulas

The MPE limits in the regulations are called out in E-field, H-field and power density or plane-wave equivalent power density. The formulas above, however, all manipulate the power-density, distance, transmitter power and antenna gain. There is a relationship between power density and the two fields that applies perfectly in the far field, and may apply reasonably well in the near field. (This is discussed in more detail in the Antenna Fundamentals chapter.)

Once S has been calculated, the E and H fields can be determined. E can be calculated in volts per meter (V/m) by the formula:

$$E = \sqrt{3770S} \qquad \text{Eq 5.12}$$

where E is in V/m and S is in mW/cm²

H can be calculated in amperes per meter (A/m) by the formula:

$$H = \sqrt{\frac{S}{37.7}} \qquad \text{Eq 5.13}$$

where H is in A/m and S is in mW/cm²

The values of S, E and H, if applicable, can be compared to the values in the MPE limit tables in the rules.

This calculation is only valid in the far field of the antenna. In the near field the relationship is not this simple. This calculation may prove useful to you as you analyze your station for compliance with the FCC MPE limits. If you know the E or H field strength at some point in the far field then you can calculate the other value at that same point.

If a steady carrier level were used in all these formula evaluations, the station being evaluated at that power level and frequency can be operated into the antenna used in the calculation at 100% duty cycle (CW key down). It will have the MPE calculated at the distance used for the calculation. Any points more distant than this point also will be in compliance with the regulations. If PEP is used in the calculations, no additional calculations need to be made for this frequency, power level, antenna and exposure locations, assuming that the point has been calculated for the nearest points of exposure. This calculation is good for any operating mode for an indefinite exposure. Repeat this calculation for other bands, power levels and antennas, assuming that the points being calculated are in the far field of the antenna in question.

The formulas also can be used for average exposure, using power averaged over the appropriate averaging time, as described elsewhere in this chapter. *Bulletin 65* contains additional formulas, including a number of them for parabolic reflectors and other aperture antennas. These formulas have *not* been reproduced in the condensation of *Bulletin 65* that appears in Chapter 6. Contact the FCC for information about how to obtain a full copy of *Bulletin 65*, or go to their Web page at **http://www.fcc.gov/oet/info/documents/bulletins/#65**.

Using Graphs to Evaluate RF Exposure

It is possible to create graphs of field strength or power density based on computer analysis or other calculations. Figure 5.13 shows one such graph. The Novice and Technician class question pools contain questions about such graphs. The figure represents a beam antenna, such as a Yagi, that you might use

with your amateur station. Some people might find it easier to read such graphs than search through the data in a table or use formulas. Each antenna type requires its own graph, so you still may have to search through many drawings to find the one that best describes your station. Graphs such as these have been included in *Bulletin 65*.

The power density of Figure 5.13 represents the signal in the main beam of this antenna. It is expressed for various levels of effective radiated power (ERP). ERP takes the antenna gain into account. For example, if you are using an antenna with 10 dBd of gain, and your transmitter produces 100-watts PEP output, then you would use the 1000-W ERP line. If you use only 10-watts PEP output with this antenna then you would use the 100-W ERP line.

Suppose you want to know the power density at a point 10 meters from your antenna when you have 1000-W ERP. Point 1 on this graph conveniently locates the 10-meter distance on the 1000-W ERP line of the graph. Now look to the axis along the left edge of the graph and read the power density. If you judged the value to be about 0.35 mW/cm² you would be pretty close.

Of course your evaluation is not complete at this point. Now you will have to determine the MPE limits for controlled and uncontrolled environments at your operating frequency. For a signal in the VHF range (30 to 300 MHz), the controlled environment power density limit is 1.0 mW/cm², so the power density at 10 meters is below this limit. For an uncontrolled environment, however, the power density limit is 0.2 mW/cm², so you will have to increase the distance to meet this limit.

To find the distance for this uncontrolled environment limit, you should find 0.2 mW/cm² on the power density axis, and look across to the right until you come to the 1000-W ERP line. You should come to point 5 on this graph. Now look down to the distance axis, and you should estimate that at about 15 meters you will meet the uncontrolled limit.

As you can see, a graph like this one can be quite helpful in evaluating the RF exposure from your station at various distances and ERP levels. They have been reproduced in Chapter 6 of this book, the partial reprint of *Bulletin 65*.

Antenna Patterns

All the evaluation methods discussed so far evaluate exposure in the main beam of the antenna, either at the height of the antenna as a worst-case, or at specific

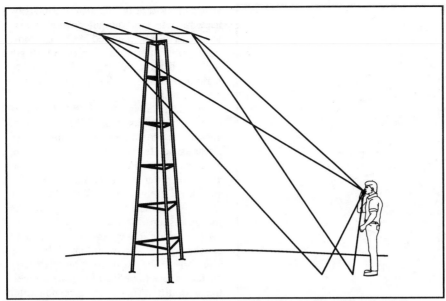

Figure 5.12—The signal on the ground results from a combination of all the signals arriving at the observer. In this case, signals from different parts of the antenna arrive directly, along with signals reflected from the ground. Each arrives in a phase relationship dependent on the relative lengths of the paths involved. These signals can add or subtract to varying degrees an any particular point.

heights in the direction the antenna is pointing. The actual field strengths will be maximum in the main beam of the antenna, and less in other directions. In most cases, amateurs will evaluate either simple antennas, such as dipoles, that are more or less omnidirectional, or rotatable antennas that can be pointed in any direction. In either case, evaluating in the main beam of the antenna is appropriate.

In other cases, though, especially with non-rotatable antennas, it may be helpful to consider how the fields vary near an antenna with the pattern of the antenna. This may help determine that a particular area has fields that are below the applicable MPE limit.

You can use the published pattern of an antenna to some degree when calculating exposure. Figure 5.14 shows the free-space radiation pattern of a 3-element Yagi antenna. At first you might believe that you should use the "above ground" pattern to evaluate the exposure potential of an antenna above ground. Unfortunately, this is not valid. Antenna patterns are derived in the far field—very far away from the antenna. At great distances, the rays from various parts of the antenna, reflected off ground, add up in or out of phase to form a pattern when the signal strength is plotted on a graph. Things are not nearly so precise in the near field, where one can be much closer to one element in an array than another. In this case the angles between the antenna and ground, and the observer and ground, are

much different than they are very far away from the antenna. The far-field pattern of that antenna would indicate that there is *no* energy below the antenna at all, a conclusion that is not borne out by computer modeling of the near field.

The free-space pattern of Figure 5.14 does demonstrate that *some* energy is di-

rected downward. Figure 5.12 shows that an observer on the ground will "see" two signals from the antenna—a direct signal and one reflected off ground. Depending on the relative path length of the two signals, they could arrive at the observer in or out of phase. If they are in phase, they will add—the reason that a *ground reflection factor* was included in all the tables. It can be seen, however, that the pattern shows that the amount of signal directed downward is not as much as is found in the main vertical lobe of the antenna. This pattern can be used with some reliability to predict the amount of energy directed downward.

Most antenna patterns use a decibel scale. The reference level is usually set to the point of maximum gain, and is usually set at 0 dB. You can look at the pattern and determine by how much the energy is reduced in a particular direction, and apply that to the evaluation process. For example, the point marked "A" on Figure 5.14 or 5.15 is about 12 dB less than it is in the main beam. If you want to evaluate exposure *in that direction,* you can reduce the amount of power used in the calculation by that amount and use the tables or formulas to estimate compliance. You can use the formulas and tables featured earlier in this chapter to determine how much to reduce the power for any particular reduction in dB. Some antenna patterns may have a decimal number scale instead of a dB scale, so look carefully.

This process does have its limitations,

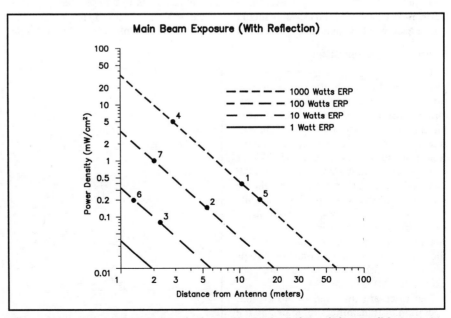

Figure 5.13—Using computer analysis or other calculations, it is possible to crate a graphical display of the field strengths and power densities for various antennas and transmitter power levels. This graph represents the power density in the main beam of an antenna such as a Yagi. Various effective radiated power (ERP) levels are given. ERP takes the antenna gain into account, referenced to a halfwave dipole in free space.

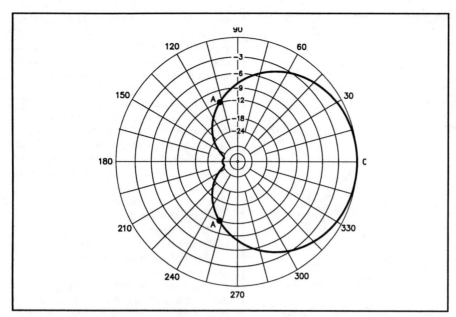

Figure 5.14—This is the free-space elevation pattern of a typical Yagi antenna. Less energy is directed downward toward the ground than in the main beam of the antenna. This is looking at the antenna from the side.

however. The patterns are derived from a far-field analysis that doesn't apply perfectly in the near field. This is especially true for the deep nulls that exist in some antenna patterns; you really can't count on their being present to that extent in the near field. Even the other areas of the pattern do not apply perfectly in the near field. However, according to the FCC in *Bulletin 65*, the patterns can be used with some degree of confidence. A good rule of thumb is that pattern nulls exceeding 15 dB or so are suspect, and probably should not be used without some modification.

Multi-transmitter Sites

The term "multi-transmitter site" applies to multi-transmitter amateur stations, such as are used in some contests, and to commercial sites, such as the mountaintop location of some amateur repeaters. Some amateur stations use multiple transmitters, such as an HF DX or contest station that also accesses a VHF PacketCluster. Other stations might be located at sites also occupied by transmitters in other radio services. Two or more transmitters could be operating at the same time, each adding to the exposure level. In these cases, the operators must take steps to ensure that the *total* exposure does not exceed the MPE level.

The rules are intended to ensure that operation of transmitters regulated by the FCC doesn't result in exposure in excess of MPE limits. It is fairly easy to make this determination for single transmitters when there are no other sources of RF to complicate things. However, many transmitters operate in proximity to other transmitters. It is entirely possible for two or more transmitters to be below their own limits, but the total exposure from all operating together to be greater than the permitted MPEs.

The FCC regulations cover this very likely situation. In most cases, all the significant RF transmitters operating at multi-transmitter sites generally must be considered when determining if the site's total exposure is in compliance. All significant emitters are jointly responsible for overall site compliance. The antenna tables elsewhere in this article cannot be used to determine actual power-density levels, as will be required to evaluate most multi-transmitter sites. The field-strength formulas in this article and in *Bulletin 65* or various antenna-modeling programs can be used instead.

At multi-transmitter sites, all significant contributions to the RF environment should be considered—not just those fields associated with one specific source. To this end, the FCC has determined that any transmitter that operates at an exposure level greater than 5% of the power density *permitted to its own operation* is jointly responsible *with all the other operators within its exposure area* who also exceed 5% for site compliance. In those areas where the exposure from the transmitter is less than 5% of the MPE level, the operator is not jointly responsible. Note that this is *not* the same as 5% of the total exposure power density, which could sometimes be unknown to any single transmitter at the site. This actually covers a lot of small stations like amateur repeaters, although a station evaluation may be required to demonstrate that the exposure is below the 5% threshold.

Categorical Exemptions Again

The FCC doesn't expect all low-power transmitters necessarily be responsible for site compliance at sites where they contribute only a tiny fraction of the total RF energy. The rules limit the responsibility of some operators at the site. In those areas where the exposure from a transmitter or system is less than 5% of the MPE level permitted to that transmitter, the operator

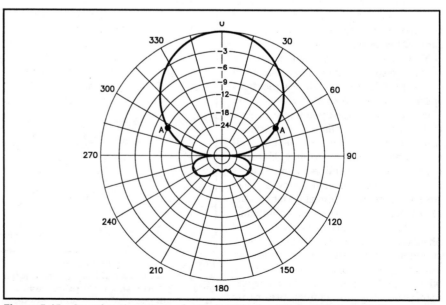

Figure 5.15—An azimuth pattern of a typical Yagi antenna. This is a bird's-eye view of the antenna.

is generally not jointly responsible with the other operators on the site for overall site compliance.

For example, the controlled power-density MPE limit for a 146-MHz transmitter is 1.0 mW/cm². If that transmitter were operating alone, the operator would have to ensure that no one was exposed to a power density greater than that, averaged over 6 minutes. (A controlled environment was selected for this example because most repeater sites are not open to the general public.) This exposure would normally occur only close to the antenna, with rooftop exposure being considerably less than this. Let's assume that the exposure on the rooftop near the amateur antenna's tower is 0.1 mW/cm², well within the limits. Twenty-two feet away from the tower base, the power density from the amateur repeater drops to 0.05 mW/cm². This is 5% of the exposure permitted for a 146-MHz transmitter.

However, if another transmitter starts operating at the site, things may change. Let's assume that three different 156-MHz commercial stations also share the site. The controlled limit for this frequency also is 1.0 mW/cm². Let's assume that the rooftop exposure for each station is 0.98 mW/cm². This also is just within the MPE limit, as long as only one transmitter is on at a time. If one transmitter and the amateur station are transmitting for the full 6-minute exposure period (likely with an amateur repeater), the total field would be 1.08 mW/cm². This is over the MPE limit *if people are present on the rooftop.* In this case, the amateur licensee, even though the repeater is only contributing a

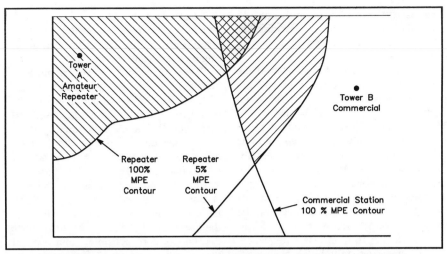

Figure 5.17—A bird's eye view of a rooftop installation. The line marked "5% contour" shows the area in which the exposure exceeds 5% of that permitted to the amateur repeater located on tower "A." The "100% contour" shows the area that is above the MPE limit for either the repeater or the transmitter on the adjacent tower "B." Under these circumstances, the amateur operator is solely responsible for the area with the diagonal cross hatch because it exceeds the MPE limits for the repeater station. The areas within the other 100%-contour boundary are out of compliance for the transmitter on tower "B." The amateur operator is, however, also jointly responsible for the overall compliance within the area with the double cross hatch because the repeater's contribution to overall exposure is greater than the 5% permitted to the repeater.

Figure 5.16—This multiple-transmitter site can be difficult to evaluate! *(Photo courtesy Robert Cleveland, FCC Office of Engineering and Technology)*

small part of the field, would be responsible for site compliance in all areas of the site where the repeater exceeds the 5% MPE level, or 0.05 mW/cm². In this case, the amateur licensee would be responsible for areas up to 22 feet from the tower base, under the conditions stipulated in the previous paragraph. Even if other transmitters on the site made the areas farther away even more non-compliant, each licensee is responsible only for their 5% areas.

Calculating Total Site Exposure Levels

The example just cited was an easy one; both transmitters operated between 30 and 300 MHz, where the controlled MPE limit is constant at 1.0 mW/cm². In this case, one can simply add up the MPE levels and obtain the total exposure. In many cases, though, the involved transmitters could be operating on frequencies with different MPE limits, such as an amateur repeater used in the earlier example on 146 MHz sharing a site with a TV transmitter on 600 MHz. In this case, the controlled MPE limit for the 146 MHz transmitter is 1.0 mW/cm²; the controlled MPE limit for the 600 MHz transmitter is 2.0 mW/cm². (The MPE limit increases for frequencies higher than 300 MHz.)

Even in cases where transmitters are operating on different frequencies, with different MPE limits, it is relatively easy to calculate total exposure at multi-transmitter sites. The antenna tables elsewhere in this article cannot be used to determine actual power-density levels. The field-

strength formulas in this article and in *Bulletin 65* or various antenna-modeling programs can be used instead. For any point being evaluated, determine what percentage of the permitted MPE will actually be encountered for each transmitter. Then, add up the percentages for any transmitters that could be in operation simultaneously. If the total percentage exceeds 100%, the site is not in compliance. For example, if a 2-meter transmitter creates exposure at 40% of what is permitted on that frequency, and a simultaneous transmission is occurring by a 1.5-GHz commercial transmitter at the same site at 70% of the limit, the total is 110%. This site is out of compliance, even though each transmitter is being operated below its own limit.

To determine overall exposure when different frequencies are involved, first convert the exposure to a percentage. In the case of the 146-MHz repeater and the 600-MHz TV station, assume that at the base of the tower, the 146-MHz exposure is 10% of its permitted MPE. If the TV station creates an exposure of 1.9 mW/cm² at the base of the tower, this is 95% of the permitted MPE for that transmitter. If you add up the two percentages, you have the total exposure. In this case, the total is 105%, and the area below the tower is not compliant if both transmitters are on *and* people remain in that area for the 6-minute controlled environment averaging period. The 146-MHz MPE in the area next to the tower is 10% of what is permitted on 146 MHz. This exceeds 5%, so the repeater operator is jointly responsible for site compliance.

This would be equally true even if the 600-MHz TV transmitter were creating a power density of 10 mW/cm^2, which would be at 500% of the permitted limit. The 146-MHz transmitter operator would still be responsible for areas where its own MPE was greater than 0.05 mW/cm^2—5% of the MPE permitted to a 146-MHz transmitter. If in this case the repeater was operating on a 1.2 GHz repeater with a power density of 0.1 mW/cm^2 at the base of the tower, the MPE from the 1.2 GHz repeater is 2.5% of the MPE level permitted at that frequency. Strictly speaking, only the TV station operator is responsible for site compliance. The amateur should certainly help out, if possible. If a site were missing compliance by only a few percent and the amateur could move the repeater antenna higher up the tower, that would certainly be a "neighborly" gesture. Likewise, the amateur should share the results of the station evaluation with other operators on the site, to help them determine if the overall site is not in compliance.

Not Included

In general, all major emitters at a site should be considered when determining overall site compliance. However, the FCC has clarified that in most cases, those stations whose MPE levels are less than 5% of the permitted level need not be considered when determining overall site compliance. Likewise, those stations that are categorically exempt from evaluation generally do not need to be considered, either. In both cases, the stations that are exempt, or less than 5%, are *presumed* not to be a factor.

In the case of the 146-MHz repeater discussed earlier as an example, the power density from the repeater does not need to be considered past 22 feet from the base of the tower. At this point the exposure level drops below 5% of the MPE level permitted on 146 MHz. In the case of the 1.2 GHz repeater, its exposure does not need to be included in any calculations on the rooftop, because its exposure level is below the 5% level.

However, some types of stations, such as amateur repeaters using less than 500 W ERP, do not need to be evaluated. In addition, however, those stations that are not required to be evaluated generally are *presumed* not to be responsible for site compliance. Amateur repeaters using less than 500 W effective radiated power (ERP) and those whose antennas are not mounted on buildings and are located

32.8 feet (10 meters) or higher above ground, generally do not need to be evaluated. These stations are not usually included in determining overall site compliance.

These exclusions, however, must be considered in the overall context of the FCC's main goal—that people not be exposed to RF energy above the limits. If there were 20 stations all operating at 5% of the limit on a particular site, and another operating at 10% of the limit, the total would be 110%. If each of the "5%" stations was not considered, and the 10% station claimed that the site was at 10%, the error would be quite large. Although the specifics of the rules would indicate that no one is responsible, other parts of the rules do permit the FCC to require stations that are otherwise exempt to conduct evaluations. Amateurs should consider carefully whether circumstances might make it helpful to evaluate an operation that is otherwise categorically exempt.

Keep in mind, too, that although some of our examples show that rooftop exposure was below 5%, as one gets closer and closer to an amateur antenna, the 5% threshold (and the MPE limit threshold) sooner or later may be crossed. In most cases, an amateur repeater will have *some* areas of responsibility on any site, even if that responsibility extends only to areas on the tower. In many cases involving an Amateur Radio transmitter, only a very small area would be encompassed by that 5%. Joint responsibility might only exist in the immediate vicinity of the amateur antenna.

A repeater trustee, for example, might have that 5% level extend only to those areas up to 10 feet below the antenna, and thus be responsible for overall site compliance only to that area. In this case, the responsibility may be only to radio service personnel climbing the tower, and generally a controlled exposure environment would apply. However if tower maintenance people (who may or may not be trained about RF exposure) are present, an uncontrolled environment may be more appropriate.

The FCC can require any operator to conduct an evaluation if they believe that there could be a problem. *Bulletin 65* clarifies that these stations are presumed to be in compliance with their own individual MPE limits and generally do not need to be included when calculating overall site compliance. These are generally presumed not to be jointly responsible for site compliance. However, these are *not* iron-clad assumptions. FCC rules, in Section 1.1312(a) stipulate that the FCC can require that any station file an Environmental Assessment (EA) or conduct a routine

environmental evaluation to demonstrate that it is not necessary to request an EA—even those covered by specific categorical exemptions.

The FCC will make these determinations on a case by case basis, but in cases where a station that is categorically exempt from evaluation, or a station that is creating exposure that is less that 5% of what is permitted to it, the FCC could determine that the particular station needs to share responsibility for site compliance. Clearly, if an amateur station shares space with a high-power broadcast station, the "5% rule" is pretty straightforward, but if a number of low-power transmitters share a site, even minor emitters might have to make changes to their station if the overall site compliance is more than the MPE limits allow. It is quite possible for some sites to have literally hundreds of transmitters, most of which are operating below the 5% level, even though the overall site's RF exposure could be greater than the MPE limits. The best approach is to err on the side of caution and cooperate with other operators on the site, if there is a compliance problem. There is, of course, no substitute for your own good judgment; use it as it appears to be appropriate in "gray" areas. This may prevent the FCC from having to make a case out of your station.

The Unknowns at Multitransmitter Sites

In some cases, amateurs may not be able to obtain full information about the other transmitters on the site. If you find yourself in this situation, you should attempt to secure information from the site owner. If that isn't available, make the best estimates possible of other transmitter powers and antenna gains on the site to determine compliance. In most cases, the repeater operator will need to cooperate with other site users to determine the overall exposure of all the transmitters on the site.

Bulletin 65 and Multi-Transmitter Sites

While *Bulletin 65* does not include simple tables for hams to use to evaluate their stations, it does have an extensive section on compliance at multiple transmitter sites. Hams who operate multi-multi contest stations (or repeater operators) may want to read the entire bulletin, to get a head start on understanding the issues involved at multiple transmitter sites.

Real World Considerations in Doing Evaluations

Of course, the real world is not quite as neat as the formulas and tables would like it to be! Ground has slope, antennas have nearby conductors changing their patterns, feed lines sometimes radiate and Murphy can strike: Things do go wrong. Knowing how and when to apply these factors sometimes requires the sound technical judgment of the station operator!

Nearby Conductors and Antennas

Antennas can and do interact with nearby conductors. Conductors located near an antenna can usually pick up and reradiate some of the signal, which can complicate analysis. Such nearby conductors can sometimes conduct signals away from the antenna and reradiate them closer to areas of exposure. An example of the latter phenomenon would be an antenna, located within several feet of the phone line, running back into the operator's house. In cases where the phone line is located very close to areas of exposure, the MPEs could be exceeded under some circumstances. Such nearby conductors can be an unintended integral part of the antenna system. This can complicate antenna modeling, because these nearby conductors should be accurately entered into the model.

If you have a considerable safety margin in your evaluation, there is little risk that additional reradiation from nearby conductors will result in local fields that are higher than the permitted MPEs. You may want to consider whether tower structures, guy wires, nearby utility wires or large metal objects could be affecting your results. In general, those objects near the antenna, or near the area being evaluated, will have the greatest potential effect.

Grounding, Feed Line Radiation, Transmitter Leakage

In a well-designed station, virtually all the RF energy is radiated by the antenna. The formulas, tables and modeling software described in *Bulletin 65* all assume that all the power comes from the antenna system. In most cases, this is a reasonable assumption. Even the operator of the station probably receives more energy from the antenna than that inadvertently radiated from other sources. This is virtually certain to be true for most situations, where the people being exposed are not *much* closer to the source of incidental radiation than they are to the antenna. However, it is possible in some circumstances, especially for the operator, that people could be very close to the feed line or some other source of incidental radiation.

However, short of making actual field-strength measurements (with all the inherent problems in doing so), this incidental radiation can be virtually impossible to predict. Neither the FCC regulations nor *Bulletin 65* can fully address this possibility. All the evaluation methods consider only the RF coming from the antenna. Normally, these incidental radiators will not be considered during a routine evaluation. They cannot, however, be completely ignored.

Incidental radiators will not be evaluated quantitatively, but subjectively. Amateurs should be familiar with the circumstances under which excessive incidental radiation can occur and ensure that those circumstances are not present in the well-designed amateur station. The following problems can result in excessive incidental RF radiation:

Figure 5.18—The tower, guy wires and utility wires near this antenna can affect the level of the fields near the antenna and the other conductors.

- End fed antennas whose connection occurs directly in the shack
- Feed line radiation caused by antenna-system imbalance
- Excessive feed line leakage caused by broken or missing shield connections on coaxial cables
- Excessive feed line leakage caused by inferior grade coaxial cables
- Improper grounding of station equipment
- Improper shielding of station equipment
- Improperly fastened or damaged waveguide connectors
- Other "RF in the shack" problems

Many of these station problems can be traced to defects in the installation or maintenance of the station. These problems should normally be corrected as a routine part of designing and operating an effective and safe amateur station.

A poorly designed antenna system may have an unbalanced feed line connected to a balanced antenna, a feed line that runs at an acute angle to the antenna (see Figure 5.19), an inferior grade of coaxial cable that results in excessive feed line leakage or some defect or problem with the shield integrity on a coaxial cable.

A full discussion of grounding is beyond the scope of this book. However, properly grounding a transmitting installation can minimize problems with "RF in the shack," an unpleasant situation where small RF burns can be felt whenever the operator touches any station apparatus. RF in the shack is usually caused by antenna-system defects. The most effective cure is to locate the cause of the problem, but RF in the shack can sometimes be cured with station grounding. *The ARRL Handbook*

for Radio Amateurs and *Radio Frequency Interference: How to Find It and Fix It* (also published by the ARRL) both feature information about grounding.

In a well-designed transmitter, all the RF energy is contained inside the transmitter until it is sent out of the output connector to the antenna. The transmitter chassis is usually well shielded, with RF bypass leads keeping the RF where it belongs. If you are using a commercial transmitter, the chances are excellent that it is *not* the source of unwanted RF emissions. However, things can sometimes go wrong. Bypass components can fail, or shielding can be removed. (If you service your transmitter and remove a shield cover with 47 separate sheet-metal screws, it may be tempting to use only 4 screws to put it back together, but this will probably decrease the effectiveness of the shield.)

Near the End!

One other factor to consider is that the total RF energy radiated from the ends of the conductors used in antennas like dipoles or Yagi arrays is generally less than the energy radiated from the center. This is because by the time the RF energy gets to the end, some of the energy has been radiated away. If you are doing exact modeling, you will be able to determine that you can generally be closer to the ends of an antenna than you can be to the center, or the "hot" end of a longwire. This could be especially helpful to evaluate an antenna like an inverted V, where you could be closer to the end than the center.

Figure 5.20 shows the electric field directly under a half-wave dipole that is 30 feet in the air. The graph shows the field in the axis of the wire at ground level, as if the person being exposed were starting at the center and walking toward the ends.

Attenuation by Buildings

It is difficult to estimate the amount of attenuation of the transmitted field strength that may result from buildings, vegetation, etc. The amount of attenuation will depend on factors such as frequency and the construction material used. A conservative evaluation generally does not include additional attenuation for buildings. However, *Bulletin 65* does conclude that for most rooftop installations, 10 to 20 dB of attenuation by the building might be expected for people located on lower floors.

PAPERWORK

Once an Amateur Radio operator determines that a station complies by doing the station evaluation (or determines that no evaluation is required), the station may be put into immediate operation. There's no need for FCC approval before operating. The FCC does not require you to keep any records of your routine RF radiation exposure evaluation. However it is a good idea to keep them. They may prove useful if the FCC would ask for documentation to demonstrate that an evaluation has been performed. The Commission recommends that each amateur keep a record of the station evaluation procedure and its results, in case questions arise.

Other than a short certification on Form 610 station applications, the regulations do not normally require hams to file proof of evaluation with the FCC. The FCC will ask you to demonstrate that you have read

Figure 5.19—This feed line is very asymmetrical with respect to the antenna. This configuration could result in excessive feed line radiation—possibly a problem for the station operator or persons located near the feed line.

and understood the FCC Rules about RF-radiation exposure by indicating that understanding on FCC Form 610 (Reprinted in Appendix C of this book) when you apply for your license.

Actually, the regulations do contain a provision that would allow an amateur to file an Environmental Assessment (EA) with the Environmental Protection Agency, however, the costs and time delays associated with an EA are usually prohibitive, especially for an amateur station. The Commission expects that it is highly unlikely that an amateur will be taking such an action. EAs are *not* normally required for amateur stations. An EA is required for any station that wants to continue to operate even though they exceed the MPE limits. It is not likely that an amateur would choose to file an EA in lieu of making changes to his or her station, to be in compliance with the MPE limits. The regulations will require that hams affirm on their station applications that they have read the regulations and that they are in compliance with them.

CORRECTING PROBLEMS

An antenna that is higher and farther away from people also reduces the strength of the radiated fields that anyone will be exposed to. If you can raise your antenna higher in the air or move it farther from your neighbor's property line you will reduce exposure.

A half-wavelength dipole antenna that is only 5 meters above the ground would generally create a stronger RF field on the ground beneath the antenna than many other antennas. For example, a three-element Yagi antenna or a three-element quad antenna both have significantly more gain than a dipole. Yet at a height of 30 meters both of these antennas would produce a smaller RF field strength on the ground beneath the antenna than would the low dipole. As a general rule, place your antenna at least as high as necessary to ensure that you meet the FCC radiation exposure guidelines.

When routine evaluation of an Amateur Radio station indicates that the RF radiation could be in excess of the FCC-specified limits, the station licensee must take action to bring the station operation into compliance with the regulations. The vast majority of stations will pass their evaluations handily. But some stations whose antennas are close to areas of exposure may not meet the MPE limits.

The FCC gives amateurs considerable flexibility in correcting problems. They are relying on the demonstrated technical ability of amateurs and their familiarity with their own stations and operating environments to make the appropriate changes to their stations or their operation to be in compliance with the MPE limits.

The following list offers some guidance on the types of changes that could be made to a station. It is not intended to be all inclusive:

- Relocate transmitting antennas to result in less exposure to people
- Choose a different antenna type to result in less exposure to people
- Control the pointing of directional antennas to reduce exposure to people
- Reduce transmitter operating power to reduce exposure
- Use a different operating mode that results in lower average transmitter power and exposure
- Reduce operating time to reduce average transmitter power and exposure
- Change the operating frequency to use a frequency where the MPE limit is higher
- Controlling access / signs
- Combinations of some or all of these

Relocating Antennas

This can be one of the easiest and most effective changes to make. In general, if you can locate your antenna farther away from people, their exposure will be less. Because an RF field diminishes rapidly with increasing distance between the measurement point and the source of RF energy, relocating the station's antenna(s) can reduce the field strength below the MPE limits. An antenna that is high and in the clear is usually going to have a field that is much reduced from a low antenna located near areas of possible exposure. Relocating a low antenna so that it is high and in the clear will have a second benefit; it will usually improve the DX performance of your station, giving you more low-angle radiation for HF DXing or VHF.

Antennas that have gain usually result in a concentration of energy, even in the near field. This can be an advantage or disadvantage. If the antenna can be located such that the gain is primarily away from areas of possible exposure, either in the horizontal or vertical plane, this could provide another means of meeting the regulations.

Moving a vertical antenna farther away from a house or nearby property also can significantly reduce exposure. Those pesky indoor and apartment-balcony antennas are particularly troublesome; if you can move them away from the building, they will work better for you and result in less exposure.

You must always take care to position your amateur antennas in a manner so they cannot harm you or anyone else. The simplest way to do this is to always install them high and in the clear, away from buildings or other locations where people might be close to them. To prevent **RF burns** you must be sure no one can touch the antenna while you are transmitting into it. It doesn't matter what type of antenna it is, or how much power you are running. If you or someone else can touch the antenna, it is

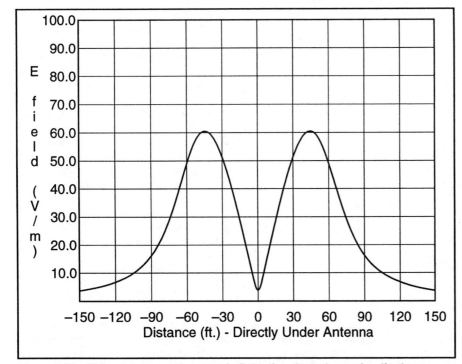

Figure 5.20—This plot shows the electric field directly under a wire dipole antenna.

too close.

Of course the one exception to this is the antenna on a hand-held radio. You aren't likely to receive an RF burn from touching the antenna on your hand-held radio because the transmitter power is quite low. You should still keep the antenna as far from you or anyone else as possible, to minimize your exposure to the RF electromagnetic fields from the radiated signal.

Choose a Different Antenna Type

There is no magic antenna that will solve all your RF-exposure woes, but the selection of an antenna can influence exposure, in both directions. In general, a large antenna usually results in a smaller field at any particular near-field point than a small antenna! This is because if one is near a small antenna, one is near the *entire* antenna, where with a large antenna, portions of the antenna may be far away.

A directional antenna, such as a Yagi array, can minimize exposure to areas off the sides and back. This comes with a price; exposure in the direction the antenna is pointing is often higher than it would be with an antenna with less gain. End-fed wires worked against earth ground almost always result in more exposure in the shack or nearby rooms than would an antenna located farther away, fed with feed line. On VHF and UHF, high-gain vertical antennas located up high often result in less exposure on the ground than would result from a simple ground plane at the same height.

In general, most gain antennas (such as Yagi arrays) radiate most of their energy toward the horizon or at low angles above the horizon as seen at the height of the antenna on the supporting tower.

The RF field at ground level is usually less (and sometimes much less) than the energy in the main beam of the antenna. This general rule usually does not apply to vertical antennas located at ground level.

It is Not Polite to Point

This old adage serves to remind us that the exposure from a gain antenna is maximum in one (or more) directions and minimum in others. It sounds too good to be true, but it is true; if you determine that your station exceeds the MPEs in a particular direction toward a particular house, the FCC considers it perfectly acceptable that you, as control operator, do not point your antenna at full power in that direction if someone is present in that direction at the time. You also can use the directional patterns of antennas to good effect; locate the antenna such that the nulls in the pattern fall toward areas where people are present, especially on the higher bands.

For example, if an amateur were to determine that his or her station was in compliance at full power to all surrounding uncontrolled areas except for one corner of a neighboring property when the antenna was aimed in that direction, one way of complying would be to avoid pointing a directional antenna in that direction if people are present on that part of the neighboring property.

In addition to using the free-space pattern of your antenna to calculate exposure (this was discussed earlier), you also can use the radiation pattern of the antenna to your advantage in controlling exposure. For example, if you position your dipole antenna (with maximum radiation off the sides of the antenna and minimum radiation off the ends) so the ends are pointed at your neighbor's house (or your house), you will reduce the exposure. A beam antenna can have an even more dramatic effect on reducing the exposure. Simply do not point the antenna in the direction where people will most likely be located.

QRP, Modes and Time

ARRL is not recommending that all stations run QRP (although there are a few avid QRPers on the ARRL staff, along with avid DXers, avid big-gun and little-pistol contesters), but reducing power is certainly an option. Higher transmitter power will produce stronger radiated RF fields. So using the minimum power necessary to carry out your communications will minimize the exposure of anyone near your station. Reducing power is one effective way of meeting the FCC MPE limits. You may find that you are not in compliance at 1500 watts, but at 1100 watts, you are just under the limit.

Some modes result in more average power than others. FM, RTTY or other digital modes have a duty factor of 100%, Morse CW has a duty factor of about 40% and voice SSB ranges from 20% to 40%. If you are running 1000 watts on RTTY and choose to use SSB instead, your average power during the time you are transmitting will drop from 1000 watts to about 200 watts. This can make a *big* difference in your exposure and the necessary compliance distances.

You also can adjust your on and off operating times to reduce your average power during the averaging period. For example, if an amateur were to discover that the MPE limits had been exceeded for uncontrolled exposure after 25 minutes of transmitting, the FCC would consider it perfectly acceptable to take a 5-minute break after 25 minutes. Thus, if necessary, an amateur may tailor the operating pattern of the station (on/off times) to meet the MPE requirements. It will then be the responsibility of the control operator and station licensee to ensure that the maximum time used for these calculations is not exceeded at any time during station operation if people could be exposed. It would be easy to forget during a long ragchew that no more than 4 minutes out of any 6-minute period are allowed, as an example—for controlled exposure.

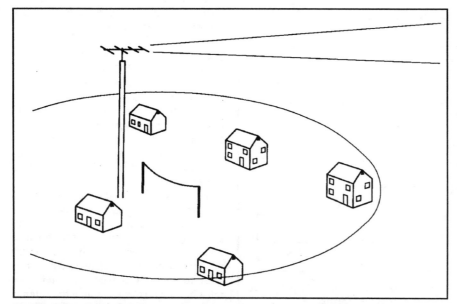

Figure 5.21—Antennas that are located up high are generally located far away from people. To the untrained, it may appear that the small antenna located between the houses will create less RF nearby than the big tower, but the antenna that is up in the air will create a smaller field on the ground.

Frequency

Even your choice of operating frequency can have an affect. Humans absorb less RF energy at some frequencies (and the MPE is higher at those frequencies). You can reduce exposure by selecting an operating frequency with a higher MPE.

The MPE limits vary with frequency. If your operation on 160 through 10 meters resulted in 0.4 mW/cm² uncontrolled exposure, you would have to reduce your average power on 10 meters by half, but you could use full power on 15 meters!

Controlling Access

Amateurs may be able to exercise control over access to areas that might have exposure that exceeds the MPE. As examples, if an amateur has authorized control over a private area, such as his or her own backyard, the areas that might have excessive exposure could be fenced in, or signs could be posted that indicate that the area may contain RF energy and is not authorized for entry for the general public, although this may invite more questions than some amateurs want to answer. Access can be controlled with fences, locked doors or any other reasonable means. Controlling access to areas where high RF energy may be present is probably the easiest method of controlling exposure.

It is important to note that for general population/uncontrolled exposures it is not often possible to control access or otherwise limit exposure duration to the extent that averaging times can be applied. In those situations, it is often necessary to assume continuous exposure to whatever exposure duration could be expected to occur with the on/off cycles of the transmitter.

Signs

The FCC accepts posted signs as a means of controlling exposure. If an amateur repeater were located on a rooftop and the exposure exceeded the MPEs after three minutes of continuous operation, a sign could be posted that indicates that RF is present and that it is not permissible to remain in the area for more than three minutes. This applies easily to occupational exposure areas.

Suitable signs are available from a number of sources. The National Association of Broadcasters, EMED Co., Inc. and Richard Tell Associates all sell such signs. See Appendix E, Resources, for contact information.

The Magic Combination

These various solutions can be combined. One could relocate an antenna and reduce power in combination to bring the exposure into compliance. One could reduce operating times whenever necessary, perhaps when a neighboring dwelling is known to be occupied.

EVALUATIONS AND THE FCC

The FCC has always relied on the Amateur Radio Service to follow the rules. Although Amateur Radio does have a few bad apples, overall, hams can be very proud of our rules-compliance record. The FCC expects that most hams will follow the requirements of these rules, too. For the most part, they expect that they will not need to become involved in the day-to-day management of individual amateur stations. The FCC may receive inquiries from neighbors of radio operators about the RF exposure from that station. In that case, it is possible that radio operators will receive an informal "inquiry" from the FCC in response. This inquiry will ask about the station, its frequency, power, modes and antennas. They also will ask if the station required an evaluation and ask for a summary of the results. For the most part, the FCC will assume that the evaluation was done correctly, and inform the inquiring neighbor that the station is operating in accordance with FCC rules. Although the FCC does retain the right to show up at your door and measure your fields, this would normally not be done, except under unusual circumstances. An example might be the ham who indicates that his 1500-watt CW station with a five-element, 10-meter band Yagi 10 feet from a neighbor's second-story bedroom window was in compliance at full duty cycle. (This unlikely sounding station was described to an ARRL employee by an FCC staffer!)

The FCC Worksheet

The FCC has included a worksheet in *Supplement B*. This optional worksheet has instructions on how to include the various factors necessary to do a station evaluation and provides a handy way to maintain a record of the evaluation. It runs step by step through the procedures outlined in this article, using the methods outlined in *Supplement B*. The worksheet describes the methods to calculate power to the antenna using feed line losses, and how to calculate ERP using both feed line losses and antenna gain. This is another example of how the FCC has made the evaluation process as clear and easy as possible for the Amateur Radio Service.

ARRL Worksheet

The FCC worksheet is comprehensive, guiding hams through a number of steps for evaluation thresholds for single transmitters and repeaters, and a comprehensive evaluation procedure. The ARRL has

---

Why Should We Even Bother?

No doubt many of you are shaking your heads and muttering, "Why should I even bother to do an evaluation? The FCC will never enforce these rules. This is a waste of time!"

There are a number of important reasons why amateurs should follow *all* FCC rules, including these. The Amateur Radio Service has a tradition of compliance with FCC regulations; Part 97 is our bible! The ARRL has worked hard to help the FCC fine tune these rules for the Amateur Radio Service. If we hope for more cooperation in the future, we must set the best example possible. The FCC (and our Amateur Radio supporters on Capitol Hill) must be assured that the majority of hams follow all the rules "by the book."

Safety also is a concern. While RF energy isn't known to cause major health problems, the research is still continuing. The levels that have been set by various standards bodies and the FCC are our best assurance that no ill effects on human health are expected from the normal operation of radio transmitters. Being in compliance buys peace of mind for you and your family. As the old saying goes, "better safe than sorry."

Your neighbors may also have questions and concerns. (The ARRL has already received quite a few questions on this subject from neighbors of hams.) Many of these concerns can be easily addressed by explaining the requirements to your neighbors and showing them the results of your station evaluation. The new rules even offer us a significant advantage; if our neighbors do have concerns, we are much better off being able to demonstrate that there are rules governing our conduct and that we have done what the rules require.

In most cases, these evaluations are not hard! They can usually be done by looking at a table, or spending a few minutes with some free software or a calculator. There is not much to lose, and a lot to gain. So, hams should complete their station evaluations and point to them with pride!—*Chris D. Imlay, W3KD, ARRL General Counsel*

developed a simplified worksheet and instructions that will be helpful for amateurs that only need to do some parts of the evaluation. See Chapter 1 of this book for a copy of this worksheet.

RFX AND RFI

Radio Frequency Interference and Radio Frequency Exposure are *not* the same. One concerns interference to or from electronics equipment; the other concerns human exposure to RF energy. The topics are worlds apart, although hams have been overheard talking about the new "RFI" rules. The levels involved are generally worlds apart, too. Most consumer electronics equipment has immunity to fields of about 3 volts per meter. The *lowest* level of exposure in the new rules is a level of 27.5 volts per meter, about 20 dB higher than the level that causes RFI! The highest permissible exposure level is a whopping 614 volts per meter—some 46 dB higher! This often gives us a clear indication that we are not exceeding the MPE levels in neighboring homes—most neighbors of hams do not have RFI problems, indicating that the fields are not substantially greater than about 3 volts per meter.

CONCLUSION

This chapter told you how to do the required station evaluation. But much like the provisions of the National Electrical Code and house wiring, the provisions of the law are not intended to replace safety and common sense. In addition to the RF-exposure provisions in Part 97, hams should continue to practice RF-safety techniques. The earlier chapters of this book discussed the principles of "prudent avoidance." Don't let your enthusiasm for learning about your station evaluation cause you to skip the important fundamentals in the earlier chapters.

Overall, these regulations are not difficult for the Amateur Radio Service. Most hams don't have to do an evaluation at all. Most of those who are required to do an evaluation can do so using relatively easy methods. Once the evaluation is complete, hams can go back to their favorite hamming, answering their own questions about RF exposure and hopefully addressing any neighbors' concerns. All in all, it seems like it is not a bad trade-off.

Evaluating Compliance with FCC Guidelines for Human Exposure to Radiofrequency Electromagnetic Fields

OET Bulletin 65

Edition 97-01

August 1997

Evaluating Compliance with FCC Guidelines for Human Exposure to Radiofrequency Electromagnetic Fields

OET BULLETIN 65
Edition 97-01

August 1997

AUTHORS
Robert F. Cleveland, Jr.
David M. Sylvar
Jerry L. Ulcek

Standards Development Branch
Allocations and Standards Division
Office of Engineering and Technology
Federal Communications Commission
Washington, D.C. 20554

The first edition of this bulletin was issued as OST Bulletin No. 65 in October 1985. This is a revised version of that original bulletin.

NOTE: Mention of commercial products does not constitute endorsement by the Federal Communications Commission or by the authors.

ACKNOWLEDGEMENTS

The following individuals and organizations from outside the FCC reviewed an early draft of this bulletin. Their valuable comments and suggestions greatly enhanced the accuracy and usefulness of this document, and their assistance is gratefully acknowledged.

Joseph A. Amato, Maxwell RF Radiation Safety, Ltd.
Edward Aslan, Lockheed Martin Microwave (Narda)
Ameritech Mobile Communications, Inc.
Dr. Tadeusz M. Babij, Florida International University
Dr. Quirano Balzano, Motorola
David Baron, P.E., Holaday Industries, Inc.
Howard I. Bassen, U.S. Food and Drug Administration
Clarence M. Beverage, Communications Technologies, Inc.
Dr. Donald J. Bowen, AT&T Laboratories
Cellular Telecommunications Industry Association
Dr. C.K. Chou, City of Hope National Medical Center
Jules Cohen, P.E., Consulting Engineer
Dr. David L. Conover, National Institute for Occupational Safety & Health
Cohen, Dippell and Everist, P.C.
Robert D. Culver, Lohnes and Culver
Fred J. Dietrich, Ph.D., Globalstar
Electromagnetic Energy Association
Professor Om P. Gandhi, University of Utah
Robert Gonsett, Communications General Corp.
Hammett & Edison, Inc.
Norbert Hankin, U.S. Environmental Protection Agency
James B. Hatfield, Hatfield & Dawson
Robert Johnson
Dr. John A. Leonowich
Dr. W. Gregory Lotz, National Institute for Occupational Safety & Health
Frederick O. Maia, National Volunteer Examiners (Amateur Radio Service)
Ed Mantiply, U.S. Environmental Protection Agency
Robert Moore
Dr. Daniel Murray, Okanagan University College
Dr. John M. Osepchuk, Full Spectrum Consulting
Professor Wayne Overbeck, California State University, Fullerton
Personal Communications Industry Association
Ronald C. Petersen, Lucent Technologies
David B. Popkin
Kazimierz Siwiak, P.E.
Richard A. Tell, Richard Tell Associates, Inc.
Rory Van Tuyl, Hewlett-Packard Laboratories
Louis A. Williams, Jr., Louis A. Williams, Jr. and Associates

Contributions from the following FCC staff members are also acknowledged:
Kwok Chan, Errol Chang, William Cross, Richard Engelman, Bruce Franca and Jay Jackson

i

FIGURES

iii

INTRODUCTION

This revised OET Bulletin 65 has been prepared to provide assistance in determining whether proposed or existing transmitting facilities, operations or devices comply with limits for human exposure to radiofrequency (RF) fields adopted by the Federal Communications Commission (FCC). The bulletin offers guidelines and suggestions for evaluating compliance. *However, it is not intended to establish mandatory procedures, and other methods and procedures may be acceptable if based on sound engineering practice.*

In 1996, the FCC adopted new guidelines and procedures for evaluating environmental effects of RF emissions. The new guidelines incorporate two tiers of exposure limits based on whether exposure occurs in an occupational or "controlled" situation or whether the general population is exposed or exposure is in an "uncontrolled" situation. In addition to guidelines for evaluating fixed transmitters, the FCC adopted new limits for evaluating exposure from mobile and portable devices, such as cellular telephones and personal communications devices. The FCC also revised its policy with respect to categorically excluding certain transmitters and services from requirements for routine evaluation for compliance with the guidelines.

This bulletin is a revision of the FCC's OST Bulletin 65, originally issued in 1985. Although certain technical information in the original bulletin is still valid, this revised version updates other information and provides additional guidance for evaluating compliance with the the new FCC policies and guidelines. The bulletin is organized into the following sections: Introduction, Definitions and Glossary, Background Information, Prediction Methods, Measuring RF Fields, Controlling Exposure to RF Fields, References and Appendices. Appendix A provides a summary of the new FCC guidelines and the requirements for routine evaluation. Additional information specifically for use in evaluating compliance for radio and television broadcast stations is included in a supplement to this bulletin (Supplement A). A supplement for the Amateur Radio Service will also be issued (Supplement B), and future supplements may be issued to provide additional information for other services. This bulletin and its supplements may be revised, as needed.

In general, the information contained in this bulletin is intended to enable an applicant to make a reasonably quick determination as to whether a proposed or existing facility is in compliance with the limits. In addition to calculations and the use of tables and figures, Section 4, dealing with controlling exposure, should be consulted to ensure compliance, especially with respect to occupational/controlled exposures. In some cases, such as multiple-emitter locations, measurements or a more detailed analysis may be required. In that regard, Section 3 on measuring RF fields provides basic information and references on measurement procedures and instrumentation.

For further information on any of the topics discussed in this bulletin, you may contact the FCC's RF safety group at: +1 202 418-2464. Questions and inquiries can also be e-mailed to: rfsafety@fcc.gov. The FCC's World Wide Web Site provides information on FCC decision documents and bulletins relevant to the RF safety issue. The address is: www.fcc.gov/oet/rfsafety.

1

DEFINITIONS AND GLOSSARY OF TERMS

The following specific words and terms are used in this bulletin. These definitions are adapted from those included in the American National Standards Institute (ANSI) 1992 RF exposure standard [Reference 1], from NCRP Report No. 67 [Reference 19] and from the FCC's Rules (47 CFR § 2.1 and § 1.1310).

Average (temporal) power. The time-averaged rate of energy transfer.

Averaging time. The appropriate time period over which exposure is averaged for purposes of determining compliance with RF exposure limits (discussed in more detail in Section 1).

Continuous exposure. Exposure for durations exceeding the corresponding averaging time.

Decibel (dB). Ten times the logarithm to the base ten of the ratio of two power levels.

Duty factor. The ratio of pulse duration to the pulse period of a periodic pulse train. Also, may be a measure of the temporal transmission characteristic of an intermittently transmitting RF source such as a paging antenna by dividing average transmission duration by the average period for transmissions. A duty factor of 1.0 corresponds to continuous operation.

Effective radiated power (ERP) (in a given direction). The product of the power supplied to the antenna and its gain relative to a half-wave dipole in a given direction.

Equivalent Isotropically Radiated Power (EIRP). The product of the power supplied to the antenna and the antenna gain in a given direction relative to an isotropic antenna.

Electric field strength (E). A field vector quantity that represents the force (**F**) on an infinitesimal unit positive test charge (**q**) at a point divided by that charge. Electric field strength is expressed in units of volts per meter (V/m).

Energy density (electromagnetic field). The electromagnetic energy contained in an infinitesimal volume divided by that volume.

Exposure. Exposure occurs whenever and wherever a person is subjected to electric, magnetic or electromagnetic fields other than those originating from physiological processes in the body and other natural phenomena.

Exposure, partial-body. Partial-body exposure results when RF fields are substantially nonuniform over the body. Fields that are nonuniform over volumes comparable to the human body may occur due to highly directional sources, standing-waves, re-radiating sources or in the near field. See **RF "hot spot"**.

Far-field region. That region of the field of an antenna where the angular field distribution is essentially independent of the distance from the antenna. In this region (also called the free space region), the field has a predominantly plane-wave character, i.e., locally uniform distribution of electric field strength and magnetic field strength in planes transverse to the direction of propagation.

Gain (of an antenna). The ratio, usually expressed in decibels, of the power required at the input of a loss-free reference antenna to the power supplied to the input of the given antenna to produce, in a given direction, the same field strength or the same power density at the same distance. When not specified otherwise, the gain refers to the direction of maximum radiation. Gain may be considered for a specified polarization. Gain may be referenced to an isotropic antenna (dBi) or a half-wave dipole (dBd).

General population/uncontrolled exposure. For FCC purposes, applies to human exposure to RF fields when the general public is exposed or in which persons who are exposed as a consequence of their employment may not be made fully aware of the potential for exposure or cannot exercise control over their exposure. Therefore, members of the general public always fall under this category when exposure is not employment-related.

Hertz (Hz). The unit for expressing frequency, (f). One hertz equals one cycle per second.

Magnetic field strength (H). A field vector that is equal to the magnetic flux density divided by the permeability of the medium. Magnetic field strength is expressed in units of amperes per meter (A/m).

Maximum permissible exposure (MPE). The rms and peak electric and magnetic field strength, their squares, or the plane-wave equivalent power densities associated with these fields to which a person may be exposed without harmful effect and with an acceptable safety factor.

Near-field region. A region generally in proximity to an antenna or other radiating structure, in which the electric and magnetic fields do not have a substantially plane-wave character, but vary considerably from point to point. The near-field region is further subdivided into the reactive near-field region, which is closest to the radiating structure and that contains most or nearly all of the stored energy, and the radiating near-field region where the radiation field predominates over the reactive field, but lacks substantial plane-wave character and is complicated in structure. For most antennas, the outer boundary of the reactive near field region is commonly taken to exist at a distance of one-half wavelength from the antenna surface.

Occupational/controlled exposure. For FCC purposes, applies to human exposure to RF fields when persons are exposed as a consequence of their employment and in which those persons who are exposed have been made fully aware of the potential for exposure and can exercise control over their exposure. Occupational/controlled exposure limits also apply where exposure is of a transient nature as a result of incidental passage through a location where exposure levels may be above general population/uncontrolled limits (see definition above), as long as the exposed person has been made fully aware of the potential for exposure and can exercise control over his or her exposure by leaving the area or by some other appropriate means.

Peak Envelope Power (PEP). The average power supplied to the antenna transmission line by a radio transmitter during one radiofrequency cycle at the crest of the modulation envelope taken under normal operating conditions.

Power density, average (temporal). The instantaneous power density integrated over a source repetition period.

Power density (S). Power per unit area normal to the direction of propagation, usually expressed in units of watts per square meter (W/m^2) or, for convenience, units such as milliwatts per square centimeter (mW/cm^2) or microwatts per square centimeter ($\mu W/cm^2$). For plane waves, power density, electric field strength (E) and magnetic field strength (H) are related by the impedance of free space, i.e., 377 ohms, as discussed in Section 1 of this bulletin. Although many survey instruments indicate power density units ("far-field equivalent" power density), the actual quantities measured are E or E^2 or H or H^2.

Power density, peak. The maximum instantaneous power density occurring when power is transmitted.

Power density, plane-wave equivalent or far-field equivalent. A commonly-used terms associated with any electromagnetic wave, equal in magnitude to the power density of a plane wave having the same electric (E) or magnetic (H) field strength.

Radiofrequency (RF) spectrum. Although the RF spectrum is formally defined in terms of frequency as extending from 0 to 3000 GHz, for purposes of the FCC's exposure guidelines, the frequency range of interest in 300 kHz to 100 GHz.

Re-radiated field. An electromagnetic field resulting from currents induced in a secondary, predominantly conducting, object by electromagnetic waves incident on that object from one or more primary radiating structures or antennas. Re-radiated fields are sometimes called "reflected" or more correctly "scattered fields." The scattering object is sometimes called a "re-radiator" or "secondary radiator".

RF "hot spot." A highly localized area of relatively more intense radio-frequency radiation that manifests itself in two principal ways:

> (1) The presence of intense electric or magnetic fields immediately adjacent to conductive objects that are immersed in lower intensity ambient fields (often referred to as re-radiation), and

> (2) Localized areas, not necessarily immediately close to conductive objects, in which there exists a concentration of RF fields caused by reflections and/or narrow beams produced by high-gain radiating antennas or other highly directional sources. In both cases, the fields are characterized by very rapid changes in field strength with distance. RF hot spots are normally associated with very nonuniform exposure of the body (partial body exposure). This is not to be confused with an actual thermal hot spot within the absorbing body.

Root-mean-square (rms). The effective value, or the value associated with joule heating, of a periodic electromagnetic wave. The rms value is obtained by taking the square root of the mean of the squared value of a function.

Scattered radiation. An electromagnetic field resulting from currents induced in a secondary, conducting or dielectric object by electromagnetic waves incident on that object from one or more primary sources.

Short-term exposure. Exposure for durations less than the corresponding averaging time.

Specific absorption rate (SAR). A measure of the rate of energy absorbed by (dissipated in) an incremental mass contained in a volume element of dielectric materials such as biological tissues. SAR is usually expressed in terms of watts per kilogram (W/kg) or milliwatts per gram (mW/g). Guidelines for human exposure to RF fields are based on SAR thresholds where adverse biological effects may occur. When the human body is exposed to an RF field, the SAR experienced is proportional to the squared value of the electric field strength induced in the body.

Wavelength (λ). The wavelength (λ) of an electromagnetic wave is related to the frequency (f) and velocity (v) by the expression $v = f\lambda$. In free space the velocity of an electromagnetic wave is equal to the speed of light, i.e., approximately 3×10^8 m/s.

Section 1: BACKGROUND INFORMATION

FCC Implementation of NEPA

The National Environmental Policy Act of 1969 (NEPA) requires agencies of the Federal Government to evaluate the effects of their actions on the quality of the human environment.[1] To meet its responsibilities under NEPA, the Commission has adopted requirements for evaluating the environmental impact of its actions.[2] One of several environmental factors addressed by these requirements is human exposure to RF energy emitted by FCC-regulated transmitters and facilities.

The FCC's Rules provide a list of various Commission actions which may have a significant effect on the environment. If FCC approval to construct or operate a facility would likely result in a significant environmental effect included in this list, the applicant for such a facility must submit an "Environmental Assessment" or "EA" of the environmental effect including information specified in the FCC Rules. It is the responsibility of the applicant to make an initial determination as to whether it is necessary to submit an EA.

If it is necessary for an applicant to submit an EA that document would be reviewed by FCC staff to determine whether the next step in the process, the preparation of an Environmental Impact Statement or "EIS," is necessary. An EIS is only prepared if there is a staff determination that the action in question will have a significant environmental effect. If an EIS is prepared, the ultimate decision as to approval of an application could require a full vote by the Commission, and consideration of the issues involved could be a lengthy process. Over the years since NEPA implementation, there have been relatively few EIS's filed with the Commission. This is because most environmental problems are resolved in the process well prior to EIS preparation, since this is in the best interest of all and avoids processing delays.

Many FCC application forms require that applicants indicate whether their proposed operation would constitute a significant environmental action under our NEPA procedures. When an applicant answers this question on an FCC form, in some cases documentation or an explanation of how an applicant determined that there would ***not*** be a significant environmental effect may be requested by the FCC operating bureau or office. This documentation may take the form of an environmental statement or engineering statement that accompanies the application. Such a statement is ***not*** an EA, since an EA is only submitted if there is evidence for a significant environmental effect. In the overwhelming number of cases, applicants attempt to mitigate any potential for a significant environmental effect before submission of either an environmental statement or an EA. This may involve informal

---

[1] National Environmental Policy Act of 1969, 42 U.S.C. Section 4321, <u>et seq.</u>

[2] See 47 CFR § 1.1301, <u>et seq.</u>

consultation with FCC staff, either prior to the filing of an application or after an application has been filed, over possible means of avoiding or correcting an environmental problem.

FCC Guidelines for Evaluating Exposure to RF Emissions

In 1985, the FCC first adopted guidelines to be used for evaluating human exposure to RF emissions.[3] The FCC revised and updated these guidelines on August 1, 1996, as a result of a rule-making proceeding initiated in 1993.[4] The new guidelines incorporate limits for Maximum Permissible Exposure (MPE) in terms of electric and magnetic field strength and power density for transmitters operating at frequencies between 300 kHz and 100 GHz. Limits are also specified for localized ("partial body") absorption that are used primarily for evaluating exposure due to transmitting devices such as hand-held portable telephones. Implementation of the new guidelines for mobile and portable devices became effective August 7, 1996. For other applicants and licensees a transition period was established before the new guidelines would apply.[5]

The FCC's MPE limits are based on exposure limits recommended by the National Council on Radiation Protection and Measurements (NCRP)[6] and, over a wide range of frequencies, the exposure limits developed by the Institute of Electrical and Electronics Engineers, Inc., (IEEE) and adopted by the American National Standards Institute (ANSI) to

---

[3] *See Report and Order,* GEN Docket No. 79-144, 100 FCC 2d 543 (1985); and *Memorandum Opinion and Order,* 58 RR 2d 1128 (1985). The guidelines originally adopted by the FCC were the 1982 RF protection guides issued by the American National Standards Institute (ANSI).

[4] See *Report and Order*, ET Docket 93-62, FCC 96-326, adopted August 1, 1996, 61 Federal Register 41,006 (1996), 11 FCC Record 15,123 (1997). The FCC initiated this rule-making proceeding in 1993 in response to the 1992 revision by ANSI of its earlier guidelines for human exposure. The Commission responded to seventeen petitions for reconsideration filed in this docket in two separate Orders: *First Memorandum Opinion and Order,* FCC 96-487, adopted December 23, 1996, 62 Federal Register 3232 (1997), 11 FCC Record 17,512 (1997); and *Second Memorandum Opinion and Order and Notice of Proposed Rulemaking,* adopted August 25, 1997.

[5]   This transition period was recently extended. With the exception of the Amateur Radio Service, the date now established for the end of the transition period is October 15, 1997. See *Second Memorandum Opinion and Order and Notice of Proposed Rule Making,* ET Docket 93-62, adopted August 25, 1997. Therefore, the new guidelines will apply to applications filed on or after this date. For the Amateur Service only, the new guidelines will apply to applications filed on or after January 1, 1998. In addition, the Commission has adopted a date certain of September 1, 2000, by which time all existing facilities and devices must be in compliance with the new guidelines (see *Second Memorandum Opinion and Order*).

[6] *See Reference 20*, "Biological Effects and Exposure Criteria for Radiofrequency Electromagnetic Fields," NCRP Report No. 86 (1986), National Council on Radiation Protection and Measurements (NCRP), Bethesda, MD. The NCRP is a non-profit corporation chartered by the U.S. Congress to develop information and recommendations concerning radiation protection.

replace the 1982 ANSI guidelines.[7] Limits for localized absorption are based on recommendations of both ANSI/IEEE and NCRP. The FCC's new guidelines are summarized in Appendix A.

In reaching its decision on adopting new guidelines the Commission carefully considered the large number of comments submitted in its rule-making proceeding, and particularly those submitted by the U.S. Environmental Protection Agency (EPA), the Food and Drug Administration (FDA) and other federal health and safety agencies. The new guidelines are based substantially on the recommendations of those agencies, and it is the Commission's belief that they represent a consensus view of the federal agencies responsible for matters relating to public safety and health.

The FCC's limits, and the NCRP and ANSI/IEEE limits on which they are based, are derived from exposure criteria quantified in terms of specific absorption rate (SAR).[8] The basis for these limits is a whole-body averaged SAR threshold level of 4 watts per kilogram (4 W/kg), as averaged over the entire mass of the body, above which expert organizations have determined that potentially hazardous exposures may occur. The new MPE limits are derived by incorporating safety factors that lead, in some cases, to limits that are more conservative than the limits originally adopted by the FCC in 1985. Where more conservative limits exist they do not arise from a fundamental change in the RF safety criteria for whole-body averaged SAR, but from a precautionary desire to protect subgroups of the general population who, potentially, may be more at risk.

The new FCC exposure limits are also based on data showing that the human body absorbs RF energy at some frequencies more efficiently than at others. As indicated by Table 1 in Appendix A, the most restrictive limits occur in the frequency range of 30-300 MHz where whole-body absorption of RF energy by human beings is most efficient. At other frequencies whole-body absorption is less efficient, and, consequently, the MPE limits are less restrictive.

MPE limits are defined in terms of power density (units of milliwatts per centimeter squared: mW/cm^2), electric field strength (units of volts per meter: V/m) and magnetic field strength (units of amperes per meter: A/m). In the far-field of a transmitting antenna, where the electric field vector (E), the magnetic field vector (H), and the direction of propagation

---

[7] *See Reference 1*, ANSI/IEEE C95.1-1992, "Safety Levels with Respect to Human Exposure to Radio Frequency Electromagnetic Fields, 3 kHz to 300 GHz." Copyright 1992, The Institute of Electrical and Electronics Engineers, Inc., New York, NY. The 1992 ANSI/IEEE exposure guidelines for field strength and power density are similar to those of NCRP Report No. 86 for most frequencies except those above 1.5 GHz.

[8] Specific absorption rate is a measure of the rate of energy absorption by the body. SAR limits are specified for both whole-body exposure and for partial-body or localized exposure (generally specified in terms of spatial peak values).

can be considered to be all mutually orthogonal ("plane-wave" conditions), these quantities are related by the following equation.[9]

$$S = \frac{E^2}{3770} = 37.7 H^2$$

(1)

where: S = power density (mW/cm^2)
E = electric field strength (V/m)
H = magnetic field strength (A/m)

In the near-field of a transmitting antenna the term "far-field equivalent" or "plane-wave equivalent" power density is often used to indicate a quantity calculated by using the near-field values of E^2 or H^2 as if they were obtained in the far-field. As indicated in Table 1 of Appendix A, for near-field exposures the values of plane-wave equivalent power density are given in some cases for reference purposes only. These values are sometimes used as a convenient comparison with MPEs for higher frequencies and are displayed on some measuring instruments.

The FCC guidelines incorporate two separate tiers of exposure limits that are dependent on the situation in which the exposure takes place and/or the status of the individuals who are subject to exposure. The decision as to which tier applies in a given situation should be based on the application of the following definitions.

Occupational/controlled exposure limits apply to situations in which persons are exposed as a consequence of their employment and in which those persons who are exposed have been made fully aware of the potential for exposure and can exercise control over their exposure. Occupational/controlled exposure limits also apply where exposure is of a transient nature as a result of incidental passage through a location where exposure levels may be above general population/uncontrolled limits (see below), as long as the exposed person has been made fully aware of the potential for exposure and can exercise control over his or her exposure by leaving the area or by some other appropriate means. As discussed later, the occupational/controlled exposure limits also apply to amateur radio operators and members of their immediate household.

General population/uncontrolled exposure limits apply to situations in which the general public may be exposed or in which persons who are exposed as a consequence of their employment may not be made fully aware of the potential for exposure or cannot exercise control over their exposure. Therefore, members of the general public would always be considered under this category when exposure is not employment-related, for example, in the case of a telecommunications tower that exposes persons in a nearby residential area.

---

[9]   Note that this equation is written so that power density is expressed in units of mW/cm^2. The impedance of free space, 377 ohms, is used in deriving the equation.

For purposes of applying these definitions, awareness of the potential for RF exposure in a workplace or similar environment can be provided through specific training as part of an RF safety program. Warning signs and labels can also be used to establish such awareness as long as they provide information, in a prominent manner, on risk of potential exposure and instructions on methods to minimize such exposure risk.[10] However, warning labels placed on low-power consumer devices such as cellular telephones are not considered sufficient to achieve the awareness necessary to qualify these devices as operating under the occupational/controlled category. In those situations the general population/uncontrolled exposure limits will apply.

A fundamental aspect of the exposure guidelines is that they apply to power densities or the squares of the electric and magnetic field strengths that are spatially averaged over the body dimensions. Spatially averaged RF field levels most accurately relate to estimating the whole-body averaged SAR that will result from the exposure and the MPEs specified in Table 1 of Appendix A are based on this concept. This means that local values of exposures that exceed the stated MPEs may not be related to non-compliance if the spatial average of RF fields over the body does not exceed the MPEs. Further discussion of spatial averaging as it relates to field measurements can be found in Section 3 of this bulletin and in the ANSI/IEEE and NCRP reference documents noted there.

Another feature of the exposure guidelines is that exposures, in terms of power density, E^2 or H^2, may be averaged over certain periods of time with the average not to exceed the limit for continuous exposure.[11] As shown in Table 1 of Appendix A, the averaging time for occupational/controlled exposures is 6 minutes, while the averaging time for general population/uncontrolled exposures is 30 minutes. It is important to note that for general population/uncontrolled exposures it is often not possible to control exposures to the extent that averaging times can be applied. In those situations, it is often necessary to assume continuous exposure.

As an illustration of the application of time-averaging to occupational/controlled exposure consider the following. The relevant interval for time-averaging for occupational/controlled exposures is six minutes. This means, for example, that during any given six-minute period a worker could be exposed to two times the applicable power density limit for three minutes as long as he or she were not exposed at all for the preceding or following three minutes. Similarly, a worker could be exposed at three times the limit for two minutes as long as no exposure occurs during the preceding or subsequent four minutes, and so forth.

---

[10] For example, a sign warning of RF exposure risk and indicating that individuals should not remain in the area for more than a certain period of time could be acceptable. Reference [3] provides information on acceptable warning signs.

[11] Note that although the FCC did not explicitly adopt limits for *peak* power density, guidance on these types of exposures can be found in Section 4.4 of the ANSI/IEEE C95.1-1992 standard.

This concept can be generalized by considering Equation (2) that allows calculation of the allowable time(s) for exposure at [a] given power density level(s) during the appropriate time-averaging interval to meet the exposure criteria of Table 1 of Appendix A. The sum of the products of the exposure levels and the allowed times for exposure must equal the product of the appropriate MPE limit and the appropriate time-averaging interval.

$$\sum S_{exp} t_{exp} = S_{limit} t_{avg} \tag{2}$$

where:
S_{exp} = power density level of exposure (mW/cm^2)
S_{limit} = appropriate power density MPE limit (mW/cm^2)
t_{exp} = allowable time of exposure for S_{exp}
t_{avg} = appropriate MPE averaging time

For the example given above, if the MPE limit is 1 mW/cm^2, then the right-hand side of the equation becomes 6 mW-min/cm^2 (1 mW/cm^2 X 6 min). Therefore, if an exposure level is determined to be 2 mW/cm^2, the allowed time for exposure at this level during any six-minute interval would be a total of 3 minutes, since the left side of the equation must equal 6 (2 mW/cm^2 X 3 min). Of course, many other combinations of exposure levels and times may be involved during a given time-averaging interval. However, as long as the sum of the products on the left side of the equation equals the right side, the *average* exposure will comply with the MPE limit. It is very important to remember that time-averaging applies to *any* interval of t_{avg}. Therefore, in the above example, consideration would have to be given to the exposure situation both before and after the allowed three-minute exposure. The time-averaging interval can be viewed as a "sliding" period of time, six minutes in this case.

Another important point to remember concerning the FCC's exposure guidelines is that they constitute *exposure* limits (not *emission* limits), and they are relevant only to locations that are *accessible* to workers or members of the public. Such access can be restricted or controlled by appropriate means such as the use of fences, warning signs, etc., as noted above. For the case of occupational/controlled exposure, procedures can be instituted for working in the vicinity of RF sources that will prevent exposures in excess of the guidelines. An example of such procedures would be restricting the time an individual could be near an RF source or requiring that work on or near such sources be performed while the transmitter is turned off or while power is appropriately reduced. In the case of broadcast antennas, the use of auxiliary antennas could prevent excessive exposures to personnel working on or near the main antenna site, depending on the separation between the main and auxiliary antennas. Section 4 of this bulletin should be consulted for further information on controlling exposure to comply with the FCC guidelines.

Applicability of New Guidelines

The FCC's environmental rules regarding RF exposure identify particular categories of existing and proposed transmitting facilities, operations and devices for which licensees and applicants are required to conduct an initial environmental evaluation, and prepare an Environmental Assessment if the evaluation indicates that the transmitting facility, operation or device exceeds or will exceed the FCC's RF exposure guidelines. For transmitting facilities, operations and devices not specifically identified, the Commission has determined, based on calculations, measurement data and other information, that such RF sources offer little potential for causing exposures in excess of the guidelines. Therefore, the Commission "categorically excluded" applicants and licensees from the requirement to perform routine, initial environmental evaluations of such sources to demonstrate compliance with our guidelines. However, the Commission still retains the authority to request that a licensee or an applicant conduct an environmental evaluation and, if appropriate, file environmental information pertaining to an otherwise categorically excluded RF source if it is determined that there is a possibility for significant environmental impact due to RF exposure.[12]

In that regard, all transmitting facilities and devices regulated by this Commission that are the subject of an FCC decision or action (e.g., grant of an application or response to a petition or inquiry) are expected to comply with the appropriate RF radiation exposure guidelines, or, if not, to file an Environmental Assessment (EA) for review under our NEPA procedures, if such is required. It is important to emphasize that the categorical exclusions are *not* exclusions from *compliance* but, rather, exclusions from performing routine evaluations to demonstrate compliance. Normally, the exclusion from performing a routine evaluation will be a sufficient basis for assuming compliance, unless an applicant or licensee is otherwise notified by the Commission or has reason to believe that the excluded transmitter or facility encompasses exceptional characteristics that could cause non-compliance.

It should also be stressed that even though a transmitting source or facility may not be categorically excluded from routine evaluation, no further environmental processing is required once it has been demonstrated that exposures are within the guidelines, as specified in Part 1 of our rules. These points have been the source of some confusion in the past among FCC licensees and applicants, some of whom have been under the impression that filing an EA is always required.

In adopting its new exposure guidelines, the Commission also adopted new rules indicating which transmitting facilities, operations and devices will be categorically excluded from performing routine, initial evaluations. The new exclusion criteria are based on such factors as type of service, antenna height, and operating power. The new criteria were adopted in an attempt to obtain greater consistency and scientific rigor in determining requirements for RF evaluation across the various FCC-regulated services.

---

[12]   *See* 47 CFR §§ 1.1307(c) and (d).

Routine environmental evaluation for RF exposure is required for transmitters, facilities or operations that are included in the categories listed in Table 2 of Appendix A or in FCC rule parts 2.1091 and 2.1093 (for portable and mobile devices). This requirement applies to some, but not necessarily all, transmitters, facilities or operations that are authorized under the following parts of our rules: 5, 15, 21 (Subpart K), 22 (Subpart E), 22 (Subpart H), 24, 25, 26, 27, 73, 74 (Subparts A, G, I, and L), 80 (ship earth stations), 90 (paging operations and Specialized Mobile Radio), 97 and 101 (Subpart L). Within a specific service category, conditions are listed in Table 2 of Appendix A to determine which transmitters will be subject to routine evaluation. These conditions are generally based on one or more of the following variables: (1) operating power, (2) location, (3) height above ground of the antenna and characteristics of the antenna or mode of transmission. In the case of Part 15 devices, only devices that transmit on millimeter wave frequencies and unlicensed Personal Communications Service (PCS) devices are covered, as noted in rule parts 2.1091 and 2.1093 (see section on mobile and portable devices of Appendix A).

Transmitters and facilities not included in the specified categories are excluded from routine evaluation for RF exposure. We believe that such transmitting facilities generally pose little or no risk for causing exposures in excess of the guidelines. However, as noted above, in exceptional cases the Commission may, on its own merit or as the result of a petition, require environmental evaluation of transmitters or facilities even though they are otherwise excluded from routine evaluation. Also, at multiple-transmitter sites applications for non-excluded transmitters should consider significant contributions of other co-located transmitters (see discussion of multiple-transmitter evaluation in Section 2).

If a transmitter operates using relatively high power, and there is a possibility that workers or the public could have access to the transmitter site, such as at a rooftop site, then routine evaluation is justified. In Table 2 of Appendix A, an attempt was made to identify situations in the various services where such conditions could prevail. In general, at rooftop transmitting sites evaluation will be required if power levels are above the values indicated in Table 2 of Appendix A. These power levels were chosen based on generally "worst-case" assumptions where the most stringent uncontrolled/general population MPE limit might be exceeded within several meters of transmitting antennas at these power levels. In the case of paging antennas, the likelihood that duty factors, although high, would not normally be expected to be 100% was also considered. Of course, if procedures are in place at a site to limit accessibility or otherwise control exposure so that the safety guidelines are met, then the site is in compliance and no further environmental processing is necessary under our rules.

Tower-mounted ("non-rooftop") antennas that are used for cellular telephone, PCS, and Specialized Mobile Radio (SMR) operations warrant a somewhat different approach for evaluation. While there is no evidence that typical installations in these services cause ground-level exposures in excess of the MPE limits, construction of these towers has been a topic of ongoing public controversy on environmental grounds, and we believe it necessary to ensure that there is no likelihood of excessive exposures from these antennas. Although we believe there is no need to require routine evaluation of towers where antennas are mounted high above the ground, out of an abundance of caution the FCC requires that tower-mounted installations be evaluated if antennas are mounted lower than 10 meters above ground and the total power of all channels being used is over 1000 watts effective radiated power (ERP), or 2000 W ERP for broadband PCS.[13] These height and power combinations were chosen as thresholds recognizing that a theoretically "worst case" site could use many channels and several thousand watts of power. At such power levels a height of 10 meters above ground is not an unreasonable distance for which an evaluation generally would be advisable. For antennas mounted higher than 10 meters, measurement data for cellular facilities have indicated that ground-level power densities are typically hundreds to thousands of times below the new MPE limits.

In view of the expected proliferation of these towers in the future and possible use of multiple channels and power levels at these installations, and to ensure that tower installations are properly evaluated when appropriate, we have instituted these new requirements for this limited category of tower-mounted antennas in these services. For consistency we have instituted similar requirements for several other services that could use relatively high power levels with antennas mounted on towers lower than 10 meters above ground.

Paging systems operated under Part 22 (Subpart E) and Part 90 of our rules previously have been categorically exempted from routine RF evaluation requirements. However, the potential exists that the new, more restrictive limits may be exceeded in accessible areas by relatively high-powered paging transmitters with rooftop antennas.[14] These transmitters may operate with high duty factors in densely populated urban environments. The record and our own data indicate the need for ensuring appropriate evaluation of such facilities, especially at multiple transmitter sites. Accordingly, paging stations authorized under Part 22 (Subpart E) and Part 90 are also subject to routine environmental evaluation for RF exposure if an antenna is located on a rooftop and if its ERP exceeds 1000 watts.

Mobile and Portable Devices

As noted in Appendix A, mobile and portable transmitting devices that operate in the Cellular Radiotelephone Service, the Personal Communications Services (PCS), the General Wireless Communications Service, the Wireless Communication Service, the Satellite Communications services, the Maritime Services (ship earth stations only) and Specialized Mobile Radio Service authorized, respectively, under Part 22 (Subpart H), Part 24, Part 25, Part 26, Part 27, Part 80, and Part 90 of the FCC's Rules are subject to routine environmental evaluation for RF exposure prior to equipment authorization or use. Unlicensed PCS, NII and millimeter wave devices are also subject to routine environmental evaluation for RF exposure

---

[13] For broadband PCS, 2000 W is used as a threshold, instead of 1000 W, since at these operating frequencies the exposure criteria are less restrictive by about a factor of two.

[14] For example, under Part 90, paging operations in the 929-930 MHz band may operate with power levels as high as 3500 W ERP.

prior to equipment authorization or use. All other mobile, portable, and unlicensed transmitting devices are normally categorically excluded from routine environmental evaluation for RF exposure (see Section 2 and Appendix A for further details).

For purposes of these requirements mobile devices are defined by the FCC as transmitters designed to be used in other than fixed locations and to generally be used in such a way that a separation distance of at least 20 centimeters is normally maintained between radiating structures and the body of the user or nearby persons. These devices are normally evaluated for exposure potential with relation to the MPE limits given in Table 1 of Appendix A.

The FCC defines portable devices, for purposes of these requirements, as transmitters whose radiating structures are designed to be used within 20 centimeters of the body of the user. As explained later, in Section 2 and in Appendix A, portable devices are to be evaluated with respect to limits for specific absorption rate (SAR).

Operations in the Amateur Radio Service

In the FCC's recent *Report and Order*, certain amateur radio installations were made subject to routine evaluation for compliance with the FCC's RF exposure guidelines.[15] Also, amateur licensees will be expected to demonstrate their knowledge of the FCC guidelines through examinations. Applicants for new licenses and renewals also will be required to demonstrate that they have read and that they understand the applicable rules regarding RF exposure. Before causing or allowing an amateur station to transmit from any place where the operation of the station could cause human exposure to RF radiation levels in excess of the FCC guidelines amateur licensees are now required to take certain actions. A routine RF radiation evaluation is required if the transmitter power of the station exceeds the levels shown in Table 1 and specified in 47 CFR § 97.13(c)(1).[16] Otherwise the operation is categorically excluded from routine RF radiation evaluation, except as a result of a specific motion or petition as specified in Sections 1.1307(c) and (d) of the FCC's Rules, (see earlier discussion in Section 1 of this bulletin).

The Commission's *Report and Order* instituted a requirement that operator license examination question pools will include questions concerning RF safety at amateur stations. An additional five questions on RF safety will be required within each of three written examination elements. The Commission also adopted the proposal of the American Radio

---

[15] *See* para. 160 of *Report and Order*, ET Dkt 93-62. *See also*, 47 CFR § 97.13, as amended.

[16] These levels were chosen to roughly parallel the frequency of the MPE limits of Table 1 in Appendix A. These levels were modified from the Commission's original decision establishing a flat 50 W power threshold for routine evaluation of amateur stations (*see Second Memorandum Opinion and Order,* ET Docket 93-62, FCC 97-303, adopted August 25, 1997).

TABLE 1. Power thresholds for routine evaluation of amateur radio stations.

Wavelength Band	Transmitter Power (watts)
MF	
160 m	500
HF	
80 m	500
75 m	500
40 m	500
30 m	425
20 m	225
17 m	125
15 m	100
12 m	75
10 m	50
VHF (all bands)	50
UHF	
70 cm	70
33 cm	150
23 cm	200
13 cm	250
SHF (all bands)	250
EHF (all bands)	250

Relay League (ARRL) that amateur operators should be required to certify, as part of their license application process, that they have read and understand our bulletins and the relevant FCC rules.

When routine evaluation of an amateur station indicates that exposure to RF fields could be in excess of the exposure limits specified by the FCC (see Appendix A), the licensee must take action to correct the problem and ensure compliance (see Section 4 of this bulletin on controlling exposure). Such actions could be in the form of modifying patterns of operation, relocating antennas, revising a station's technical parameters such as frequency, power or emission type or combinations of these and other remedies.

In complying with the Commission's *Report and Order*, amateur operators should follow a policy of systematic avoidance of excessive RF exposure. The Commission has said that it will continue to rely upon amateur operators, in constructing and operating their stations, to take steps to ensure that their stations comply with the MPE limits for both occupational/controlled and general public/uncontrolled situations, as appropriate. In that regard, amateur radio operators and members of their immediate household are considered to be in a "controlled environment" and are subject to the occupational/controlled MPE limits. Neighbors who are not members of an amateur operator's household are considered to be members of the general public, since they cannot reasonably be expected to exercise control over their exposure. In those cases general population/uncontrolled exposure MPE limits will apply.

In order to qualify for use of the occupational/controlled exposure criteria, appropriate restrictions on access to high RF field areas must be maintained and educational instruction in RF safety must be provided to individuals who are members of the amateur operator's household. Persons who are not members of the amateur operator's household but who are present temporarily on an amateur operator's property may also be considered to fall under the occupational/controlled designation provided that appropriate information is provided them about RF exposure potential if transmitters are in operation and such persons are exposed in excess of the general population/uncontrolled limits.

Amateur radio facilities represent a special case for determining exposure, since there are many possible antenna types that could be designed and used for amateur stations. However, several relevant points can be made with respect to analyzing amateur radio antennas for potential exposure that should be helpful to amateur operators in performing evaluations.

First of all, the generic equations described in this bulletin can be used for analyzing fields due to almost all antennas, although the resulting estimates for power density may be overly-conservative in some cases. Nonetheless, for general radiators and for aperture antennas, if the user is knowledgeable about antenna gain, frequency, power and other relevant factors, the equations in this section can be used to estimate field strength and power density as described earlier. In addition, other resources are available to amateur radio operators for analyzing fields near their antennas. The ARRL Radio Amateur Handbook contains an excellent section on analyzing amateur radio facilities for compliance with RF guidelines (Reference [4]). Also, the FCC and the EPA conducted a study of several amateur radio stations in 1990 that provides a great deal of measurement data for many types of antennas commonly used by amateur operators (Reference [10]).

Amateur radio organizations and licensees are encouraged to develop their own more detailed evaluation models and methods for typical antenna configurations and power/frequency combinations. The FCC is working with the amateur radio community to develop a supplement to this bulletin that will be designed specifically for evaluating amateur radio installations. For example, the supplement will contain information on projected minimum exclusion distances from typical amateur antenna installations. The supplement should be completed soon after release of this bulletin. Once the amateur radio supplement is released by the FCC it will be made available for downloading at the FCC's World Wide Web Site for "RF safety." Amateur radio applicants and licensees are encouraged to monitor the Web Site for release of the supplement. The address is: www.fcc.gov/oet/rfsafety. Information on availability of the supplement, as well as other RF-related questions, can be directed to the FCC's "RF Safety Program" at: (202) 418-2464 or to: rfsafety@fcc.gov.

Section 2: PREDICTION METHODS

The material in this section is designed to provide assistance in determining whether a given facility would be in compliance with guidelines for human exposure to RF radiation. The calculational methods discussed below should be helpful in evaluating a particular exposure situation. However, for certain transmitting facilities, such as radio and television broadcast stations, a specific supplement to this bulletin has been developed containing information and compliance guidelines specific to those stations.[17] Therefore, applicants for radio and television broadcast facilities may wish to first consult this supplement that concentrates on AM radio, FM radio and television broadcast antennas. Applicants for many broadcast facilities should be able to determine whether a given facility would be in compliance with FCC guidelines by simply consulting the tables and figures in this supplement. However, in addition, with respect to occupational/controlled exposure, all applicants should consult Section 4 of this bulletin concerning controlling exposures that may occur during maintenance or other procedures carried out at broadcast and other telecommunications sites.

Applicants may consult the relevant sections below, which describe how to estimate field strength and power density levels from typical, general radiators as well as from aperture

---

[17]   *Supplement A to* OET Bulletin 65, Version 97-01, *Additional Information for Radio and Television Broadcast Stations.* This supplement can be downloaded from the FCC's RF Safety World Wide Web Site: www.fcc.gov/oet/rfsafety. For further information contact the RF safety program at: +1 (202) 418-2464.

antennas such as microwave and satellite dish antennas. The general equations given below can be used for predicting field strength and power density in the vicinity of most antennas, including those used for paging and in the commercial mobile radio service (CMRS). They can also be used for making conservative predictions of RF fields in the vicinity of antennas used for amateur radio transmissions, as discussed earlier.

Equations for Predicting RF Fields

Calculations can be made to predict RF field strength and power density levels around typical RF sources. For example, in the case of a single radiating antenna, a prediction for power density in the far-field of the antenna can be made by use of the general Equations (3) or (4) below [for conversion to electric or magnetic field strength see Equation (1) in Section 1]. These equations are generally accurate in the far-field of an antenna but will over-predict power density in the near field, where they could be used for making a "worst case" or conservative prediction.

$$S = \frac{PG}{4\pi R^2} \qquad (3)$$

where: S = power density (in appropriate units, e.g. mW/cm^2)
P = power input to the antenna (in appropriate units, e.g., mW)
G = power gain of the antenna in the direction of interest relative to an isotropic radiator
R = distance to the center of radiation of the antenna (appropriate units, e.g., cm)

or:

$$S = \frac{EIRP}{4\pi R^2} \qquad (4)$$

where: EIRP = equivalent (or effective) isotropically radiated power

When using these and other equations care must be taken to use the *correct units* for all variables. For example, in Equation (3), if power density in units of mW/cm^2 is desired then power should be expressed in milliwatts and distance in cm. Other units may be used, but care must be taken to use correct conversion factors when necessary. Also, it is important to note that the power gain factor, *G*, in Equation (3) is normally *numeric* gain. Therefore, when power gain is expressed in logarithmic terms, i.e., dB, a conversion is required using the relation:

$$G = 10^{\frac{dB}{10}}$$

For example, a logarithmic power gain of 14 dB is equal to a numeric gain of 25.12.

In some cases operating power may be expressed in terms of "effective radiated power" or "ERP" instead of EIRP. ERP is power referenced to a half-wave dipole radiator instead of to an isotropic radiator. Therefore, if ERP is given it is necessary to convert ERP into EIRP in order to use the above equations. This is easily done by multiplying the ERP by the factor of 1.64, which is the gain of a half-wave dipole relative to an isotropic radiator. For example, if ERP is used in Equation (4) the relation becomes:

$$S = \frac{EIRP}{4\pi R^2} = \frac{1.64\ ERP}{4\pi R^2} = \frac{0.41\ ERP}{\pi R^2} \tag{5}$$

For a truly worst-case prediction of power density at or near a surface, such as at ground-level or on a rooftop, 100% reflection of incoming radiation can be assumed, resulting in a potential doubling of predicted field strength and a four-fold increase in (far-field equivalent) power density. In that case Equations (3) and (4) can be modified to:

$$S = \frac{(2)^2 PG}{4\pi R^2} = \frac{PG}{\pi R^2} = \frac{EIRP}{\pi R^2} \tag{6}$$

In the case of FM radio and television broadcast antennas, the U.S. Environmental Protection Agency (EPA) has developed models for predicting ground-level field strength and power density [Reference 11]. The EPA model recommends a more realistic approximation for ground reflection by assuming a maximum 1.6-fold increase in field strength leading to an increase in power density of 2.56 (1.6 X 1.6). Equation (4) can then be modified to:

$$S = \frac{2.56 \; EIRP}{4\pi R^2} = \frac{0.64 \; EIRP}{\pi R^2} \qquad (7)$$

If ERP is used in Equation (7), the relation becomes:

$$S = \frac{0.64 \; EIRP}{\pi R^2} = \frac{(0.64)(1.64) \; ERP}{\pi R^2} = \frac{1.05 \; ERP}{\pi R^2} \qquad (8)$$

It is sometimes convenient to use units of microwatts per centimeter squared ($\mu W/cm^2$) instead of mW/cm^2 in describing power density. The following simpler form of Equation (8) can be derived if power density, **S**, is to be expressed in units of $\mu W/cm^2$:

$$S = \frac{33.4 \; ERP}{R^2} \qquad (9)$$

where: S = power density in $\mu W/cm^2$
            ERP = power in watts
            R = distance in meters

An example of the use of the above equations follows. A station is transmitting at a frequency of 100 MHz with a total nominal ERP (including all polarizations) of 10 kilowatts (10,000 watts) from a tower-mounted antenna. The height to the center of radiation is 50 meters above ground-level. Using the formulas above, what would be the calculated "worst-case" power density that could be expected at a point 2 meters above ground (approximate head level) and at a distance of 20 meters from the base of the tower? Note that this type of analysis *does not* take into account the vertical radiation pattern of the antenna, i.e., no information on directional characteristics of signal propagation is considered. Use of actual vertical radiation pattern data for the antenna would most likely significantly reduce ground-level exposure predictions from those calculated below (see later discussion), resulting in a more realistic estimate of the actual exposure levels.

From simple trigonometry the distance **R** can be calculated to be 52 meters [square root of: $(48)^2 + (20)^2$], assuming essentially flat terrain. Therefore, using Equation (9), the calculated conservative "worst case" power density is:

$$S = \frac{33.4\ (10{,}000\ watts)}{(52\ m)^2} = about\ 124\ \mu W/cm^2$$

By consulting Table 1 of Appendix A it can be determined that the limit for general population/uncontrolled exposure at 100 MHz is 0.2 mW/cm^2 or 200 μW/cm^2. Therefore, this calculation shows that even under worst-case conditions this station would comply with the general population/uncontrolled limits, at least at a distance of 20 meters from the tower. Similar calculations could be made to ensure compliance at other locations, such as at the base of the tower where the shortest direct line distance, R, to the ground would occur.

Relative Gain and Main-Beam Calculations

The above-described equations can be used to calculate fields from a variety of radiating antennas, such as omni-directional radiators, dipole antennas and antennas incorporating directional arrays. However, in many cases the use of equations such as Equations (3) and (4) will result in an overly conservative "worst case" prediction of the field at a given point. Alternatively, if information concerning an antenna's vertical radiation pattern is known, a relative field factor (relative gain) derived from such a pattern can be incorporated into the calculations to arrive at a more accurate representation of the field at a given point of interest. For example, in the case of an antenna pointing toward the horizon, if the relative gain in the main beam is 1.0, then in other directions downward from horizontal the field may be significantly less than 1.0. Therefore, radiation from the antenna directly toward the ground may be significantly reduced from the omni-directional case and a more realistic prediction of the field can be obtained for the point of interest.

For example, in the calculation above, it can be shown from trigonometry that the depression angle below horizontal of the vector corresponding to the distance, R, is about 68°. For purposes of illustration, assume that the antenna in this example has its main beam pointed approximately toward the horizon and, at a depression angle of 68°, the field relative to the main beam (relative gain) is −6 dB (a factor of 0.5 in terms of field strength and 0.25 in terms of power density). In that case the calculation above can be modified giving a more accurate representation of the power density at the ground-level point of interest, as follows.

$$S = \frac{33.4 \; F^2 \; ERP}{R^2} = \frac{33.4 \; (0.5)^2 \; (10,000 \; watts)}{(52 \; m)^2} = \text{about } 31 \; \mu W/cm^2$$

where: F = the relative field factor (relative numeric gain)

In general, Equation (9) can be modified to:

$$S = \frac{33.4 \; (F^2) \; ERP}{R^2} \qquad\qquad (10)$$

where: S = power density in $\mu W/cm^2$
F = relative field factor (relative numeric gain)
ERP = power in watts
R = distance in meters

When the point of interest where exposure may occur is in or near the main radiated beam of an antenna, Equation (3) or its derivatives can be used. In other words, the factor, F, in such cases would be assumed to be 1.0. Such cases occur when, for example, a nearby building or rooftop may be in the main beam of a radiator. For convenience in determining exposures in such situations, Equation (3) has been used to derive Figures 1 and 2. These figures allow a quick determination of the power density at a given distance from an antenna in its main beam for various levels of ERP.[18] Intermediate ERPs can be estimated by interpolation, or the next highest ERP level can be used as a worst case approximation.

Figure 1 assumes no reflection off of a surface. However, at a rooftop location where the main-beam may be directed parallel and essentially along or only slightly above the surface of the roof, there may be reflected waves that would contribute to exposure. Therefore, Figure 2 was derived for the latter case using the EPA-recommended reflection factor of $(1.6)^2 = 2.56$ (see earlier discussion), and the values shown are more conservative. When using Figures 1 or 2 a given situation should be considered on its own merits to determine which figure is more appropriate. For rooftop locations it is also important to note that exposures *inside* a building can be expected to be reduced by at least 10-20 dB due to attenuation caused by building materials in the walls and roof.

---

[18]   To convert to EIRP use the relation: EIRP = ERP X 1.64.

Main-Beam Exposure (No Reflection)

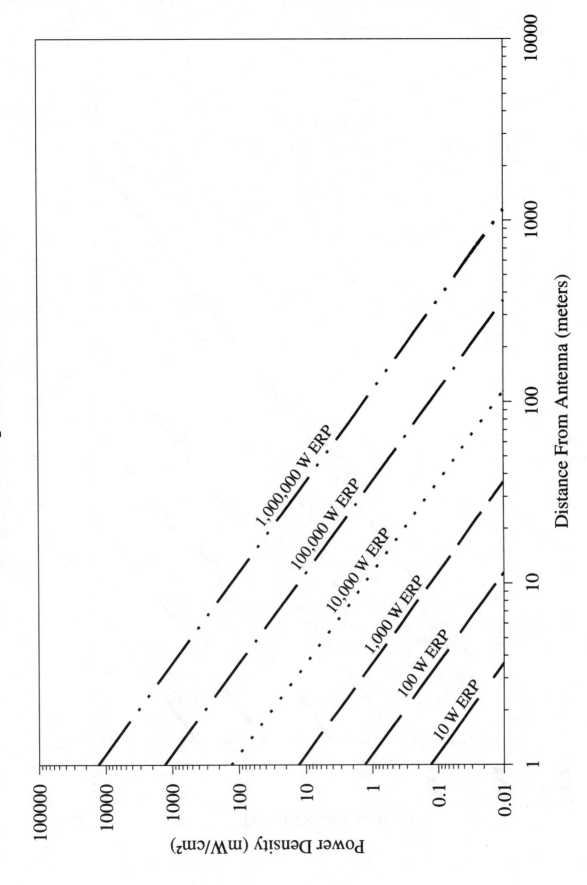

FIGURE 1. Power Density vs. Distance (assumes no surface reflection).

Main-Beam Exposure (With Reflection)

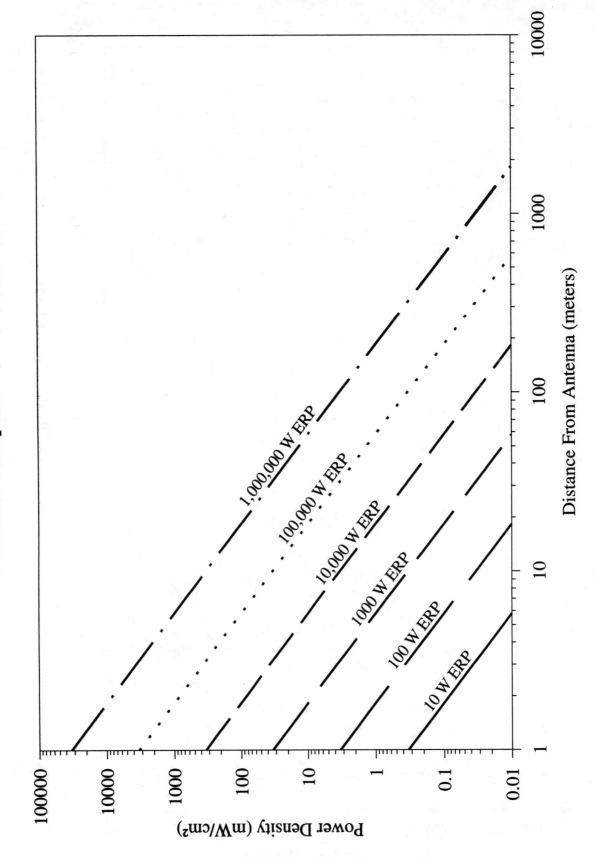

FIGURE 2. Power Density vs. Distance (assumes surface reflection).

Multiple-Transmitter Sites and Complex Environments

It is common for multiple RF emitters to be co-located at a given site. Antennas are often clustered together at sites that may include a variety of RF sources such as radio and television broadcast towers, CMRS antennas and microwave antennas. The FCC's exposure guidelines are meant to apply to any exposure situation caused by transmitters regulated by the FCC. Therefore, at multiple-transmitter sites, all significant contributions to the RF environment should be considered, not just those fields associated with one specific source. When there are multiple transmitters at a given site collection of pertinent technical information about them will be necessary to permit an analysis of the overall RF environment by calculation or computer modeling. However, if this is not practical a direct measurement survey may prove to be more expedient for assessing compliance (see Section 3 of this bulletin that deals with measurements for more information).

The rules adopted by the FCC specify that, in general, at multiple transmitter sites actions necessary to bring the area into compliance with the guidelines are the shared responsibility of all licensees whose transmitters produce field strengths or power density levels at the area in question in excess of 5% of the exposure limit (in terms of power density or the square of the electric or magnetic field strength) applicable to their particular transmitter.[22] When performing an evaluation for compliance with the FCC's RF guidelines *all* significant contributors to the ambient RF environment should be considered, including those otherwise excluded from performing routine RF evaluations, and applicants are expected to make a good-faith effort to consider these other transmitters. For purposes of such consideration, significance can be taken to mean *any* transmitter producing more than 5% of the applicable exposure limit (in terms of power density or the square of the electric or magnetic field strength) at accessible locations. The percentage contributions are then added to determine whether the limits are (or would be) exceeded. If the MPE limits are exceeded, then the responsible party or parties, as described below, must take action to either bring the area into compliance or submit an EA.

Applicants and licensees should be able to calculate, based on considerations of frequency, power and antenna characteristics the distance from their transmitter where their signal produces an RF field equal to, or greater than, the 5% threshold limit. The applicant or licensee then shares responsibility for compliance in any accessible area or areas within this 5% "contour" where the appropriate limits are found to be exceeded.

The following policy applies in the case of an application for a proposed transmitter, facility or modification (not otherwise excluded from performing a routine RF evaluation) that would *cause non-compliance* at an accessible area previously in compliance. In such a case, it is the responsibility of the applicant to either ensure compliance or submit an EA if emissions from the applicant's transmitter or facility would result in an exposure level at the non-complying area that exceeds 5% of the exposure limits applicable to that transmitter or facility in terms of power density or the square of the electric or magnetic field strength.

For a renewal applicant whose transmitter or facility (not otherwise excluded from routine evaluation) contributes to the RF environment at an accessible area *not in compliance* with the guidelines the following policy applies. The renewal applicant must submit an EA if emissions from the applicant's transmitter or facility, at the area in question, result in an exposure level that exceeds 5% of the exposure limits applicable to that particular transmitter

---

[22]   *See* 47 C.F.R. 1.1307(b)(3), as amended.

in terms of power density or the square of the electric or magnetic field strength. In other words, although the renewal applicant may only be responsible for a fraction of the total exposure (greater than 5%), the applicant (along with any other licensee undergoing renewal at the same time) will trigger the EA process, unless suitable corrective measures are taken to prevent non-compliance before preparation of an EA is necessary. In addition, in a renewal situation if a determination of non-compliance is made, other co-located transmitters contributing more than the 5% threshold level must share responsibility for compliance, regardless of whether they are categorically excluded from routine evaluation or submission of an EA.

Therefore, at multiple-transmitter sites the various responsibilities for evaluating the RF environment, taking actions to ensure compliance or submitting an EA may lie either with a newcomer to the site, with a renewal applicant (or applicants) or with all significant users, depending on the situation. In general, an applicant or licensee for a transmitter at a multiple-transmitter site should seek answers to the following questions in order to determine compliance responsibility.

(1) New transmitter proposed for a multiple-transmitter site.

- Is the transmitter in question already categorically excluded from routine evaluation?

- If *yes*, routine evaluation of the application is not required.

- If *not excluded*, is the site in question already in compliance with the FCC guidelines?

- If *no,* the applicant must submit an EA with its application notifying the Commission of the non-compying situation, unless measures are to be taken to ensure compliance. Compliance is the responsibility of licensees of all transmitters that contribute to non-complying area(s) in excess of the applicable 5% threshold at the existing site. If the existing site is subsequently brought into compliance *without* consideration of the new applicant then the next two questions below apply.

- If *yes*, would the proposed transmitter cause non-compliance at the site in question?
- If *yes*, the applicant must submit an EA (or submit a new EA in the situation described above) with its application notifying the Commission of the potentially non-complying situation, unless measures will be taken by the applicant to ensure compliance. In this situation, it is the responsibility of the applicant to ensure compliance, since the existing site is already in compliance.

- If *no*, no further environmental evaluation is required and the applicant certifies compliance.

(2) Renewal applicant at a multiple-transmitter site

- Is the transmitter in question already categorically excluded from routine evaluation?

- If *yes*, routine evaluation of the application is not required.

- If *not excluded*, is the site in question already in compliance with the FCC guidelines?

- If *no*, the applicant must submit an EA with its application notifying the Commission of the non-compying situation, unless measures are taken to ensure compliance. Compliance is the responsibility of licensees of all transmitters that contribute to non-complying area(s) in excess of the applicable 5% threshold.

- If *yes,* no further environmental evaluation is necessary and the applicant certifies compliance.

The Commission expects its licensees and applicants to cooperate in resolving problems involving compliance at multiple-transmitter sites. Also, owners of transmitter sites are expected to allow applicants and licensees to take reasonable steps to comply with the FCC's requirements. When feasible, site owners should also encourage co-location and common solutions for controlling access to areas that may be out of compliance. In situations where disputes arise or where licensees cannot reach agreement on necessary compliance actions, a licensee or applicant should notify the FCC licensing bureau. The bureau may then determine whether appropriate FCC action is necessary to facilitate a resolution of the dispute.

The FCC's MPE limits vary with frequency. Therefore, in mixed or broadband RF fields where several sources and frequencies are involved, the fraction of the recommended limit (in terms of power density or square of the electric or magnetic field strength) incurred within each frequency interval should be determined, and the sum of all fractional contributions should not exceed 1.0, or 100% in terms of percentage. For example, consider an antenna farm with radio and UHF television broadcast transmitters. At a given location that is accessible to the general public it is determined that FM radio station X contributes 100 μW/cm^2 to the total power density (which is 50% of the applicable 200 μW/cm^2 MPE limit for the FM frequency band). Also, assume that FM station Y contributes an additional 50 μW/cm^2 (25% of its limit) and that a nearby UHF-TV station operating on Channel 35 (center frequency = 599 MHz) contributes 200 μW/cm^2 at the same location (which is 50% of the applicable MPE limit for this frequency of 400 μW/cm^2). The sum of all of the percentage contributions then equals 125%, and the location is not in compliance with the MPE limits for the general public. Consequently, measures must be taken to bring the site into compliance such as restricting access to the area (see Section 4 of this bulletin on controlling exposure).

As noted above, in such situations it is the shared responsibility of site occupants to take whatever actions are necessary to bring a site into compliance. In the above case, the allocation of responsibility could be generally based on each station's percentage contribution to the overall power density at the problem location, although such a formula for allocating responsibility is not an FCC requirement, and other formulas may be used, as appropriate.

When attempting to predict field strength or power density levels at multiple transmitter sites the general equations discussed in this section of the bulletin can be used at many sites, depending on the complexity of the site. Individual contributions can often be determined at a given location using these prediction methods, and then power densities (or squares of field strength values) can be added together for the total predicted exposure level.
In addition, time-averaging of exposures may be possible, as explained in Section 1 of this bulletin. For sites involving radio and television broadcast stations, the methods described in Supplement A for broadcast stations can be used in some circumstances when a site is not overly complex. Also, for wireless communications sites, some organizations have developed commercially-available software for modeling sites for compliance purposes.[23]

When considering the contributions to field strength or power density from other RF sources, care should be taken to ensure that such variables as reflection and re-radiation are considered. In cases involving very complex sites predictions of RF fields may not be possible, and a measurement survey may be necessary (see Section 3 of this bulletin).

The following example illustrates a simple situation involving multiple antennas. The process for determining compliance for other situations can be similarly accomplished using the techniques described in this section and in Supplement A to this bulletin that deals with radio and television broadcast operations. However, as mentioned above, at very complex sites measurements may be necessary.

In the simple example shown in Figure 4 it is desired to determine the power density at a given location **X** meters from the base of a tower on which are mounted two antennas. One antenna is a CMRS antenna with several channels, and the other is an FM broadcast antenna. The system parameters that must be known are the total ERP for each antenna and the operating frequencies (to determine which MPE limits apply). The heights above ground level for each antenna, **H1** and **H2,** must be known in order to calculate the distances, **R1** and **R2**, from the antennas to the point of interest. The methods described in this section (and in Supplement A for FM antennas) can be used to determine the power density contributions of each antenna at the location of interest, and the percentage contributions (compared to the applicable MPE limit for that frequency) are added together as described above to determine if the location complies with the applicable exposure guidelines. If the location is accessible to the public, the general/population limits apply. Otherwise occupational/controlled limits should be used.

---

[23] For example, the following two U.S. companies have recently begun marketing such software: (1) Richard Tell Associates, Inc., telephone: (702) 645-3338; and (2) UniSite, telephone: (972) 348-7632.

Another type of complex environment is a site with multiple towers. The same general process may be used to determine compliance as described above, if appropriate. Distances from each transmitting antenna to the point of interest must be calculated, and RF levels should be calculated at the point of interest due to emissions from each transmitting antenna using the most accurate model. Limits, percentages and cumulative percent of the limit may then be determined in the same manner as for Figure 4. Figure 5 illustrates such a situation.

Another situation may involve a single antenna that creates significant RF levels at more than one type of location. Figure 6 illustrates such a situation where exposures on a rooftop as well as on the ground are possible. The same considerations apply here as before and can be applied to predict RF levels at the points of interest. As mentioned previously, with respect to rooftop environments, it is also important to remember that building attenuation can be expected to reduce fields inside of the building by approximately 10-20 dB.

Situations where tower climbing is involved may be complicated and may require reduction of power or shutting down of transmitters during maintenance tasks (also see Section 4 of this bulletin on controlling exposure). Climbing of AM towers involves exposure due to RF currents induced in the body of the climber, and guidelines are available for appropriate power reduction (see Supplement A, Section 1, dealing with AM broadcast stations). For FM, TV and other antennas that may be mounted on towers, the highest exposures will be experienced near the active elements of each antenna and may require shutting off or greatly reducing power when a worker passes near the elements.

The equations in this section can also be used to calculate worst-case RF levels either below or above antennas that are side-mounted on towers. In the example shown in Figure 7, a more complicated situation arises when a worker is climbing an AM tower on which are side-mounted two other antennas. In this case the safest and most conservative approach would be to consult Supplement A, Section 1, for the appropriate AM power level to use and then to ensure that the transmitters for the other antennas are shut down when the climber passes near each side-mounted antenna's elements.

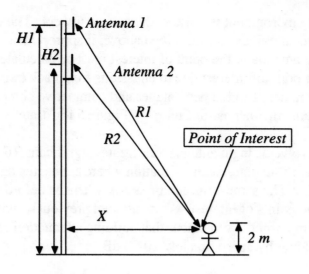

Figure 4. Single tower, co-located antennas, ground-level exposure (at 2 m).

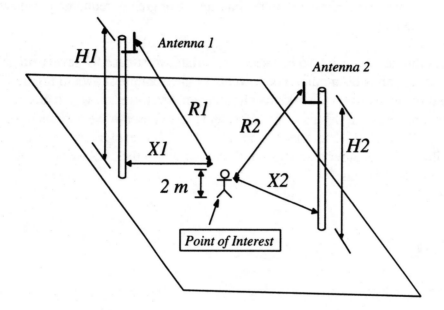

FIGURE 5. Antennas on multiple towers contributing to RF field at point of interest.

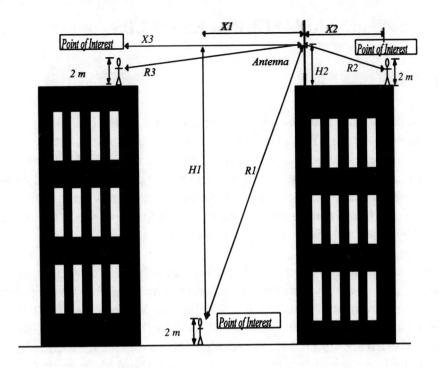

FIGURE 6. Single roof-top antenna, various exposure locations.

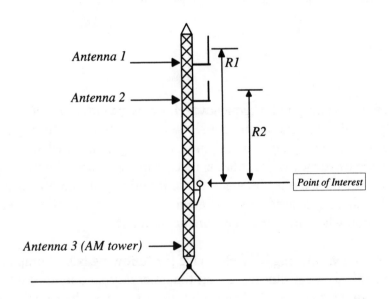

FIGURE 7. Single tower, co-located antennas, on-tower exposure.

Section 3: MEASURING RF FIELDS

Reference Material

In some cases the prediction methods described in Section 2 of this bulletin cannot be used, and actual measurements of the RF field may be necessary to determine whether there is a potential for human exposure in excess of the MPE limits specified by the FCC. For example, in a situation such as an antenna farm, with multiple users the models discussed previously would not always be applicable. Measurements may also be desired for cases in which predictions are slightly greater or slightly less than the threshold for excessive exposure or when fields are likely to be seriously distorted by objects in the field, e.g., conductive structures.

Techniques and instrumentation are available for measuring the RF environment near broadcast and other transmitting sources. In addition, references are available which provide detailed information on measurement procedures, instrumentation, and potential problems. Two excellent references in this area have been published by the IEEE and by the NCRP. The ANSI/IEEE document (ANSI/IEEE C95.3-1992) is entitled, "Recommended Practice for the Measurement of Potentially Hazardous Electromagnetic Fields - RF and Microwave," (Reference [2]) and the NCRP publication (NCRP Report No. 119) is entitled, "A Practical Guide to the Determination of Human Exposure to Radiofrequency Fields" (Reference [21]). Both of these documents contain practical guidelines and information for performing field measurements in broadcast and other environments, and the FCC strongly encourages their use. Other selected references are given in the reference section of this bulletin.

Instrumentation

Instruments used for measuring radiofrequency fields may be either broadband or narrowband devices. A typical broadband instrument responds essentially uniformly and instantaneously over a wide frequency range and requires no tuning. A narrowband instrument may also operate over a wide frequency range, but the instantaneous bandwidth may be limited to only a few kilohertz, and the device must be tuned to the frequency of interest. Each type of instrument has certain advantages and certain disadvantages, and the choice of which instrument to use depends on the situation where measurements are being made.

All instruments used for measuring RF fields have the following basic components: (1) an antenna to sample the field, (2) a detector to convert the time-varying output of the antenna to a steady-state or slowly varying signal, (3) electronic circuitry to process the signal, and (4) a readout device to display the measured field parameter in appropriate units.

The antennas most commonly used with broadband instruments are either dipoles that respond to the electric field (E) or loops that respond to the magnetic field (H). Surface area or displacement-current sensors that respond to the E-field are also used. In order to achieve a uniform response over the indicated frequency range, the size of the dipole or loop must be small compared to the wavelength of the highest frequency to be measured. Isotropic broadband probes contain three mutually orthogonal dipoles or loops whose outputs are *summed* so that the response is independent of orientation of the probe. The output of the dipoles or loops is converted to a proportional steady-state voltage or current by diodes or thermocouples, so that the measured parameter can be displayed on the readout device.

As described in the first edition of this bulletin, there are certain characteristics which are desirable in a broadband survey instrument. The major ones are as follows:

(1) The response of the instrument should be essentially isotropic, i.e., independent of orientation, or rotation angle, of the probe.

(2) The frequency range of the instrument and the instruments response over that range should be known. Generally this is given in terms of the error of response between certain frequency limits, e.g., ± 0.5 dB from 3 to 500 MHz.

(3) Out-of-band response characteristics of the instrument should be specified by the manufacturer to assist the user in selecting an instrument for a particular application.

For example, regions of enhanced response, or resonance, at frequencies outside of the band of interest could result in error in a measurement, if signals at the resonant frequency(ies) are present during the measurement.

(4) The dynamic range of the instrument should be at least ± 10 dB of the applicable exposure guideline.

(5) The instrument's readout device should be calibrated in units that correspond to the quantity actually being measured. An electric field probe responds to E or E^2, and a magnetic field probe responds to H or H^2, equally well in both the near-field and far-field. However, a readout device calibrated in units of power density does not read true power density if measurements are made in the near-field. This is because under plane-wave conditions, in which E, H, and power density are related by a constant quantity (the wave impedance which, for free space, is equal to 377 ohms), do not exist in the near-field where the wave impedance is complex and generally not known. Readout devices calibrated in "power density" actually read "far-field equivalent" power density or "plane-wave equivalent" power density (see discussion of MPE limits in Section 1 of this bulletin).

(6) The probe and the attached cables should only respond to the parameter being measured, e.g., a loop antenna element should respond to the magnetic field and should not interact significantly with the electric field.

(7) Shielding should be incorporated into the design of the instrument to reduce or eliminate electromagnetic interference.

(8) There should be some means, e.g., an alarm or test switch to establish that the probe is operating correctly and that none of the elements are burned out. Also, a means should be provided to alert the user if the measured signal is overloading the device.

(9) When the amplitude of the field is changing while measurements are being made, a "peak-hold" circuit may be useful. Such a change in amplitude could result either from variation in output from the source or from moving the probe through regions of the field that are non-uniform.

(10) For analog-type meters, the face of the meter should be coated with a transparent, conductive film to prevent false readings due to the accumulation of static charge in the meter itself. Also, the outer surface of the probe assembly of electric-field survey instruments should be covered with a high-resistance material to minimize errors due to static charge buildup.

(11) The instrument should be battery operated with easily replaceable or rechargeable batteries. A test switch or some other means should be provided to determine whether the batteries are properly charged. The instrument should be capable of operating

within the stated accuracy range for a time sufficient to accomplish the desired measurements without recharging or replacing the batteries.

(12) The user should be aware of the response time of the instrument, i.e., the time required for the instrument to reach a stable reading.

(13)   The device should be stable enough so that frequent readjustment to zero ("rezeroing") is not necessary. If not equipped with automatic zeroing capability, devices must be zeroed with the probe out of the field, either by shielding them or turning off the RF source(s). Either method is time consuming, making stability an especially desirable feature.

(14) If the instrument is affected by temperature, humidity, pressure, etc., the extent of the effect should be known and taken into account.

(15) The sensor elements should be sufficiently small and the device should be free from spurious responses so that the instrument responds correctly to the parameter being measured, both in the near-field and in the far-field. It should be emphasized that an instrument with a readout expressed in terms of power density will only be correct in the far-field. However, the term "far-field equivalent" or "plane-wave equivalent" power density is sometimes used in this context and would be acceptable as long as its meaning is understood and it is appropriately applied to the situation of interest (see discussion in Section 1).

(16) The instrument should respond to the average (rms) values of modulated fields independent of modulation characteristics. With respect to measurements of pulsed sources such as radar transmitters, many commercially-available survey instruments cannot measure high peak-power pulsed fields accurately. In such cases, the instrument should be chosen carefully to enable fields close to the antenna to be accurately measured.

(17) The instrument should be durable and able to withstand shock and vibration associated with handling in the field or during shipping. A storage case should be provided.

(18) The accuracy of the instrument should not be affected by exposure to light or other forms of ambient RF and low-frequency electromagnetic fields.

(19) The markings on the meter face should be sufficiently large to be easily read at arm's length.

(20) Controls should be clearly labeled and kept to a minimum, and operating procedures should be relatively simple.

(21) Typical meters use high-resistance leads that can be particularly susceptible to flexure noise when measuring fields at relatively low intensities. Therefore, when a broadband isotropic meter is used for measuring power density levels that fall into the lower range of detectability of the instrument (e.g., a few $\mu W/cm^2$), the meter should exhibit low noise levels if such measurements are to have any meaning.

(22) When measuring fields in multiple-emitter environments, the ability of many commonly available RF broadband survey meters to accurately measure multiple signals of varying frequencies may be limited by how the meter sums the outputs of its diode detectors. This can lead to over-estimates of the total RF field that may be significant. Although such estimates can represent a "worst case," and are allowable for compliance purposes, users of these meters should be aware of this possible source of error.

A useful characteristic of broadband probes used in multiple-frequency RF environments is a frequency-dependent response that corresponds to the variation in MPE limits with frequency. Broadband probes having such a "shaped" response permit direct assessment of compliance at sites where RF fields result from antennas transmitting over a wide range of frequencies. Such probes can express the composite RF field as a percentage of the applicable MPEs.

Another practical characteristic of some RF field instruments is their ability to automatically determine spatial averages of RF fields. Because the MPEs for exposure are given in terms of spatial averages, it is helpful to simplify the measurement of spatially variable fields via data averaging as the survey is being performed. Spatial averaging can be achieved via the use of "data loggers" attached to survey meters or circuitry built into the meter.

Narrowband devices may also be used to characterize RF fields for exposure assessment. In contrast to broadband devices, narrowband instruments may have bandwidths of only a few hundred kilohertz or less. Narrowband instruments, such as field-strength meters and spectrum analyzers, must be tuned from frequency to frequency, and the field level at each frequency measured. Spectrum analyzers can be scanned over a band of frequencies, and the frequency and peak-amplitude information can be stored and printed for later analysis. The results of all narrowband measurements may then be combined to determine the total field.

As with broadband instruments, narrowband devices consist of basically four components: an antenna, cables to carry the signal from the antenna, electronic circuitry to process the output from the antenna and convert it to a steady-state signal proportional to the parameter being measured, and a readout device. Narrowband instruments may use various antennas, such as rods (monopoles), loops, dipoles, biconical, conical log spiral antennas or aperture antennas such as pyramidal horns or parabolic reflectors. A knowledge of the gain, the antenna factor, or the effective area for a particular antenna provides a means for determining the appropriate field parameter from a measurement of voltage or power. Cable loss also should be taken into account. Tunable field strength meters and spectrum analyzers are appropriate narrowband instruments to use for measuring antenna terminal voltage or power at selected frequencies. Each has certain advantages and disadvantages.

Field Measurements

Before beginning a measurement survey it is important to characterize the exposure situation as much as possible. An attempt should be made to determine:

(1) The frequency and maximum power of the RF source(s) in question, as well as any nearby sources.

(2) Duty factor, if applicable, of the source(s).

(3) Areas that are accessible to either workers or the general public.

(4) The location of any nearby reflecting surfaces or conductive objects that could produce regions of field intensification ("hot spots").

(5) For pulsed sources, such as radar, the pulse width and repetition rate and the antenna scanning rate.

(6) If appropriate, antenna gain and vertical and horizontal radiation patterns.

(7) Type of modulation of the source(s).

(8) Polarization of the antenna(s).

(9) Whether measurements are to be made in the near-field, in close proximity to a leakage source, or under plane-wave conditions. The type of measurement needed can influence the type of survey probe, calibration conditions and techniques used.

If possible, one should estimate the maximum expected field levels, in order to facilitate the selection of an appropriate survey instrument. For safety purposes, the electric field (or the far-field equivalent power density derived from the E-field) should be measured first because the body absorbs more energy from the electric field, and it is potentially more hazardous. In many cases it may be best to begin by using a broadband instrument capable of accurately measuring the total field from all sources in all directions. If the total field does not exceed the relevant exposure guideline in accessible areas, and if the measurement technique employed is sufficiently accurate, such a determination would constitute a showing of compliance with that particular guideline, and further measurements would be unnecessary.

When using a broadband survey instrument, spatially-averaged exposure levels may be determined by slowly moving the probe while scanning over an area approximately equivalent to the vertical cross-section (projected area) of the human body. An average can be estimated by observing the meter reading during this scanning process or be read directly on those meters that provide spatial averaging. Spatially averaging exposure is discussed in more detail in the ANSI/IEEE and NCRP documents referenced above. A maximum field reading may also be desirable, and, if the instrument has a "peak hold" feature, can be obtained by observing the peak reading according to the instrument instructions. Otherwise, the maximum reading can be determined by simply recording the peak during the scanning process.

The term "hot spots" has been used to describe locations where peak readings occur. Often such readings are found near conductive objects, and the question arises as to whether it is valid to consider such measurements for compliance purposes. According to the ANSI C95.3 guidelines (Reference [2]) measurements of field strength to determine compliance are to be made, "at distances 20 cm or greater from any object." Therefore, as long as the 20 cm criterion is satisfied, such peak readings should be considered as indicative of the field *at that point.* However, as far as *average* exposure is concerned such localized readings may not be relevant if accessibility to the location is restricted or time spent at the location is limited (see Section 4 of this bulletin on controlling exposure). It should be noted that most broadband survey instruments already have a 5 cm separation built into the probe.

In many situations there may be several RF sources. For example, a broadcast antenna farm or multiple-use tower could have several types of RF sources including AM, FM, and TV, as well as CMRS and microwave antennas. Also, at rooftop sites many different types of CMRS antennas are commonly present. In such situations it is generally useful to use both broadband and narrowband instrumentation to fully characterize the electromagnetic environment. Broadband instrumentation could be used to determine what the overall field levels appeared to be, while narrowband instrumentation would be required to determine the relative contributions of each signal to the total field if the broadband measurements exceed the most restrictive portion of the applicable MPEs. The "shaped" probes mentioned earlier will also provide quantification of the total field in terms of percentage of the MPE limits.

In cases where personnel may have close access to intermittently active antennas, for example at rooftop locations, measurement surveys should attempt to minimize the uncertainty associated with the duty cycle of the various communications transmitters at the site to arrive at a conservative estimate of maximum possible exposure levels.

At broadcast sites it is important to determine whether stations have auxiliary, or stand-by, antennas at a site in addition to their main antennas. In such cases, either the main antenna or the auxiliary antenna, which may be mounted lower to the ground, may result in the highest RF field levels in accessible areas, and contributions from both must be properly evaluated.

At frequencies above about 300 MHz it is usually sufficient to measure only the electric field (E) or the mean-squared electric field. For frequencies equal to or less than 30 MHz, for example frequencies in the AM broadcast band, measurements for determining compliance with MPE limits require independent measurement of *both* E field and the magnetic field (H). For frequencies between 30 and 300 MHz it may be possible through analysis to show that measurement of only one of the two fields, not both, is sufficient for determining compliance. Further discussion of this topic can be found in Sections 4.3(2) and 6.6 of Reference [1]. At sites with higher frequency sources, such as UHF-TV stations, only E-field measurements should be attempted since the loop antennas used in H-field probes are subject to out-of-band resonances at these frequencies.

In many situations a relatively large sampling of data will be necessary to spatially resolve areas of field intensification that may be caused by reflection and multipath interference. Areas that are normally occupied by personnel or are accessible to the public should be examined in detail to determine exposure potential.

If narrowband instrumentation and a linear antenna are used, field intensities at three mutually orthogonal orientations of the antenna must be obtained at each measurement point. The values of E^2 or H^2 will then be equal to the sum of the squares of the corresponding, orthogonal field components.

If an aperture antenna is used, unless the test antenna responds uniformly to all polarizations in a plane, e.g., a conical log-spiral antenna, it should be rotated in both azimuth and elevation until a maximum is obtained. The antenna should then be rotated about its longitudinal axis and the measurement repeated so that both horizontally and vertically polarized field components are measured. It should be noted that when using aperture antennas in reflective or near-field environments, significant negative errors may be obtained.

When making measurements, procedures should be followed which minimize possible sources of error. For example, when the polarization of a field is known, all cables associated with the survey instrument should be held perpendicular to the electric field in order to minimize pickup. Ideally, non-conductive cable, e.g., optical fiber, should be used, since substantial error can be introduced by cable pick-up.

Interaction of the entire instrument (probe plus readout device) with the field can be a significant problem below approximately 10 MHz, and it may be desirable to use a self-contained meter or a fiber-optically coupled probe for measuring electric field at these frequencies. Also, at frequencies below about 1 MHz, the body of the person making the measurement may become part of the antenna, and error from probe/cable pickup and instrument/body interaction may be reduced by supporting the probe and electronics on a dielectric structure made of wood, styrofoam, etc. In all cases, it is desirable to remove all unnecessary personnel from an area where a survey is being conducted in order to minimize errors due to reflection and field perturbation.

In areas with relatively high fields, it is a good idea to occasionally hold the probe fixed and rotate the readout device and move the connecting cable while observing the meter reading. Alternatively, cover the entire sensor of the probe with metal foil and observe the meter reading. Any significant change usually indicates pickup in the leads and interference problems. When a field strength meter or spectrum analyzer is used in the above environments, the antenna cable should occasionally be removed and replaced with an impedance matched termination. Any reading on the device indicates pickup or interference.

As noted previously, substantial errors may be introduced due to zero drift. If a device is being used which requires zeroing, it should frequently be checked for drift. This should be done with the probe shielded with metal foil, with the probe removed from the field or, ideally, with the source(s) shut off.

With regard to compliance with the FCC's guidelines in mixed or broadband fields where several sources and frequencies are involved, the fraction or percentage of the recommended limit for power density (or square of the field strength) incurred within each frequency interval should be determined, and the sum of all contributions should not exceed 1.0 or 100% (see discussion of this topic in Section 1 of this bulletin). As mentioned before, probes with "shaped" responses may be useful in these environments.

Federal Communications Commission
Office of Engineering & Technology

Evaluating Compliance with FCC Guidelines for Human Exposure to Radiofrequency Electromagnetic Fields

Additional Information for Amateur Radio Stations

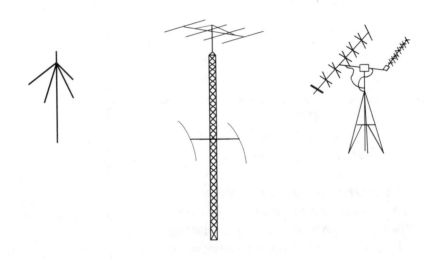

Supplement B
(Edition 97-01)
to
OET Bulletin 65 *(Edition 97-01)*

Evaluating Compliance with FCC Guidelines for Human Exposure to Radiofrequency Electromagnetic Fields

Additional Information for Amateur Radio Stations

SUPPLEMENT B
Edition 97-01
to
OET BULLETIN 65
Edition 97-01

November 1997

AUTHORS
Jerry L. Ulcek
Robert F. Cleveland, Jr.

Standards Development Branch
Allocations and Standards Division
Office of Engineering and Technology
Federal Communications Commission
Washington, D.C. 20554

IMPORTANT NOTE

This supplement is designed to be used in connection with the FCC's OET Bulletin 65, Version 97-01. The information in this supplement provides additional detailed information that can be used for evaluating compliance of amateur radio stations with FCC guidelines for exposure to radiofrequency electromagnetic fields. However, users of this supplement should also consult Bulletin 65 for complete information on FCC policies, guidelines and compliance-related issues. Definitions of terms used in this supplement are given in Bulletin 65. Bulletin 65 can be viewed and downloaded from the FCC's Office of Engineering and Technology's World Wide Web Internet Site: http://www.fcc.gov/oet/rfsafety.

ACKNOWLEDGMENTS

The following individuals and organizations reviewed an early draft of this supplement. Their valuable comments and suggestions greatly enhanced the accuracy and usefulness of this document, and their assistance is gratefully acknowledged.

American Radio Relay League (ARRL) HQ staff and outside advisors:

Paul Danzer, N1II, ARRL Assistant Technical Editor
Walter L Furr III, K4UNC
Robert E. Gold, M.D., WB0KIZ, ARRL RF Safety Committee
Gerald Griffin, M.D., K6NN, ARRL RF Safety Committee
Arthur W. Guy, Ph.D., W7PO, ARRL RF Safety Committee
Ed Hare, W1RFI, ARRL Laboratory Supervisor
John Hennessee, N1KB, ARRL Regulatory Information Specialist
Tom Hogerty, KC1J, ARRL Regulatory Information Branch Supervisor
Howard Huntington, K9KM, ARRL Central Division Vice Director
Wayne Irwin, W1KI, ARRL VEC Assistant to the Manager
Bart Jahnke, W9JJ, ARRL VEC Manager
Zack Lau, W1VT, ARRL Senior Laboratory Engineer
Roy Lewellan, P.E., W7EL, ARRL Technical Advisor
Jim Maxwell, Ph.D., W6CF, ARRL Pacific Division Vice Director
Gary E. Myers, K9CZB, ARRL RF Safety Committee
William J. Raskoff, M.D., K6SQL, Chairman, ARRL RF Safety Committee
Peter Richeson, KA5COI,
Kazimierz Siwiak, P.E, KE4PT, ARRL RF Safety Committee
R. Dean Straw, N6BV, ARRL Senior Assistant Technical Editor
Larry Wolfgang, WR1B, ARRL Senior Assistant Technical Editor

Jay Jackson, Wireless Telecommunications Bureau, FCC
John B. Johnston, Wireless Telecommunications Bureau, FCC
Fred O. Maia, W5YI Group, Inc., and W5YI-VEC, Inc.
Professor Wayne Overbeck, California State University, Fullerton
Edwin Mantiply, U.S. Environmental Protection Agency

i

ii

Introduction

In 1996, the FCC adopted new guidelines and procedures for evaluating human exposure to environmental radiofrequency (RF) electromagnetic fields from FCC-regulated transmitters. The new guidelines replaced those adopted by the FCC in 1985 (the 1982 RF protection guides of the American National Standards Institute, ANSI).[1] The FCC's guidelines are used for evaluating exposure from fixed station transmitters and from mobile and portable transmitting devices, such as cellular telephones and personal communications devices, in accordance with FCC responsibilities under the National Environmental Policy Act of 1969 (NEPA).[2] These rule changes set new limits on maximum permissible exposure (MPE) levels that apply to all transmitters and licensees regulated by the FCC.

The FCC also revised its policy regarding transmitters, facilities and operations for which routine evaluation for compliance is required before granting an application. A routine evaluation is a determination as to whether the station conforms to the RF exposure requirements. For amateur stations, the new policy requires that the station be subject to routine evaluation when it will be operated above certain power levels. In the past, although amateur stations were expected to comply with the FCC's guidelines, routine station evaluation was not required.

In August, 1997, the FCC issued a revised technical bulletin, OET Bulletin 65[3], that provides assistance and guidance to applicants and licensees in determining whether proposed or existing transmitting facilities, operations or devices comply with FCC-adopted limits for human exposure to RF fields. Although Bulletin 65 provides basic information concerning evaluation for compliance, it is recognized that additional specific guidance and information may be helpful for certain specialized categories of stations and transmitters such as radio and television stations and amateur stations. Therefore, supplements to Bulletin 65 have been prepared to provide this additional information. Supplement A was developed for radio and television broadcasting stations and this supplement (Supplement B) has been prepared for amateur stations. Users of this supplement are also strongly advised to consult Bulletin 65 itself for complete information and guidance related to RF guideline compliance. It should also be noted that, although Bulletin 65 and this supplement offer guidelines and suggestions for evaluating compliance, they are not intended to establish mandatory procedures, and other methods and procedures may be acceptable if based on sound engineering practice.

---

[1]   *See Report and Order*, ET Docket 93-62, FCC 96-326, adopted August 1, 1996, 61 *Federal Register* 41006, 11 FCC Rcd 15123 (1997). The FCC initiated this rule-making proceeding in 1993 in response to the 1992 revision by ANSI of its earlier guidelines for human exposure.

[2]   See 47 CFR § 1.1301, et seq.

[3]   To view and download OET 65, the website address is: http://www.fcc.gov/oet/

In general, the information contained in Bulletin 65 and in this supplement is intended to enable the applicant or amateur to make a reasonably quick determination as to whether a proposed or existing amateur station is in compliance with the exposure guidelines and if not, the steps that can be taken to bring it into compliance.[4] Bulletin 65 and this supplement include information on calculational methods, tables and figures that can be used in determining compliance. In addition, amateurs are encouraged to consult Section 4 of Bulletin 65 that deals with controlling exposure. In some cases, e.g., some multiple-emitter locations such as amateur repeater sites and multi-transmitter contest-style stations, measurements or a more detailed analysis may be required. In that regard, the part of Section 2 of Bulletin 65 dealing with multiple transmitter sites and, also, Section 3 of Bulletin 65 dealing with measurements and instrumentation provide basic information and references.

The new FCC limits for exposure incorporate two tiers of exposure limits based on whether exposure occurs in an occupational or "controlled" situation or whether the general population is exposed or exposure is in an "uncontrolled" situation. A detailed discussion of the guidelines and adopted limits are included in Bulletin 65.

As mentioned, in the FCC's recent *Report and Order,* certain amateur radio installations were made subject to routine evaluation for compliance with the FCC's RF exposure guidelines, effective January 1, 1997 (this date was later extended).[5] Section 97.13 of the Commission's Rules, 47 C.F.R. § 97.13, requires the licensee to take certain actions before causing or allowing an amateur station to transmit from any place where the operation of the station would cause human exposure to levels of RF fields that are in excess of the FCC guidelines. The licensee must perform the routine evaluation if the transmitter power of the station exceeds the levels specified in 47 CFR § 97.13(c)(1) and repeated in Table 1.[6] Amateurs may use the optional worksheet shown in Appendix B of this supplement to help in determining whether a routine evaluation is required.

All mobile amateur stations are categorically excluded from this requirement. Such mobile stations are presumed to be used only for very infrequent intermittent two-way operation. They are, however, required to comply with the exposure guidelines. Otherwise the operation is categorically excluded from routine RF radiation evaluation except as specified in Sections 1.1307(c) and (d) of the FCC's Rules.

---

[4] As is the case with all other FCC rules, an amateur station licensee or grantee is responsible for compliance with the FCC's rules for RF exposure.

[5] *See* para. 152 of Report and Order, ET Docket 93-62, (footnote 4). *See also*, 47 CFR 97.13, as amended. In the FCC's *First Memorandum Opinion and Order* in this docket, FCC 96-487, released December 24, 1996, the Commission extended the implementation date of the new guidelines for the amateur radio service to January 1, 1998. *See* 62 Federal Register 3232 (January 22, 1997).

[6] These levels were chosen to roughly parallel the frequency of the MPE limits of Table 1 in Appendix A of this supplement. These levels were modified from the Commission's original decision establishing a flat 50 W power threshold for routine evaluation of amateur stations (see *Second Memorandum Opinion and Order*, ET Docket 93-62, FCC 97-303, adopted August 25, 1997).

Table 1. Power Thresholds for Routine Evaluation of Amateur Radio Stations

Wavelength Band	Evaluation Required if Power* (watts) Exceeds:
MF	
160 m	500
HF	
80 m	500
75 m	500
40 m	500
30 m	425
20 m	225
17 m	125
15 m	100
12 m	75
10 m	50
VHF (all bands)	50
UHF	
70 cm	70
33 cm	150
23 cm	200
13 cm	250
SHF (all bands)	250
EHF (all bands)	250
Repeater stations (all bands)	non-building-mounted antennas: height above ground level to lowest point of antenna < 10 m <u>and</u> power > 500 W ERP building-mounted antennas: power > 500 W ERP

* Transmitter power = PEP input to antenna. For repeater stations *only,* power exclusion based on ERP (effective radiated power).

No station is exempt from *compliance* with the FCC's rules and with the MPE limits. However, many amateur stations are categorically exempt from the requirement to perform a *routine station evaluation* for compliance. Stations operating at or below the power levels given in Table 1, are not required by the FCC to perform a routine evaluation for compliance. Also, stations using mobile and portable (hand-held) transmitters (as defined by the FCC's rules) are not required to be routinely evaluated.[7] Amateur repeater stations transmitting with 500 W ERP or less whose antennas are not mounted on buildings, but rather on stand alone towers, and which are located at least 10 meters above ground are also categorically exempt from performing an evaluation. In the case of building-mounted repeater station antennas, the exemption applies regardless of height if the ERP is 500 W or less.

Many classes of amateur stations are categorically exempt from the need to do a station evaluation. This is because the circumstances under which exempt stations are usually operated are such that the station is presumed to be in compliance with the MPEs. Under some circumstances, such as an antenna that is located unusually near people such as an indoor antenna in a living space or a balcony mounted antenna a foot or so away from a neighbor's balcony, the FCC could require a station evaluation or take other action. FCC rule parts 1.1307 (c) and 1.1307 (d) could require that in cases where a station is categorically exempt, the FCC can require additional action, including a station evaluation, be taken by the station licensee if the FCC believes there is reason to believe that the exposure levels are being exceeded.

Although not required by the FCC's rules, it is advisable that mobile stations also be considered for potential exposure before an amateur automatically applies the categorical exemption. As an example, a 500-watt, 10-meter mobile installation with a vehicle mounted antenna would certainly merit a closer look. On VHF, the use of a high-power amplifier could also present problems in some cases. In general, it is recommended that in these higher powered installations, the antenna be located such that the vehicle occupants will be shielded from the antenna during normal use. One good location is in the center of an all-metal roof. Locations to be avoided for high-power operation would be a trunk-mounted antenna, or installation on a vehicle with a fiberglass roof. In general, mobile installations, even higher-powered ones, should not exceed the MPEs if sound installation guidelines are followed. The ARRL *Handbook* and ARRL antenna books, available from the ARRL, have additional material on mobile installations and antennas (see footnote 9).

---

[7] The FCC has defined "mobile" devices as those designed to be used in other than fixed locations and to be used in such a way that a separation distance of at least 20 cm is normally maintained between the transmitter's radiating structure(s) and the body of the user or nearby persons. The FCC defines "portable" devices as those designed to be used so that the radiating structure(s) of the device is/are within 20 cm of the body of the user. For example, this definition would apply to handheld cellular phones. Although amateur mobile and portable (handheld) PTT devices are categorically exempt from routine evaluation, users are cautioned to be aware that relatively high-powered mobile or portable devices can expose persons in their immediate vicinity to significant RF fields under conditions of relatively continuous transmission. An example might be a 100-110 W vehicle-mounted mobile antenna that is mounted in such a way (e.g., on a rear window) so that RF fields are created inside the vehicle. An example of this was noted in the FCC's measurement survey of typical amateur radio stations that is cited in Footnote 10.

Even if the regulations do not require an evaluation, there could be a number of reasons to conduct one anyway. At a minimum, such an evaluation would be good practice for the time when a station change is made that would require an evaluation. In addition, the results of an evaluation will certainly demonstrate to the amateur and his or her neighbors that the station's operation is well within the guidelines and is not a cause for concern. In the case of some of the unusual circumstances described earlier, the FCC's rules could require an evaluation of a station otherwise categorically exempt. In all cases, regardless of categorical exemption, the FCC's rules require compliance with the MPE limits. In most cases, the FCC will rely on amateurs to determine for themselves how the evaluation requirements apply to their stations, but under the rules, the FCC does have the flexibility to ask that an evaluation be performed on any transmitter regulated by the FCC.

The Commission's *Report and Order* instituted a requirement that amateur license examination question pools will include questions concerning RF environmental safety at amateur stations. Five questions on RF safety are required within each of the first three levels of written examination elements. Applicants for new amateur licenses must demonstrate their knowledge of the FCC Guidelines through the examinations prepared and administered by the volunteer examiners. The Commission also adopted the proposal of the American Radio Relay League (ARRL) that amateurs should be required to certify, as part of their license application process, that they have read and understand our bulletins and the relevant FCC rules. In addition, applicants for new, renewed and modified primary, club, military recreation and radio amateur civil emergency service (RACES) station licenses and applicants for a reciprocal permit for alien amateur licenses are also required to certify that they have read and understood the applicable rules regarding RF exposure.

When routine evaluation of an amateur station indicates that exposure to RF fields could be in excess of the limits specified by the FCC, the licensee must take action to correct the problem and ensure compliance (see Section 4 of OET Bulletin 65 on controlling exposure). Such actions could be in the form of modifying patterns of operation, relocating antennas, revising a station's technical parameters such as frequency, power or emission type or combinations of these and other remedies. For example, assume an amateur applicant or licensee determined that his or her station was in compliance at full power with all relevant FCC limits in all surrounding areas except for one corner of a neighboring property when a certain antenna was aimed in that direction. In such a case, one way of complying would be to simply avoid pointing the antenna in that direction when people are present at that location.

Amateur station licensees are also expected to follow a policy of systematic avoidance of excessive RF exposure. In its *Report and Order* the Commission said that it will continue to rely upon amateurs, in constructing and operating their stations, to take steps to ensure that their stations comply with the MPE limits for both occupational/controlled and general public/uncontrolled situations, as appropriate. In that regard, for a typical amateur station located at a residence, the amateur station licensee and members of his or her immediate household are considered to be in a "controlled environment" and as such are subject to the occupational/controlled MPE limits. All persons, with particular emphasis on neighbors, who are not members of an amateur station licensee's household are considered to be members of the general public, because they cannot reasonably be expected to exercise control over their exposure. In those cases, general population/uncontrolled exposure MPE limits apply. Similar considerations apply to amateur stations located at places other than a residence.[8]

To qualify for use of the occupational/controlled exposure criteria, appropriate restrictions on access to high RF field areas must be maintained and educational instruction in RF safety must be provided to individuals who are members of the amateur's household. Persons who are not members of the amateur's household but who are present temporarily on an amateur's property may also be considered to fall under the occupational/controlled designation provided that appropriate information is provided them about RF exposure potential if transmitters are in operation and such persons are exposed in excess of the general population/uncontrolled limits. As one example of educational materials, the 1998 *ARRL Handbook for Radio Amateurs* has a section on RF safety. The ARRL also publishes other materials on RF safety and RF exposure. Much of this material is available for viewing or downloading from the ARRL's World Wide Web site[9].

Amateur stations represent a unique case for determining exposure because there are many possible transmitting antenna types that could be designed and used for amateur service. However, several relevant points can be made with respect to analyzing amateur radio antennas for potential exposure that should be helpful to amateur licensees in performing evaluations.

First, the generic equations described in OET Bulletin 65 and in this supplement can be used for analyzing fields due to almost all antennas, although the resulting estimates for power density may be overly-conservative in some cases. Nonetheless, for general radiators and for aperture antennas, if the user is knowledgeable about antenna gain, frequency, power and other relevant factors, the equations in this section can be used to estimate field strength and power density as described earlier. In addition, other resources are available to amateurs for analyzing fields near their antennas. For example, as mentioned above, the ARRL provides excellent material available to help amateurs analyze their radio facilities for compliance with RF guidelines. Also, in 1996 the FCC released the final report of a 1990 study conducted by the FCC and the Environmental Protection Agency (EPA) of several amateur radio stations that provides a great deal of measurement data for many types of

---

[8] The definitions of these exposure criteria are discussed in more detail in OET Bulletin 65 and in the Commission's *Report and Order*.

[9] Contact: American Radio Relay League, Inc., QST Magazine, 225 Main St., Newington, CT 06111 Voice: 860-594-0200, FAX: 860-594-0294, Email: pubsales@arrl.org, Tech info: tis@arrl.org, Web Site: http://www.arrl.org/news/rfsafety/. The ARRL has developed the ARRL RF Exposure Package, and this material has been reproduced at the ARRL Web site. Paper copies are available from the ARRL Technical Information Service. In addition, recent articles have appeared in amateur publications that discuss amateur compliance with the FCC's RF rules. Two examples are: (1) *"The FCC's New RF-Exposure Regulations*, by Ed Hare, KA1CV, in *QST*, January 1997; and (2) *"Complying with the FCC's New RF Safety Rules*, by Wayne Overbeck, N6NB, in *CQ VHF*, January 1997. CQ Communications, Inc. 76 North Broadway, Hicksville, NY, 11801-2953. Tel: (516) 681-2922 FAX: (516) 681-2926 Email: CQVHF@aol.com; 72127.745@compuserve.com; cqcomm@delphi.com; Web Site: http://members.aol.com/cqvhf/ .

antennas commonly used by amateurs.[10] The FCC/EPA study concluded that, for most of the stations surveyed, RF protection guidelines were not exceeded in most accessible areas. However, the report also indicated that at higher power levels or with different facility configurations, higher exposure levels could not be completely ruled out.

This supplement contains information that should allow amateur licensees to predict RF field levels at their station site and determine distances that should be maintained from transmitting antennas in order to comply with the FCC's guidelines. The tables in this supplement represent the more commonly used types of amateur station antennas. For those types not covered by the tables, it may be necessary for the licensee to calculate the fields that are present by means of equations from this supplement, Bulletin 65, computer modelling, or direct measurements.[11] Material from the ARRL contains additional charts and tables developed with the same methods used to create the information in this supplement.

The FCC is relying on the demonstrated technical skills of amateurs to comply with these rules, select an evaluation method and to conduct their own station evaluations. The methods outlined in Bulletin 65 and this supplement can be used, but amateurs are free to select alternative methods as long as they are technically valid. If an amateur station is evaluated and found to be in compliance with the rules, no paperwork need be filed with the FCC, other than any required certifications as part of the Form 610 station application, and the station may be immediately put into operation.

Amateur radio organizations and licensees are encouraged to develop their own more detailed evaluation models and methods for typical antenna configurations and power/frequency combinations.[12] Such models and methods have been utilized in developing the material in this supplement. In addition, FCC staff will continue to work with the amateur radio community to assist licensees and applicants in evaluating compliance.

Information on RF safety issues is generally available at the FCC's World Wide Web Site. OET bulletins and supplements, such as this one, and other relevant FCC orders and documents can be downloaded from the specific web site for "RF safety." For example, information on the biological effects and potential hazards of RF radiation are discussed in an FCC publication (OET Bulletin 56), entitled "Questions and Answers about Biological Effects and Potential Hazards of Radiofrequency Radiation." This document can be downloaded from

---

[10] Federal Communications Commission (FCC), "Measurements of Environmental Electromagnetic Fields at Amateur Radio Stations," FCC Report No. FCC/OET ASD-9601, February 1996. FCC, Office of Engineering and Technology (OET), Washington, D.C. 20554. Copies can be ordered from the National Technical Information Service (NTIS), 1 800-553-6847 (Order No. PB96-145016), or the report can be downloaded from OET's Home Page on the World Wide Web at: http://www.fcc.gov/oet/.

[11] See Bulletin 65 for a discussion of measurement techniques and instrumentation.

[12] For example, a power density "calculator" has been developed by Kenneth Harker, KM5FA, and can be accessed at the following World Wide Web site: http://www.utexas.edu/students/utarc/. This program is based on a C version of a public domain BASIC program written by Prof. Wayne Overbeck that appeared in the January, 1997, issue of *CQ VHF*. The source code for this program may be downloaded at: ftp://members.aol.com/cqvhf/97issues/rfsafety.bas

the web site, or copies can be requested from the FCC's RF safety program. The FCC's home page address is: www.fcc.gov. The web site address for the RF safety program is: www.fcc.gov/oet/rfsafety. Information on RF safety issues can also be directed to the FCC's RF safety program at: (202) 418-2464 [FAX: (202) 418-1918] or by calling the FCC's toll-free number: 1 (888) CALL FCC [1 (888) 225-5322].

Section 1
What is Radiofrequency Radiation?

Radiofrequency (RF) energy is one type of electromagnetic energy. Electromagnetic waves and associated phenomena can be discussed in terms of energy, radiation or fields. Electromagnetic "radiation" can be defined as waves of electric and magnetic energy moving together (i.e., radiating) through space. These waves are generated by the movement of electrical charges. For example, the movement of charge in a radio station antenna (the alternating current) creates electromagnetic waves that radiate away from the antenna and can be intercepted by receiving antennas. Electromagnetic "field" refers to the electric and magnetic environment existing at some location due to a radiating source such as an antenna.

An electromagnetic wave is characterized by its wavelength and frequency. The wavelength is the distance covered by one complete wave cycle. The frequency is the number of waves passing a point in a second. For example, a typical radio wave transmitted by a 2-meter VHF station has a wavelength of about 2 meters and a frequency of about 145 million cycles per second (145 million hertz): one cycle/second = one hertz, abbreviated Hz.

Electromagnetic waves travel through space at the speed of light. Wavelength and frequency are inversely related by a simple equation: (frequency) times (wavelength) = the speed of light, or $f \ \text{x} \ \lambda = c$. Since the speed of light is a constant quantity, high-frequency electromagnetic waves have short wavelengths and low-frequency waves have long wavelengths. Frequency bands used for amateur radio transmissions are usually characterized by their approximate corresponding wavelengths, e.g., 12, 15, 17, 20 meters, etc.

The electromagnetic "spectrum" includes all of the various forms of electromagnetic energy ranging from extremely low frequency (ELF) energy (with very long wavelengths) to all the way up to X-rays and gamma rays which have very high frequencies and correspondingly short wavelengths. In between these extremes lie radio waves, microwaves, infrared radiation, visible light and ultraviolet radiation, respectively. The RF part of the electromagnetic spectrum can generally be defined as that part of the spectrum where electromagnetic waves have frequencies that range from about 3 kilohertz (kHz) to 300 gigahertz (GHz). Figure 1 illustrates the electromagnetic spectrum and the approximate relationship between the various forms of electromagnetic energy. Further information on RF

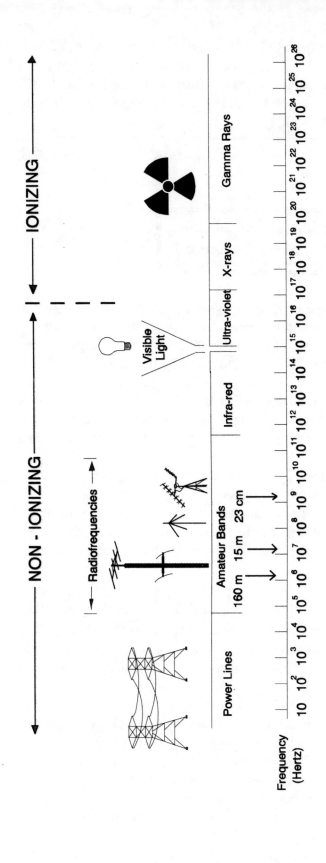

Figure 1. The Electromagnetic Spectrum

electromagnetic field exposure and potential biological effects can be found in the FCC's OET Bulletin 56.[13]

Section 2
FCC Exposure Guidelines and Their Application

The FCC's guidelines for Maximum Permissible Exposure (MPE) are defined in terms of power density (units of milliwatts per centimeter squared: mW/cm^2), electric field strength (units of volts per meter: V/m) and magnetic field strength (units of amperes per meter: A/m). In the far-field, *in free space* of a transmitting antenna, where the electric field vector (E), the magnetic field vector (H), and the direction of propagation can be considered to be all mutually orthogonal ("plane-wave" conditions), these quantities are related by the following equation.[14]

$$S = \frac{E^2}{3770} = 37.7H^2 \qquad (1)$$

where: S = power density (mW/cm^2)
E = electric field strength (V/m)
H = magnetic field strength (A/m)

In the near-field of a transmitting antenna, the term "far-field equivalent" or "plane-wave equivalent" power density is often used to indicate a quantity calculated by using the near-field values of E^2 or H^2 as if they were obtained in the far-field. As indicated in Table 1 of Appendix A for near-field exposures the values of plane-wave equivalent power density are given in some cases for reference purposes only. These values are sometimes used as a convenient comparison with MPEs for higher frequencies and are displayed on some measuring instruments.

---

[13] *"Questions and Answers about Biological Effects and Potential Hazards of Radiofrequency Radiation,"* OET Bulletin No. 56, Third Edition, January 1989. This bulletin can be viewed and downloaded at the FCC's OET World Wide Web site: http://www.fcc.gov/oet/rfsafety. Also, note that this bulletin is being revised, and a new version should be available in early 1998.

[14] Note that this equation is written so that power density is expressed in units of mW/cm^2. The impedance of free space, 377 ohms, is used in deriving the equation.

Exposure Environments

The FCC guidelines incorporate two separate tiers of exposure limits that are dependent on the situation in which the exposure takes place and/or the status of the individuals who are subject to exposure. The decision as to which tier applies in a given situation should be based on the application of the following definitions.

Occupational/controlled exposure limits apply to situations in which persons are exposed as a consequence of their employment and in which those persons who are exposed have been made fully aware of the potential for exposure and can exercise control over their exposure. Occupational/controlled exposure limits also apply where exposure is of a transient nature as a result of incidental passage through a location where exposure levels may be above general population/uncontrolled limits (see below), as long as the exposed person has been made fully aware of the potential for exposure and can exercise control over his or her exposure by leaving the area or by some other appropriate means. As discussed previously, occupational/controlled exposure limits apply to amateur licensees and members of their immediate household (but not their neighbors - see below). In general, a controlled environment is one for which access is controlled or restricted. In the case of an amateur station, the licensee or grantee is the person responsible for controlling access and providing the necessary information and training as described above.

General population/uncontrolled exposure limits apply to situations in which the general public may be exposed or in which persons who are exposed as a consequence of their employment may not be made fully aware of the potential for exposure or cannot exercise control over their exposure. Therefore, members of the general public always fall under this category when exposure is not employment-related, as in the case of residents in an area near a broadcast tower. Neighbors of amateurs and other non-household members would normally be subject to the general population/uncontrolled exposure limits.

For purposes of applying these definitions, awareness of the potential for RF exposure in a controlled or similar environment can be provided through specific training. Warning signs and labels can also be used to establish such awareness as long as they provide information, in a prominent manner, on risk of potential exposure and instructions on methods to minimize such exposure risk.[15]

Time and Spatial Averaging

A fundamental aspect of the exposure guidelines is that they apply to power densities or the squares of the electric and magnetic field strengths that are spatially averaged over the body dimensions. Spatially averaged RF field levels most accurately relate to estimating the whole-body averaged specific absorption rate (SAR) that will result from the exposure and the

---

[15] For example, a sign warning of RF exposure risk and indicating that individuals should not remain in the area for more than a certain period of time could be acceptable. Bulletin 65 provides more information on warning signs.

MPEs specified in Table 1 of Appendix A are based on this concept. This means that local values of exposures that exceed the stated MPEs do not imply non-compliance if the spatial average of RF fields over the body does not exceed the MPEs. Further discussion of spatial averaging as it relates to field measurements can be found in Section 3 of Bulletin 65 and in the ANSI/IEEE and NCRP reference documents noted there.

Another feature of the exposure guidelines is that exposures, in terms of power density, E^2 or H^2, may be averaged over certain periods of time with the average not to exceed the limit for continuous exposure. As shown in Table 1 of Appendix A, the averaging time for occupational/controlled exposures is 6 minutes, while the averaging time for general population/uncontrolled exposures is 30 minutes. It is important to note that for general population/uncontrolled exposures it is usually not possible or practical to control access or otherwise limit exposure duration to the extent that averaging times can be applied. In those situations, it would normally be necessary to assume continuous exposure to RF fields that would be created by the on/off cycles of the radiating source.

As an illustration of the application of time-averaging to occupational/controlled exposure (such as would occur at an amateur station) consider the following. The relevant interval for time-averaging for occupational/controlled exposures is six minutes. This means, for example, that during any given six-minute period an amateur or a worker could be exposed to two times the applicable power density limit for three minutes as long as he or she were not exposed at all for the preceding or following three minutes. Similarly, a worker could be exposed at three times the limit for two minutes as long as no exposure occurs during the preceding or subsequent four minutes, and so forth.

This concept can be generalized by considering Equation (2) that allows calculation of the allowable time(s) for exposure at [a] given power density level(s) during the appropriate time-averaging interval to meet the exposure criteria of Table 1 of Appendix A. The sum of the products of the exposure levels and the allowed times for exposure must equal the product of the appropriate MPE limit and the appropriate time-averaging interval.

$$\sum S_{exp} t_{exp} = S_{limit} t_{avg} \qquad (2)$$

where: S_{exp} = power density level of exposure (mW/cm^2)
S_{limit} = appropriate power density MPE limit (mW/cm^2)
t_{exp} = allowable time of exposure for S_{exp}
t_{avg} = appropriate MPE averaging time

For the example given above, if the MPE limit is 1 mW/cm^2, then the right-hand side of the equation becomes 6 mW-min/cm^2 (1 mW/cm^2 X 6 min). Therefore, if an exposure level is determined to be 2 mW/cm^2, the allowed time for exposure at this level during any six-minute interval would be a total of 3 minutes, since the left side of the equation must equal 6 (2 mW/cm^2 X 3 min). Of course, many other combinations of exposure levels and times may be involved during a given time-averaging interval. However, as long as the sum of the products on the left side of the equation equals the right side, the *average* exposure will meet the MPE limit. It is very important to remember that time-averaging applies to *any* t_{avg}. Therefore, in the above example, consideration would have to be given to the exposure situation both before and after the allowed three-minute exposure. The time-averaging interval can be viewed as a "sliding" period of time, six minutes in this case.

Another important point to remember concerning the FCC's exposure guidelines is that they constitute *exposure* limits (not *emission* limits), and they are relevant only to locations that are *accessible* to workers (or members of an amateur's household) or members of the public. Such access can be restricted or controlled by appropriate means such as the use of fences, warning signs, etc., as noted above. For the case of occupational/controlled exposure, procedures can be instituted for working in the vicinity of RF sources that will prevent exposures in excess of the guidelines. An example of such procedures would be restricting the time an individual could be near an RF source or requiring that work on or near such sources be performed while the transmitter is turned off or while power is appropriately reduced. Section 4 of Bulletin 65 should be consulted for further information on controlling exposure to comply with the FCC guidelines.

The concept of power averaging includes both on and off times, and the "duty factor" of the transmitting mode being used. Various modes of operation have their own duty factor that is representative of the ratio between average and peak power. Table 2 shows the duty factors for several modes commonly in use by amateurs. To obtain an easy estimate of average power, multiply the transmitter peak envelope power by the duty factor, then multiply that result by the **worst-case** percentage of time the station would be on the air in, e.g., a 6-minute period (the averaging time for controlled exposure) or a 30-minute period (the averaging time for uncontrolled exposure). This is an example of "source-based" time averaging.

For example, if a 1,500-watt PEP amateur single-sideband station (with no speech processing) transmits ("worst case") two minutes on, two minutes off then two minutes on again in any six-minute period (the averaging time period for controlled exposure), then for controlled exposure situations the effective power would be:

1,500 W X 0.2 (20% from Table 2) X ⅔ (4 of 6 minutes) = 200 W

For uncontrolled exposures the averaging time is 30 minutes and the total transmission time during any 30-minute period would be 20 minutes out of 30. The result would then also be:

1,500 W X 0.2 X ⅔ (20 of 30 minutes) = 200 W

On the other hand, if the transmission cycle were, say, 7 minutes on, 7 minutes off, the average power would be higher, since there would be continuous exposure over a six-minute period (controlled and uncontrolled time-averaging periods specify **any** six or thirty minute period, respectively). In this case the average power (for controlled exposure) becomes:

1,500 W X 0.2 X 1.0 (6 of 6 minutes) = 300 W

For uncontrolled/general population exposure average power becomes:

1,500 W   X 0.2 X .53 (16 of 30 minutes) = 159 W

Another example might be a 500-watt CW station that is used in a DX pileup, transmitting 15 seconds every two minutes (45 seconds for six minutes) the result would be the same for either controlled or uncontrolled exposure:

500 W X 0.4 (40% from Table 2) X 0.125 (45 of 360 seconds) = 25 W (controlled)

500 W X 0.4 X 0.125 (225 of 1800 seconds) = 25 W (uncontrolled)

For the case of a 250-watt FM base station used to talk for 5 minutes on, 5 minutes off, 5 minutes on (worst case) calculated power becomes (since worst case is 5 minutes transmission during any six-minute period or 15 minutes during any 30-minute period):

250 W X 1.0 (100% from Table 2) X 0.833 (5 of 6 minutes) = 208.3 W (controlled)

250 W X 1 X 0.5 (15 out of 30 minutes) = 125 W (uncontrolled)

Table 2. Duty Factor of Modes Commonly Used by Amateurs

Mode	Duty Factor	Notes
Conversational SSB	20%	Note 1
Conversational SSB	50%	Note 2
Voice FM	100%	
FSK or RTTY	100%	
AFSK SSB	100%	
Conversational CW	40%	
Carrier	100%	Note 3

Note 1: Includes voice characteristics and syllabic duty factor. No speech processing.
Note 2: Includes voice characteristics and syllabic duty factor. Heavy speech processor employed.
Note 3: A full carrier is commonly used for tune-up purposes

Section 3
Methods of Predicting Human Exposure

Amateurs can select from a number of technically valid methods that can be useful in performing the required station evaluations. In general, it will be appropriate to use one of the following methods:

o *Estimated compliance distances using tables developed from field-strength equations*
o *Estimated compliance distances using tables derived from antenna modeling*
o *Estimated compliance distances using antenna modeling (NEC, MININEC, etc.)*
o *Estimated compliance distances using field-strength equations*
o *Estimated compliance distances using software developed from field-strength equations*
o *Estimated compliance distances using calibrated field-strength measurements*

In addition, methods for controlling exposure outlined in this supplement and in Section 4 of OET Bulletin 65 should be consulted for information on various means of ensuring compliance.

Tables Using Far-Field Formulas

Most amateurs will use various tables to estimate compliance distances for MPE limits. The simplest of these tables was developed using a far-field equation and assuming ground reflection of electromagnetic waves from the RF source. This model, although simplified, has been verified to be a reasonable approximation against a number of dipole, ground-plane and Yagi antennas, based on computer modeling (see later discussion) carried out by the ARRL. The ARRL reports that this model does not necessarily apply to all antenna types. Computer models of small HF loops, for example, yield RF fields very near the antenna that are much higher than the far-field formula predicts. In most cases, however, the tables derived from this far-field approximation give conservative results that over-predict exposure levels. Tables 4 a. and 4 b. are probably the easiest of the tables to use. They are followed by a number of tables based on specific antenna types.

The first step an amateur should take is to select the simple table that best applies to your station and determine the estimated compliance distance(s) for the relevant operating band(s). If a compliance distance is less than the actual distance to an exposure location, the station "passes" and the evaluation is complete. It can be that simple. Remember that these distances are for the absolute distance from the antenna at any angle. Remember also, that the FCC's limits are *exposure* limits, not *emission* limits. Therefore, if high RF levels are present at a given location, but no one will be exposed at that location, this does not mean the station is out of compliance.

Tables Derived from NEC Modeling

In many cases, actual exposure below an antenna can be significantly less than that indicated by the tables based on far-field considerations. If a station "passes" using the far-field tables, this could be a moot point, although some licensees may still need to demonstrate actual predicted exposure levels. There are no easy answers to actual near-field predictions for actual antennas over real ground. The ARRL has, however, used Numeric Electromagnetic Code (NEC4) antenna modeling software to predict fields from a number of actual antennas and ground conditions. The results are summarized in a series of tables. These tables are located in Section 4 of this supplement, beginning with Table 18. Amateurs who desire a more accurate estimate of the RF fields expected near their antennas are encouraged to refer to this section. In many cases, a station that may not pass based on "worst-case" predictions could easily be shown to be in compliance using these tables. The ARRL tables offered in this supplement are only a few examples of a large number of tables prepared by that organization using this method.

Antenna Modeling

The same methods used to derive the NEC-modeled tables can be applied to any antenna situation. Amateurs are known to use many unique and varied types of antennas, and it is not possible to develop tables for every possible antenna type or combination. Some amateurs may want to evaluate the effect of multiple antennas or other conductors in proximity to their antennas in order to have a more accurate estimate of exposure than could be obtained from other calculational methods. For example, many amateurs may wish to use antenna-modeling software for this purpose.

Many antenna-modeling programs are based on NEC or MININEC analysis. These programs often yield very accurate results. An amateur enters his or her antenna dimensions and ground characteristics into the antenna model, and the program is then executed to calculate electric and magnetic field strengths near the antenna. These programs do require some amount of user skill, but the average amateur should not experience too much difficulty in using them. The ARRL Web page maintains a list of software vendors who sell antenna modeling software (http://www.arrl.org/news/rfsafety).[16]

---

[16] Note: Brian Beezley, K6STI, has made a scaled-down version of his Antenna Optimizer software available. Download NF.ZIP at: http://oak.oakland.edu:8080/pub/hamradio/arrl/bbs/programs/. Contact: Brian Beezley, K6STI, 3532 Linda Vista Drive, San Marcos, CA 92069, Phone: 619-599-4962, Email: k6sti@n2.net. Also, The equations used for the simple, far-field tables in this supplement have been used to develop a program written in Basic by Wayne Overbeck, N6NB. This program has been made available for downloading from ftp://members.aol.com/cqvhf/97issues/rfsafety.bas. It has also been rewritten into a power density calculator by Ken Harker, KM5FA, and can be accessed at http://www.utexas.edu/students/utarc/. Roy Lewellan, W7EL, sells ELNEC and EZNEC antenna modeling software, based on MININEC or NEC2. ELNEC is based on MININEC, but does not have near-field capability. EZNEC is based on NEC2 and can be used to predict the near field strength. This software is available from W7EL Software, PO Box 6658, Beaverton, OR 97007, Phone: 503-646-2885, Fax: 503-671-9046, Email: w7el@teleport.com, Web Site: ftp://ftp.teleport.com/vendors/w7el/

Prediction Methods and Derivation of Tables

The tables, figures and graphs provided in this supplement should allow most amateur station licensees and applicants to easily determine the steps necessary to ensure that their stations will comply with the FCC's guidelines. By using the appropriate table or figure for a given antenna type, the station licensee should be able to obtain the necessary compliance information. As an example, to ensure compliance for a station using a certain antenna type and transmitter power level, the minimum separation distance between a person and an antenna is given in the appropriate table. Since continuous exposure is assumed for convenience, and because time-averaging of exposure is allowed, these distances will be conservative (most amateur station transmissions are two-way and thus not continuous for significant periods of time).

The tables and figures are based both on far-field equations (see Bulletin 65) and also on data obtained from computer programs such as the Numeric Electromagnetic Code (NEC) and MININEC. Much of this information has been provided by individual amateur licensees and amateur radio organizations. When this is the case, the source of the table or information has been provided. The tables are provided for a sample of the most commonly-used amateur station antennas. For other antennas or system configurations, amateurs may have to perform their own calculations or other evaluation based on the information in Bulletin 65.

As discussed in Bulletin 65, calculations can be made to predict RF field strength and power density levels around typical RF sources. For example, in the case of a non-directional antenna, a prediction for power density in the far-field of the antenna can be made by use of the general Equations (3) or (4) below [for conversion to electric or magnetic field strength see Equation (1) above]. These equations are generally accurate in the far-field of an antenna but will over-predict power density in the near field, where it could be used for making a "worst case" or conservative prediction.

$$S = \frac{PG}{4\pi R^2} \qquad (3)$$

where: S = power density (in appropriate units, e.g. mW/cm^2)
P = power input to the antenna (in appropriate units, e.g., mW)
G = power gain of the antenna in the direction of interest relative to an isotropic radiator (dBi)
R = distance to the center of radiation of the antenna (appropriate units, e.g., cm)

or:

$$S = \frac{EIRP}{4\pi R^2} \qquad (4)$$

where: EIRP = equivalent (or effective) isotropically radiated power

When using these and other equations care must be taken to use the **correct units** for all variables. For example, in Equation (3), if power density in units of mW/cm^2 is desired then power should be expressed in milliwatts and distance in cm. Other units may be used, but care must be taken to use correct conversion factors when necessary. Also, it is important to note that the power gain factor, **G**, in Equation (3) is normally **numeric** gain. Therefore, when power gain is expressed in logarithmic terms, i.e., dB, a conversion is required using the relation:

$$G = 10^{\frac{dB}{10}}$$

For example, a logarithmic power gain of 14 dB is equal to a numeric gain of 25.1. Table 3 gives factors that can be used for converting logarithmic and numerical gain.

Table 3. Gain Conversion

Gain (dBi)	Numeric Gain	Gain (dBi)	Numeric Gain
1	1.3	11	12.6
2	1.6	12	15.9
3	2.0	13	20.0
4	2.5	14	25.1
5	3.2	15	31.6
6	4.0	16	39.8
7	5.0	18	63.1
8	6.3	20	100.0
9	7.9	25	316.2
10	10.0	30	1000.0

In many cases, operating power may be expressed in terms of "effective radiated power" or "ERP" instead of EIRP. ERP is referenced to a half-wave dipole radiator instead of an isotropic radiator. Therefore, if ERP is given it is necessary to convert ERP into EIRP in order to use the above equations. This is easily done by multiplying the ERP by the factor of 1.64, the gain of a half-wave dipole relative to an isotropic radiator. Conversely, divide

EIRP by 1.64 to obtain ERP. For example, if ERP is used in Equation (4) the relation becomes:

$$S = \frac{EIRP}{4\pi R^2} = \frac{1.64 \; ERP}{4\pi R^2} = \frac{0.41 \; ERP}{\pi R^2} \tag{5}$$

For a truly worst-case prediction of power density at or near a surface, such as at ground-level or on a rooftop, 100% reflection of incoming radiation could be assumed, resulting in a potential doubling of predicted field strength and a four-fold increase in (far-field equivalent) power density. In that case Equations (3) and (4) can be modified as follows to:

$$S = \frac{(2)^2 PG}{4\pi R^2} = \frac{PG}{\pi R^2} = \frac{EIRP}{\pi R^2} \tag{6}$$

As discussed in Bulletin 65, for the case of FM radio and television broadcast antennas, the U.S. Environmental Protection Agency (EPA) developed models for predicting ground-level field strength and power density.   The EPA model recommended a more realistic approximation for ground reflection by assuming a maximum 1.6-fold increase in field strength leading to an increase in power density of 2.56 (1.6 X 1.6). Equation (4) then becomes:

$$S = \frac{2.56 \; EIRP}{4\pi R^2} = \frac{0.64 \; EIRP}{\pi R^2} \tag{7}$$

If ERP is used in Equation (7), the relation becomes:

$$S = \frac{0.64 \; EIRP}{\pi R^2} = \frac{(0.64)(1.64) \; ERP}{\pi R^2} = \frac{1.05 \; ERP}{\pi R^2} \tag{8}$$

It is often convenient to use units of microwatts per centimeter squared (μW/cm^2) instead of mW/cm^2 in describing power density. The following simpler form of Equation (8) can be derived if power density, **S**, is to be expressed in units of μW/cm^2:

$$S = \frac{33.4 \; ERP}{R^2} \tag{9}$$

where: S = power density in μW/cm^2
        ERP = power in watts
        R = distance in meters

An example of the use of the above equations follows. A repeater station is transmitting at a frequency of 146.94 MHz with a total nominal ERP (including all polarizations) of 1 kilowatt (1,000 watts) from a tower-mounted antenna. The height to the center of radiation is 10 meters above ground-level. Using the formulas above, what would be the calculated "worst-case" power density that could be expected at a point 2 meters above ground (approximate head level) and at a distance of 20 meters from the base of the tower (e.g., at a neighbor's property line where the more restrictive general population exposure limits would apply)? Note that this type of analysis *does not* take into account the specific radiation pattern of the antenna, i.e., no information on directionality of propagation is considered. Use of actual radiation pattern data would likely significantly reduce actual ground-level exposures from those calculated below, but often this is unnecessary when using "worst case" approximations or for amateur stations where operating powers may not be that high (see Bulletin 65 for further discussion)

From simple trigonometry the distance **R** can be calculated to be about 21.5 meters [square root of: $(8)^2 + (20)^2$]. Therefore, using Equation (9), the calculated power density is:

$$S = \frac{33.4 \; (1,000 \; watts)}{(21.5 \; m)^2} = about \; 72 \; \mu W/cm^2$$

By consulting Table 1 in Appendix A, it can be determined that the limit for general population/uncontrolled exposure at 146.94 MHz is 0.2 mW/cm^2 (200 μW/cm^2). Therefore, this calculation shows that even under "worst-case" conditions this station would easily comply with the general population/uncontrolled limits at the neighbor's property line. Similar calculations could be made to ensure compliance at other locations, such as at the base of the tower where the shortest direct line distance, R, to the ground would occur and where worst-case exposure of the amateur's household members might occur.

Measurements

The equations and calculational methods described here and in OET Bulletin 65 have been used to develop the tables, figures and graphs in this supplement. In addition, direct measurement of RF fields can be performed, and this topic is also discussed in Bulletin 65. Bulletin 65 includes an extensive section on the topic of performing measurements of RF field strength and power density. However, in general, most amateurs will not have access to the appropriate calibrated equipment to make such measurements. The field-strength meters in common use by amateurs operators and inexpensive hand-held field strength meters do not provide the accuracy necessary for reliable measurements, especially when different frequencies may be encountered at a given measurement location. As discussed in Bulletin 65, repeatability and accuracy of more than a few dB is often difficult to achieve even with the best available instrumentation and expertise.

Section 4
Estimated Compliance Distances from Typical Transmitting Antennas

---

Tables Based on Far-Field Equations

The following tables are based on use of the far-field equations for power density given above (Equations 3 and following) assuming the reflection factor used by the EPA. These tables represent "worst case" estimates of the far-field equivalent power density. These tables should be used unless the exposure situation of interest is in the main beam or lobe of the antenna being considered. In the latter case, surface reflection would not necessarily be of major concern.

Tables 4-17 are not height-specific. To use these tables it is necessary to match the characteristics of the antenna in question as closely as possible to those of the appropriate table and locate the distance to the appropriate environment boundary. For example, for a 500 watt, 21 MHz, horizontal, half-wave dipole antenna refer to Table 7. In order to comply with the occupational/controlled environment, a line-of-sight distance of 2.8 meters would have to be maintained from the antenna for conditions of continuous transmission. The distance required to comply with the limit for the general public/uncontrolled exposure criteria distance would be a minimum of 6.3 meters for continuous transmission. If the antenna in question is operated with a power level in between two of the levels given in a table it is possible to interpolate the distance given between the actual power level and the next highest power level.

For example, consider the following situation. Using Table 5 it is desired to find the distance necessary to comply with the occupational/controlled limit for a three-element, tri-band Yagi antenna transmitting at approximately 14 MHz with 700 watts of power (peak envelope power or PEP). Since a specific entry for 700 watts is not given in Table 5, the appropriate distance must be determined from those given. There are two ways to do this. The first and simplest approach is to simply use the distance corresponding to the next highest power level. This approach will lead to a more conservative distance, but may not be a problem if the actual separation distance is more anyway. For this approach, using the entry for 1,000 watts results in a distance requirement of 4.5 meters in order to be assured of meeting the controlled/occupational criteria.

The second approach for this case is to interpolate between the entries for 500 watts and 1,000 watts, respectively. This requires solving for the value x in the following relation and adding the value obtained for x to the distance requirement for 500 watts (3.1 m). To set up this calculation, 200 watts is obtained from 700-500 watts, 500 watts is obtained from 1000 - 500 watts and 1.4 is obtained from 4.5 - 3.1.

Solving for x

$$\frac{200 \ watts}{500 \ watts} = \frac{x}{1.4}$$

Solving for x yields approximately 0.6. Adding this to 3.1 results in a value of about 3.7 m. Therefore, using this method, at 700 watts and 14 MHz, the required "worst-case" distance is determined to be about 3.7 meters from the antenna in order to comply with the occupational/controlled limit.

NOTE: Some of the tables in this section use the following abbreviations:

✓ **con = occupational/controlled exposure limit(s)**
✓ **unc = general population/uncontrolled exposure limit(s)**
✓ **f = transmitter frequency**
✓ **HAG = antenna height above ground level**
✓ **m = meter(s)**

Examples Using Models

The following two examples illustrate how tables such as Tables 24 and 26 can be used.

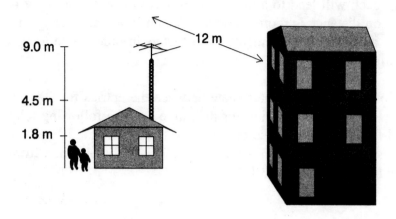

FIGURE 2. Illustration of use of Table 24.

In Figure 2, an amateur station located at a residence is transmitting using a three-element Yagi antenna (20 meter/14.35 MHz) that is located approximately 9 m above ground level. Maximum *average* operating power is 1,000 watts. From Table 24 it is apparent that a person standing at ground level (taken as the 1.8 meters level based on a person's height) would always be exposed below the guidelines, regardless of whether they are considered under the occupational/controlled or the general population/uncontrolled tiers of exposure limits. If only single story residences were located near this amateur station then the station would be assumed to be in compliance with FCC exposure guidelines. However, in the case shown in Figure 2 a three-story apartment building is located adjacent to the amateur station. People living in this building would have to be considered under the general population/uncontrolled exposure guidelines. Since the antenna is the same height (9 meters) as the third story of this building, the amateur would have to ensure that the transmitting antenna is at least 8.8 meters from the apartment building. Since the actual distance in this case is 12 meters, the amateur station can be assumed to be in compliance. However, if the distance were not at least 8.8 meters, the amateur station may not comply but there would still be several options for actions that could ensure compliance. These include (but are not necessarily limited to) raising the center of radiation of the antenna to an appropriate height above the apartment building, moving the antenna to the other side of his property, or possibly incorporating duty cycle considerations into determining exposure levels.

9.1 m

4.6 m
3.7 m

1.8 m

FIGURE 3. Illustration of use of Table 26.

In Figure 3, an amateur station is using a 40 meter/ 7.3 MHz horizontal half-wave dipole antenna that extends from outside a second floor window to a nearby tree. The antenna is approximately 4.6 meters off the ground, and average transmitter power is 1,500 watts. From Table 26 the station would be in compliance with FCC RF guidelines if the amateur or members of his/her immediate household (occupational/controlled exposure) remained directly below the antenna (see 1.8 m column in the table).   However, in this example, a household member on the second floor of the house would have to maintain a minimum distance of 2.7 m from the antenna (see 3.7 m and 4.6 m columns for occupational/controlled exposure) in order to ensure compliance.   Note also that from Table 26 compliance distances required for a height of 1.8 m are 2 m (general population/uncontrolled). Neighbors of the amateur or persons who do not fit the category of occupational/controlled must stay at least 2 m from the antenna, while at ground level, in order to ensure compliance for continuous exposure. Since the antenna is approximately 4.6 m. off the ground, a person of around 1.8 m. tall would be 2.8 meters from the antenna while they were standing at ground level. Therefore, this station would be in compliance with uncontrolled limits using the parameters listed above

Also, for the case shown in Figure 3, the amateur station is using a ten-meter, three-element Yagi antenna mounted on the roof of the house that is operated with 100 watts of average power. This power level was chosen because the second floor of the house is located between 3.7-6.1 meters above ground (see Table 19).   Since the antenna is mounted approximately 9 meters above ground level, the amateur has decided to operate without any duty factor or time-averaging restrictions that might be necessary if higher power levels were used. As shown in Table 19, the station would be in compliance with the RF guidelines for both occupational/controlled and general population/uncontrolled categories for ground-level and 2nd floor (3.7 and 6.1 m. heights) exposure. If the amateur in this case were to choose to transmit using both antennas simultaneously it would be necessary to consider the total contributions of both antennas to field strength or power density levels at possible exposure locations. This topic is discussed in detail in Bulletin 65, Section 2 (multiple transmitter environments).

Section 5
Controlling Exposure to RF Fields

The FCC's guidelines for exposure to radiofrequency electromagnetic fields incorporate two tiers of limits, one for "general population/uncontrolled" exposure and another for "occupational/controlled" exposure. Amateurs and members of their immediate household are considered by the FCC under the "occupational/controlled" exposure limits. Neighbors, guests, people walking by on the street, delivery people, maintenance people coming to work on the property where an amateur station is located, etc., are normally considered to fall under the "general population/uncontrolled" exposure category. However, under some conditions persons transient through the station property may be considered under the occupation/controlled criteria as discussed in Bulletin 65.

In order for an amateur to perform an evaluation of his or her station for RF compliance, the following questions should first be asked:

> *(1) Which category of exposure applies at the location(s) in question?*
> *(2) What type(s) of transmitting antenna is/are being used?*
> *(3) What transmitting power levels will be used?*
> *(4) How far is the area being evaluated from the antenna(s) in question?*

The tables in this supplement can then be used to help determine compliance with exposure guidelines. If this supplement does not contain a table that is relevant to the particular station parameters, Bulletin 65 should be consulted for alternative methods of determining compliance (e.g., calculations, measurements, etc.)

After an evaluation is performed, if a determination is made that a potential problem exists, Section 4 of Bulletin 65 should be consulted for a discussion of recommended methods for reducing or controlling exposure. Such methods could include one or more of the following:

1. *Restricting access to high RF-field areas*
2. *Operating at reduced power when people are present in high RF-field areas*
3. *Transmitting at times when people are not present in high RF-field areas*
3. *Considering duty factor of transmissions*
4. *Time-averaging exposure*
5. *Relocating antennas or raising antenna height*
6. *Incorporating shielding techniques*
7. *Using monitoring or protective devices*
8. *Erecting warning/notification signage*

Limiting access may be the easiest method to reduce exposure. If an antenna is in an area where access is generally restricted (such as a fenced-in yard) it may be sufficient to simply control access to the yard when transmissions are in progress (assuming exposure levels exceed the guidelines in the yard). An antenna could also be placed high enough on a tower or mast so that access to high RF levels is generally impossible.

Reducing transmitting power can also significantly reduce exposure levels. The power output of a transmitter has a linear relationship with the power density exposure level that could be experienced by a person near the transmitting antenna. For example, if power output is reduced by 20% then power density at a given location will also be reduced by 20%.

An often overlooked method of reducing exposure is by utilizing the inherent duty factor of the transmissions from an amateur station. The worst-case duty factor, 100%, occurs during continuous or "key down" transmissions. However, most amateur service two-way transmissions are more likely to be of the "key on, key off" type, resulting in more typical duty factors of, say, 50%.

Consider the following example. An amateur station transmits on-off keyed telegraphy emission type A1A on 28.05 MHz (10 Meter band). The station antenna is a half-wave dipole mounted outside a nearby window (about 9 meters above ground). The station transmits with 1,500 watts PEP. The amateur needs to know how far he or she should be from the antenna to comply with the RF safety guidelines. Table 18 indicates that there must be a distance of about 4.3 meters from the antenna during transmissions. However, the antenna is located closer than 4.3 m to the control point. Assume that with an emission type A1A transmission, the antenna would be energized about 50% of the time. In this case, first consider exposure of the station operator. In such a case limits for the occupational/controlled criteria apply, and the averaging time for occupational/controlled exposure is 6 minutes (see last column in Table 1, Appendix A). This means that the station operator can be exposed at or below 100% of the limit indefinitely. However, exposure in excess of 100% of the limits is permissible if the **time-averaged** exposure (over 6 minutes) is 100% or less of the MPE limit. For example, should the operator only send telegraph approximately 3 minutes out of every 6 minutes, compliance could be based on a 50% duty factor (50% reduction in power level used in determining exposure potential). This means that the distance values in tables such as Table 18 can be reduced by the multiplication factors shown in Table 32. In this example, the distance requirement for compliance for continuous exposure can be calculated to be:

(0.71)(4.3 meters) = 3 meters

Table 32. Duty Factor Conversion

Duty factor percentage	Multiplication factor
75%	0.87
66%	0.82
50%	0.71
33%	0.58
25%	0.5
10%	0.32

Conclusion

As of January 1, 1998, amateur licensees and grantees will be expected to routinely evaluate their stations for potential human exposure to RF fields that may exceed the FCC-adopted limits for maximum permissible exposure (MPE). If such an evaluation shows that potential exposure will exceed the MPE limits, the amateur licensee must take appropriate corrective action to bring the station into compliance before transmission occurs (see 47 CFR § 97.13(c), as amended.

The Commission has always relied on the skills and demonstrated abilities of amateurs to comply with its technical rules, and it will continue to do so. The Commission believes that amateur licensees and applicants should be sufficiently qualified to conduct their own evaluations and act accordingly. In OET Bulletin 65 and in this supplement we attempt to provide the amateur community with as much information as possible to accomplish these tasks. In addition, Commission staff will continue to be available to answer questions and provide further information if requested. The Commission will also continue to work with amateur organizations such as the ARRL to improve the usefulness, accuracy and inclusiveness of this supplement.

Future editions of this supplement (as well as of Bulletin 65) may be issued as needed to update the data and information provided here or to make any major corrections that may be necessary. In that regard, the Commission invites amateurs to provide input to FCC staff relating to evaluating RF exposure and the contents of the Bulletin 65 and its supplements. We also encourage the amateur community to continue its activities in developing its own methods and information for performing RF environmental evaluations. We believe that these efforts will result in an improved and safe amateur service that will benefit both amateur licensees and those persons residing or working near amateur facilities.

Appendix A
Exposure Criteria Adopted by the FCC

Table 1 lists the exposure criteria adopted by the FCC for various transmitting frequencies and for the categories of "controlled/occupational" and "general population/uncontrolled" exposures. The limits are defined in terms of electric field strength, magnetic field strength and power density. Intervals for time averaging of exposures are also given. For further information and more detail consult OET Bulletin 65.

Table 1. *FCC Limits for Maximum Permissible Exposure (MPE)*

(A) Limits for Occupational/Controlled Exposure

Frequency Range (MHz)	Electric Field Strength (E) (V/m)	Magnetic Field Strength (H) (A/m)	Power Density (S) (mW/cm^2)	Averaging Time $|E|^2$, $|H|^2$ or S (minutes)
0.3-3.0	614	1.63	(100)*	6
3.0-30	1842/f	4.89/f	(900/f^2)*	6
30-300	61.4	0.163	1.0	6
300-1500	--	--	f/300	6
1500-100,000	--	--	5	6

(B) Limits for General Population/Uncontrolled Exposure

Frequency Range (MHz)	Electric Field Strength (E) (V/m)	Magnetic Field Strength (H) (A/m)	Power Density (S) (mW/cm^2)	Averaging Time $|E|^2$, $|H|^2$ or S (minutes)
0.3-1.34	614	1.63	(100)*	30
1.34-30	824/f	2.19/f	(180/f^2)*	30
30-300	27.5	0.073	0.2	30
300-1500	--	--	f/1500	30
1500-100,000	--	--	1.0	30

f = frequency in MHz   *Plane-wave equivalent power density

NOTE 1: *Occupational/controlled* limits apply in situations in which persons are exposed as a consequence of their employment provided those persons are fully aware of the potential for exposure and can exercise control over their exposure. Limits for occupational/controlled exposure also apply in situations when an individual is transient through a location where occupational/controlled limits apply provided he or she is made aware of the potential for exposure. These limits apply to amateur station licensees and members of their immediate household as discussed in the text.

NOTE 2: *General population/uncontrolled* exposures apply in situations in which the general public may be exposed, or in which persons that are exposed as a consequence of their employment may not be fully aware of the potential for exposure or can not exercise control over their exposure. As discussed in the text, these limits apply to neighbors living near amateur radio stations.

Appendix B
Optional Worksheet and Record of Compliance

This optional worksheet can be used to determine whether routine evaluation of an amateur station is required by the FCC's rules. It also can be used as an aid in determining compliance. However, use of this worksheet is not required by the FCC.

Optional Worksheet and Record of Compliance
with FCC Guidelines for Human Exposure
to Radiofrequency Electromagnetic Fields
for Amateur Radio Stations

Instructions

Introduction. This optional worksheet is intended to be helpful when determining whether any particular combination of transmitting apparatus at an amateur radio station ("a setup") is in compliance with the FCC rules (47 C.F.R. § 1.1310) concerning human exposure to radiofrequency (RF) electromagnetic fields.

The purpose of the first section of this worksheet is to help determine, for any given setup of the amateur station, whether the routine RF evaluation prescribed by FCC rules (47 C.F.R. § 1.1307(b)) must be performed before the setup can be used for transmitting. In the event that a routine RF evaluation must be performed, that requirement may be satisfied by completing the second section of the worksheet, by using methods outlined in OET Bulletin 65 or by employing another technically valid method.

The person responsible for making the determination is the person named on the data base license grant as the primary station licensee or as the club, military recreation or RACES station license trustee, and any alien whose amateur radio station is transmitting from a place where the service is regulated by the FCC under the authority derived from a reciprocal arrangement. When completed, this worksheet may be retained in the station records so that if and when the setup is changed, it may more easily be re-evaluated. **Do not send the completed worksheets to the FCC.**

If the amateur station is to be operated on more than one wavelength band, or with several different antennas or combinations of apparatus, each is considered to be a separate setup. It might be helpful, therefore, to complete a separate worksheet for each setup. For an amateur radio station where two or more transmitters are used with the same antenna on the same wavelength band, it is only necessary to consider the setup that uses the highest power to the antenna input.

Top of each page. At the top of each page are blanks to fill in the amateur station call sign (item 1), the wavelength band under consideration (item 2), and a number or identifier that will identify the each setup (item 3). The purpose for repeating these items on each page is so that the various pages of a particular completed worksheet could be reassembled if they were to become separated. Additionally, at the top of Page 1 of the worksheet, there are blanks for the location of the station (item 4), the name of the person completing the worksheet (item 5), and the date (item 6).

Section I
Items 7 and 8. Fill in the manufacturer and model of the transmitter or transceiver and any RF power amplifier, or a brief description of these if they are home built.

Item 9. Fill in the Peak Envelope Power (PEP) output of the transmitter (use the PEP of the external amplifier, if one is to be used), in Watts (A). Many commercially manufactured transmitters and RF power amplifiers have a built-in power meter that can provide a measurement of PEP with reasonable accuracy for this purpose. Also, commercially manufactured external PEP reading power meters are available for stations that use common coaxial cables as feed lines. If there isn't any capability to measure the PEP output, the maximum PEP capability specified by the manufacturer may be used, or a reasonable estimate, based on factors such as measured power input, the maximum capability of the final amplifier devices or the power supply, may be used.

Optional Worksheet and Record of Compliance with FCC Guidelines
for Human Exposure to Radiofrequency Electromagnetic Fields
for Amateur Radio Stations

Instructions

Check the PEP output against Table 1. Because the PEP input to the antenna (H) can't be more than the PEP output (A), it's worthwhile at this point to take a quick look at Table 1 on page 3 of Supplement B to OET Bulletin 65. If the PEP output (A) does not exceed the value listed for the wavelength band under consideration, neither will the PEP input to the antenna (H). If that is the case, a routine RF evaluation is not required for this setup, and it isn't necessary to complete the rest of the worksheet. Otherwise, continue as follows.

Item 10. Fill in the PEP output used in item 9, converted to dBW. The power unit dBW expresses the ratio of the power in question to 1 Watt, in deciBels. The following chart can be used to convert common PEP levels in Watts to dBW. For power levels that fall in between the levels given, use the next higher power.

Watts	dBW
1	0
2	3
3	5
5	7
10	10
15	12
20	13
25	14
30	15
40	16
50	17
80	19
100	20
150	22
200	23
500	27
1000	30
1200	31
1500	32

Alternatively, the following mathematical formula can be used to do the conversion:

$$power_{dBW} = 10 \times \log (power_{Watts})$$

Items 11 and 12. Fill in the feed line type and loss (attenuation) specification (C). The attenuation or loss of a feed line is higher for higher frequencies. Therefore, the wavelength band of operation must be taken into account when determining what the feed line loss specification is. Manufacturers of coaxial cables develop tables showing the attenuation of various types of cables at various frequencies. There are also graphs and charts showing feed line attenuation versus frequency in readily available amateur radio handbooks and publications. The conservative approximate loss specifications for commonly used feed line type, given in the table on the next page, can also be used. In terms of feed line loss, a "conservative" estimate means that the feed line is very unlikely to have a lower loss than the estimate, although it may easily have a higher loss than estimated.

Item 13. Fill in the length of the feed line in feet (D).

Item 14. Fill in the calculated feed line loss (E) in dB. Calculate the feed line loss (E) by multiplying the feed line loss specification (C) by the feed line length (D). Inherent feed line loss often increases as the feed line ages. Also, feed line loss is considerably larger if the antenna impedance is not matched to the feed line impedance (causing a high SWR). However, for the purposes of this work sheet, do not consider or rely upon any additional feed line loss attributable to feed line aging or mismatch.

Optional Worksheet and Record of Compliance with FCC Guidelines for Human Exposure to Radiofrequency Electromagnetic Fields for Amateur Radio Stations

Instructions

Feed Line Loss Specification for Commonly Used Feed Lines (dB/100 feet)

Band	RG-58	RG-8X	RG-8A, RG-213	RG-8 Foam	"9913" & eqv	½" 50Ω "hardline"	"Ladder line"
160 m	0.5	0.4	0.3	0.2	0.2	0	0
80 m & 75 m	0.7	0.5	0.4	0.3	0.2	0.1	0
40 m	1.1	0.7	0.5	0.4	0.3	0.2	0
30 m	1.4	0.9	0.6	0.5	0.4	0.2	0
20 m	1.7	1.1	0.8	0.6	0.5	0.3	0
17 m	2.0	1.2	0.9	0.7	0.6	0.3	0.1
15 m	2.2	1.3	1.0	0.7	0.6	0.3	0.1
12 m	2.4	1.4	1.1	0.8	0.6	0.3	0.2
10 m	2.5	1.5	1.3	0.9	0.7	0.4	0.2
6 m	3.5	2.1	1.7	1.2	0.9	0.5	0.3
2 m	6.5	3.6	3.0	2.0	1.6	1.0	0.7
1¼ m	8.4	4.6	4.0	2.6	2.0	1.3	
70 cm	12	6.5	5.8	3.6	2.8	1.9	
33 cm	19	9.6	9.0	5.4	4.0	3.0	
23 cm	23	12	11	6.4	4.6	3.7	
13 cm		15	15	8.8	6.4	5.2	

This table provides conservative approximations for common types of feed lines. It is not meant to represent the actual attenuation performance of any particular product made by any particular manufacturer. The actual attenuation of any particular sample of a feed line type may vary somewhat from other samples of the same type because of differences in materials or manufacturing. If the feed line manufacturer's specification is available, use that instead of the values listed in this table. The term "hardline", as used above means commercial grade coaxial cable with a solid center conductor, foam dielectric, and solid or corrugated jacket. The term "ladder line", as used above, means 450Ω insulated window line with parallel conductors.

**Optional Worksheet and Record of Compliance with FCC Guidelines
for Human Exposure to Radiofrequency Electromagnetic Fields
for Amateur Radio Stations**

Instructions

<u>Items 15 and 16</u>. There may be other loss causing components in the feed line between the transmitter or external amplifier output and the antenna input. For example, there may be antenna switches or relays, directional couplers, duplexers, cavities or other filters. Usually the losses introduced by these components are so small as to be negligible. However, for setups operating in the VHF and higher frequency bands, the losses introduced by feed line components can be substantial. If this is the case, fill in a brief description of what these components are in item 15, and a conservative estimate of the total loss in dB in item 16, feed line components loss (F). Otherwise, write 0 (zero) in item 16. In terms of feed line component loss, a "conservative" estimate means that the feed line components are very unlikely to have a lower loss than the estimate, although they may easily have a higher loss than estimated. If the feed line component loss is not known, write 0 (zero) in item 16.

<u>Item 17</u>. Fill in the PEP input to the antenna, in dBW (G). Calculate this by subtracting the calculated feed line loss (E) and the feed line components loss (F) from the PEP output in dBW (B). Expressed as a mathematical equation, this is:

$$G = B - E - F$$

If G is less than 17 dBW, a routine RF evaluation is not required for this setup, and it isn't necessary to complete the rest of the worksheet. Otherwise, continue as follows.

<u>Item 18</u>. Fill in the PEP input to the antenna used in item 17, converted to Watts. The following table can be used to convert PEP levels in dBW to Watts. The entries in this table correspond to the power levels in Table 1 in OET Bulletin 65, Supplement B. For power levels that fall in between the levels given, use the next higher power.

dBW	Watts
17.0	50
18.5	70
18.8	75
20.0	100
21.0	125
21.8	150
23.0	200
23.5	225
24.0	250
26.3	425
27.0	500

Alternatively, the following mathematical formula can be used to do the conversion:

$$power_{Watts} = 10^{\frac{power_{dBW}}{10}}$$

<u>Item 19</u>. If the setup under consideration is an amateur radio repeater station, skip over this item and go directly to item 20. Otherwise, proceed as follows: Compare the PEP input to the antenna in Watts (H) to the power level listed in Table 1 in OET Bulletin 65, Supplement B, for the wavelength band to be used.

If the PEP input to the antenna in Watts (H) <u>is less than or equal to</u> the power level listed in Table 1 of OET Bulletin 65, Supplement B, for the wavelength band to be used, put a check mark in the first box. This means that the FCC rules do not require that a routine RF evaluation of the amateur radio setup be performed before it

**Optional Worksheet and Record of Compliance with FCC Guidelines
for Human Exposure to Radiofrequency Electromagnetic Fields
for Amateur Radio Stations**

Instructions

can be operated. It is not necessary to complete the rest of the worksheet.

On the other hand, if the PEP input to the antenna in Watts (H) <u>exceeds</u> the power level listed in Table 1 in OET Bulletin 65, Supplement B, for the wavelength band to be used, put a check mark in the second box. This means that a routine RF evaluation of this setup must be performed before it may be used to transmit. This requirement may be satisfied by completing the second section of the worksheet, by using methods outlined in OET Bulletin 65 or by employing any other technically valid method.

<u>Note</u>: Items 20 through 26 are only for amateur radio repeater setups.

<u>Item 20</u>. Fill in the manufacturer and model of the transmitting antenna for the amateur repeater setup, or a brief description of the antenna type (e.g. vertical collinear array).

<u>Item 21</u>. Check the appropriate box to indicate whether or not the repeater antenna is mounted on a building.

<u>Item 22</u>. Fill in the height above ground level of the lowest radiating part of the repeater antenna, in meters (I). One meter equals 3.28 feet.

<u>Item 23</u>. Fill in the maximum gain of the repeater antenna, in dBd (J). The term maximum gain means the highest antenna gain the antenna exhibits in any direction, not just in the direction of nearby places where people could be exposed to RF electromagnetic fields. The unit "dBd" means that the gain is expressed as a ratio between the power flux density ("pfd") that the antenna in question produces and the pfd

that a lossless half-wave dipole antenna would produce in free space (when both antennas have the same input power. Antenna gain of commercially manufactured antennas mounted in various typical arrangements is generally measured by the manufacturer on an antenna test range. The manufacturer may specify maximum antenna gain in dBd or dBi or both. If the gain is specified in dBi, for the purpose of this item simply subtract 2.15 dB from the dBi specification to obtain the dBd. Take into account, if possible, any increase in the gain resulting from the mounting arrangement (e.g. if the antenna is side-mounted on a tower). If the it is a home built antenna, estimate the maximum gain likely to be realized for an antenna of that type. Although antenna gain includes antenna efficiency, assume the efficiency is 100% for the purpose of this item.

<u>Item 24</u>. Fill in the maximum effective radiated power (ERP), in dBW (K). Calculate this by adding the PEP input to the antenna in dBW (G) and the estimated maximum repeater antenna gain (J). Expressed as a mathematical equation, this is:

$$K = G + J$$

<u>Item 25</u>. Fill in the maximum ERP used in item 24, converted to Watts (L), using the same methods as in the instruction for item 18.

<u>Item 26</u>. If L is less than or equal to 500 Watts (or K is less than 27 dBW), a routine RF evaluation is not required for this amateur radio repeater setup. Furthermore, even if L exceeds 500 Watts (i.e. K equals or exceeds 27 dBW), <u>provided that</u> the antenna is <u>not</u> located on a building <u>and</u> is

**Optional Worksheet and Record of Compliance with FCC Guidelines
for Human Exposure to Radiofrequency Electromagnetic Fields
for Amateur Radio Stations**

Instructions

installed such that the lowest point of the antenna is at least 10 meters (33 feet) above the ground level, a routine RF evaluation of this amateur radio repeater setup is not required. In either case, put a check mark in the first box. This indicates that a routine RF evaluation of the amateur radio repeater setup is not required before it can be operated.

In all other cases, put a check mark in the second box. This means that a routine RF evaluation of this amateur radio repeater setup must be performed before it can be operated. This requirement may be satisfied by completing the second section of the worksheet, by using methods outlined in OET Bulletin 65 or by employing any other technically valid method.

Section II

Item 27. Fill in a brief description of the antenna. If it is a commercially made antenna, indicate the manufacturer and type.

Item 28. Fill in the height above ground level of the lowest radiating part of the antenna, in meters (M). One meter equals 3.28 feet.

Item 29. Fill in the antenna gain in dBi (N). The term "antenna gain" generally refers to the field intensity at a given distance radiated by the antenna with a given power input, relative to an ideal lossless reference antenna type such as a half-wave dipole (dBd) or an isotropic radiator (dBi), fed with the same power and measured at the same distance. Antenna gain is a result of the directivity (i.e. that more energy is radiated in some directions than in others) and the efficiency (that some portion of the energy is not radiated as electromagnetic fields, but is instead converted to heat as a result of electrical resistance in the antenna materials and its surroundings). For this item, consider only the directivity of the antenna. The efficiency factor is considered in items 35-36.

Check Table 4. At this point, refer to Table 4 in Supplement B to OET Bulletin 65 (the W5YI table). For the wavelength band indicated in item 2, and using the PEP input to the antenna (H) and the antenna gain (N) from the worksheet, find the minimum necessary separation distances in meters from the antenna for uncontrolled and controlled environments. Pencil these distances in item 38 as (T) and (U) respectively. For power levels and antenna gains between those provided in the table, use the next higher values. This table is for a worst case analysis. Proceed now to the instruction for item 39, understanding that, if the worst case distances derived using Table 4 are not met in reality, they can be erased from item 38 and the evaluation can then proceed into further detail with the next instruction.

Item 30. Fill in the emission type used (e.g. SSB, CW, FM, FSK, AFSK, etc.).

Item 31. Fill in an emission type factor (O). The following table may be used.

CW Morse telegraphy	0.4
SSB voice	0.2
SSB voice, heavy speech processing	0.5
SSB AFSK	1.0
SSB SSTV	1.0
FM voice or data	1.0
FSK	1.0
AM voice, 50% modulation	0.5
AM voice, 100% modulation	0.3
ATV, video portion, image	0.6
ATV, video portion, black screen	0.8

Optional Worksheet and Record of Compliance with FCC Guidelines
for Human Exposure to Radiofrequency Electromagnetic Fields
for Amateur Radio Stations

Instructions

This emission type factor accounts for the fact that, for some modulated emission types that have a non-constant envelope, the PEP can be considerably larger than the average power. See also Table 2 in Supplement B of OET Bulletin 65 which provides examples of duty factors for modes commonly used by amateur radio operators.

Items 32 and 33. Fill in the transmit duty cycle and duty cycle factor. The duty cycle is the percentage of time in a given time interval (6 or 30 minutes) that the amateur radio station is in a transmitting condition, including instants where a transmission is in progress, but there is momentarily no power input to the antenna (e.g. the spaces between the "dits" and "dahs" of Morse telegraphy, the pauses between words of SSB telephony). The duty cycle factor is simply this percentage expressed in decimal form. For example, 20% becomes 0.2.

This transmit duty cycle is one of the parameters that is most easily controlled by the amateur radio station operator. As an example, with directed net or list operation, consideration should be given to whether the station is a net control station (relatively more transmit time) or a check-in (lots of listening time, relatively less transmission). When transmissions are carried through a repeater, the repeater timer may serve as a reminder to limit the length of continuous transmissions. With casual two way conversations, the transmit duty cycle could be approximated as 50%. A more detailed discussion, with examples, is contained in Supplement B to OET Bulletin 65 under the heading of "Time and Spacial Averaging".

Item 34. Fill in the average power input to the antenna (Q), in Watts. This is calculated by multiplying the PEP input to the antenna, in Watts (H), by the emission type factor (O) and the duty cycle factor (P). Expressed as a mathematical equation, this is:

$$Q = H \times O \times P$$

Check Tables 5-17 and/or 18-31. At this point, refer to Tables 5 through 17 (the Overbeck/Siwiak/FCC tables) and/or Tables 18 through 31 in Supplement B to OET Bulletin 65 (the ARRL tables). For the wavelength band indicated in item 2, and using the average power input to the antenna (Q) and selecting the appropriate table for the type of antenna, find the minimum necessary separation distances in meters from the antenna for uncontrolled and controlled environments. Note the limitations on appropriate use of these tables set forth in the bulletin. Write the distances found in item 38 as (T) and (U) respectively. For power levels and antenna gains between those provided in the table, use the next higher values.

Items 35 and 36. This item can be used for calculating the power flux density in accordance with the methods outlined in OET Bulletin 65, where antenna efficiency is a significant factor. Fill in the antenna efficiency and antenna efficiency factor (R). The antenna efficiency is the percentage of the input power that is radiated as electromagnetic energy. The antenna efficiency factor is simply this percentage expressed in decimal form. For example, 20% becomes 0.2. For most antennas, the efficiency is high enough to be negligible. For some antennas, however, particularly shortened vertical ground plane antennas, mobile whips, resistor broadbanded antennas, and small loops, the radiation resistance of

Optional Worksheet and Record of Compliance with FCC Guidelines
for Human Exposure to Radiofrequency Electromagnetic Fields
for Amateur Radio Stations

Instructions

the antenna may be so low that a significant portion of the energy is lost as heat in the antenna and it's ground system. Consult available amateur radio publications literature for more details. Otherwise, assume that the antenna efficiency is 100% and the antenna efficiency factor (R) is 1.0.

Item 37. Fill in the average power radiated (S). This is calculated by multiplying the average power input to the antenna (Q) by the antenna efficiency factor (R). Expressed as a mathematical equation, this is:

$$S = Q \times R$$

Item 38. This item is for filling in the distances, in meters, obtained from the various tables in Supplement B to OET Bulletin 65. It is also a good idea to jot the table number down next to this item so that the source of the distances indicated is known.

Item 39. Fill in the actual estimated, calculated or measured shortest physical distances, in meters, between the radiating part of the station antenna and the nearest place where the public or a person unaware of RF fields could be present, and the nearest place where a person who is aware of the RF fields could be present, (V) and (U) respectively.

Item 40. This item is a table where the evaluator may fill in calculated or measured power flux densities at locations where persons may be present. Power flux density may be calculated by methods outlined in Section 3 of Supplement B to OET Bulletin 65. If valid measurements are made at a reduced power level (that would comply

with exposure guidelines), it can be assumed that these measurements may be adjusted proportionally to predict field levels at a higher power.

Conclusions Section

At the end of the work sheet is a page where the evaluator can indicate his or her finding that the evaluated amateur radio setup is in compliance with FCC rules. A setup that does not comply must not be used for transmission until it is brought into compliance.

The evaluator should check the boxes [   ] next to any and all statements that apply to the evaluated amateur radio setup. The blank lines can also be used to elaborate on circumstances that support the conclusion.

The first four check boxes are for the situation where, for any of various reasons, it is very unlikely or simply not possible for any person to be in a location where he or she would be exposed to radiofrequency electromagnetic fields that are strong enough to exceed the levels prescribed in the FCC Guidelines for Human Exposure. The second four boxes are for the situation where a person could be in a location where he or she could be briefly exposed to radiofrequency electromagnetic fields that are strong enough to exceed the levels prescribed, but that other considerations ensure that a person will not remain in that location long enough to receive exposure in excess of that allowed by the FCC Guidelines for Human Exposure.

Optional Worksheet and Record of Compliance with FCC Guidelines for Human Exposure to Radiofrequency Electromagnetic Fields for Amateur Radio Stations

1. Call sign: _____ 2. Wavelength band: _____ 3. Setup #: _____

4. Station location: _____

5. Evaluated by: _____ 6. Date: _____

I. Initial Determination as to whether a Routine Evaluation is required by FCC Rule Section 97.13 for this amateur radio station setup.

7. Transmitter description: _____

8. External amplifier description: _____

        9. Peak Envelope Power (PEP) output, in Watts: (A)_____Watts

        10. PEP output, converted to dBW: (B)_____ dBW

11. Feed line type: _____   12. Feed line loss specification: (C)_____ dB/100 feet

        13. Feed line length: (D)_____ feet

        14. Calculated feed line loss: (E)_____ dB

15. Other feed line components, if any: _____

        16. Feed line components loss: (F)_____ dB

        17. PEP input to antenna, in dBW: (G)_____ dBW

        18. PEP input to antenna, converted to Watts: (H)_____ Watts

19. INITIAL DETERMINATION FOR STATIONS OTHER THAN REPEATERS:
    (for repeater stations go to the next page)

[   ] Based on the peak envelope power input to the antenna (H) calculated above, **a routine evaluation is NOT required** by FCC rules for operation as described of this setup in the stated wavelength band. It is not necessary to complete the rest of this worksheet.

[   ] Based on the peak envelope power input to the antenna (H) calculated above, **a routine evaluation is required** by FCC rules for operation as described of this setup in the stated wavelength band. The licensee may satisfy the requirement for a routine evaluation by completing the rest of this worksheet.

Optional Worksheet and Record of Compliance with FCC Guidelines for Human Exposure to Radiofrequency Electromagnetic Fields

1. Call sign: _____ 2. Wavelength band: _____ 3. Setup #:_____

---

20. Repeater antenna description: _____

21. Repeater antenna location: [ ] mounted on a building   [ ] not on a building

    22. Minimum repeater antenna height above ground level: (I) _____ meters

    23. Estimated maximum repeater antenna gain: (J)_____ dBd

    24. Maximum Effective Radiated Power (ERP), in dBW: (K)_____ dBW

    25. Maximum ERP, converted to Watts: (L)_____ Watts

26. INITIAL DETERMINATION FOR AMATEUR REPEATER STATIONS:

[ ] Based on the effective radiated power (L) calculated above and the antenna height (I) and location of the antenna, **a routine evaluation is NOT required** by FCC rules for operation as described of this amateur radio repeater station in the stated wavelength band. It is not necessary to complete the rest of this worksheet.

[ ] Based on the effective radiated power (L) calculated above and the antenna height (I) and location of the antenna, **a routine evaluation is required** by FCC rules for operation as described of this amateur radio repeater station in the stated wavelength band. The licensee may satisfy the requirement for a routine evaluation by completing the rest of this worksheet.

Reminders:

• A routine evaluation is not required for vehicular mobile or hand-held amateur radio setups. However, amateur radio operators should be aware of the potential for exposure to radiofrequency electromagnetic fields from these setups, and take measures (such as reducing transmitting power to the minimum necessary, positioning the radiating antenna as far from humans as practical, and limiting continuous transmitting time) accordingly to protect themselves and the occupants of their vehicles.

• The operation of each amateur radio setup must not exceed the FCC's guidelines for human exposure to radiofrequency electromagnetic fields, regardless of whether or not a routine evaluation is required.

• Although a particular amateur radio setup may by itself be in compliance with the FCC's guidelines for human exposure to radiofrequency electromagnetic fields, the cumulative effect of all simultaneously operating amateur radio setups (and any other operating transmitters in other services) at the same location or in the immediate vicinity must also be considered.

Optional Worksheet and Record of Compliance with FCC Guidelines
for Human Exposure to Radiofrequency Electromagnetic Fields

1. Call sign: _____   2. Wavelength band: _____   3. Setup #:_____

II. Routine Evaluation of amateur radio station setup.

27. Antenna description: _____

                            28. Antenna height above ground level: (M)_____ meters

                    29. Lossless antenna gain (directivity only): (N)_____ dBi

30. Emission type: _____   31. Emission type factor: (O)_____

32. Transmit duty cycle: _____%   33. Duty cycle factor: (P)_____

                    34. Average power input to the antenna: (Q)_____ Watts

35. Antenna efficiency: _____%   36. Antenna efficiency factor: (R)_____

                          37. Average power radiated: (S) _____ Watts

38. Minimum necessary distance from radiating part of antenna to place where:

                    - public may be present (uncontrolled): (T)_____ meters

         - amateur radio operator may be present (controlled): (U)_____ meters

39. Actual distance from radiating part of antenna to nearest place where:

                    - public may be present (uncontrolled): (V)_____ meters

         - amateur radio operator may be present (controlled): (W)_____ meters

40. Calculated power flux density:

Location	Power Flux Density

Optional Worksheet and Record of Compliance with FCC Guidelines for Human Exposure to Radiofrequency Electromagnetic Fields

1. Call sign: _____   2. Wavelength band: _____   3. Setup #:_____

CONCLUSIONS

Based on this routine evaluation, operation of this amateur radio station setup in accordance with the technical parameters entered above complies with the FCC's guidelines for human exposure to radiofrequency (RF) electromagnetic fields. The following statements provide the basis for this conclusion.

[ ]   It is physically impossible or extremely unlikely under normal circumstances for any person to be in any location where their exposure to RF electromagnetic fields would exceed the FCC guidelines, because:

    [ ]   the antenna is installed high enough on a tower or tree or other antenna support structure, such that it is not possible under normal circumstances for persons to get close enough to the antenna to be where the strength of the RF electromagnetic fields exceed the levels in the applicable FCC guidelines.

    [ ]   fences, locked gates and/or doors prevent persons who are unaware of the possibility of RF exposure from normally gaining access to locations where the strength of the RF electromagnetic fields exceed the levels in the applicable FCC guidelines.

    [ ]   _____

    _____

    _____

[ ]   Although persons could normally be in location(s) where the RF fields from the evaluated setup exceed the guideline levels, the following factors ensure that FCC human exposure guidelines will not be exceeded:

    [ ]   Signs have been installed that alert persons to the presence of RF electromagnetic fields and warn them not to remain for an extended period.

    [ ]   The locations where RF electromagnetic fields may exceed the guideline levels are roadways or other areas where human presence is transient.

    [ ]   _____

    _____

    _____

8

Antenna Tables

The tables in this chapter show the required minimum compliance distance for many common antenna types. The description of each antenna, along with the information in Chapter 5, will help you choose the one that best matches your antenna.

There are many, many possible antenna configurations used in the Amateur Radio Service. It is not possible to come up with a table that represents every possible combination. The first table in this chapter covers several pages! It is a generic table, showing the calculated "worst-case" compliance distances, in feet, for different frequency, antenna gain and power combinations. Chapter 5 explains in detail how to use this table. In many cases, this table will represent the easiest way to estimate compliance distance. If you "pass" using this table, it is not necessary to consider height above ground or other factors —your evaluation is complete.

In other cases, it is reasonable to use a table that represents an antenna similar, but not identical, to the one you want to evaluate, and apply it conservatively to your antenna installation. If, for example, you have a random wire that zigzags around your property, but you determine that all areas of exposure are greater than the distances for verticals, dipoles, G5RVs and ground plane antennas at similar heights, you can presume with some confidence that your antenna also passes. ARRL members who would like some assistance determining how to apply these tables to their stations can contact the ARRL Technical Information Service at ARRL Headquarters for assistance (email: tis@arrl.org, tel: 860-594-0214). The ARRL is continuing to refine and develop new models; they may be able to offer some specific information about antenna types not featured in this Chapter.

General-Purpose Table for HF Bands (Revised from Table 4A from Supplement B)—Part 1

Freq MHz	Ant. (dBi)	5 W con	5 W unc	25 W con	25 W unc	50 W con	50 W unc	100 W con	100 W unc	200 W con	200 W unc	300 W con	300 W unc
2	0	0.1	0.2	0.2	0.3	0.3	0.5	0.5	0.7	0.7	1.0	0.8	1.2
	1	0.1	0.2	0.3	0.4	0.4	0.6	0.5	0.8	0.7	1.1	0.9	1.4
	2	0.1	0.2	0.3	0.4	0.4	0.6	0.6	0.9	0.8	1.2	1.0	1.5
	3	0.1	0.2	0.3	0.5	0.5	0.7	0.7	1.0	0.9	1.4	1.1	1.7
	4	0.2	0.2	0.4	0.6	0.5	0.8	0.7	1.1	1.0	1.6	1.3	1.9
	5	0.2	0.3	0.4	0.6	0.6	0.9	0.8	1.2	1.2	1.8	1.4	2.2
	6	0.2	0.3	0.5	0.7	0.7	1.0	0.9	1.4	1.3	2.0	1.6	2.4
4	0	0.1	0.3	0.3	0.7	0.4	1.0	0.6	1.4	0.9	2.0	1.1	2.4
	1	0.2	0.4	0.4	0.8	0.5	1.1	0.7	1.6	1.0	2.2	1.2	2.7
	2	0.2	0.4	0.4	0.9	0.6	1.2	0.8	1.8	1.1	2.5	1.4	3.0
	3	0.2	0.4	0.4	1.0	0.6	1.4	0.9	2.0	1.2	2.8	1.5	3.4
	4	0.2	0.5	0.5	1.1	0.7	1.6	1.0	2.2	1.4	3.1	1.7	3.8
	5	0.2	0.6	0.6	1.2	0.8	1.8	1.1	2.5	1.6	3.5	1.9	4.3
	6	0.3	0.6	0.6	1.4	0.9	2.0	1.2	2.8	1.8	3.9	2.2	4.8
7.3	0	0.3	0.6	0.6	1.3	0.8	1.8	1.1	2.5	1.6	3.6	2.0	4.4
	1	0.3	0.6	0.6	1.4	0.9	2.0	1.3	2.9	1.8	4.0	2.2	5.0
	2	0.3	0.7	0.7	1.6	1.0	2.3	1.4	3.2	2.0	4.5	2.5	5.6
	3	0.4	0.8	0.8	1.8	1.1	2.5	1.6	3.6	2.3	5.1	2.8	6.2
	4	0.4	0.9	0.9	2.0	1.3	2.9	1.8	4.0	2.6	5.7	3.1	7.0
	5	0.5	1.0	1.0	2.3	1.4	3.2	2.0	4.5	2.9	6.4	3.5	7.8
	6	0.5	1.1	1.1	2.5	1.6	3.6	2.3	5.1	3.2	7.2	3.9	8.8
10.15	0	0.4	0.8	0.8	1.8	1.1	2.5	1.6	3.5	2.2	5.0	2.7	6.1
	1	0.4	0.9	0.9	2.0	1.3	2.8	1.8	4.0	2.5	5.6	3.1	6.9
	2	0.4	1.0	1.0	2.2	1.4	3.2	2.0	4.5	2.8	6.3	3.5	7.7
	3	0.5	1.1	1.1	2.5	1.6	3.5	2.2	5.0	3.2	7.1	3.9	8.7
	4	0.6	1.3	1.3	2.8	1.8	4.0	2.5	5.6	3.6	7.9	4.3	9.7
	5	0.6	1.4	1.4	3.1	2.0	4.5	2.8	6.3	4.0	8.9	4.9	10.9
	6	0.7	1.6	1.6	3.5	2.2	5.0	3.2	7.1	4.5	10.0	5.5	12.2
14.35	0	0.5	1.1	1.1	2.5	1.6	3.5	2.2	5.0	3.2	7.1	3.9	8.7
	1	0.6	1.3	1.3	2.8	1.8	4.0	2.5	5.6	3.6	7.9	4.4	9.7
	2	0.6	1.4	1.4	3.2	2.0	4.5	2.8	6.3	4.0	8.9	4.9	10.9
	3	0.7	1.6	1.6	3.5	2.2	5.0	3.2	7.1	4.5	10.0	5.5	12.3
	4	0.8	1.8	1.8	4.0	2.5	5.6	3.6	7.9	5.0	11.2	6.1	13.7
	5	0.9	2.0	2.0	4.5	2.8	6.3	4.0	8.9	5.6	12.6	6.9	15.4
	6	1.0	2.2	2.2	5.0	3.2	7.1	4.5	10.0	6.3	14.1	7.7	17.3
	7	1.1	2.5	2.5	5.6	3.5	7.9	5.0	11.2	7.1	15.9	8.7	19.4
	8	1.3	2.8	2.8	6.3	4.0	8.9	5.6	12.6	8.0	17.8	9.7	21.8
	9	1.4	3.2	3.2	7.1	4.5	10.0	6.3	14.1	8.9	20.0	10.9	24.4
	10	1.6	3.5	3.5	7.9	5.0	11.2	7.1	15.8	10.0	22.4	12.3	27.4
	11	1.8	4.0	4.0	8.9	5.6	12.6	7.9	17.8	11.2	25.1	13.8	30.8
	12	2.0	4.5	4.5	10.0	6.3	14.1	8.9	19.9	12.6	28.2	15.4	34.5
18.168	0	0.6	1.4	1.4	3.2	2.0	4.5	2.8	6.3	4.0	9.0	4.9	11.0
	1	0.7	1.6	1.6	3.6	2.2	5.0	3.2	7.1	4.5	10.1	5.5	12.3
	2	0.8	1.8	1.8	4.0	2.5	5.6	3.6	8.0	5.0	11.3	6.2	13.8
	3	0.9	2.0	2.0	4.5	2.8	6.3	4.0	9.0	5.7	12.7	6.9	15.5
	4	1.0	2.2	2.2	5.0	3.2	7.1	4.5	10.1	6.4	14.2	7.8	17.4
	5	1.1	2.5	2.5	5.6	3.6	8.0	5.0	11.3	7.1	15.9	8.7	19.5
	6	1.3	2.8	2.8	6.3	4.0	8.9	5.7	12.7	8.0	17.9	9.8	21.9
	7	1.4	3.2	3.2	7.1	4.5	10.0	6.3	14.2	9.0	20.1	11.0	24.6
	8	1.6	3.6	3.6	8.0	5.0	11.3	7.1	15.9	10.1	22.5	12.3	27.6
	9	1.8	4.0	4.0	8.9	5.7	12.6	8.0	17.9	11.3	25.3	13.8	31.0
	10	2.0	4.5	4.5	10.0	6.3	14.2	9.0	20.1	12.7	28.4	15.5	34.7
	11	2.2	5.0	5.0	11.2	7.1	15.9	10.1	22.5	14.2	31.8	17.4	39.0
	12	2.5	5.6	5.6	12.6	8.0	17.9	11.3	25.2	16.0	35.7	19.6	43.7

General-Purpose Table for HF Bands (Revised from Table 4A from Supplement B)—Part 2

Freq MHz	Ant. (dBi)	400 W con	400 W unc	500 W con	500 W unc	600 W con	600 W unc	750 W con	750 W unc	1000 W con	1000 W unc	1500 W con	1500 W unc
2	0	0.9	1.4	1.0	1.6	1.1	1.7	1.3	1.9	1.5	2.2	1.8	2.7
	1	1.1	1.6	1.2	1.8	1.3	1.9	1.4	2.1	1.7	2.5	2.0	3.0
	2	1.2	1.8	1.3	2.0	1.4	2.2	1.6	2.4	1.9	2.8	2.3	3.4
	3	1.3	2.0	1.5	2.2	1.6	2.4	1.8	2.7	2.1	3.1	2.6	3.8
	4	1.5	2.2	1.7	2.5	1.8	2.7	2.0	3.0	2.3	3.5	2.9	4.3
	5	1.7	2.5	1.9	2.8	2.0	3.0	2.3	3.4	2.6	3.9	3.2	4.8
	6	1.9	2.8	2.1	3.1	2.3	3.4	2.6	3.8	3.0	4.4	3.6	5.4
4	0	1.2	2.8	1.4	3.1	1.5	3.4	1.7	3.8	2.0	4.4	2.4	5.4
	1	1.4	3.1	1.6	3.5	1.7	3.8	1.9	4.3	2.2	5.0	2.7	6.1
	2	1.6	3.5	1.8	3.9	1.9	4.3	2.2	4.8	2.5	5.6	3.0	6.8
	3	1.8	3.9	2.0	4.4	2.2	4.8	2.4	5.4	2.8	6.2	3.4	7.6
	4	2.0	4.4	2.2	4.9	2.4	5.4	2.7	6.1	3.1	7.0	3.8	8.6
	5	2.2	5.0	2.5	5.6	2.7	6.1	3.0	6.8	3.5	7.9	4.3	9.6
	6	2.5	5.6	2.8	6.2	3.1	6.8	3.4	7.6	3.9	8.8	4.8	10.8
7.3	0	2.3	5.1	2.5	5.7	2.8	6.2	3.1	7.0	3.6	8.1	4.4	9.9
	1	2.6	5.7	2.9	6.4	3.1	7.0	3.5	7.8	4.0	9.0	5.0	11.1
	2	2.9	6.4	3.2	7.2	3.5	7.9	3.9	8.8	4.5	10.1	5.6	12.4
	3	3.2	7.2	3.6	8.0	3.9	8.8	4.4	9.9	5.1	11.4	6.2	13.9
	4	3.6	8.1	4.0	9.0	4.4	9.9	4.9	11.1	5.7	12.8	7.0	15.6
	5	4.1	9.1	4.5	10.1	5.0	11.1	5.5	12.4	6.4	14.3	7.8	17.5
	6	4.5	10.2	5.1	11.4	5.6	12.5	6.2	13.9	7.2	16.1	8.8	19.7
10.15	0	3.2	7.1	3.5	7.9	3.9	8.7	4.3	9.7	5.0	11.2	6.1	13.7
	1	3.6	7.9	4.0	8.9	4.4	9.7	4.9	10.9	5.6	12.6	6.9	15.4
	2	4.0	8.9	4.5	10.0	4.9	10.9	5.5	12.2	6.3	14.1	7.7	17.3
	3	4.5	10.0	5.0	11.2	5.5	12.3	6.1	13.7	7.1	15.8	8.7	19.4
	4	5.0	11.2	5.6	12.6	6.2	13.8	6.9	15.4	7.9	17.8	9.7	21.7
	5	5.6	12.6	6.3	14.1	6.9	15.4	7.7	17.3	8.9	19.9	10.9	24.4
	6	6.3	14.1	7.1	15.8	7.7	17.3	8.7	19.4	10.0	22.4	12.2	27.4
14.35	0	4.5	10.0	5.0	11.2	5.5	12.3	6.1	13.7	7.1	15.8	8.7	19.4
	1	5.0	11.2	5.6	12.6	6.2	13.8	6.9	15.4	7.9	17.8	9.7	21.8
	2	5.6	12.6	6.3	14.1	6.9	15.4	7.7	17.3	8.9	19.9	10.9	24.4
	3	6.3	14.1	7.1	15.8	7.8	17.3	8.7	19.4	10.0	22.4	12.3	27.4
	4	7.1	15.9	7.9	17.8	8.7	19.4	9.7	21.7	11.2	25.1	13.7	30.7
	5	8.0	17.8	8.9	19.9	9.8	21.8	10.9	24.4	12.6	28.2	15.4	34.5
	6	8.9	20.0	10.0	22.3	10.9	24.5	12.2	27.4	14.1	31.6	17.3	38.7
	7	10.0	22.4	11.2	25.1	12.3	27.5	13.7	30.7	15.9	35.5	19.4	43.4
	8	11.3	25.2	12.6	28.1	13.8	30.8	15.4	34.5	17.8	39.8	21.8	48.7
	9	12.6	28.2	14.1	31.6	15.5	34.6	17.3	38.7	20.0	44.6	24.4	54.7
	10	14.2	31.7	15.8	35.4	17.4	38.8	19.4	43.4	22.4	50.1	27.4	61.3
	11	15.9	35.5	17.8	39.7	19.5	43.5	21.8	48.7	25.1	56.2	30.8	68.8
	12	17.8	39.9	19.9	44.6	21.8	48.8	24.4	54.6	28.2	63.1	34.5	77.2
18.168	0	5.7	12.7	6.3	14.2	6.9	15.5	7.8	17.4	9.0	20.1	11.0	24.6
	1	6.4	14.2	7.1	15.9	7.8	17.4	8.7	19.5	10.1	22.5	12.3	27.6
	2	7.1	16.0	8.0	17.9	8.7	19.6	9.8	21.9	11.3	25.2	13.8	30.9
	3	8.0	17.9	9.0	20.0	9.8	21.9	11.0	24.5	12.7	28.3	15.5	34.7
	4	9.0	20.1	10.1	22.5	11.0	24.6	12.3	27.5	14.2	31.8	17.4	38.9
	5	10.1	22.6	11.3	25.2	12.4	27.6	13.8	30.9	15.9	35.7	19.5	43.7
	6	11.3	25.3	12.7	28.3	13.9	31.0	15.5	34.6	17.9	40.0	21.9	49.0
	7	12.7	28.4	14.2	31.7	15.6	34.8	17.4	38.9	20.1	44.9	24.6	55.0
	8	14.2	31.9	15.9	35.6	17.4	39.0	19.5	43.6	22.5	50.4	27.6	61.7
	9	16.0	35.7	17.9	40.0	19.6	43.8	21.9	48.9	25.3	56.5	31.0	69.2
	10	17.9	40.1	20.1	44.8	22.0	49.1	24.6	54.9	28.4	63.4	34.7	77.7
	11	20.1	45.0	22.5	50.3	24.6	55.1	27.6	61.6	31.8	71.1	39.0	87.1
	12	22.6	50.5	25.2	56.4	27.7	61.8	30.9	69.1	35.7	79.8	43.7	97.8

General-Purpose Table for HF Bands (Revised from Table 4A from Supplement B)—Part 3

Freq MHz	Ant. (dBi)	5 W con	5 W unc	25 W con	25 W unc	50 W con	50 W unc	100 W con	100 W unc	200 W con	200 W unc	300 W con	300 W unc
21.45	0	0.7	1.7	1.7	3.7	2.4	5.3	3.3	7.5	4.7	10.6	5.8	13.0
	1	0.8	1.9	1.9	4.2	2.7	5.9	3.8	8.4	5.3	11.9	6.5	14.5
	2	0.9	2.1	2.1	4.7	3.0	6.7	4.2	9.4	6.0	13.3	7.3	16.3
	3	1.1	2.4	2.4	5.3	3.3	7.5	4.7	10.6	6.7	15.0	8.2	18.3
	4	1.2	2.7	2.7	5.9	3.8	8.4	5.3	11.9	7.5	16.8	9.2	20.6
	5	1.3	3.0	3.0	6.7	4.2	9.4	6.0	13.3	8.4	18.8	10.3	23.1
	6	1.5	3.3	3.3	7.5	4.7	10.6	6.7	14.9	9.4	21.1	11.6	25.9
	7	1.7	3.7	3.7	8.4	5.3	11.9	7.5	16.8	10.6	23.7	13.0	29.0
	8	1.9	4.2	4.2	9.4	5.9	13.3	8.4	18.8	11.9	26.6	14.6	32.6
	9	2.1	4.7	4.7	10.6	6.7	14.9	9.4	21.1	13.3	29.8	16.3	36.5
	10	2.4	5.3	5.3	11.8	7.5	16.7	10.6	23.7	15.0	33.5	18.3	41.0
	11	2.7	5.9	5.9	13.3	8.4	18.8	11.9	26.6	16.8	37.6	20.6	46.0
	12	3.0	6.7	6.7	14.9	9.4	21.1	13.3	29.8	18.9	42.2	23.1	51.6
24.99	0	0.9	2.0	2.0	4.4	2.8	6.2	3.9	8.7	5.5	12.3	6.8	15.1
	1	1.0	2.2	2.2	4.9	3.1	6.9	4.4	9.8	6.2	13.8	7.6	17.0
	2	1.1	2.5	2.5	5.5	3.5	7.8	4.9	11.0	6.9	15.5	8.5	19.0
	3	1.2	2.8	2.8	6.2	3.9	8.7	5.5	12.3	7.8	17.4	9.5	21.3
	4	1.4	3.1	3.1	6.9	4.4	9.8	6.2	13.8	8.7	19.5	10.7	23.9
	5	1.6	3.5	3.5	7.8	4.9	11.0	6.9	15.5	9.8	21.9	12.0	26.9
	6	1.7	3.9	3.9	8.7	5.5	12.3	7.8	17.4	11.0	24.6	13.5	30.1
	7	2.0	4.4	4.4	9.8	6.2	13.8	8.7	19.5	12.3	27.6	15.1	33.8
	8	2.2	4.9	4.9	11.0	6.9	15.5	9.8	21.9	13.9	31.0	17.0	37.9
	9	2.5	5.5	5.5	12.3	7.8	17.4	11.0	24.6	15.5	34.8	19.0	42.6
	10	2.8	6.2	6.2	13.8	8.7	19.5	12.3	27.6	17.4	39.0	21.4	47.8
	11	3.1	6.9	6.9	15.5	9.8	21.9	13.8	30.9	19.6	43.8	24.0	53.6
	12	3.5	7.8	7.8	17.4	11.0	24.6	15.5	34.7	22.0	49.1	26.9	60.1
29.7	0	1.0	2.3	2.3	5.2	3.3	7.3	4.6	10.4	6.6	14.7	8.0	18.0
	1	1.2	2.6	2.6	5.8	3.7	8.2	5.2	11.6	7.4	16.4	9.0	20.1
	2	1.3	2.9	2.9	6.5	4.1	9.2	5.8	13.1	8.3	18.5	10.1	22.6
	3	1.5	3.3	3.3	7.3	4.6	10.4	6.5	14.6	9.3	20.7	11.3	25.4
	4	1.6	3.7	3.7	8.2	5.2	11.6	7.3	16.4	10.4	23.2	12.7	28.5
	5	1.8	4.1	4.1	9.2	5.8	13.0	8.2	18.4	11.7	26.1	14.3	31.9
	6	2.1	4.6	4.6	10.3	6.5	14.6	9.2	20.7	13.1	29.3	16.0	35.8
	7	2.3	5.2	5.2	11.6	7.3	16.4	10.4	23.2	14.7	32.8	18.0	40.2
	8	2.6	5.8	5.8	13.0	8.2	18.4	11.6	26.0	16.5	36.8	20.2	45.1
	9	2.9	6.5	6.5	14.6	9.2	20.7	13.1	29.2	18.5	41.3	22.6	50.6
	10	3.3	7.3	7.3	16.4	10.4	23.2	14.7	32.8	20.7	46.4	25.4	56.8
	11	3.7	8.2	8.2	18.4	11.6	26.0	16.4	36.8	23.3	52.0	28.5	63.7
	12	4.1	9.2	9.2	20.6	13.1	29.2	18.5	41.3	26.1	58.4	32.0	71.5

General-Purpose Table for HF Bands (Revised from Table 4A from Supplement B)—Part 4

Freq MHz	Ant. (dBi)	400 W con	400 W unc	500 W con	500 W unc	600 W con	600 W unc	750 W con	750 W unc	1000 W con	1000 W unc	1500 W con	1500 W unc
21.45	0	6.7	15.0	7.5	16.7	8.2	18.3	9.2	20.5	10.6	23.7	13.0	29.0
	1	7.5	16.8	8.4	18.8	9.2	20.6	10.3	23.0	11.9	26.6	14.5	32.5
	2	8.4	18.9	9.4	21.1	10.3	23.1	11.5	25.8	13.3	29.8	16.3	36.5
	3	9.5	21.2	10.6	23.6	11.6	25.9	13.0	29.0	15.0	33.4	18.3	41.0
	4	10.6	23.7	11.9	26.5	13.0	29.1	14.5	32.5	16.8	37.5	20.6	46.0
	5	11.9	26.6	13.3	29.8	14.6	32.6	16.3	36.5	18.8	42.1	23.1	51.6
	6	13.4	29.9	14.9	33.4	16.4	36.6	18.3	40.9	21.1	47.2	25.9	57.9
	7	15.0	33.5	16.8	37.5	18.4	41.1	20.5	45.9	23.7	53.0	29.0	64.9
	8	16.8	37.6	18.8	42.1	20.6	46.1	23.0	51.5	26.6	59.5	32.6	72.8
	9	18.9	42.2	21.1	47.2	23.1	51.7	25.8	57.8	29.8	66.7	36.5	81.7
	10	21.2	47.4	23.7	52.9	25.9	58.0	29.0	64.8	33.5	74.9	41.0	91.7
	11	23.8	53.1	26.6	59.4	29.1	65.1	32.5	72.7	37.6	84.0	46.0	102.9
	12	26.7	59.6	29.8	66.6	32.6	73.0	36.5	81.6	42.2	94.3	51.6	115.4
24.99	0	7.8	17.4	8.7	19.5	9.6	21.4	10.7	23.9	12.3	27.6	15.1	33.8
	1	8.8	19.6	9.8	21.9	10.7	24.0	12.0	26.8	13.8	30.9	17.0	37.9
	2	9.8	22.0	11.0	24.6	12.0	26.9	13.4	30.1	15.5	34.7	19.0	42.5
	3	11.0	24.6	12.3	27.5	13.5	30.2	15.1	33.7	17.4	39.0	21.3	47.7
	4	12.4	27.6	13.8	30.9	15.1	33.9	16.9	37.9	19.5	43.7	23.9	53.5
	5	13.9	31.0	15.5	34.7	17.0	38.0	19.0	42.5	21.9	49.0	26.9	60.1
	6	15.6	34.8	17.4	38.9	19.1	42.6	21.3	47.7	24.6	55.0	30.1	67.4
	7	17.5	39.1	19.5	43.7	21.4	47.8	23.9	53.5	27.6	61.7	33.8	75.6
	8	19.6	43.8	21.9	49.0	24.0	53.7	26.8	60.0	31.0	69.3	37.9	84.9
	9	22.0	49.2	24.6	55.0	26.9	60.2	30.1	67.3	34.8	77.7	42.6	95.2
	10	24.7	55.2	27.6	61.7	30.2	67.6	33.8	75.5	39.0	87.2	47.8	106.8
	11	27.7	61.9	30.9	69.2	33.9	75.8	37.9	84.8	43.8	97.9	53.6	119.9
	12	31.1	69.4	34.7	77.6	38.0	85.1	42.5	95.1	49.1	109.8	60.1	134.5
29.7	0	9.3	20.7	10.4	23.2	11.4	25.4	12.7	28.4	14.7	32.8	18.0	40.1
	1	10.4	23.3	11.6	26.0	12.7	28.5	14.2	31.9	16.4	36.8	20.1	45.0
	2	11.7	26.1	13.1	29.2	14.3	32.0	16.0	35.7	18.5	41.3	22.6	50.5
	3	13.1	29.3	14.6	32.7	16.0	35.9	17.9	40.1	20.7	46.3	25.4	56.7
	4	14.7	32.9	16.4	36.7	18.0	40.2	20.1	45.0	23.2	52.0	28.5	63.6
	5	16.5	36.9	18.4	41.2	20.2	45.2	22.6	50.5	26.1	58.3	31.9	71.4
	6	18.5	41.4	20.7	46.2	22.7	50.7	25.3	56.6	29.3	65.4	35.8	80.1
	7	20.8	46.4	23.2	51.9	25.4	56.8	28.4	63.6	32.8	73.4	40.2	89.9
	8	23.3	52.1	26.0	58.2	28.5	63.8	31.9	71.3	36.8	82.3	45.1	100.8
	9	26.1	58.4	29.2	65.3	32.0	71.6	35.8	80.0	41.3	92.4	50.6	113.2
	10	29.3	65.6	32.8	73.3	35.9	80.3	40.1	89.8	46.4	103.7	56.8	127.0
	11	32.9	73.6	36.8	82.2	40.3	90.1	45.0	100.7	52.0	116.3	63.7	142.5
	12	36.9	82.5	41.3	92.3	45.2	101.1	50.5	113.0	58.4	130.5	71.5	159.8

General-Purpose Table for VHF and UHF Bands (Revised from Table 4B from Supplement B)—Part 1

Freq MHz	Ant. (dBi)	5 W con	5 W unc	25 W con	25 W unc	50 W con	50 W unc	100 W con	100 W unc	200 W con	200 W unc	300 W con	300 W unc
50	0	1.0	2.3	2.3	5.2	3.3	7.4	4.7	10.5	6.6	14.8	8.1	18.1
	1	1.2	2.6	2.6	5.9	3.7	8.3	5.3	11.7	7.4	16.6	9.1	20.3
	2	1.3	2.9	2.9	6.6	4.2	9.3	5.9	13.2	8.3	18.6	10.2	22.8
	3	1.5	3.3	3.3	7.4	4.7	10.5	6.6	14.8	9.4	20.9	11.5	25.6
	4	1.7	3.7	3.7	8.3	5.2	11.7	7.4	16.6	10.5	23.5	12.9	28.7
	5	1.9	4.2	4.2	9.3	5.9	13.2	8.3	18.6	11.8	26.3	14.4	32.3
	6	2.1	4.7	4.7	10.4	6.6	14.8	9.3	20.9	13.2	29.5	16.2	36.2
	7	2.3	5.2	5.2	11.7	7.4	16.6	10.5	23.4	14.8	33.2	18.2	40.6
	8	2.6	5.9	5.9	13.2	8.3	18.6	11.8	26.3	16.6	37.2	20.4	45.6
	9	3.0	6.6	6.6	14.8	9.3	20.9	13.2	29.5	18.7	41.7	22.9	51.1
	10	3.3	7.4	7.4	16.6	10.5	23.4	14.8	33.1	20.9	46.8	25.6	57.4
	11	3.7	8.3	8.3	18.6	11.7	26.3	16.6	37.2	23.5	52.5	28.8	64.3
	12	4.2	9.3	9.3	20.8	13.2	29.5	18.6	41.7	26.4	59.0	32.3	72.2
	15	5.9	13.2	13.2	29.4	18.6	41.6	26.3	58.9	37.2	83.3	45.6	102.0
	18	8.3	18.6	18.6	41.6	26.3	58.8	37.2	83.2	52.6	117.6	64.4	144.1
	21	11.7	26.3	26.3	58.7	37.2	83.1	52.5	117.5	74.3	166.1	91.0	203.5
	24	16.6	37.1	37.1	83.0	52.5	117.3	74.2	166.0	105.0	234.7	128.5	287.4
144	0	1.0	2.3	2.3	5.2	3.3	7.4	4.7	10.5	6.6	14.8	8.1	18.1
	1	1.2	2.6	2.6	5.9	3.7	8.3	5.3	11.7	7.4	16.6	9.1	20.3
	2	1.3	2.9	2.9	6.6	4.2	9.3	5.9	13.2	8.3	18.6	10.2	22.8
	3	1.5	3.3	3.3	7.4	4.7	10.5	6.6	14.8	9.4	20.9	11.5	25.6
	4	1.7	3.7	3.7	8.3	5.2	11.7	7.4	16.6	10.5	23.5	12.9	28.7
	5	1.9	4.2	4.2	9.3	5.9	13.2	8.3	18.6	11.8	26.3	14.4	32.3
	6	2.1	4.7	4.7	10.4	6.6	14.8	9.3	20.9	13.2	29.5	16.2	36.2
	7	2.3	5.2	5.2	11.7	7.4	16.6	10.5	23.4	14.8	33.2	18.2	40.6
	8	2.6	5.9	5.9	13.2	8.3	18.6	11.8	26.3	16.6	37.2	20.4	45.6
	9	3.0	6.6	6.6	14.8	9.3	20.9	13.2	29.5	18.7	41.7	22.9	51.1
	10	3.3	7.4	7.4	16.6	10.5	23.4	14.8	33.1	20.9	46.8	25.6	57.4
	11	3.7	8.3	8.3	18.6	11.7	26.3	16.6	37.2	23.5	52.5	28.8	64.3
	12	4.2	9.3	9.3	20.8	13.2	29.5	18.6	41.7	26.4	59.0	32.3	72.2
	15	5.9	13.2	13.2	29.4	18.6	41.6	26.3	58.9	37.2	83.3	45.6	102.0
	18	8.3	18.6	18.6	41.6	26.3	58.8	37.2	83.2	52.6	117.6	64.4	144.1
	21	11.7	26.3	26.3	58.7	37.2	83.1	52.5	117.5	74.3	166.1	91.0	203.5
	24	16.6	37.1	37.1	83.0	52.5	117.3	74.2	166.0	105.0	234.7	128.5	287.4
222	0	1.0	2.3	2.3	5.2	3.3	7.4	4.7	10.5	6.6	14.8	8.1	18.1
	1	1.2	2.6	2.6	5.9	3.7	8.3	5.3	11.7	7.4	16.6	9.1	20.3
	2	1.3	2.9	2.9	6.6	4.2	9.3	5.9	13.2	8.3	18.6	10.2	22.8
	3	1.5	3.3	3.3	7.4	4.7	10.5	6.6	14.8	9.4	20.9	11.5	25.6
	4	1.7	3.7	3.7	8.3	5.2	11.7	7.4	16.6	10.5	23.5	12.9	28.7
	5	1.9	4.2	4.2	9.3	5.9	13.2	8.3	18.6	11.8	26.3	14.4	32.3
	6	2.1	4.7	4.7	10.4	6.6	14.8	9.3	20.9	13.2	29.5	16.2	36.2
	7	2.3	5.2	5.2	11.7	7.4	16.6	10.5	23.4	14.8	33.2	18.2	40.6
	8	2.6	5.9	5.9	13.2	8.3	18.6	11.8	26.3	16.6	37.2	20.4	45.6
	9	3.0	6.6	6.6	14.8	9.3	20.9	13.2	29.5	18.7	41.7	22.9	51.1
	10	3.3	7.4	7.4	16.6	10.5	23.4	14.8	33.1	20.9	46.8	25.6	57.4
	11	3.7	8.3	8.3	18.6	11.7	26.3	16.6	37.2	23.5	52.5	28.8	64.3
	12	4.2	9.3	9.3	20.8	13.2	29.5	18.6	41.7	26.4	59.0	32.3	72.2
	15	5.9	13.2	13.2	29.4	18.6	41.6	26.3	58.9	37.2	83.3	45.6	102.0
	18	8.3	18.6	18.6	41.6	26.3	58.8	37.2	83.2	52.6	117.6	64.4	144.1
	21	11.7	26.3	26.3	58.7	37.2	83.1	52.5	117.5	74.3	166.1	91.0	203.5
	24	16.6	37.1	37.1	83.0	52.5	117.3	74.2	166.0	105.0	234.7	128.5	287.4

General-Purpose Table for VHF and UHF Bands (Revised from Table 4B from Supplement B)—Part 2

Freq MHz	Ant. (dBi)	400 W con	400 W unc	500 W con	500 W unc	600 W con	600 W unc	750 W con	750 W unc	1000 W con	1000 W unc	1500 W con	1500 W unc
50	0	9.4	20.9	10.5	23.4	11.5	25.6	12.8	28.7	14.8	33.1	18.1	40.6
	1	10.5	23.5	11.7	26.3	12.9	28.8	14.4	32.2	16.6	37.2	20.3	45.5
	2	11.8	26.4	13.2	29.5	14.4	32.3	16.1	36.1	18.6	41.7	22.8	51.1
	3	13.2	29.6	14.8	33.1	16.2	36.2	18.1	40.5	20.9	46.8	25.6	57.3
	4	14.8	33.2	16.6	37.1	18.2	40.7	20.3	45.4	23.5	52.5	28.7	64.3
	5	16.7	37.2	18.6	41.6	20.4	45.6	22.8	51.0	26.3	58.9	32.3	72.1
	6	18.7	41.8	20.9	46.7	22.9	51.2	25.6	57.2	29.5	66.1	36.2	80.9
	7	21.0	46.9	23.4	52.4	25.7	57.4	28.7	64.2	33.2	74.1	40.6	90.8
	8	23.5	52.6	26.3	58.8	28.8	64.4	32.2	72.0	37.2	83.2	45.6	101.9
	9	26.4	59.0	29.5	66.0	32.3	72.3	36.1	80.8	41.7	93.3	51.1	114.3
	10	29.6	66.2	33.1	74.0	36.3	81.1	40.6	90.7	46.8	104.7	57.4	128.2
	11	33.2	74.3	37.2	83.1	40.7	91.0	45.5	101.7	52.5	117.5	64.3	143.9
	12	37.3	83.4	41.7	93.2	45.7	102.1	51.1	114.2	59.0	131.8	72.2	161.4
	15	52.7	117.8	58.9	131.7	64.5	144.2	72.1	161.3	83.3	186.2	102.0	228.1
	18	74.4	166.3	83.2	186.0	91.1	203.7	101.9	227.8	117.6	263.0	144.1	322.1
	21	105.1	235.0	117.5	262.7	128.7	287.8	143.9	321.7	166.1	371.5	203.5	455.0
	24	148.4	331.9	166.0	371.1	181.8	406.5	203.3	454.5	234.7	524.8	287.4	642.7
144	0	9.4	20.9	10.5	23.4	11.5	25.6	12.8	28.7	14.8	33.1	18.1	40.6
	1	10.5	23.5	11.7	26.3	12.9	28.8	14.4	32.2	16.6	37.2	20.3	45.5
	2	11.8	26.4	13.2	29.5	14.4	32.3	16.1	36.1	18.6	41.7	22.8	51.1
	3	13.2	29.6	14.8	33.1	16.2	36.2	18.1	40.5	20.9	46.8	25.6	57.3
	4	14.8	33.2	16.6	37.1	18.2	40.7	20.3	45.4	23.5	52.5	28.7	64.3
	5	16.7	37.2	18.6	41.6	20.4	45.6	22.8	51.0	26.3	58.9	32.3	72.1
	6	18.7	41.8	20.9	46.7	22.9	51.2	25.6	57.2	29.5	66.1	36.2	80.9
	7	21.0	46.9	23.4	52.4	25.7	57.4	28.7	64.2	33.2	74.1	40.6	90.8
	8	23.5	52.6	26.3	58.8	28.8	64.4	32.2	72.0	37.2	83.2	45.6	101.9
	9	26.4	59.0	29.5	66.0	32.3	72.3	36.1	80.8	41.7	93.3	51.1	114.3
	10	29.6	66.2	33.1	74.0	36.3	81.1	40.6	90.7	46.8	104.7	57.4	128.2
	11	33.2	74.3	37.2	83.1	40.7	91.0	45.5	101.7	52.5	117.5	64.3	143.9
	12	37.3	83.4	41.7	93.2	45.7	102.1	51.1	114.2	59.0	131.8	72.2	161.4
	15	52.7	117.8	58.9	131.7	64.5	144.2	72.1	161.3	83.3	186.2	102.0	228.1
	18	74.4	166.3	83.2	186.0	91.1	203.7	101.9	227.8	117.6	263.0	144.1	322.1
	21	105.1	235.0	117.5	262.7	128.7	287.8	143.9	321.7	166.1	371.5	203.5	455.0
	24	148.4	331.9	166.0	371.1	181.8	406.5	203.3	454.5	234.7	524.8	287.4	642.7
222	0	9.4	20.9	10.5	23.4	11.5	25.6	12.8	28.7	14.8	33.1	18.1	40.6
	1	10.5	23.5	11.7	26.3	12.9	28.8	14.4	32.2	16.6	37.2	20.3	45.5
	2	11.8	26.4	13.2	29.5	14.4	32.3	16.1	36.1	18.6	41.7	22.8	51.1
	3	13.2	29.6	14.8	33.1	16.2	36.2	18.1	40.5	20.9	46.8	25.6	57.3
	4	14.8	33.2	16.6	37.1	18.2	40.7	20.3	45.4	23.5	52.5	28.7	64.3
	5	16.7	37.2	18.6	41.6	20.4	45.6	22.8	51.0	26.3	58.9	32.3	72.1
	6	18.7	41.8	20.9	46.7	22.9	51.2	25.6	57.2	29.5	66.1	36.2	80.9
	7	21.0	46.9	23.4	52.4	25.7	57.4	28.7	64.2	33.2	74.1	40.6	90.8
	8	23.5	52.6	26.3	58.8	28.8	64.4	32.2	72.0	37.2	83.2	45.6	101.9
	9	26.4	59.0	29.5	66.0	32.3	72.3	36.1	80.8	41.7	93.3	51.1	114.3
	10	29.6	66.2	33.1	74.0	36.3	81.1	40.6	90.7	46.8	104.7	57.4	128.2
	11	33.2	74.3	37.2	83.1	40.7	91.0	45.5	101.7	52.5	117.5	64.3	143.9
	12	37.3	83.4	41.7	93.2	45.7	102.1	51.1	114.2	59.0	131.8	72.2	161.4
	15	52.7	117.8	58.9	131.7	64.5	144.2	72.1	161.3	83.3	186.2	102.0	228.1
	18	74.4	166.3	83.2	186.0	91.1	203.7	101.9	227.8	117.6	263.0	144.1	322.1
	21	105.1	235.0	117.5	262.7	128.7	287.8	143.9	321.7	166.1	371.5	203.5	455.0
	24	148.4	331.9	166.0	371.1	181.8	406.5	203.3	454.5	234.7	524.8	287.4	642.7

General-Purpose Table for VHF and UHF Bands (Revised from Table 4B from Supplement B)—Part 3

Freq MHz	Ant. (dBi)	5 W con	5 W unc	25 W con	25 W unc	50 W con	50 W unc	100 W con	100 W unc	200 W con	200 W unc	300 W con	300 W unc
420	0	0.9	2.0	2.0	4.4	2.8	6.3	4.0	8.8	5.6	12.5	6.9	15.3
	1	1.0	2.2	2.2	5.0	3.1	7.0	4.4	9.9	6.3	14.0	7.7	17.2
	2	1.1	2.5	2.5	5.6	3.5	7.9	5.0	11.1	7.0	15.8	8.6	19.3
	3	1.3	2.8	2.8	6.3	4.0	8.8	5.6	12.5	7.9	17.7	9.7	21.7
	4	1.4	3.1	3.1	7.0	4.4	9.9	6.3	14.0	8.9	19.8	10.9	24.3
	5	1.6	3.5	3.5	7.9	5.0	11.1	7.0	15.7	10.0	22.3	12.2	27.3
	6	1.8	3.9	3.9	8.8	5.6	12.5	7.9	17.7	11.2	25.0	13.7	30.6
	7	2.0	4.4	4.4	9.9	6.3	14.0	8.9	19.8	12.5	28.0	15.3	34.3
	8	2.2	5.0	5.0	11.1	7.0	15.7	9.9	22.2	14.1	31.4	17.2	38.5
	9	2.5	5.6	5.6	12.5	7.9	17.6	11.2	24.9	15.8	35.3	19.3	43.2
	10	2.8	6.3	6.3	14.0	8.8	19.8	12.5	28.0	17.7	39.6	21.7	48.5
	11	3.1	7.0	7.0	15.7	9.9	22.2	14.0	31.4	19.9	44.4	24.3	54.4
	12	3.5	7.9	7.9	17.6	11.1	24.9	15.8	35.2	22.3	49.8	27.3	61.0
	15	5.0	11.1	11.1	24.9	15.7	35.2	22.3	49.8	31.5	70.4	38.5	86.2
	18	7.0	15.7	15.7	35.1	22.2	49.7	31.4	70.3	44.5	99.4	54.4	121.8
	21	9.9	22.2	22.2	49.6	31.4	70.2	44.4	99.3	62.8	140.4	76.9	172.0
	24	14.0	31.4	31.4	70.1	44.4	99.2	62.7	140.3	88.7	198.4	108.6	242.9
902	0	0.6	1.4	1.4	3.0	1.9	4.3	2.7	6.0	3.8	8.5	4.7	10.5
	1	0.7	1.5	1.5	3.4	2.1	4.8	3.0	6.8	4.3	9.6	5.2	11.7
	2	0.8	1.7	1.7	3.8	2.4	5.4	3.4	7.6	4.8	10.8	5.9	13.2
	3	0.9	1.9	1.9	4.3	2.7	6.0	3.8	8.5	5.4	12.1	6.6	14.8
	4	1.0	2.1	2.1	4.8	3.0	6.8	4.3	9.6	6.1	13.5	7.4	16.6
	5	1.1	2.4	2.4	5.4	3.4	7.6	4.8	10.7	6.8	15.2	8.3	18.6
	6	1.2	2.7	2.7	6.0	3.8	8.5	5.4	12.0	7.6	17.0	9.3	20.9
	7	1.4	3.0	3.0	6.8	4.3	9.6	6.0	13.5	8.6	19.1	10.5	23.4
	8	1.5	3.4	3.4	7.6	4.8	10.7	6.8	15.2	9.6	21.5	11.7	26.3
	9	1.7	3.8	3.8	8.5	5.4	12.0	7.6	17.0	10.8	24.1	13.2	29.5
	10	1.9	4.3	4.3	9.5	6.0	13.5	8.5	19.1	12.1	27.0	14.8	33.1
	11	2.1	4.8	4.8	10.7	6.8	15.2	9.6	21.4	13.6	30.3	16.6	37.1
	12	2.4	5.4	5.4	12.0	7.6	17.0	10.8	24.0	15.2	34.0	18.6	41.6
	15	3.4	7.6	7.6	17.0	10.7	24.0	15.2	34.0	21.5	48.0	26.3	58.8
	18	4.8	10.7	10.7	24.0	15.2	33.9	21.5	48.0	30.3	67.8	37.2	83.1
	21	6.8	15.2	15.2	33.9	21.4	47.9	30.3	67.8	42.9	95.8	52.5	117.4
	24	9.6	21.4	21.4	47.9	30.3	67.7	42.8	95.7	60.5	135.3	74.1	165.8
1240	0	0.5	1.2	1.2	2.6	1.6	3.6	2.3	5.2	3.3	7.3	4.0	8.9
	1	0.6	1.3	1.3	2.9	1.8	4.1	2.6	5.8	3.7	8.2	4.5	10.0
	2	0.6	1.4	1.4	3.2	2.1	4.6	2.9	6.5	4.1	9.2	5.0	11.2
	3	0.7	1.6	1.6	3.6	2.3	5.1	3.3	7.3	4.6	10.3	5.6	12.6
	4	0.8	1.8	1.8	4.1	2.6	5.8	3.7	8.2	5.2	11.5	6.3	14.1
	5	0.9	2.0	2.0	4.6	2.9	6.5	4.1	9.2	5.8	13.0	7.1	15.9
	6	1.0	2.3	2.3	5.1	3.2	7.3	4.6	10.3	6.5	14.5	8.0	17.8
	7	1.2	2.6	2.6	5.8	3.6	8.2	5.2	11.5	7.3	16.3	8.9	20.0
	8	1.3	2.9	2.9	6.5	4.1	9.1	5.8	12.9	8.2	18.3	10.0	22.4
	9	1.5	3.2	3.2	7.3	4.6	10.3	6.5	14.5	9.2	20.5	11.2	25.1
	10	1.6	3.6	3.6	8.1	5.2	11.5	7.3	16.3	10.3	23.0	12.6	28.2
	11	1.8	4.1	4.1	9.1	5.8	12.9	8.2	18.3	11.6	25.8	14.2	31.7
	12	2.1	4.6	4.6	10.3	6.5	14.5	9.2	20.5	13.0	29.0	15.9	35.5
	15	2.9	6.5	6.5	14.5	9.2	20.5	13.0	29.0	18.3	41.0	22.4	50.2
	18	4.1	9.1	9.1	20.5	12.9	28.9	18.3	40.9	25.9	57.9	31.7	70.9
	21	5.8	12.9	12.9	28.9	18.3	40.9	25.8	57.8	36.5	81.7	44.8	100.1
	24	8.2	18.3	18.3	40.8	25.8	57.7	36.5	81.6	51.6	115.4	63.2	141.4

General-Purpose Table for VHF and UHF Bands (Revised from Table 4B from Supplement B)—Part 4

Freq MHz	Ant. (dBi)	400 W con	400 W unc	500 W con	500 W unc	600 W con	600 W unc	750 W con	750 W unc	1000 W con	1000 W unc	1500 W con	1500 W unc
420	0	7.9	17.7	8.8	19.8	9.7	21.7	10.8	24.2	12.5	28.0	15.3	34.3
	1	8.9	19.9	9.9	22.2	10.9	24.3	12.2	27.2	14.0	31.4	17.2	38.5
	2	10.0	22.3	11.1	24.9	12.2	27.3	13.6	30.5	15.8	35.2	19.3	43.1
	3	11.2	25.0	12.5	28.0	13.7	30.6	15.3	34.2	17.7	39.5	21.7	48.4
	4	12.5	28.1	14.0	31.4	15.4	34.4	17.2	38.4	19.8	44.4	24.3	54.3
	5	14.1	31.5	15.7	35.2	17.2	38.5	19.3	43.1	22.3	49.8	27.3	60.9
	6	15.8	35.3	17.7	39.5	19.3	43.3	21.6	48.4	25.0	55.8	30.6	68.4
	7	17.7	39.6	19.8	44.3	21.7	48.5	24.3	54.3	28.0	62.7	34.3	76.7
	8	19.9	44.5	22.2	49.7	24.4	54.4	27.2	60.9	31.4	70.3	38.5	86.1
	9	22.3	49.9	24.9	55.8	27.3	61.1	30.5	68.3	35.3	78.9	43.2	96.6
	10	25.0	56.0	28.0	62.6	30.7	68.5	34.3	76.6	39.6	88.5	48.5	108.4
	11	28.1	62.8	31.4	70.2	34.4	76.9	38.5	86.0	44.4	99.3	54.4	121.6
	12	31.5	70.5	35.2	78.8	38.6	86.3	43.1	96.5	49.8	111.4	61.0	136.4
	15	44.5	99.5	49.8	111.3	54.5	121.9	60.9	136.3	70.4	157.4	86.2	192.7
	18	62.9	140.6	70.3	157.2	77.0	172.2	86.1	192.5	99.4	222.3	121.8	272.2
	21	88.8	198.6	99.3	222.0	108.8	243.2	121.6	271.9	140.4	314.0	172.0	384.6
	24	125.4	280.5	140.3	313.6	153.6	343.6	171.8	384.1	198.4	443.5	242.9	543.2
902	0	5.4	12.1	6.0	13.5	6.6	14.8	7.4	16.5	8.5	19.1	10.5	23.4
	1	6.1	13.6	6.8	15.2	7.4	16.6	8.3	18.6	9.6	21.4	11.7	26.2
	2	6.8	15.2	7.6	17.0	8.3	18.6	9.3	20.8	10.8	24.0	13.2	29.4
	3	7.6	17.1	8.5	19.1	9.3	20.9	10.4	23.4	12.1	27.0	14.8	33.0
	4	8.6	19.1	9.6	21.4	10.5	23.4	11.7	26.2	13.5	30.3	16.6	37.1
	5	9.6	21.5	10.7	24.0	11.8	26.3	13.2	29.4	15.2	34.0	18.6	41.6
	6	10.8	24.1	12.0	26.9	13.2	29.5	14.8	33.0	17.0	38.1	20.9	46.7
	7	12.1	27.0	13.5	30.2	14.8	33.1	16.6	37.0	19.1	42.8	23.4	52.4
	8	13.6	30.3	15.2	33.9	16.6	37.2	18.6	41.5	21.5	48.0	26.3	58.7
	9	15.2	34.0	17.0	38.1	18.6	41.7	20.8	46.6	24.1	53.8	29.5	65.9
	10	17.1	38.2	19.1	42.7	20.9	46.8	23.4	52.3	27.0	60.4	33.1	74.0
	11	19.2	42.9	21.4	47.9	23.5	52.5	26.2	58.7	30.3	67.8	37.1	83.0
	12	21.5	48.1	24.0	53.8	26.3	58.9	29.4	65.8	34.0	76.0	41.6	93.1
	15	30.4	67.9	34.0	75.9	37.2	83.2	41.6	93.0	48.0	107.4	58.8	131.5
	18	42.9	95.9	48.0	107.3	52.5	117.5	58.7	131.4	67.8	151.7	83.1	185.8
	21	60.6	135.5	67.8	151.5	74.2	166.0	83.0	185.6	95.8	214.3	117.4	262.4
	24	85.6	191.4	95.7	214.0	104.8	234.4	117.2	262.1	135.3	302.7	165.8	370.7
1240	0	4.6	10.3	5.2	11.5	5.6	12.6	6.3	14.1	7.3	16.3	8.9	19.9
	1	5.2	11.6	5.8	12.9	6.3	14.2	7.1	15.8	8.2	18.3	10.0	22.4
	2	5.8	13.0	6.5	14.5	7.1	15.9	7.9	17.8	9.2	20.5	11.2	25.1
	3	6.5	14.6	7.3	16.3	8.0	17.8	8.9	19.9	10.3	23.0	12.6	28.2
	4	7.3	16.3	8.2	18.3	8.9	20.0	10.0	22.4	11.5	25.8	14.1	31.6
	5	8.2	18.3	9.2	20.5	10.0	22.4	11.2	25.1	13.0	29.0	15.9	35.5
	6	9.2	20.6	10.3	23.0	11.3	25.2	12.6	28.1	14.5	32.5	17.8	39.8
	7	10.3	23.1	11.5	25.8	12.6	28.2	14.1	31.6	16.3	36.5	20.0	44.7
	8	11.6	25.9	12.9	28.9	14.2	31.7	15.8	35.4	18.3	40.9	22.4	50.1
	9	13.0	29.0	14.5	32.5	15.9	35.6	17.8	39.8	20.5	45.9	25.1	56.2
	10	14.6	32.6	16.3	36.4	17.8	39.9	19.9	44.6	23.0	51.5	28.2	63.1
	11	16.3	36.5	18.3	40.9	20.0	44.8	22.4	50.0	25.8	57.8	31.7	70.8
	12	18.3	41.0	20.5	45.8	22.5	50.2	25.1	56.2	29.0	64.8	35.5	79.4
	15	25.9	57.9	29.0	64.8	31.7	70.9	35.5	79.3	41.0	91.6	50.2	112.2
	18	36.6	81.8	40.9	91.5	44.8	100.2	50.1	112.0	57.9	129.4	70.9	158.4
	21	51.7	115.6	57.8	129.2	63.3	141.6	70.8	158.3	81.7	182.7	100.1	223.8
	24	73.0	163.3	81.6	182.5	89.4	199.9	100.0	223.5	115.4	258.1	141.4	316.1

NEC TABLES

Most of the following tables are the result of NEC 4.1 runs of antennas modeled over average ground. There are tables for half-wave dipoles, G5RV-type antennas, Yagis, quarter-wave verticals, half-wave verticals, ground planes and other miscellaneous antennas.

These tables show the horizontal compliance distance from the antenna, modeled at an exposure height of 6 feet, 12 feet, 20 feet and at the height of the antenna. This corresponds generally to ground-level, first-story and second-story exposure, with the figure at the height of the antenna representing a typical exposure distance near the maximum for that type of antenna. A result of "0" feet indicates that the exposure below (or above in a few cases) the antenna at that particular exposure height is below the limit at that height for all areas surrounding the antenna, including immediately below the antenna. See the discussion in Chapter 5 for information on how to apply these tables to amateur station configurations.

In each class of antennas, the antenna type has been modeled at heights of 10, 20, 30, 40, 50 and 60 feet. These tables feature those antennas for which the horizontal compliance distance was greater than 0 feet somewhere below the antenna. The horizontal compliance distance for antennas located higher than the ones shown in these tables was calculated to be 0 feet, except at the height of the antenna. A distance of "0" feet indicates that the exposure below (or above in some cases) the antenna was less than the MPE limit at all points at that particular exposure height.

The distances at the height of the antenna give a reasonable indication of the maximum compliance distances to be expected for each antenna type.

HALF-WAVE HORIZONTAL DIPOLES

The following tables show the horizontal compliance distance for half-wave horizontal dipoles. The model calculates the required distance from the antenna at various heights, modeled at the center of the antenna. The exposure at the ends is generally less, so this is a conservative estimate for the minimum distance from any part of the antenna.

This model also serves reasonably well for inverted V antennas, using the minimum height of the inverted V as the antenna height. The 10-meter dipole models are a reasonable estimate of the compliance distance to be expected for VHF dipoles located at the same height.

160-meter band horizontal, half-wave dipole, Frequency = 2.0 MHz, Height above ground = 10 feet

Horizontal distance (feet) from any part of the antenna for compliance with
occupational/controlled or general population/uncontrolled exposure limits*

Power** (watts)	6 feet con.	6 feet unc.	12 feet con.	12 feet unc.	20 feet con.	20 feet unc.	10 feet con.	10 feet unc.
10	0	0	0	0	0	0	.5	.5
25	0	0	0	0	0	0	.5	.5
50	0	0	0	0	0	0	.5	.5
100	0	0	0	0	0	0	.5	1
200	0	0	0	0	0	0	1	1
250	0	0	0	0	0	0	1	1.5
300	0	0	0	0	0	0	1	1.5
400	0	0	0	0	0	0	1	1.5
500	0	0	0	0	0	0	1	1.5
600	0	0	0	0	0	0	1.5	2
750	0	0	0	0	0	0	1.5	2
1000	0	0	0	0	0	0	1.5	2.5
1250	0	0	0	1	0	0	2	2.5
1500	0	0	0	1.5	0	0	2	3

Height above ground (feet) where exposure occurs

80-meter band horizontal, half-wave dipole, Frequency = 4.0 MHz, Height above ground = 10 feet

Horizontal distance (feet) from any part of the antenna for compliance with
occupational/controlled or general population/uncontrolled exposure limits*

Power** (watts)	6 feet con.	6 feet unc.	12 feet con.	12 feet unc.	20 feet con.	20 feet unc.	10 feet con.	10 feet unc.
10	0	0	0	0	0	0	.5	.5
25	0	0	0	0	0	0	.5	1
50	0	0	0	0	0	0	.5	1
100	0	0	0	0	0	0	1	1.5
200	0	0	0	0	0	0	1	2
250	0	0	0	.5	0	0	1	2.5
300	0	0	0	1	0	0	1.5	2.5
400	0	0	0	2	0	0	1.5	3
500	0	0	0	2.5	0	0	1.5	3.5
600	0	0	0	2.5	0	0	1.5	3.5
750	0	1.5	0	3	0	0	2	4
1000	0	3	0	4	0	0	2	4.5
1250	0	4	.5	4.5	0	0	2.5	5
1500	0	4.5	1	5	0	0	2.5	5.5

Height above ground (feet) where exposure occurs

---

* 0 feet indicates that the exposure at the height in the column above or below the antenna is in compliance.
** Power = Average power input to the antenna. See Chapter 5.

40-meter band horizontal, half-wave dipole, Frequency = 7.3 MHz, Height above ground = 10 feet

Horizontal distance (feet) from any part of the antenna for compliance with occupational/controlled or general population/uncontrolled exposure limits*

Height above ground (feet) where exposure occurs

Power**	6 feet		12 feet		20 feet		10 feet	
(watts)	con.	unc.	con.	unc.	con.	unc.	con.	unc.
10	0	0	0	0	0	0	.5	1
25	0	0	0	0	0	0	1	1.5
50	0	0	0	0	0	0	1	2
100	0	0	0	1.5	0	0	1.5	2.5
200	0	0	0	3	0	0	2	3.5
250	0	2	0	3.5	0	0	2	4
300	0	3	0	4	0	0	2	4.5
400	0	4	1	4.5	0	0	2.5	5
500	0	5	1.5	5	0	0	2.5	5.5
600	0	5.5	2	5.5	0	0	3	6
750	0	6.5	2.5	6	0	0	3.5	7
1000	0	8	3	7	0	0	3.5	8
1250	2	9	3.5	8	0	0	4	8.5
1500	3	9.5	4	8.5	0	0	4.5	9.5

40-meter band horizontal, half-wave dipole, Frequency = 7.3 MHz, Height above ground = 20 feet

Horizontal distance (feet) from any part of the antenna for compliance with occupational/controlled or general population/uncontrolled exposure limits*

Height above ground (feet) where exposure occurs

Power**	6 feet		12 feet		20 feet		30 feet	
(watts)	con.	unc.	con.	unc.	con.	unc.	con.	unc.
10	0	0	0	0	.5	1	0	0
25	0	0	0	0	1	1.5	0	0
50	0	0	0	0	1	2	0	0
100	0	0	0	0	1.5	2.5	0	0
200	0	0	0	0	1.5	3.5	0	0
250	0	0	0	0	2	4	0	0
300	0	0	0	0	2	4	0	0
400	0	0	0	0	2.5	5	0	0
500	0	0	0	0	2.5	5.5	0	0
600	0	0	0	0	2.5	6	0	0
750	0	0	0	0	3	6.5	0	0
1000	0	0	0	1.5	3.5	7.5	0	0
1250	0	0	0	4.5	4	8	0	0
1500	0	0	0	6	4	9	0	0

30-meter band horizontal, half-wave dipole, Frequency = 10.15 MHz, Height above ground = 10 feet

Horizontal distance (feet) from any part of the antenna for compliance with occupational/controlled or general population/uncontrolled exposure limits*

Height above ground (feet) where exposure occurs

Power**	6 feet		12 feet		20 feet		10 feet	
(watts)	con.	unc.	con.	unc.	con.	unc.	con.	unc.
10	0	0	0	0	0	0	.5	1.5
25	0	0	0	0	0	0	1	2
50	0	0	0	1	0	0	1.5	2.5
100	0	0	0	2.5	0	0	1.5	3.5
200	0	3.5	0	4	0	0	2.5	5
250	0	4.5	1	4.5	0	0	2.5	5.5
300	0	5	1.5	5	0	0	3	6
400	0	6.5	2	6	0	0	3	6.5
500	0	7	2.5	7	0	0	3.5	7.5
600	1	8	3	7.5	0	0	4	8
750	2.5	9	3.5	8.5	0	0	4	9
1000	3.5	10.5	4	9.5	0	0	5	10
1250	4.5	11.5	4.5	10.5	0	3	5.5	11
1500	5	12.5	5	11.5	0	5.5	6	12

* 0 feet indicates that the exposure at the height in the column above or below the antenna is in compliance.
** Power = Average power input to the antenna. See Chapter 5.

30-meter band horizontal, half-wave dipole, Frequency = 10.15 MHz, Height above ground = 20 feet

Horizontal distance (feet) from any part of the antenna for compliance with occupational/controlled or general population/uncontrolled exposure limits*

Height above ground (feet) where exposure occurs

Power** (watts)	6 feet con.	6 feet unc.	12 feet con.	12 feet unc.	20 feet con.	20 feet unc.	30 feet con.	30 feet unc.
10	0	0	0	0	.5	1	0	0
25	0	0	0	0	1	1.5	0	0
50	0	0	0	0	1	2	0	0
100	0	0	0	0	1.5	3	0	0
200	0	0	0	0	2	4	0	0
250	0	0	0	0	2	4.5	0	0
300	0	0	0	0	2.5	5	0	0
400	0	0	0	0	3	6	0	0
500	0	0	0	0	3	6.5	0	0
600	0	0	0	0	3.5	7	0	0
750	0	0	0	2.5	3.5	8	0	0
1000	0	0	0	5.5	4	9	0	0
1250	0	0	0	7.5	4.5	10	0	3.5
1500	0	1	0	8.5	5	11	0	6

20-meter band horizontal, half-wave dipole, Frequency = 14.35 MHz, Height above ground = 10 feet

Horizontal distance (feet) from any part of the antenna for compliance with occupational/controlled or general population/uncontrolled exposure limits*

Height above ground (feet) where exposure occurs

Power** (watts)	6 feet con.	6 feet unc.	12 feet con.	12 feet unc.	20 feet con.	20 feet unc.	10 feet con.	10 feet unc.
10	0	0	0	0	0	0	1	1.5
25	0	0	0	.5	0	0	1	2.5
50	0	0	0	2.5	0	0	1.5	3
100	0	3	0	4	0	0	2	4.5
200	0	5.5	2	5.5	0	0	3	6
250	0	6.5	2.5	6.5	0	0	3	7
300	0	7	2.5	7	0	0	3.5	7.5
400	2	8.5	3.5	8	0	0	4	8.5
500	3	9.5	4	9	0	0	4.5	9.5
600	3.5	10	4.5	10	0	3.5	5	10
750	4.5	11.5	5	11	0	6.5	5.5	11
1000	5.5	13	5.5	12.5	0	9.5	6	12.5
1250	6.5	14	6.5	14	0	11.5	7	14
1500	7	15	7	15	0	13	7.5	15

20-meter band horizontal, half-wave dipole, Frequency = 14.35 MHz, Height above ground = 20 feet

Horizontal distance (feet) from any part of the antenna for compliance with occupational/controlled or general population/uncontrolled exposure limits*

Height above ground (feet) where exposure occurs

Power** (watts)	6 feet con.	6 feet unc.	12 feet con.	12 feet unc.	20 feet con.	20 feet unc.	30 feet con.	30 feet unc.
10	0	0	0	0	1	1.5	0	0
25	0	0	0	0	1	2	0	0
50	0	0	0	0	1.5	3	0	0
100	0	0	0	0	2	4	0	0
200	0	0	0	0	2.5	5.5	0	0
250	0	0	0	0	3	6	0	0
300	0	0	0	0	3	6.5	0	0
400	0	0	0	0	3.5	7.5	0	0
500	0	0	0	1	4	8.5	0	0
600	0	0	0	4	4	9.5	0	0
750	0	0	0	6	4.5	10.5	0	5.5
1000	0	2	0	8.5	5.5	12	0	8.5
1250	0	7	0	10.5	6	14	0	11
1500	0	9.5	0	12	6.5	15	0	13

* 0 feet indicates that the exposure at the height in the column above or below the antenna is in compliance.
** Power = Average power input to the antenna. See Chapter 5.

20-meter band horizontal, half-wave dipole, Frequency = 14.35 MHz, Height above ground = 30 feet

Horizontal distance (feet) from any part of the antenna for compliance with
occupational/controlled or general population/uncontrolled exposure limits*

Height above ground (feet) where exposure occurs

Power**	6 feet		12 feet		20 feet		30 feet	
(watts)	con.	unc.	con.	unc.	con.	unc.	con.	unc.
10	0	0	0	0	0	0	1	1.5
25	0	0	0	0	0	0	1	2
50	0	0	0	0	0	0	1.5	3
100	0	0	0	0	0	0	2	4
200	0	0	0	0	0	0	2.5	5.5
250	0	0	0	0	0	0	3	6.5
300	0	0	0	0	0	0	3	7
400	0	0	0	0	0	0	3.5	8
500	0	0	0	0	0	0	4	9
600	0	0	0	0	0	0	4.5	10
750	0	0	0	0	0	0	5	11
1000	0	0	0	0	0	4.5	5.5	13
1250	0	0	0	0	0	7.5	6.5	14.5
1500	0	0	0	0	0	9.5	7	16

17-meter band horizontal, half-wave dipole, Frequency = 18.168 MHz, Height above ground = 10 feet

Horizontal distance (feet) from any part of the antenna for compliance with
occupational/controlled or general population/uncontrolled exposure limits*

Height above ground (feet) where exposure occurs

Power**	6 feet		12 feet		20 feet		10 feet	
(watts)	con.	unc.	con.	unc.	con.	unc.	con.	unc.
10	0	0	0	0	0	0	1	2
25	0	0	0	2	0	0	1.5	3
50	0	1	0	3.5	0	0	2	4
100	0	4	1.5	5	0	0	2.5	5.5
200	0	7	3	7	0	0	3.5	7.5
250	1	8	3.5	8	0	0	4	8.5
300	2	8.5	3.5	9	0	0	4	9
400	3.5	10	4.5	10	0	5.5	5	10
500	4	11	5	11.5	0	8	5.5	11.5
600	5	12	5.5	12.5	0	10	6	12
750	5.5	13	6	13.5	0	12	6.5	13.5
1000	7	14.5	7	15.5	0	14.5	7.5	15
1250	8	16	8	17	0	17	8.5	16.5
1500	8.5	17	9	18.5	0	18.5	9	18

17-meter band horizontal, half-wave dipole, Frequency = 18.168 MHz, Height above ground = 20 feet

Horizontal distance (feet) from any part of the antenna for compliance with
occupational/controlled or general population/uncontrolled exposure limits*

Height above ground (feet) where exposure occurs

Power**	6 feet		12 feet		20 feet		30 feet	
(watts)	con.	unc.	con.	unc.	con.	unc.	con.	unc.
10	0	0	0	0	1	1.5	0	0
25	0	0	0	0	1.5	2.5	0	0
50	0	0	0	0	1.5	3.5	0	0
100	0	0	0	0	2.5	5	0	0
200	0	0	0	0	3	7	0	0
250	0	0	0	0	3.5	8	0	0
300	0	0	0	0	4	8.5	0	0
400	0	0	0	3.5	4.5	10	0	0
500	0	0	0	6	5	11.5	0	5.5
600	0	0	0	7.5	5.5	12.5	0	7.5
750	0	4.5	0	9.5	6	14	0	10
1000	0	9.5	0	12.5	7	16.5	0	13.5
1250	0	12.5	0	15	8	18.5	0	16.5
1500	0	14.5	0	17	8.5	20.5	0	19

* 0 feet indicates that the exposure at the height in the column above or below the antenna is in compliance.
** Power = Average power input to the antenna. See Chapter 5.

17-meter band horizontal, half-wave dipole, Frequency = 18.168 MHz, Height above ground = 30 feet

Horizontal distance (feet) from any part of the antenna for compliance with
occupational/controlled or general population/uncontrolled exposure limits*

Height above ground (feet) where exposure occurs

Power**	6 feet		12 feet		20 feet		30 feet	
(watts)	con.	unc.	con.	unc.	con.	unc.	con.	unc.
10	0	0	0	0	0	0	1	2
25	0	0	0	0	0	0	1.5	3
50	0	0	0	0	0	0	2	4
100	0	0	0	0	0	0	2.5	5.5
200	0	0	0	0	0	0	3.5	8
250	0	0	0	0	0	0	4	8.5
300	0	0	0	0	0	0	4.5	9.5
400	0	0	0	0	0	2	5	11
500	0	0	0	0	0	5.5	5.5	12
600	0	0	0	0	0	7.5	6	13.5
750	0	0	0	0	0	10	7	15
1000	0	0	0	0	0	13	8	17.5
1250	0	0	0	9	0	16.5	8.5	19.5
1500	0	0	0	14.5	0	21	9.5	21.5

15-meter band horizontal, half-wave dipole, Frequency = 21.45 MHz, Height above ground = 10 feet

Horizontal distance (feet) from any part of the antenna for compliance with
occupational/controlled or general population/uncontrolled exposure limits*

Height above ground (feet) where exposure occurs

Power**	6 feet		12 feet		20 feet		10 feet	
(watts)	con.	unc.	con.	unc.	con.	unc.	con.	unc.
10	0	0	0	0	0	0	1	2
25	0	0	0	2.5	0	0	1.5	3
50	0	2.5	0	4	0	0	2	4.5
100	0	5	2	6	0	0	3	6
200	.5	8	3.5	8.5	0	0	4	8.5
250	2.5	9	4	9.5	0	4	4.5	9.5
300	3	9.5	4.5	10.5	0	6.5	5	10.5
400	4	11	5.5	12	0	9.5	5.5	12
500	5	12	6	13.5	0	11.5	6	13
600	5.5	13	6.5	14.5	0	13.5	6.5	14
750	6.5	14.5	7.5	16	0	15.5	7.5	15.5
1000	8	16	8.5	18	0	19	8.5	17.5
1250	9	17.5	9.5	20	4	21	9.5	19
1500	9.5	19	10.5	21.5	6.5	23.5	10.5	20.5

15-meter band horizontal, half-wave dipole, Frequency = 21.45 MHz, Height above ground = 20 feet

Horizontal distance (feet) from any part of the antenna for compliance with
occupational/controlled or general population/uncontrolled exposure limits*

Height above ground (feet) where exposure occurs

Power**	6 feet		12 feet		20 feet		30 feet	
(watts)	con.	unc.	con.	unc.	con.	unc.	con.	unc.
10	0	0	0	0	1	2	0	0
25	0	0	0	0	1.5	3	0	0
50	0	0	0	0	2	4.5	0	0
100	0	0	0	0	3	6	0	0
200	0	0	0	0	4	8.5	0	0
250	0	0	0	0	4.5	10	0	0
300	0	0	0	4	5	10.5	0	0
400	0	0	0	7	5.5	12.5	0	5.5
500	0	0	0	9.5	6	14	0	8.5
600	0	6.5	0	11.5	7	16	0	10.5
750	0	10	0	15	7.5	18	0	13.5
1000	0	14	0	19	8.5	21.5	0	17.5
1250	0	17	0	22	10	25.5	0	21
1500	0	19.5	4	24.5	10.5	29	0	25

* 0 feet indicates that the exposure at the height in the column above or below the antenna is in compliance.
** Power = Average power input to the antenna. See Chapter 5.

15-meter band horizontal, half-wave dipole, Frequency = 21.45 MHz, Height above ground = 30 feet

Horizontal distance (feet) from any part of the antenna for compliance with
occupational/controlled or general population/uncontrolled exposure limits*

| Power** (watts) | Height above ground (feet) where exposure occurs | | | | | | | |
| | 6 feet | | 12 feet | | 20 feet | | 30 feet | |
	con.	unc.	con.	unc.	con.	unc.	con.	unc.
10	0	0	0	0	0	0	1	2.5
25	0	0	0	0	0	0	1.5	3.5
50	0	0	0	0	0	0	2.5	4.5
100	0	0	0	0	0	0	3	6.5
200	0	0	0	0	0	0	4.5	9
250	0	0	0	0	0	3	4.5	10
300	0	0	0	0	0	5.5	5	11
400	0	0	0	0	0	8.5	6	12.5
500	0	0	0	0	0	10.5	6.5	14
600	0	0	0	0	0	12.5	7	15.5
750	0	0	0	6.5	0	14.5	8	17
1000	0	0	0	16	0	18	9	19.5
1250	0	0	0	21	3	22	10	22
1500	0	10.5	0	25	5.5	27.5	11	24

15-meter band horizontal, half-wave dipole, Frequency = 21.45 MHz, Height above ground = 40 feet

Horizontal distance (feet) from any part of the antenna for compliance with
occupational/controlled or general population/uncontrolled exposure limits*

| Power** (watts) | Height above ground (feet) where exposure occurs | | | | | | | |
| | 6 feet | | 12 feet | | 20 feet | | 40 feet | |
	con.	unc.	con.	unc.	con.	unc.	con.	unc.
10	0	0	0	0	0	0	1	2
25	0	0	0	0	0	0	1.5	3
50	0	0	0	0	0	0	2	4.5
100	0	0	0	0	0	0	3	6
200	0	0	0	0	0	0	4	8.5
250	0	0	0	0	0	0	4.5	9.5
300	0	0	0	0	0	0	5	10.5
400	0	0	0	0	0	0	5.5	12
500	0	0	0	0	0	0	6	13.5
600	0	0	0	0	0	0	6.5	14.5
750	0	0	0	0	0	0	7.5	16.5
1000	0	0	0	0	0	3.5	8.5	19
1250	0	0	0	0	0	10	9.5	21
1500	0	0	0	0	0	13.5	10.5	23

12-meter band horizontal, half-wave dipole, Frequency = 24.99 MHz, Height above ground = 10 feet

Horizontal distance (feet) from any part of the antenna for compliance with
occupational/controlled or general population/uncontrolled exposure limits*

| Power** (watts) | Height above ground (feet) where exposure occurs | | | | | | | |
| | 6 feet | | 12 feet | | 20 feet | | 10 feet | |
	con.	unc.	con.	unc.	con.	unc.	con.	unc.
10	0	0	0	1	0	0	1	2.5
25	0	0	0	3	0	0	1.5	3.5
50	0	3	1	4.5	0	0	2.5	5
100	0	6	2.5	7	0	0	3	7
200	2	9	4	10	0	5	4.5	10
250	3	10	4.5	11	0	8	5	11
300	4	11	5.5	12.5	0	10	5.5	12
400	5	12.5	6	14	0	13	6.5	13.5
500	6	13.5	7	15.5	0	15	7	15
600	6.5	14.5	7.5	17	0	17	7.5	16.5
750	7.5	16	8.5	19	0	19.5	8.5	18
1000	9	18	10	21.5	5	23	10	20
1250	10	19.5	11	23.5	8	26	11	22
1500	11	21	12.5	25	10	28	12	23.5

* 0 feet indicates that the exposure at the height in the column above or below the antenna is in compliance.
** Power = Average power input to the antenna. See Chapter 5.

12-meter band horizontal, half-wave dipole, Frequency = 24.99 MHz, Height above ground = 20 feet

Horizontal distance (feet) from any part of the antenna for compliance with occupational/controlled or general population/uncontrolled exposure limits*

Height above ground (feet) where exposure occurs

Power**	6 feet		12 feet		20 feet		30 feet	
(watts)	con.	unc.	con.	unc.	con.	unc.	con.	unc.
10	0	0	0	0	1.5	2.5	0	0
25	0	0	0	0	2	4	0	0
50	0	0	0	0	2.5	5.5	0	0
100	0	0	0	0	3.5	7.5	0	0
200	0	0	0	4	5	10.5	0	0
250	0	0	0	7	5.5	12	0	2
300	0	0	0	10	6	13	0	5
400	0	2	0	14.5	7	15.5	0	8.5
500	0	9	0	17.5	7.5	18	0	11
600	0	12	0	20	8	20	0	13
750	0	15.5	0	23	9	24.5	0	16
1000	0	19.5	4	27.5	10.5	30.5	0	21
1250	0	22.5	7	30.5	12	35.5	2	26
1500	0	25.5	10	33	13	39	5	32

12-meter band horizontal, half-wave dipole, Frequency = 24.99 MHz, Height above ground = 30 feet

Horizontal distance (feet) from any part of the antenna for compliance with occupational/controlled or general population/uncontrolled exposure limits*

Height above ground (feet) where exposure occurs

Power**	6 feet		12 feet		20 feet		30 feet	
(watts)	con.	unc.	con.	unc.	con.	unc.	con.	unc.
10	0	0	0	0	0	0	1	2.5
25	0	0	0	0	0	0	2	3.5
50	0	0	0	0	0	0	2.5	5
100	0	0	0	0	0	0	3.5	7
200	0	0	0	0	0	4	4.5	10
250	0	0	0	0	0	6.5	5	11
300	0	0	0	0	0	8.5	5.5	12
400	0	0	0	0	0	11	6.5	14
500	0	0	0	0	0	13	7	15.5
600	0	0	0	4.5	0	15	8	16.5
750	0	0	0	14	0	17	8.5	18.5
1000	0	0	0	22	4	20.5	10	21
1250	0	11.5	0	27.5	6.5	23.5	11	23
1500	0	16.5	0	31.5	8.5	26.5	12	25

12-meter band horizontal, half-wave dipole, Frequency = 24.99 MHz, Height above ground = 40 feet

Horizontal distance (feet) from any part of the antenna for compliance with occupational/controlled or general population/uncontrolled exposure limits*

Height above ground (feet) where exposure occurs

Power**	6 feet		12 feet		20 feet		40 feet	
(watts)	con.	unc.	con.	unc.	con.	unc.	con.	unc.
10	0	0	0	0	0	0	1.5	2.5
25	0	0	0	0	0	0	2	4
50	0	0	0	0	0	0	2.5	5.5
100	0	0	0	0	0	0	3.5	7.5
200	0	0	0	0	0	0	5	10.5
250	0	0	0	0	0	0	5.5	11.5
300	0	0	0	0	0	0	6	12.5
400	0	0	0	0	0	0	6.5	14.5
500	0	0	0	0	0	0	7.5	16.5
600	0	0	0	0	0	5	8	18
750	0	0	0	0	0	11	9	20.5
1000	0	0	0	0	0	16	10.5	25
1250	0	0	0	15.5	0	20	11.5	28.5

* 0 feet indicates that the exposure at the height in the column above or below the antenna is in compliance.
** Power = Average power input to the antenna. See Chapter 5.

12-meter band horizontal, half-wave dipole, Frequency = 24.99 MHz, Height above ground = 50 feet

Horizontal distance (feet) from any part of the antenna for compliance with
occupational/controlled or general population/uncontrolled exposure limits*

Height above ground (feet) where exposure occurs

Power**	6 feet		12 feet		20 feet		50 feet	
(watts)	con.	unc.	con.	unc.	con.	unc.	con.	unc.
10	0	0	0	0	0	0	1	2.5
25	0	0	0	0	0	0	2	3.5
50	0	0	0	0	0	0	2.5	5
100	0	0	0	0	0	0	3.5	7
200	0	0	0	0	0	0	4.5	10
250	0	0	0	0	0	0	5	11
300	0	0	0	0	0	0	5.5	12
400	0	0	0	0	0	0	6.5	14
500	0	0	0	0	0	0	7	15.5
600	0	0	0	0	0	0	8	17
750	0	0	0	0	0	0	8.5	19
1000	0	0	0	0	0	0	10	21.5
1250	0	0	0	0	0	4	11	24
1500	0	0	0	0	0	13	12	26

10-meter band horizontal, half-wave dipole, Frequency = 29.7, Antenna height = 10 feet

Horizontal distance (feet) from any part of the antenna for compliance with
occupational/controlled or general population/uncontrolled exposure limits*

Height above ground (feet) where exposure occurs

Power**	6 feet		12 feet		20 feet		10 feet	
(watts)	con.	unc.	con.	unc.	con.	unc.	con.	unc.
10	0	0	0	2	0	0	1.5	2.5
25	0	0	0	4	0	0	2	4
50	0	4	2	5.5	0	0	2.5	6
100	0	7	3	8.5	0	0	3.5	8.5
200	3	10.5	5	12.5	0	9	5	12
250	4	11.5	5.5	14	0	11.5	6	13.5
300	4.5	12.5	6.5	15	0	13.5	6.5	14.5
400	6	14.5	7.5	17.5	0	17	7.5	16.5
500	7	16	8.5	19.5	0	20	8.5	18.5
600	7.5	17	9.5	21	0	22.5	9	20
750	8.5	19	10.5	23	5	25.5	10	22
1000	10.5	21	12	26	9	29.5	12	24.5
1250	11.5	23	14	28	11.5	33	13	26.5
1500	12.5	24.5	15	30	13.5	35.5	14.5	28.5

10-meter band horizontal, half-wave dipole, Frequency = 29.7, Antenna Height = 20 feet

Horizontal distance (feet) from any part of the antenna for compliance with
occupational/controlled or general population/uncontrolled exposure limits*

Height above ground (feet) where exposure occurs

Power**	6 feet		12 feet		20 feet		30 feet	
(watts)	con.	unc.	con.	unc.	con.	unc.	con.	unc.
10	0	0	0	0	1.5	3	0	0
25	0	0	0	0	2	4.5	0	0
50	0	0	0	0	3	6.5	0	0
100	0	0	0	2	4	9	0	0
200	0	0	0	10.5	6	12.5	0	6
250	0	5.5	0	14.5	6.5	14	0	8
300	0	9	0	17.5	7	15.5	0	9.5
400	0	13.5	0	22	8	18.5	0	12
500	0	16.5	2	25.5	9	21.5	0	14
600	0	19	4.5	28.5	10	25.5	0	15.5
750	0	22	6.5	32	11	33	1.5	18
1000	0	26.5	10.5	36.5	12.5	41	6	22
1250	5.5	29.5	14.5	40.5	14	47	8	28
1500	9	32.5	17.5	43.5	15.5	51.5	9.5	39.5

* 0 feet indicates that the exposure at the height in the column above or below the antenna is in compliance.
** Power = Average power input to the antenna. See Chapter 5.

10-meter band horizontal, half-wave dipole, Frequency = 29.7 MHz, Height above ground = 30 feet

Horizontal distance (feet) from any part of the antenna for compliance with occupational/controlled or general population/uncontrolled exposure limits*

Height above ground (feet) where exposure occurs

Power** (watts)	6 feet		12 feet		20 feet		30 feet	
	con.	unc.	con.	unc.	con.	unc.	con.	unc.
10	0	0	0	0	0	0	1.5	3
25	0	0	0	0	0	0	2	4.5
50	0	0	0	0	0	0	3	6
100	0	0	0	0	0	0	4	8.5
200	0	0	0	0	0	7.5	5.5	12
250	0	0	0	0	0	9.5	6	13.5
300	0	0	0	0	0	11.5	6.5	14.5
400	0	0	0	0	0	14.5	7.5	17
500	0	0	0	0	0	16.5	8.5	18.5
600	0	0	0	14	0	18.5	9	20.5
750	0	12	0	22.5	3.5	21	10	22.5
1000	0	19	0	31	7.5	24.5	12	25.5
1250	0	23.5	0	37	9.5	28	13.5	27.5
1500	0	27	0	41.5	11.5	31.5	14.5	29.5

10-meter band horizontal, half-wave dipole, Frequency = 29.7, Height above ground = 40 feet

Horizontal distance (feet) from any part of the antenna for compliance with occupational/controlled or general population/uncontrolled exposure limits*

Height above ground (feet) where exposure occurs

Power** (watts)	6 feet		12 feet		20 feet		40 feet	
	con.	unc.	con.	unc.	con.	unc.	con.	unc.
10	0	0	0	0	0	0	1.5	3
25	0	0	0	0	0	0	2	4.5
50	0	0	0	0	0	0	3	6
100	0	0	0	0	0	0	4	8.5
200	0	0	0	0	0	0	5.5	12
250	0	0	0	0	0	0	6	13.5
300	0	0	0	0	0	0	6.5	14.5
400	0	0	0	0	0	0	7.5	16.5
500	0	0	0	0	0	5	8.5	18.5
600	0	0	0	0	0	11.5	9.5	20.5
750	0	0	0	0	0	17	10.5	23
1000	0	0	0	0	0	22.5	12	27.5
1250	0	16.5	0	28.5	0	26.5	13.5	36
1500	0	23	0	37.5	0	29.5	14.5	41

10-meter band horizontal, half-wave dipole, Frequency = 29.7, Height above ground = 50 feet

Horizontal distance (feet) from any part of the antenna for compliance with occupational/controlled or general population/uncontrolled exposure limits*

Height above ground (feet) where exposure occurs

Power** (watts)	6 feet		12 feet		20 feet		50 feet	
	con.	unc.	con.	unc.	con.	unc.	con.	unc.
10	0	0	0	0	0	0	1.5	3
25	0	0	0	0	0	0	2	4.5
50	0	0	0	0	0	0	3	6
100	0	0	0	0	0	0	4	8.5
200	0	0	0	0	0	0	5.5	12
250	0	0	0	0	0	0	6	13.5
300	0	0	0	0	0	0	6.5	15
400	0	0	0	0	0	0	7.5	17
500	0	0	0	0	0	0	8.5	20
600	0	0	0	0	0	0	9.5	22
750	0	0	0	0	0	0	10.5	24.5
1000	0	0	0	0	0	0	12	27.5
1250	0	0	0	0	0	16.5	13.5	30
1500	0	0	0	0	0	23.5	15	32.5

* 0 feet indicates that the exposure at the height in the column above or below the antenna is in compliance.
** Power = Average power input to the antenna. See Chapter 5.

"G5RV-TYPE" HORIZONTAL DIPOLES

The following tables show the horizontal compliance distance for horizontal full-size G5RV antennas, or 80-meter half-wave dipoles fed with ladder line. The model calculates the required distance from the antenna at various heights, modeled at the center of the antenna. The exposure at the ends is generally less, so this is a conservative estimate for the minimum distance from any part of the antenna.

This model also serves reasonably well for G5RV-type inverted V antennas, using the minimum height of the inverted V as the antenna height. It can also provide a reasonable estimate of the minimum distance required from any part of a longwire antenna of similar dimensions. The 10-meter dipole models are a reasonable estimate of the compliance distance to be expected for VHF antennas of similar configuration.

115-foot, center-fed dipole, Frequency = 2.0 MHz, Height above ground = 10 feet

Horizontal distance (feet) from any part of the antenna for compliance with occupational/controlled or general population/uncontrolled exposure limits*

Height above ground (feet) where exposure occurs

| Power** | 6 feet | | 12 feet | | 20 feet | | 30 feet | |
(watts)	con.	unc.	con.	unc.	con.	unc.	con.	unc.
10	0	0	0	0	0	0	0	0
25	0	0	0	0	0	0	0	0
50	0	0	0	0	0	0	0	0
100	0	0	0	0	0	0	0	0
200	0	0	0	0	0	0	0	0
250	0	0	0	1.5	0	0	0	0
300	0	0	0	2	0	0	0	0
400	0	0	0	2.5	0	0	0	0
500	0	0	1.5	2.5	0	0	0	0
600	0	0	2	3	0	0	0	0
750	0	0	2	3	0	0	0	0
1000	0	0	2.5	3.5	0	0	0	0
1250	0	2	2.5	4	0	0	0	0
1500	0	3	3	4	0	0	0	0

115-foot, center-fed dipole, Frequency = 2.0 MHz, Height above ground = 20 feet

Horizontal distance (feet) from any part of the antenna for compliance with occupational/controlled or general population/uncontrolled exposure limits*

Height above ground (feet) where exposure occurs

| Power** | 6 feet | | 12 feet | | 20 feet | | 30 feet | |
(watts)	con.	unc.	con.	unc.	con.	unc.	con.	unc.
10	0	0	0	0	1	1	0	0
25	0	0	0	0	1	1	0	0
50	0	0	0	0	1	1.5	0	0
100	0	0	0	0	1.5	1.5	0	0
200	0	0	0	0	1.5	2	0	0
250	0	0	0	0	1.5	2.5	0	0
300	0	0	0	0	2	2.5	0	0
400	0	0	0	0	2	3	0	0
500	0	0	0	0	2.5	3.5	0	0
600	0	0	0	0	2.5	4	0	0
750	0	0	0	0	3	4.5	0	0
1000	0	0	0	0	3.5	5	0	0
1250	0	0	0	0	4	5.5	0	0
1500	0	0	0	0	4	6	0	0

* 0 feet indicates that the exposure at the height in the column above or below the antenna is in compliance.
** Power = Average power input to the antenna. See Chapter 5.

115-foot, center-fed dipole, Frequency = 7.3 MHz, Height above ground = 10 feet

Horizontal distance (feet) from any part of the antenna for compliance with occupational/controlled or general population/uncontrolled exposure limits*

Power**	6 feet		12 feet		20 feet		30 feet	
(watts)	con.	unc.	con.	unc.	con.	unc.	con.	unc.
10	0	0	0	0	0	0	0	0
25	0	0	0	0	0	0	0	0
50	0	0	0	0	0	0	0	0
100	0	0	0	0	0	0	0	0
200	0	0	0	0	0	0	0	0
250	0	0	0	0	0	0	0	0
300	0	0	0	0	0	0	0	0
400	0	0	0	2	0	0	0	0
500	0	0	0	2.5	0	0	0	0
600	0	0	0	3	0	0	0	0
750	0	0	0	3.5	0	0	0	0
1000	0	0	0	4.5	0	0	0	0
1250	0	2	0	5	0	0	0	0
1500	0	3	0	5.5	0	0	0	0

115-foot, center-fed dipole, Frequency = 7.3 MHz, Height above ground = 20 feet

Horizontal distance (feet) from any part of the antenna for compliance with occupational/controlled or general population/uncontrolled exposure limits*

Power**	6 feet		12 feet		20 feet		30 feet	
(watts)	con.	unc.	con.	unc.	con.	unc.	con.	unc.
10	0	0	0	0	0	0	0	0
25	0	0	0	0	0	1	0	0
50	0	0	0	0	0	1	0	0
100	0	0	0	0	1	1	0	0
200	0	0	0	0	1	2	0	0
250	0	0	0	0	1	2	0	0
300	0	0	0	0	1	2.5	0	0
400	0	0	0	0	1	3	0	0
500	0	0	0	0	1	3.5	0	0
600	0	0	0	0	1.5	4	0	0
750	0	0	0	0	1.5	4.5	0	0
1000	0	0	0	0	2	5	0	0
1250	0	0	0	0	2	5.5	0	0
1500	0	0	0	0	2.5	6	0	0

115-foot, center-fed dipole, Frequency = 10.15 MHz, Height above ground = 10 feet

Horizontal distance (feet) from any part of the antenna for compliance with occupational/controlled or general population/uncontrolled exposure limits*

Power**	6 feet		12 feet		20 feet		30 feet	
(watts)	con.	unc.	con.	unc.	con.	unc.	con.	unc.
10	0	0	0	0	0	0	0	0
25	0	0	0	0	0	0	0	0
50	0	0	0	0	0	0	0	0
100	0	0	0	0	0	0	0	0
200	0	0	0	1.5	0	0	0	0
250	0	0	0	2	0	0	0	0
300	0	0	0	2.5	0	0	0	0
400	0	0	0	3.5	0	0	0	0
500	0	0	0	4	0	0	0	0
600	0	0	0	4.5	0	0	0	0
750	0	1.5	.5	5	0	0	0	0
1000	0	3	1.5	5.5	0	0	0	0
1250	0	4	2	6.5	0	0	0	0
1500	0	4.5	2.5	7	0	0	0	0

* 0 feet indicates that the exposure at the height in the column above or below the antenna is in compliance.
** Power = Average power input to the antenna. See Chapter 5.

115-foot, center-fed dipole, Frequency = 10.15 MHz, Height above ground = 20 feet

Horizontal distance (feet) from any part of the antenna for compliance with
occupational/controlled or general population/uncontrolled exposure limits*

Power** (watts)	6 feet		12 feet		20 feet		30 feet	
	con.	unc.	con.	unc.	con.	unc.	con.	unc.
10	0	0	0	0	0	1	0	0
25	0	0	0	0	1	1.5	0	0
50	0	0	0	0	1	1.5	0	0
100	0	0	0	0	1	2	0	0
200	0	0	0	0	1.5	2.5	0	0
250	0	0	0	0	1.5	3	0	0
300	0	0	0	0	1.5	3.5	0	0
400	0	0	0	0	2	4	0	0
500	0	0	0	0	2	4.5	0	0
600	0	0	0	0	2	5	0	0
750	0	0	0	0	2.5	5.5	0	0
1000	0	0	0	0	2.5	6.5	0	0
1250	0	0	0	0	3	7	0	0
1500	0	0	0	0	3.5	7.5	0	0

115-foot, center-fed dipole, Frequency = 14.35 MHz, Height above ground = 10 feet

Horizontal distance (feet) from any part of the antenna for compliance with
occupational/controlled or general population/uncontrolled exposure limits*

Power** (watts)	6 feet		12 feet		20 feet		30 feet	
	con.	unc.	con.	unc.	con.	unc.	con.	unc.
10	0	0	0	0	0	0	0	0
25	0	0	0	0	0	0	0	0
50	0	0	0	0	0	0	0	0
100	0	0	0	2	0	0	0	0
200	0	0	0	3.5	0	0	0	0
250	0	2	0	3.5	0	0	0	0
300	0	3	0	4	0	0	0	0
400	0	4.5	1.5	4.5	0	0	0	0
500	0	5.5	2	5.5	0	0	0	0
600	0	6.5	2.5	6	0	0	0	0
750	0	7.5	3	7	0	0	0	0
1000	0	9	3.5	8.5	0	0	0	0
1250	2	10.5	3.5	10	0	4.5	0	0
1500	3	11.5	4	11	0	7	0	0

115-foot, center-fed dipole, Frequency = 14.35 MHz, Height above ground = 20 feet

Horizontal distance (feet) from any part of the antenna for compliance with
occupational/controlled or general population/uncontrolled exposure limits*

Power** (watts)	6 feet		12 feet		20 feet		30 feet	
	con.	unc.	con.	unc.	con.	unc.	con.	unc.
10	0	0	0	0	1	1.5	0	0
25	0	0	0	0	1	2	0	0
50	0	0	0	0	1.5	2.5	0	0
100	0	0	0	0	2	3	0	0
200	0	0	0	0	2	4	0	0
250	0	0	0	0	2.5	4	0	0
300	0	0	0	0	2.5	4.5	0	0
400	0	0	0	0	3	5	0	0
500	0	0	0	0	3	5.5	0	0
600	0	0	0	0	3	6	0	0
750	0	0	0	0	3.5	7	0	0
1000	0	0	0	1.5	4	8.5	0	0
1250	0	0	0	5.5	4	10	0	5
1500	0	3.5	0	7.5	4.5	11.5	0	8

* 0 feet indicates that the exposure at the height in the column above or below the antenna is in compliance.
** Power = Average power input to the antenna. See Chapter 5.

115-foot, center-fed dipole, Frequency = 18.168 MHz, Height above ground = 10 feet

Horizontal distance (feet) from any part of the antenna for compliance with
occupational/controlled or general population/uncontrolled exposure limits*

Power**	6 feet		12 feet		20 feet		30 feet	
(watts)	con.	unc.	con.	unc.	con.	unc.	con.	unc.
10	0	0	0	0	0	0	0	0
25	0	0	0	1	0	0	0	0
50	0	0	0	2.5	0	0	0	0
100	0	0	0	4	0	0	0	0
200	0	3	2	5	0	0	0	0
250	0	3.5	2.5	5.5	0	0	0	0
300	0	4	3	6	0	0	0	0
400	0	5	3.5	7	0	0	0	0
500	0	5.5	4	7.5	0	0	0	0
600	0	6	4	8	0	0	0	0
750	1.5	7	4.5	9	0	0	0	0
1000	3	8	5	10	0	2.5	0	0
1250	3.5	8.5	5.5	11	0	6	0	0
1500	4	9	6	11.5	0	8	0	0

115-foot, center-fed dipole, Frequency = 18.168 MHz, Height above ground = 20 feet

Horizontal distance (feet) from any part of the antenna for compliance with
occupational/controlled or general population/uncontrolled exposure limits*

Height above ground (feet) where exposure occurs

Power**	6 feet		12 feet		20 feet		30 feet	
(watts)	con.	unc.	con.	unc.	con.	unc.	con.	unc.
10	0	0	0	0	1	2	0	0
25	0	0	0	0	1.5	2.5	0	0
50	0	0	0	0	2	3	0	0
100	0	0	0	0	2.5	4	0	0
200	0	0	0	0	3	5	0	0
250	0	0	0	0	3	5.5	0	0
300	0	0	0	0	3.5	6	0	0
400	0	0	0	0	4	7	0	0
500	0	0	0	0	4	7.5	0	0
600	0	0	0	0	4.5	8	0	0
750	0	0	0	4	4.5	9	0	0
1000	0	0	0	6.5	5	10.5	0	0
1250	0	0	0	8.5	5.5	11.5	0	0
1500	0	0	0	10	6	13	0	4

115-foot, center-fed dipole, Frequency = 18.168 MHz, Height above ground = 30 feet

Horizontal distance (feet) from any part of the antenna for compliance with
occupational/controlled or general population/uncontrolled exposure limits*

Height above ground (feet) where exposure occurs

Power**	6 feet		12 feet		20 feet		30 feet	
(watts)	con.	unc.	con.	unc.	con.	unc.	con.	unc.
10	0	0	0	0	0	0	1	2
25	0	0	0	0	0	0	1.5	2.5
50	0	0	0	0	0	0	2	3.5
100	0	0	0	0	0	0	2.5	4
200	0	0	0	0	0	0	3	5
250	0	0	0	0	0	0	3.5	5.5
300	0	0	0	0	0	0	3.5	6
400	0	0	0	0	0	0	4	6.5
500	0	0	0	0	0	0	4	7
600	0	0	0	0	0	0	4.5	7.5
750	0	0	0	0	0	0	4.5	8
1000	0	0	0	0	0	0	5	9
1250	0	0	0	0	0	4	5.5	10
1500	0	0	0	0	0	7	6	10.5

* 0 feet indicates that the exposure at the height in the column above or below the antenna is in compliance.
** Power = Average power input to the antenna. See Chapter 5.

115-foot, center-fed dipole, Frequency = 21.45 MHz, Height above ground = 10 feet

Horizontal distance (feet) from any part of the antenna for compliance with
occupational/controlled or general population/uncontrolled exposure limits*

Height above ground (feet) where exposure occurs

Power** (watts)	6 feet		12 feet		20 feet		30 feet	
	con.	unc.	con.	unc.	con.	unc.	con.	unc.
10	0	0	0	0	0	0	0	0
25	0	0	0	1	0	0	0	0
50	0	0	0	2.5	0	0	0	0
100	0	3	0	4	0	0	0	0
200	0	5.5	2	6	0	0	0	0
250	0	6.5	2.5	7	0	0	0	0
300	0	7	3	7.5	0	0	0	0
400	2	8	3.5	8.5	0	0	0	0
500	3	9	4	9.5	0	2	0	0
600	3.5	9.5	4.5	10	0	5	0	0
750	4.5	10.5	5.5	11	0	7	0	0
1000	5.5	12	6	12.5	0	9	0	0
1250	6.5	13	7	13.5	0	11	0	0
1500	7	13.5	7.5	14.5	0	12	0	0

115-foot, center-fed dipole, Frequency = 21.45 MHz, Height above ground = 20 feet

Horizontal distance (feet) from any part of the antenna for compliance with
occupational/controlled or general population/uncontrolled exposure limits*

Height above ground (feet) where exposure occurs

Power** (watts)	6 feet		12 feet		20 feet		30 feet	
	con.	unc.	con.	unc.	con.	unc.	con.	unc.
10	0	0	0	0	1	1.5	0	0
25	0	0	0	0	1	2.5	0	0
50	0	0	0	0	1.5	3.5	0	0
100	0	0	0	0	2	4.5	0	0
200	0	0	0	0	3	6.5	0	0
250	0	0	0	0	3.5	7	0	0
300	0	0	0	0	3.5	8	0	0
400	0	0	0	3	4	9	0	0
500	0	0	0	5	4.5	10	0	0
600	0	0	0	6.5	5	10.5	0	3.5
750	0	0	0	8	5.5	11.5	0	6
1000	0	0	0	10	6.5	13	0	8.5
1250	0	5	0	11	7	14	0	10
1500	0	7	0	12.5	8	15	0	11.5

115-foot, center-fed dipole, Frequency = 21.45 MHz, Height above ground = 30 feet

Horizontal distance (feet) from any part of the antenna for compliance with
occupational/controlled or general population/uncontrolled exposure limits*

Height above ground (feet) where exposure occurs

Power** (watts)	6 feet		12 feet		20 feet		30 feet	
	con.	unc.	con.	unc.	con.	unc.	con.	unc.
10	0	0	0	0	0	0	1	1.5
25	0	0	0	0	0	0	1	2.5
50	0	0	0	0	0	0	1.5	3.5
100	0	0	0	0	0	0	2.5	5
200	0	0	0	0	0	0	3	7
250	0	0	0	0	0	0	3.5	7.5
300	0	0	0	0	0	0	4	8
400	0	0	0	0	0	0	4.5	9.5
500	0	0	0	0	0	0	5	10.5
600	0	0	0	0	0	4	5.5	11
750	0	0	0	0	0	6	6	12.5
1000	0	0	0	0	0	9	7	14
1250	0	0	0	0	0	10.5	7.5	15
1500	0	0	0	0	0	12	8	16.5

* 0 feet indicates that the exposure at the height in the column above or below the antenna is in compliance.
** Power = Average power input to the antenna. See Chapter 5.

115-foot, center-fed dipole, Frequency = 24.99 MHz, Height above ground = 10 feet

Horizontal distance (feet) from any part of the antenna for compliance with
occupational/controlled or general population/uncontrolled exposure limits*

Height above ground (feet) where exposure occurs

Power** (watts)	6 feet con.	6 feet unc.	12 feet con.	12 feet unc.	20 feet con.	20 feet unc.	30 feet con.	30 feet unc.
10	0	0	0	0	0	0	0	0
25	0	0	0	2	0	0	0	0
50	0	0	0	3.5	0	0	0	0
100	0	2	1.5	5	0	0	0	0
200	0	5	3	6.5	0	0	0	0
250	0	5.5	3.5	7.5	0	0	0	0
300	0	6	3.5	8	0	0	0	0
400	0	7.5	4.5	9.5	0	3.5	0	0
500	2	8	5	10.5	0	9	0	0
600	3	9	5	11.5	0	12	0	0
750	4	11	6	14	0	15.5	0	5
1000	5	14	6.5	17.5	0	19.5	0	13.5
1250	5.5	16	7.5	19.5	0	22	0	18
1500	6	17.5	8	21.5	0	24	0	21

115-foot, center-fed dipole, Frequency = 24.99 MHz, Height above ground = 20 feet

Horizontal distance (feet) from any part of the antenna for compliance with
occupational/controlled or general population/uncontrolled exposure limits*

Height above ground (feet) where exposure occurs

Power** (watts)	6 feet con.	6 feet unc.	12 feet con.	12 feet unc.	20 feet con.	20 feet unc.	30 feet con.	30 feet unc.
10	0	0	0	0	1.5	2	0	0
25	0	0	0	0	1.5	3	0	0
50	0	0	0	0	2	3.5	0	0
100	0	0	0	0	2.5	4.5	0	0
200	0	0	0	0	3.5	6	0	0
250	0	0	0	0	3.5	6.5	0	0
300	0	0	0	0	4	7	0	0
400	0	0	0	4.5	4.5	8	0	0
500	0	0	0	7	4.5	10.5	0	0
600	0	0	0	9	5	13.5	0	4.5
750	0	7	0	11.5	5.5	17	0	10
1000	0	13	0	17	6	21	0	16.5
1250	0	16.5	0	20.5	6.5	24	0	20.5
1500	0	19	0	23	7	26.5	0	23

115-foot, center-fed dipole, Frequency = 24.99 MHz, Height above ground = 30 feet

Horizontal distance (feet) from any part of the antenna for compliance with
occupational/controlled or general population/uncontrolled exposure limits*

Height above ground (feet) where exposure occurs

Power** (watts)	6 feet con.	6 feet unc.	12 feet con.	12 feet unc.	20 feet con.	20 feet unc.	30 feet con.	30 feet unc.
10	0	0	0	0	0	0	1.5	2.5
25	0	0	0	0	0	0	2	3
50	0	0	0	0	0	0	2.5	4
100	0	0	0	0	0	0	3	5
200	0	0	0	0	0	0	3.5	6.5
250	0	0	0	0	0	0	4	7
300	0	0	0	0	0	0	4	7.5
400	0	0	0	0	0	0	4.5	8.5
500	0	0	0	0	0	4	5	10
600	0	0	0	0	0	9	5.5	12.5
750	0	0	0	0	0	13	6	15.5
1000	0	0	0	0	0	17.5	6.5	19.5
1250	0	0	0	7.5	0	21	7	22.5
1500	0	8.5	0	14	0	24	7.5	25

* 0 feet indicates that the exposure at the height in the column above or below the antenna is in compliance.
** Power = Average power input to the antenna. See Chapter 5.

115-foot, center-fed dipole, Frequency = 29.7 MHz, Height above ground = 10 feet

Horizontal distance (feet) from any part of the antenna for compliance with occupational/controlled or general population/uncontrolled exposure limits*

Power** (watts)	6 feet		12 feet		20 feet		30 feet	
	con.	unc.	con.	unc.	con.	unc.	con.	unc.
10	0	0	0	0	0	0	0	0
25	0	0	0	2.5	0	0	0	0
50	0	1.5	0	4	0	0	0	0
100	0	4	2	5.5	0	0	0	0
200	0	6.5	3.5	7.5	0	0	0	0
250	1.5	7	4	8	0	0	0	0
300	2.5	7.5	4.5	8.5	0	0	0	0
400	3.5	8.5	5	9.5	0	0	0	0
500	4	9.5	5.5	10	0	2	0	0
600	5	10	6	10.5	0	4	0	0
750	5.5	10.5	6.5	11.5	0	5.5	0	0
1000	6.5	11.5	7.5	12	0	7.5	0	0
1250	7	12	8	13	0	8.5	0	0
1500	7.5	12.5	8.5	13.5	0	9	0	0

115-foot, center-fed dipole, Frequency = 29.7 MHz, Height above ground = 20 feet

Horizontal distance (feet) from any part of the antenna for compliance with occupational/controlled or general population/uncontrolled exposure limits*

Power** (watts)	6 feet		12 feet		20 feet		30 feet	
	con.	unc.	con.	unc.	con.	unc.	con.	unc.
10	0	0	0	0	1	2	0	0
25	0	0	0	0	1.5	3.5	0	0
50	0	0	0	0	2	4.5	0	0
100	0	0	0	0	3	6	0	0
200	0	0	0	0	4	7.5	0	0
250	0	0	0	2.5	4.5	8.5	0	0
300	0	0	0	4	5	9	0	0
400	0	0	0	6	5.5	9.5	0	2.5
500	0	0	0	7	6	10.5	0	4.5
600	0	0	0	7.5	6.5	11	0	6
750	0	0	0	8.5	7	11.5	0	7.5
1000	0	0	0	10	7.5	12.5	0	9.5
1250	0	0	2.5	10.5	8.5	13.5	0	10.5
1500	0	2.5	4	11.5	9	14	0	12

115-foot, center-fed dipole, Frequency = 29.7 MHz, Height above ground = 30 feet

Horizontal distance (feet) from any part of the antenna for compliance with occupational/controlled or general population/uncontrolled exposure limits*

Power** (watts)	6 feet		12 feet		20 feet		30 feet	
	con.	unc.	con.	unc.	con.	unc.	con.	unc.
10	0	0	0	0	0	0	1	2.5
25	0	0	0	0	0	0	1.5	3.5
50	0	0	0	0	0	0	2.5	4.5
100	0	0	0	0	0	0	3	6
200	0	0	0	0	0	0	4	8
250	0	0	0	0	0	0	4.5	8.5
300	0	0	0	0	0	0	5	9
400	0	0	0	0	0	0	5.5	10
500	0	0	0	0	0	0	6	11
600	0	0	0	0	0	2.5	6.5	11.5
750	0	0	0	0	0	5	7	12
1000	0	0	0	0	0	7	8	13.5
1250	0	0	0	0	0	8	8.5	14
1500	0	0	0	0	0	9	9	15

* 0 feet indicates that the exposure at the height in the column above or below the antenna is in compliance.
** Power = Average power input to the antenna. See Chapter 5.

QUARTER-WAVE, GROUND-MOUNTED VERTICALS

The following tables show the horizontal compliance distance for quarter-wave, ground-mounted verticals, using 64 radials buried 1 inch below excellent ground. In general, this represents typical performance of excellent vertical systems. The model calculates the required distance from the antenna at various heights, modeled at the center of the antenna. In some cases, the exposure height is located above the antenna. In these cases, a compliance distance of "0" feet indicates that the areas above the antenna at that exposure height are in compliance.

This model also serves reasonably well as a conservative estimate for end-fed random wires longer than a quarter wavelength, for ground-plane antennas located very close to ground and for most antennas where people can be located very close to the antenna conductors. The 10-meter models serve as a reasonable estimate for VHF ground-mounted, quarter-wa

---

160-meter band quarter-wave vertical, 64 radials, Frequency = 2.0 MHz, Height above ground = 0 feet

Horizontal distance (feet) from any part of the antenna for compliance with
occupational/controlled or general population/uncontrolled exposure limits*

Power**	6 feet		12 feet		20 feet		30 feet	
(watts)	con.	unc.	con.	unc.	con.	unc.	con.	unc.
10	.5	.5	.5	.5	.5	.5	.5	.5
25	.5	.5	.5	.5	.5	.5	.5	.5
50	.5	1	.5	1	.5	1	.5	.5
100	1	1	1	1	.5	1	.5	1
200	1	1.5	1	1.5	1	1.5	1	1.5
250	1	1.5	1	1.5	1	1.5	1	1.5
300	1	1.5	1	1.5	1	1.5	1	1.5
400	1.5	2	1.5	1.5	1	1.5	1	1.5
500	1.5	2	1.5	2	1.5	2	1.5	2
600	1.5	2	1.5	2	1.5	2	1.5	2
750	1.5	2.5	1.5	2.5	1.5	2.5	1.5	2
1000	2	2.5	2	2.5	2	2.5	2	2.5
1250	2	3	2	3	2	3	2	2.5
1500	2	3	2	3	2	3	2	3

80-meter band quarter-wave vertical, 64 radials, Frequency = 4.0 MHz, Height above ground = 0 feet

Horizontal distance (feet) from any part of the antenna for compliance with
occupational/controlled or general population/uncontrolled exposure limits*

Power**	6 feet		12 feet		20 feet		30 feet	
(watts)	con.	unc.	con.	unc.	con.	unc.	con.	unc.
10	.5	.5	.5	.5	.5	.5	.5	.5
25	.5	1	.5	1	.5	1	.5	1.5
50	.5	1.5	.5	1.5	.5	1.5	.5	1.5
100	1	2	1	1.5	1	1.5	1	2
200	1	2.5	1	2.5	1	2	1.5	2
250	1.5	2.5	1.5	2.5	1.5	2.5	1.5	2.5
300	1.5	3	1.5	3	1.5	2.5	1.5	2.5
400	1.5	3.5	1.5	3	1.5	3	1.5	3
500	2	3.5	1.5	3.5	1.5	3.5	2	3
600	2	4	2	4	1.5	3.5	2	3.5
750	2	4.5	2	4.5	2	4	2	3.5
1000	2.5	5	2.5	5	2	4.5	2	4
1250	2.5	5.5	2.5	5.5	2.5	5	2.5	4.5
1500	3	6	3	6	2.5	5.5	2.5	4.5

---

* 0 feet indicates that the exposure at the height in the column above or below the antenna is in compliance.
** Power = Average power input to the antenna. See Chapter 5.

40-meter band quarter-wave vertical, 64 radials, Frequency = 7.3 MHz, Height above ground = 0 feet

Horizontal distance (feet) from any part of the antenna for compliance with
occupational/controlled or general population/uncontrolled exposure limits*

Height above ground (feet) where exposure occurs

Power**	6 feet		12 feet		20 feet		30 feet	
(watts)	con.	unc.	con.	unc.	con.	unc.	con.	unc.
10	.5	1	.5	1	.5	1.5	.5	1.5
25	1	1.5	1	1.5	1.5	2	1.5	2
50	1	2	1	2	1.5	2.5	1.5	3
100	1.5	3	1.5	2.5	2	3	2	3.5
200	2	4	2	3.5	2	4	2.5	4.5
250	2	4.5	2	4	2.5	4.5	3	5
300	2.5	5	2	4.5	2.5	4.5	3	5
400	2.5	5.5	2.5	5	3	5.5	3.5	5.5
500	3	6.5	2.5	5.5	3	6	3.5	6
600	3	7	3	6	3.5	6.5	3.5	6.5
750	3.5	7.5	3	7	3.5	7	4	7
1000	4	9	3.5	8	4	8	4.5	8
1250	4.5	10	4	8.5	4.5	8.5	5	8.5
1500	5	10.5	4.5	9.5	4.5	9	5	9.5

30-meter band quarter-wave vertical, 64 radials, Frequency = 10.15 MHz, Height above ground = 0 feet

Horizontal distance (feet) from any part of the antenna for compliance with
occupational/controlled or general population/uncontrolled exposure limits*

Height above ground (feet) where exposure occurs

Power**	6 feet		12 feet		20 feet		30 feet	
(watts)	con.	unc.	con.	unc.	con.	unc.	con.	unc.
10	1	1.5	.5	1.5	1.5	2	0	0
25	1	2	1.5	2	1.5	2.5	0	0
50	1.5	3	1.5	2.5	2	3.5	0	0
100	2	4	2	3.5	2.5	4.5	0	0
200	2.5	5.5	2.5	4.5	3	5.5	0	0
250	3	6	2.5	5	3.5	6	0	0
300	3	6.5	3	5.5	3.5	6.5	0	0
400	3.5	7.5	3.5	6.5	4	7	0	0
500	4	8.5	3.5	7	4.5	8	0	0
600	4.5	9	4	7.5	4.5	8.5	0	0
750	4.5	10.5	4	8	5	9	0	0
1000	5.5	12	4.5	9.5	5.5	10	0	0
1250	6	13	5	10.5	6	11	0	0
1500	6.5	14.5	5.5	12	6.5	11.5	0	3.5

20-meter band quarter-wave vertical, 64 radials, Frequency = 14.35 MHz, Height above ground = 0 feet

Horizontal distance (feet) from any part of the antenna for compliance with
occupational/controlled or general population/uncontrolled exposure limits*

Height above ground (feet) where exposure occurs

Power**	6 feet		12 feet		20 feet		30 feet	
(watts)	con.	unc.	con.	unc.	con.	unc.	con.	unc.
10	1	2	1.5	2.5	0	0	0	0
25	1.5	2.5	2	3	0	0	0	0
50	2	3.5	2.5	4	0	0	0	0
100	2.5	5	3	5.5	0	0	0	0
200	3.5	7	4	7	0	2.5	0	0
250	3.5	8	4	7.5	0	3.5	0	0
300	4	8.5	4.5	8	0	4.5	0	0
400	4.5	10	5	9	0	5.5	0	0
500	5	11	5.5	9.5	0	6.5	0	0
600	5.5	12.5	5.5	10.5	0	7	0	0
750	6	13.5	6	11.5	0	8	0	0
1000	7	16	7	13	2.5	9.5	0	0
1250	8	17.5	7.5	14.5	3.5	10.5	0	0
1500	8.5	19.5	8	15.5	4.5	12	0	0

* 0 feet indicates that the exposure at the height in the column above or below the antenna is in compliance.
** Power = Average power input to the antenna. See Chapter 5.

17-meter band quarter-wave vertical, 64 radials, Frequency = 18.168 MHz, Height above ground = 0 feet

Horizontal distance (feet) from any part of the antenna for compliance with occupational/controlled or general population/uncontrolled exposure limits*

Height above ground (feet) where exposure occurs

Power** (watts)	6 feet con.	6 feet unc.	12 feet con.	12 feet unc.	20 feet con.	20 feet unc.	30 feet con.	30 feet unc.
10	1.5	2	2	3	0	0	0	0
25	2	3	2.5	3.5	0	0	0	0
50	2	4	3	4.5	0	0	0	0
100	3	6	3.5	6	0	0	0	0
200	4	8.5	4.5	7.5	0	0	0	0
250	4	9.5	4.5	8	0	0	0	0
300	4.5	10.5	5	9	0	0	0	0
400	5.5	12	5.5	10	0	0	0	0
500	6	13.5	6	11	0	0	0	0
600	6.5	15	6.5	12	0	2.5	0	0
750	7.5	16.5	7	13	0	5	0	0
1000	8.5	19	7.5	15	0	7.5	0	0
1250	9.5	21.5	8	17	0	9.5	0	0
1500	10.5	23.5	9	19	0	11	0	0

15-meter band quarter-wave vertical, 64 radials, Frequency = 21.45 MHz, Height above ground = 0 feet

Horizontal distance (feet) from any part of the antenna for compliance with occupational/controlled or general population/uncontrolled exposure limits*

Height above ground (feet) where exposure occurs

Power** (watts)	6 feet con.	6 feet unc.	12 feet con.	12 feet unc.	20 feet con.	20 feet unc.	30 feet con.	30 feet unc.
10	1.5	2.5	.5	2	0	0	0	0
25	2	4	1	3	0	0	0	0
50	2.5	5	2	4	0	0	0	0
100	3.5	6.5	3	5.5	0	0	0	0
200	4.5	9.5	4	7.5	0	0	0	0
250	5	11	4	8.5	0	0	0	0
300	5.5	12	4.5	9	0	0	0	0
400	6	14	5	10.5	0	0	0	0
500	6.5	15.5	5.5	11.5	0	0	0	0
600	7.5	17	6	13	0	0	0	0
750	8	19	6.5	14.5	0	0	0	0
1000	9.5	22	7.5	17	0	5	0	0
1250	11	24.5	8.5	19.5	0	8	0	0
1500	12	28	9	22	0	10.5	0	0

12-meter band quarter-wave vertical, 64 radials, Frequency = 24.99 MHz, Height above ground = 0 feet

Horizontal distance (feet) from any part of the antenna for compliance with occupational/controlled or general population/uncontrolled exposure limits*

Height above ground (feet) where exposure occurs

Power** (watts)	6 feet con.	6 feet unc.	12 feet con.	12 feet unc.	20 feet con.	20 feet unc.	30 feet con.	30 feet unc.
10	2	3	0	0	0	0	0	0
25	2.5	4.5	0	0	0	0	0	0
50	3	5.5	0	3	0	0	0	0
100	4	7.5	0	5	0	0	0	0
200	5	10.5	2.5	7.5	0	0	0	0
250	5.5	12	3	8.5	0	0	0	0
300	6	13.5	3.5	9.5	0	0	0	0
400	7	15.5	4.5	11	0	0	0	0
500	7.5	17.5	5	12.5	0	0	0	0
600	8	19.5	5.5	14	0	0	0	0
750	9	21.5	6.5	16.5	0	0	0	0
1000	10.5	25	7.5	20	0	0	0	0
1250	12	27.5	8.5	23	0	6	0	0
1500	13.5	30	9.5	26	0	10.5	0	0

* 0 feet indicates that the exposure at the height in the column above or below the antenna is in compliance.
** Power = Average power input to the antenna. See Chapter 5.

10-meter band quarter-wave vertical, 64 radials, Frequency = 29.7 MHz, Height above ground = 0 feet

Horizontal distance (feet) from any part of the antenna for compliance with
occupational/controlled or general population/uncontrolled exposure limits*

Power**	\		\	Height above ground (feet) where exposure occurs				
	6 feet		12 feet		20 feet		30 feet	
(watts)	con.	unc.	con.	unc.	con.	unc.	con.	unc.
10	2	3.5	0	0	0	0	0	0
25	3	5	0	0	0	0	0	0
50	3.5	6.5	0	0	0	0	0	0
100	4.5	8.5	0	4.5	0	0	0	0
200	6	12.5	0	8	0	0	0	0
250	6.5	14	0	9	0	0	0	0
300	7	15.5	2.5	10	0	0	0	0
400	8	18	3.5	12.5	0	0	0	0
500	8.5	20.5	4.5	15	0	0	0	0
600	9.5	22.5	5.5	17	0	0	0	0
750	10.5	25	6.5	20	0	0	0	0
1000	12.5	28.5	8	24	0	0	0	0
1250	14	31.5	9	28	0	7.5	0	0
1500	15.5	34.5	10	31	0	17	0	0

HALF-WAVE VERTICALS

The following tables show the horizontal compliance distance for half-wave type verticals at various heights above average ground. The models are fed at the center, but feeding at the end results in approximately the same compliance distances. This model applies well to antennas such as the Cushcraft R7000, the Gap, etc. These antennas were modeled without a radial system. For those half-wave type verticals that have a radial system, a conservative estimate would the worst case of the quarter-wave or half-wave vertical compliance distance. A compliance distance of "0" feet indicates that the areas above or below the antenna at that exposure height are in compliance.

This model also serves reasonably well as a conservative estimate for end-fed random wires longer than a quarter wavelength, for ground-plane antennas located very close to ground of for most antennas where people can be located very close to the antenna conductors. The 10-meter models serve as a reasonable estimate for VHF half-wave type antennas. Using the next higher power level serves as a reasonable estimate for 5/8-wave type antennas.

40-meter band half-wave, center-fed vertical, Frequency = 7.3 MHz, Height above ground = 10 feet

Horizontal distance (feet) from any part of the antenna for compliance with
occupational/controlled or general population/uncontrolled exposure limits*

Power**				Height above ground (feet) where exposure occurs				
	6 feet		12 feet		20 feet		30 feet	
(watts)	con.	unc.	con.	unc.	con.	unc.	con.	unc.
10	0	0	.5	1.5	.5	1.5	.5	1
25	0	0	1.5	2	.5	1.5	.5	1.5
50	0	0	1.5	2.5	1.5	2	1	1.5
100	0	0	2	3	1.5	2.5	1	2
200	0	0	2	3.5	2	3	1.5	2.5
250	0	0	2.5	4	2	3.5	1.5	3
300	0	0	2.5	4	2	4	1.5	3
400	0	0	2.5	4.5	2.5	4	1.5	3.5
500	0	0	3	5	2.5	4.5	2	4
600	0	0	3	5	2.5	5	2	4.5
750	0	0	3	5.5	3	5.5	2.5	5
1000	0	0	3.5	6	3	6	2.5	5.5
1250	0	2.5	4	6.5	3.5	6.5	3	6
1500	0	3.5	4	7	4	7	3	6.5

* 0 feet indicates that the exposure at the height in the column above or below the antenna is in compliance.
** Power = Average power input to the antenna. See Chapter 5.

30-meter band half-wave, center-fed vertical, Frequency = 10.15 MHz, Height above ground = 10 feet

Horizontal distance (feet) from any part of the antenna for compliance with
occupational/controlled or general population/uncontrolled exposure limits*

| Power** | 6 feet | | 12 feet | | 20 feet | | 30 feet | |
(watts)	con.	unc.	con.	unc.	con.	unc.	con.	unc.
10	0	0	.5	1.5	.5	1.5	.5	1
25	0	0	1.5	2	.5	2	1	1.5
50	0	0	1.5	3	1.5	2.5	1	2
100	0	0	2	3.5	1.5	3	1.5	3
200	0	0	2.5	4.5	2	4	2	4
250	0	0	3	4.5	2.5	4	2	4.5
300	0	0	3	5	2.5	4.5	2.5	5
400	0	0	3	5.5	2.5	5	2.5	5.5
500	0	0	3.5	6	3	5.5	3	6.5
600	0	0	3.5	6.5	3	6	3.5	7
750	0	2.5	4	7	3.5	6.5	3.5	7.5
1000	0	4	4.5	7.5	4	7.5	4	9
1250	0	5	4.5	8.5	4	8	4.5	10
1500	0	6	5	9	4.5	8.5	5	10.5

20-meter band half-wave, center-fed vertical, Frequency = 14.35 MHz, Height above ground = 10 feet

Horizontal distance (feet) from any part of the antenna for compliance wit
occupational/controlled or general population/uncontrolled exposure limits*

| Power** | 6 feet | | 12 feet | | 20 feet | | 30 feet | |
(watts)	con.	unc.	con.	unc.	con.	unc.	con.	unc.
10	0	0	1.5	2	.5	1.5	1	1.5
25	0	0	1.5	2.5	1.5	2	1	2
50	0	0	2	3.5	1.5	2.5	1.5	3
100	0	0	2.5	4.5	2	3.5	2	4
200	0	0	3	5.5	2.5	5	2.5	5.5
250	0	0	3.5	6	2.5	5.5	3	6.5
300	0	0	3.5	6.5	3	6	3.5	7
400	0	2	4	7	3.5	7	3.5	8
500	0	3.5	4.5	7.5	3.5	8	4	9
600	0	4.5	4.5	8	4	8.5	4.5	9.5
750	0	5.5	5	9	4.5	9.5	5	11
1000	0	7	5.5	10	5	11	5.5	12.5
1250	0	8	6	11	5.5	12.5	6.5	13.5
1500	0	9	6.5	12	6	13.5	7	15

17-meter band half-wave, center-fed vertical, Frequency = 18.168 MHz, Height above ground = 10 feet

Horizontal distance (feet) from any part of the antenna for compliance with
occupational/controlled or general population/uncontrolled exposure limits*

| Power** | 6 feet | | 12 feet | | 20 feet | | 30 feet | |
(watts)	con.	unc.	con.	unc.	con.	unc.	con.	unc.
10	0	0	1.5	2.5	1	2	1.5	2
25	0	0	2	3	1.5	2.5	1.5	2.5
50	0	0	2.5	4	2	3.5	2	3.5
100	0	0	3	5	2.5	5	2.5	4.5
200	0	0	3.5	6.5	3.5	7	3	6
250	0	0	4	7	3.5	8	3.5	6.5
300	0	2.5	4	7.5	4	8.5	3.5	7
400	0	4	4.5	8	4.5	10	4	8.5
500	0	5	5	9	5	11	4.5	9.5
600	0	6	5	9.5	5.5	12	5	10.5
750	0	7	5.5	10.5	6	13.5	5	12
1000	0	9	6.5	12.5	7	15	6	14
1250	0	10.5	7	14	8	17	6.5	16
1500	2.5	11.5	7.5	15.5	8.5	18	7	17.5

* 0 feet indicates that the exposure at the height in the column above or below the antenna is in compliance.
** Power = Average power input to the antenna. See Chapter 5.

15-meter band half-wave, center-fed vertical, Frequency = 21.45 MHz, Height above ground = 10 feet

Horizontal distance (feet) from any part of the antenna for compliance with
occupational/controlled or general population/uncontrolled exposure limits*

Height above ground (feet) where exposure occurs

Power**	6 feet		12 feet		20 feet		30 feet	
(watts)	con.	unc.	con.	unc.	con.	unc.	con.	unc.
10	0	0	1.5	2.5	1	2	1.5	2.5
25	0	0	2	3.5	1.5	3.5	2	3.5
50	0	0	2.5	4.5	2	4.5	2.5	4.5
100	0	0	3	5.5	3	6.5	3	5.5
200	0	0	4	7	4	9	4	7
250	0	3	4.5	7.5	4.5	9.5	4.5	7.5
300	0	4	4.5	8	5	10.5	4.5	8
400	0	5	5	9.5	5.5	12	5	9
500	0	6.5	5.5	10.5	6.5	13.5	5.5	10
600	0	7.5	6	11.5	7	14.5	6	11
750	0	9	6.5	13	7.5	16	6.5	12.5
1000	0	11	7	15.5	9	18	7	15
1250	3	13.5	7.5	17.5	9.5	20	7.5	17.5
1500	4	15.5	8	20	10.5	21.5	8	19.5

15-meter band half-wave, center-fed vertical, Frequency = 21.45 MHz, Height above ground = 20 feet

Horizontal distance (feet) from any part of the antenna for compliance with
occupational/controlled or general population/uncontrolled exposure limits*

Height above ground (feet) where exposure occurs

Power**	6 feet		12 feet		20 feet		30 feet	
(watts)	con.	unc.	con.	unc.	con.	unc.	con.	unc.
10	0	0	0	0	1.5	2	1	2
25	0	0	0	0	2	3	1.5	3
50	0	0	0	0	2	4	2	4.5
100	0	0	0	0	3	5	3	6
200	0	0	0	0	3.5	6.5	4	8.5
250	0	0	0	0	4	7	4.5	10
300	0	0	0	0	4	7.5	5	10.5
400	0	0	0	0	4.5	8.5	5.5	12.5
500	0	0	0	0	5	9.5	6	14
600	0	0	0	0	5	10	7	15
750	0	0	0	0	5.5	11.5	7.5	17
1000	0	0	0	0	6.5	13	8.5	20
1250	0	0	0	0	7	15	10	22.5
1500	0	0	0	4	7.5	17	10.5	24.5

12-meter band half-wave, center-fed vertical, Frequency = 24.99 MHz, Height above ground = 10 feet

Horizontal distance (feet) from any part of the antenna for compliance with
occupational/controlled or general population/uncontrolled exposure limits*

Height above ground (feet) where exposure occurs

Power**	6 feet		12 feet		20 feet		30 feet	
(watts)	con.	unc.	con.	unc.	con.	unc.	con.	unc.
10	0	0	1.5	2.5	1.5	2.5	0	1.5
25	0	0	2	3.5	2	4	0	2.5
50	0	0	2.5	4.5	2.5	5.5	1.5	3.5
100	0	0	3.5	6	3.5	7.5	2.5	5
200	0	2.5	4.5	8	5	10.5	3	7
250	0	4	4.5	8.5	5.5	11.5	3.5	7.5
300	0	5	5	9.5	6	12.5	4	8.5
400	0	6.5	5.5	11	6.5	14	4.5	9.5
500	0	8	6	12.5	7.5	15.5	5	11
600	0	9.5	6.5	14	8	17	5.5	12
750	0	11.5	7	16	9	18.5	6	13.5
1000	2.5	15	8	19	10.5	20.5	7	17
1250	4	18	8.5	21.5	11.5	22.5	7.5	20
1500	5	21.5	9.5	24	12.5	24.5	8.5	22.5

* 0 feet indicates that the exposure at the height in the column above or below the antenna is in compliance.
** Power = Average power input to the antenna. See Chapter 5.

12-meter band half-wave, center-fed vertical, Frequency = 24.99 MHz, Height above ground = 20 feet

Horizontal distance (feet) from any part of the antenna for compliance with
occupational/controlled or general population/uncontrolled exposure limits*

Height above ground (feet) where exposure occurs

Power**	6 feet		12 feet		20 feet		30 feet	
(watts)	con.	unc.	con.	unc.	con.	unc.	con.	unc.
10	0	0	0	0	1.5	2.5	1	2.5
25	0	0	0	0	2	3.5	2	3.5
50	0	0	0	0	2.5	4	2.5	5
100	0	0	0	0	3	5.5	3.5	7
200	0	0	0	0	4	7	4.5	10
250	0	0	0	0	4	8	5	11
300	0	0	0	0	4.5	8.5	5.5	12.5
400	0	0	0	0	5	10	6.5	14.5
500	0	0	0	0	5.5	11	7	16
600	0	0	0	0	6	12	8	17.5
750	0	0	0	0	6.5	14	9	20
1000	0	0	0	0	7	18	10	23.5
1250	0	0	0	6	8	20.5	11	26.5
1500	0	0	0	9.5	8.5	23	12.5	29.5

10-meter band half-wave, center-fed vertical, Frequency = 29.7 MHz, Height above ground = 10 feet

Horizontal distance (feet) from any part of the antenna for compliance with
occupational/controlled or general population/uncontrolled exposure limits*

Height above ground (feet) where exposure occurs

Power**	6 feet		12 feet		20 feet		30 feet	
(watts)	con.	unc.	con.	unc.	con.	unc.	con.	unc.
10	0	0	2	3	1.5	3	0	0
25	0	0	2.5	4	2	4	0	0
50	0	0	3	5	3	6	0	0
100	0	0	3.5	6.5	4	8.5	0	0
200	0	4	4.5	9	5.5	12	0	4.5
250	0	5.5	5	10.5	6	13.5	0	6
300	0	6.5	5.5	11.5	6.5	14.5	0	7
400	0	9	6	13.5	7.5	16.5	0	9
500	0	11	6.5	15.5	8.5	18.5	0	11
600	0	13.5	7	17.5	9.5	20	0	13
750	2	16.5	8	20	10.5	21.5	3	15.5
1000	4	21	9	23	12	24	4.5	20.5
1250	5.5	25.5	10.5	27	13.5	26	6	24.5
1500	6.5	30	11.5	28.5	14.5	35	7	27.5

10-meter band half-wave, center-fed vertical, Frequency = 29.7 MHz, Height above ground = 20 feet

Horizontal distance (feet) from any part of the antenna for compliance with
occupational/controlled or general population/uncontrolled exposure limits*

Height above ground (feet) where exposure occurs

Power**	6 feet		12 feet		20 feet		30 feet	
(watts)	con.	unc.	con.	unc.	con.	unc.	con.	unc.
10	0	0	0	0	1.5	2.5	1.5	2.5
25	0	0	0	0	2	3.5	2	4
50	0	0	0	0	2.5	4.5	2.5	6
100	0	0	0	0	3.5	6	3.5	8
200	0	0	0	0	4.5	8.5	5	11.5
250	0	0	0	0	4.5	9.5	6	13
300	0	0	0	0	5	10.5	6.5	14
400	0	0	0	0	5.5	13	7.5	16.5
500	0	0	0	0	6	15.5	8	18.5
600	0	0	0	0	6.5	18	9	20.5
750	0	0	0	3	7.5	21	10	23.5
1000	0	0	0	9	8.5	25	11.5	28
1250	0	0	0	14.5	9.5	28.5	13	32.5
1500	0	0	0	20	10.5	31	14	36.5

* 0 feet indicates that the exposure at the height in the column above or below the antenna is in compliance.
** Power = Average power input to the antenna. See Chapter 5.

10-meter band half-wave, center-fed vertical, Frequency = 29.7 MHz, Height above ground = 30 feet

Horizontal distance (feet) from any part of the antenna for compliance with
occupational/controlled or general population/uncontrolled exposure limits*

Power** (watts)	6 feet		12 feet		20 feet		30 feet	
	con.	unc.	con.	unc.	con.	unc.	con.	unc.
10	0	0	0	0	0	0	1.5	2.5
25	0	0	0	0	0	0	2	3.5
50	0	0	0	0	0	0	2.5	4.5
100	0	0	0	0	0	0	3.5	6
200	0	0	0	0	0	0	4.5	8.5
250	0	0	0	0	0	0	4.5	9.5
300	0	0	0	0	0	0	5	10.5
400	0	0	0	0	0	0	5.5	12.5
500	0	0	0	0	0	0	6	14.5
600	0	0	0	0	0	0	6.5	16.5
750	0	0	0	0	0	0	7.5	19
1000	0	0	0	0	0	0	8.5	23
1250	0	0	0	0	0	0	9.5	26.5
1500	0	0	0	0	0	9.5	10.5	30

6-meter band half-wave, center-fed vertical, Frequency = 50.0 MHz, Height above ground = 10 feet

Horizontal distance (feet) from any part of the antenna for compliance with
occupational/controlled or general population/uncontrolled exposure limits*

Power** (watts)	6 feet		12 feet		20 feet		30 feet	
	con.	unc.	con.	unc.	con.	unc.	con.	unc.
10	0	0	1.5	2.5	1	2	0	0
25	0	0	2	3.5	1.5	3	0	0
50	0	0	2.5	5	2	4	0	0
100	0	0	3.5	8	3	6	0	0
200	0	2.5	4.5	12	4	9.5	0	0
250	0	4.5	5	13.5	4	11	0	0
300	0	6.5	6	14.5	4.5	13	0	0
400	0	10.5	7	16.5	5.5	16	0	0
500	0	13	8	18	6	19	0	0
600	0	15	9	19.5	6.5	21.5	0	0
750	0	17.5	10	21.5	7.5	24.5	0	0
1000	2.5	21.5	12	24.5	9.5	28.5	0	0
1250	4.5	25	13.5	26.5	11	31	0	0
1500	6.5	28	14.5	28.5	13	33	0	25

6-meter band half-wave, center-fed vertical, Frequency = 50.0 MHz, Height above ground = 20 feet

Horizontal distance (feet) from any part of the antenna for compliance with
occupational/controlled or general population/uncontrolled exposure limits*

Power** (watts)	6 feet		12 feet		20 feet		30 feet	
	con.	unc.	con.	unc.	con.	unc.	con.	unc.
10	0	0	0	0	1.5	2.5	1	2
25	0	0	0	0	2	3.5	1.5	3
50	0	0	0	0	2.5	4.5	2	4.5
100	0	0	0	0	3	6.5	3	6
200	0	0	0	0	4	9.5	4	9.5
250	0	0	0	0	4.5	11	4.5	11
300	0	0	0	0	5	12.5	4.5	12.5
400	0	0	0	0	5.5	15	5.5	15.5
500	0	0	0	0	6.5	17.5	6	18
600	0	0	0	0	7	19	6.5	20
750	0	0	0	0	8	22	7.5	22.5
1000	0	0	0	21	9.5	26	9.5	25.5
1250	0	0	0	25.5	11	30.5	11	27.5
1500	0	0	0	28.5	12.5	37	12.5	30

* 0 feet indicates that the exposure at the height in the column above or below the antenna is in compliance.
** Power = Average power input to the antenna. See Chapter 5.

6-meter band half-wave, center-fed vertical, Frequency = 50.0 MHz, Height above ground = 30 feet

Horizontal distance (feet) from any part of the antenna for compliance with occupational/controlled or general population/uncontrolled exposure limits*

Power**	6 feet		12 feet		20 feet		30 feet	
(watts)	con.	unc.	con.	unc.	con.	unc.	con.	unc.
10	0	0	0	0	0	0	1.5	2.5
25	0	0	0	0	0	0	2	3.5
50	0	0	0	0	0	0	2.5	4.5
100	0	0	0	0	0	0	3	6.5
200	0	0	0	0	0	0	4	9.5
250	0	0	0	0	0	0	4.5	11.5
300	0	0	0	0	0	0	5	12.5
400	0	0	0	0	0	0	5.5	15
500	0	0	0	0	0	0	6.5	17.5
600	0	0	0	0	0	0	7	19.5
750	0	0	0	0	0	0	8	22
1000	0	0	0	0	0	0	9.5	26
1250	0	0	0	0	0	0	11.5	31.5
1500	0	0	0	0	0	20.5	12.5	35

2-meter band half-wave, center-fed vertical, Frequency = 144.0 MHz, Height above ground = 10 feet

Horizontal distance (feet) from any part of the antenna for compliance with occupational/controlled or general population/uncontrolled exposure limits*

Power**	6 feet		12 feet		20 feet		30 feet	
(watts)	con.	unc.	con.	unc.	con.	unc.	con.	unc.
10	0	0	1.5	3	0	0	0	0
25	0	0	2	4.5	0	0	0	0
50	0	0	3	6.5	0	0	0	0
100	0	0	4	9	0	0	0	0
200	0	8	5.5	13	0	0	0	0
250	0	10	6.5	15	0	0	0	0
300	0	14.5	7	16	0	0	0	0
400	0	18.5	8	17.5	0	0	0	0
500	0	20.5	9	19	0	12.5	0	0
600	0	22	9.5	20	0	14.5	0	0
750	0	23.5	11	22	0	16	0	0
1000	8	26	13	25	0	28	0	0
1250	10	28.5	15	36	0	30.5	0	0
1500	14.5	30	16	39.5	0	31.5	0	0

2-meter band half-wave, center-fed vertical, Frequency = 144.0 MHz, Height above ground = 20 feet

Horizontal distance (feet) from any part of the antenna for compliance with occupational/controlled or general population/uncontrolled exposure limits*

Power**	6 feet		12 feet		20 feet		30 feet	
(watts)	con.	unc.	con.	unc.	con.	unc.	con.	unc.
10	0	0	0	0	1.5	2.5	0	0
25	0	0	0	0	2	4	0	0
50	0	0	0	0	2.5	6	0	0
100	0	0	0	0	3.5	8.5	0	0
200	0	0	0	0	5	12	0	0
250	0	0	0	0	6	13.5	0	0
300	0	0	0	0	6.5	15	0	0
400	0	0	0	0	7.5	18	0	0
500	0	0	0	0	8.5	19.5	0	11.5
600	0	0	0	14	9.5	21	0	14.5
750	0	0	0	16.5	10.5	23	0	20.5
1000	0	0	0	21.5	12	26	0	23
1250	0	0	0	31.5	13.5	34	0	25
1500	0	33	0	33.5	15	35.5	0	34.5

* 0 feet indicates that the exposure at the height in the column above or below the antenna is in compliance.
** Power = Average power input to the antenna. See Chapter 5.

2-meter band half-wave, center-fed vertical, Frequency = 144.0 MHz, Height above ground = 30 feet

Horizontal distance (feet) from any part of the antenna for compliance with occupational/controlled or general population/uncontrolled exposure limits*

Power**	6 feet		12 feet		20 feet		30 feet	
(watts)	con.	unc.	con.	unc.	con.	unc.	con.	unc.
10	0	0	0	0	0	0	1.5	2.5
25	0	0	0	0	0	0	2	4
50	0	0	0	0	0	0	2.5	6
100	0	0	0	0	0	0	3.5	8.5
200	0	0	0	0	0	0	5	12
250	0	0	0	0	0	0	6	13.5
300	0	0	0	0	0	0	6.5	15
400	0	0	0	0	0	0	7.5	17
500	0	0	0	0	0	0	8.5	19
600	0	0	0	0	0	0	9.5	21
750	0	0	0	0	0	0	10.5	23.5
1000	0	0	0	0	0	18.5	12	27.5
1250	0	0	0	0	0	23.5	13.5	30
1500	0	0	0	0	0	31.5	15	32

222-MHz band half-wave, center-fed vertical, Frequency = 222.0 MHz, Height above ground = 10 feet

Horizontal distance (feet) from any part of the antenna for compliance with occupational/controlled or general population/uncontrolled exposure limits*

Power**	6 feet		12 feet		20 feet		30 feet	
(watts)	con.	unc.	con.	unc.	con.	unc.	con.	unc.
10	0	0	1.5	3	0	0	0	0
25	0	0	2	4.5	0	0	0	0
50	0	0	3	6.5	0	0	0	0
100	0	0	4	9	0	0	0	0
200	0	10.5	6	13	0	0	0	0
250	0	12	6.5	14.5	0	0	0	0
300	0	13	7	15.5	0	0	0	0
400	0	15	8	17	0	0	0	0
500	0	17.5	9	18	0	0	0	0
600	0	18.5	9.5	24.5	0	16	0	0
750	0	27	10.5	27.5	0	17	0	0
1000	10.5	31	13	29.5	0	27	0	0
1250	12	33.5	14.5	31	0	29	0	0
1500	13	36	15.5	32.5	0	30	0	0

222-MHz band half-wave, center-fed vertical, Frequency = 222.0 MHz, Height above ground = 20 feet

Horizontal distance (feet) from any part of the antenna for compliance with occupational/controlled or general population/uncontrolled exposure limits*

Power**	6 feet		12 feet		20 feet		30 feet	
(watts)	con.	unc.	con.	unc.	con.	unc.	con.	unc.
10	0	0	0	0	1.5	3	0	0
25	0	0	0	0	2	4.5	0	0
50	0	0	0	0	3	6.5	0	0
100	0	0	0	0	4	9	0	0
200	0	0	0	0	5.5	12.5	0	0
250	0	0	0	0	6.5	14	0	0
300	0	0	0	0	7	15	0	0
400	0	0	0	0	8	17.5	0	0
500	0	0	0	0	9	20	0	0
600	0	0	0	14.5	9.5	21	0	12
750	0	0	0	19.5	11	22.5	0	18
1000	0	0	0	22	12.5	29	0	24.5
1250	0	22.5	0	31	14	30.5	0	26.5
1500	0	24	0	33	15	31.5	0	34

* 0 feet indicates that the exposure at the height in the column above or below the antenna is in compliance.
** Power = Average power input to the antenna. See Chapter 5.

222-MHz band half-wave, center-fed vertical, Frequency = 222.0 MHz, Height above ground = 30 feet

Horizontal distance (feet) from any part of the antenna for compliance with
occupational/controlled or general population/uncontrolled exposure limits*

Height above ground (feet) where exposure occurs

Power** (watts)	6 feet con.	6 feet unc.	12 feet con.	12 feet unc.	20 feet con.	20 feet unc.	30 feet con.	30 feet unc.
10	0	0	0	0	0	0	1.5	3
25	0	0	0	0	0	0	2	4.5
50	0	0	0	0	0	0	3	6.5
100	0	0	0	0	0	0	4	9
200	0	0	0	0	0	0	5.5	12.5
250	0	0	0	0	0	0	6.5	14
300	0	0	0	0	0	0	7	15
400	0	0	0	0	0	0	8	17.5
500	0	0	0	0	0	0	9	19.5
600	0	0	0	0	0	0	9.5	21
750	0	0	0	0	0	0	11	23.5
1000	0	0	0	0	0	22	12.5	28
1250	0	0	0	0	0	24	14	30
1500	0	0	0	0	0	32.5	15	36

420-MHz band half-wave, center-fed vertical, Frequency = 420.0 MHz, Height above ground = 10 feet

Horizontal distance (feet) from any part of the antenna for compliance with
occupational/controlled or general population/uncontrolled exposure limits*

Height above ground (feet) where exposure occurs

Power** (watts)	6 feet con.	6 feet unc.	12 feet con.	12 feet unc.	20 feet con.	20 feet unc.	30 feet con.	30 feet unc.
10	0	0	0	0	0	0	0	0
25	0	0	0	3.5	0	0	0	0
50	0	0	0	5	0	0	0	0
100	0	0	2.5	7.5	0	0	0	0
200	0	8	4.5	10.5	0	0	0	0
250	0	9	5	12	0	0	0	0
300	0	10.5	5.5	13	0	0	0	0
400	0	15	6.5	16.5	0	0	0	0
500	0	15.5	7.5	17.5	0	0	0	0
600	0	17	8	18	0	0	0	0
750	0	17.5	9	23.5	0	0	0	0
1000	8	26	10.5	24.5	0	21.5	0	0
1250	9	27.5	12	25.5	0	23	0	0
1500	10.5	29	13	33.5	0	23.5	0	0

420-MHz band half-wave, center-fed vertical, Frequency = 420.0 MHz, Height above ground = 20 feet

Horizontal distance (feet) from any part of the antenna for compliance with
occupational/controlled or general population/uncontrolled exposure limits*

Height above ground (feet) where exposure occurs

Power** (watts)	6 feet con.	6 feet unc.	12 feet con.	12 feet unc.	20 feet con.	20 feet unc.	30 feet con.	30 feet unc.
10	0	0	0	0	1.5	2.5	0	0
25	0	0	0	0	2	4	0	0
50	0	0	0	0	2.5	5.5	0	0
100	0	0	0	0	3.5	7.5	0	0
200	0	0	0	0	5	10.5	0	0
250	0	0	0	0	5.5	12	0	0
300	0	0	0	0	6	13	0	0
400	0	0	0	0	7	15	0	0
500	0	0	0	0	7.5	16.5	0	0
600	0	0	0	0	8.5	18	0	0
750	0	0	0	15.5	9.5	20.5	0	0
1000	0	0	0	21	10.5	24.5	0	16.5
1250	0	0	0	22	12	29	0	20.5
1500	0	17.5	0	28	13	30	0	25

* 0 feet indicates that the exposure at the height in the column above or below the antenna is in compliance.
** Power = Average power input to the antenna. See Chapter 5.

1200-MHz band half-wave, center-fed vertical, Frequency = 1200.0 MHz, Height above ground = 10 feet

Horizontal distance (feet) from any part of the antenna for compliance with
occupational/controlled or general population/uncontrolled exposure limits*

Power**	6 feet		12 feet		20 feet		30 feet	
(watts)	con.	unc.	con.	unc.	con.	unc.	con.	unc.
10	0	0	0	0	0	0	0	0
25	0	0	0	0	0	0	0	0
50	0	0	0	0	0	0	0	0
100	0	0	0	3.5	0	0	0	0
200	0	0	0	6	0	0	0	0
250	0	0	0	6.5	0	0	0	0
300	0	0	0	7.5	0	0	0	0
400	0	0	0	8.5	0	0	0	0
500	0	7	3.5	10	0	0	0	0
600	0	8.5	4	11	0	0	0	0
750	0	10.5	5	12.5	0	0	0	0
1000	0	13	6	14.5	0	0	0	0
1250	0	15.5	6.5	16.5	0	0	0	0
1500	0	18.5	7.5	18.5	0	0	0	0

Quarter-Wave Ground Plane Antennas

The following tables show the horizontal compliance distance for quarter-wave ground-plane antennas with sloping radials. The heights shown are for the lowest part of the radial system. The compliance distance was calculated parallel with the radial system, representing a worst case. In some cases, the compliance height would be in physical contact with the radial system.

To estimate compliance in the areas midway between radial systems, use the tables for half-wave verticals at the same height as the lowest part of the ground-plane radials. A compliance distance of "0" feet indicates that the areas above or below the antenna at that exposure height are in compliance. Using the next higher power level serves as a reasonable estimate for 5/8-wave type antennas with sloping radials.

20-meter band ground plane, 45-degree radials, Frequency = 14.35 MHz, Height above ground = 10 feet

Horizontal distance (feet) from any part of the antenna for compliance with
occupational/controlled or general population/uncontrolled exposure limits*

Power**	6 feet		12 feet		20 feet		30 feet	
(watts)	con.	unc.	con.	unc.	con.	unc.	con.	unc.
10	0	0	10.5	11	2.5	2.5	0	2
25	0	0	10.5	11.5	2.5	3	1.5	2.5
50	0	0	11	11.5	2.5	4	2	3
100	0	0	11	12.5	3	5.5	2.5	4
200	0	0	11.5	13	4	7	3	5.5
250	0	0	11.5	13.5	4	8	3	6
300	0	0	12	13.5	4.5	8.5	3.5	6.5
400	0	0	12	14	5	9.5	3.5	7
500	0	0	12.5	14	5.5	10.5	4	8
600	0	0	12.5	14.5	6	11	4.5	9
750	0	0	12.5	15	6.5	12	5	10
1000	0	0	13	15.5	7	13.5	5.5	11.5
1250	0	3	13.5	16	8	15	6	13
1500	0	11	13.5	17	8.5	16	6	14.5

* 0 feet indicates that the exposure at the height in the column above or below the antenna is in compliance.
** Power = Average power input to the antenna. See Chapter 5.

17-meter band ground plane, 45-degree radials, Frequency = 18.168 MHz, Height above ground = 10 feet

Horizontal distance (feet) from any part of the antenna for compliance with
occupational/controlled or general population/uncontrolled exposure limits*

Height above ground (feet) where exposure occurs

Power**	6 feet		12 feet		20 feet		30 feet	
(watts)	con.	unc.	con.	unc.	con.	unc.	con.	unc.
10	0	0	8	8.5	1	2	1.5	2.5
25	0	0	8.5	9	1.5	3.5	2	3.5
50	0	0	8.5	9.5	2	4.5	2.5	4.5
100	0	0	9	10.5	3	6	3	5.5
200	0	0	9.5	11.5	4	8.5	4	7
250	0	0	9.5	11.5	4.5	9.5	4.5	7.5
300	0	0	10	12	5	10	4.5	8
400	0	0	10	12.5	5.5	11.5	5	8.5
500	0	0	10.5	13	6	12.5	5.5	9.5
600	0	0	10.5	13.5	6.5	13	6	10
750	0	7	11	14.5	7.5	14.5	6	11
1000	0	10.5	11.5	15.5	8.5	16	7	12.5
1250	0	12.5	11.5	17	9.5	17.5	7.5	14
1500	0	14	12	18.5	10	19	8	15.5

15-meter band ground plane, 45-degree radials, Frequency = 21.45 MHz, Height above ground = 10 feet

Horizontal distance (feet) from any part of the antenna for compliance with
occupational/controlled or general population/uncontrolled exposure limits*

Height above ground (feet) where exposure occurs

Power**	6 feet		12 feet		20 feet		30 feet	
(watts)	con.	unc.	con.	unc.	con.	unc.	con.	unc.
10	0	0	7	7.5	1	2.5	0	0
25	0	0	7	8	2	3.5	0	2.5
50	0	0	7.5	8.5	2.5	5	0	3.5
100	0	0	8	9.5	3.5	7	2	5
200	0	0	8.5	10.5	4.5	9.5	3	6.5
250	0	0	8.5	11	5	10.5	3.5	7
300	0	0	9	11.5	5.5	11.5	3.5	8
400	0	0	9	12.5	6.5	12.5	4.5	9
500	0	6.5	9.5	13	7	14	5	9.5
600	0	8.5	10	14	7.5	15	5	10.5
750	0	10.5	10	15.5	8.5	16.5	5.5	11.5
1000	0	13	10.5	17.5	9.5	18.5	6.5	13.5
1250	0	15.5	11	19.5	10.5	20	7	15
1500	0	18	11.5	21.5	11.5	21.5	8	17.5

15-meter band ground plane, 45-degree radials, Frequency = 21.45 MHz, Height above ground = 20 feet

Horizontal distance (feet) from any part of the antenna for compliance with
occupational/controlled or general population/uncontrolled exposure limits*

Height above ground (feet) where exposure occurs

Power**	6 feet		12 feet		20 feet		30 feet	
(watts)	con.	unc.	con.	unc.	con.	unc.	con.	unc.
10	0	0	0	0	8.5	9	1	2.5
25	0	0	0	0	9	9	1.5	3.5
50	0	0	0	0	9	9.5	2.5	5
100	0	0	0	0	9	10	3	7
200	0	0	0	0	9.5	10.5	4.5	9.5
250	0	0	0	0	9.5	11	5	10.5
300	0	0	0	0	9.5	11	5.5	11.5
400	0	0	0	0	9.5	11.5	6	13
500	0	0	0	0	10	12.5	7	14.5
600	0	0	0	0	10	13	7.5	16
750	0	0	0	0	10	13.5	8.5	17.5
1000	0	0	0	0	10.5	15	9.5	20.5
1250	0	0	0	0	11	17	10.5	22.5
1500	0	0	0	4.5	11	18.5	11.5	24.5

* 0 feet indicates that the exposure at the height in the column above or below the antenna is in compliance.
** Power = Average power input to the antenna. See Chapter 5.

12-meter band ground plane, 45-degree radials, Frequency = 24.99 MHz, Height above ground = 10 feet

Horizontal distance (feet) from any part of the antenna for compliance with
occupational/controlled or general population/uncontrolled exposure limits*

Height above ground (feet) where exposure occurs

Power**	6 feet		12 feet		20 feet		30 feet	
(watts)	con.	unc.	con.	unc.	con.	unc.	con.	unc.
10	0	0	6	6.5	1.5	2.5	0	0
25	0	0	6	7	2	4	0	0
50	0	0	6.5	8	2.5	5.5	0	0
100	0	0	7	9	3.5	7.5	0	0
200	0	0	7.5	10.5	5	10.5	0	5
250	0	0	8	11	5.5	11.5	0	6
300	0	2.5	8	11.5	6	12.5	0	6.5
400	0	7	8.5	13	7	14	0	8
500	0	9.5	9	14.5	7.5	15.5	0	9.5
600	0	11	9	15.5	8.5	17	2.5	10.5
750	0	13.5	9.5	17.5	9	18.5	3.5	12
1000	0	17	10.5	20.5	10.5	20.5	5	15
1250	0	20.5	11	23	11.5	22.5	6	18
1500	2.5	24	11.5	25	12.5	24	6.5	20.5

12-meter band ground plane, 45-degree radials, Frequency = 24.99 MHz, Height above ground = 20 feet

Horizontal distance (feet) from any part of the antenna for compliance with
occupational/controlled or general population/uncontrolled exposure limits*

Height above ground (feet) where exposure occurs

Power**	6 feet		12 feet		20 feet		30 feet	
(watts)	con.	unc.	con.	unc.	con.	unc.	con.	unc.
10	0	0	0	0	7.5	8	1.5	2.5
25	0	0	0	0	7.5	8	2	3.5
50	0	0	0	0	8	8.5	2.5	5
100	0	0	0	0	8	9	3.5	7.5
200	0	0	0	0	8.5	10	4.5	10.5
250	0	0	0	0	8.5	10.5	5	11.5
300	0	0	0	0	8.5	11	5.5	12.5
400	0	0	0	0	9	12	6.5	14.5
500	0	0	0	0	9	12.5	7.5	16
600	0	0	0	0	9	13.5	8	18
750	0	0	0	0	9.5	16	9	20
1000	0	0	0	3.5	10	19	10.5	23.5
1250	0	0	0	8.5	10.5	21.5	11.5	26.5
1500	0	0	0	13	11	24	12.5	29

12-meter band ground plane, 45-degree radials, Frequency = 24.99 MHz, Height above ground = 30 feet

Horizontal distance (feet) from any part of the antenna for compliance with
occupational/controlled or general population/uncontrolled exposure limits*

Height above ground (feet) where exposure occurs

Power**	6 feet		12 feet		20 feet		30 feet	
(watts)	con.	unc.	con.	unc.	con.	unc.	con.	unc.
10	0	0	0	0	0	0	7.5	8
25	0	0	0	0	0	0	7.5	8
50	0	0	0	0	0	0	8	8.5
100	0	0	0	0	0	0	8	9
200	0	0	0	0	0	0	8.5	10
250	0	0	0	0	0	0	8.5	10.5
300	0	0	0	0	0	0	8.5	11
400	0	0	0	0	0	0	9	12
500	0	0	0	0	0	0	9	13
600	0	0	0	0	0	0	9.5	14
750	0	0	0	0	0	0	9.5	16
1000	0	0	0	0	0	0	10	19
1250	0	0	0	0	0	4	10.5	22
1500	0	0	0	0	0	8	11	25.5

* 0 feet indicates that the exposure at the height in the column above or below the antenna is in compliance.
** Power = Average power input to the antenna. See Chapter 5.

10-meter band ground plane, 45-degree radials, Frequency = 29.7 MHz, Height above ground = 10 feet

Horizontal distance (feet) from any part of the antenna for compliance with
occupational/controlled or general population/uncontrolled exposure limits*

Power**	6 feet		12 feet		20 feet		30 feet	
(watts)	con.	unc.	con.	unc.	con.	unc.	con.	unc.
10	0	0	5	5.5	2	3	0	0
25	0	0	5	6.5	2.5	4	0	0
50	0	0	5.5	7.5	3	5.5	0	0
100	0	0	6	8.5	4	8	0	0
200	0	1.5	7	11	5	12	0	0
250	0	5.5	7.5	12	5.5	13	0	0
300	0	7.5	7.5	13	6	14.5	0	3
400	0	10.5	8	15	7	16.5	0	6.5
500	0	13	8.5	16.5	8	18	0	8.5
600	0	15.5	9	18.5	9	19.5	0	11
750	0	18.5	9.5	20.5	10	21	0	14
1000	1.5	23	11	23.5	12	23.5	0	18.5
1250	5.5	27.5	12	26.5	13	25.5	0	23
1500	7.5	32	13	31	14.5	27.5	3	26

10-meter band ground plane, 45-degree radials, Frequency = 29.7 MHz, Height above ground = 20 feet

Horizontal distance (feet) from any part of the antenna for compliance with
occupational/controlled or general population/uncontrolled exposure limits*

Height above ground (feet) where exposure occurs

Power**	6 feet		12 feet		20 feet		30 feet	
(watts)	con.	unc.	con.	unc.	con.	unc.	con.	unc.
10	0	0	0	0	6.5	7	2	3
25	0	0	0	0	6.5	7.5	2.5	4
50	0	0	0	0	7	7.5	3	5.5
100	0	0	0	0	7	8.5	4	8
200	0	0	0	0	7.5	10	5	11
250	0	0	0	0	7.5	11	5.5	12.5
300	0	0	0	0	8	12	6	14
400	0	0	0	0	8	14.5	7	16
500	0	0	0	0	8.5	17	8	18
600	0	0	0	0	9	19.5	8.5	20.5
750	0	0	0	6.5	9.5	22	9.5	23.5
1000	0	0	0	13	10	25.5	11	28
1250	0	0	0	18.5	11	28.5	12.5	32
1500	0	0	0	22.5	12	30.5	14	36

10-meter band ground plane, 45-degree radials, Frequency = 29.7 MHz, Height above ground = 30 feet

Horizontal distance (feet) from any part of the antenna for compliance with
occupational/controlled or general population/uncontrolled exposure limits*

Height above ground (feet) where exposure occurs

Power**	6 feet		12 feet		20 feet		30 feet	
(watts)	con.	unc.	con.	unc.	con.	unc.	con.	unc.
10	0	0	0	0	0	0	6.5	7
25	0	0	0	0	0	0	6.5	7.5
50	0	0	0	0	0	0	7	7.5
100	0	0	0	0	0	0	7	8.5
200	0	0	0	0	0	0	7.5	10
250	0	0	0	0	0	0	7.5	11
300	0	0	0	0	0	0	8	12
400	0	0	0	0	0	0	8	13.5
500	0	0	0	0	0	0	8.5	15.5
600	0	0	0	0	0	0	9	17.5
750	0	0	0	0	0	0	9.5	20
1000	0	0	0	0	0	3.5	10	24
1250	0	0	0	0	0	10	11	27.5
1500	0	0	0	0	0	21.5	12	30.5

* 0 feet indicates that the exposure at the height in the column above or below the antenna is in compliance.
** Power = Average power input to the antenna. See Chapter 5.

6-meter band ground plane, 45-degree radials, Frequency = 52.0 MHz, Height above ground = 10 feet

Horizontal distance (feet) from any part of the antenna for compliance with occupational/controlled or general population/uncontrolled exposure limits*

Height above ground (feet) where exposure occurs

Power** (watts)	6 feet con.	6 feet unc.	12 feet con.	12 feet unc.	20 feet con.	20 feet unc.	30 feet con.	30 feet unc.
10	0	0	2	3	0	0	0	0
25	0	0	2.5	4.5	0	0	0	0
50	0	0	3	6	0	2.5	0	0
100	0	0	4	8.5	0	5	0	0
200	0	4	5.5	12.5	2	8.5	0	0
250	0	6.5	6	13.5	2.5	10	0	0
300	0	9	6.5	15	3	12	0	0
400	0	12	7.5	16.5	4	15.5	0	0
500	0	14	8.5	18	5	18.5	0	0
600	0	16	9.5	19.5	5.5	21.5	0	0
750	0	18.5	10.5	21	6.5	24.5	0	0
1000	4	22	12.5	24	8.5	28	0	0
1250	6.5	25	13.5	26	10	30.5	0	0
1500	9	27.5	15	27.5	12	32.5	0	25

6-meter band ground plane, 45-degree radials, Frequency = 52.0 MHz, Height above ground = 20 feet

Horizontal distance (feet) from any part of the antenna for compliance with occupational/controlled or general population/uncontrolled exposure limits*

Height above ground (feet) where exposure occurs

Power** (watts)	6 feet con.	6 feet unc.	12 feet con.	12 feet unc.	20 feet con.	20 feet unc.	30 feet con.	30 feet unc.
10	0	0	0	0	4	4.5	0	0
25	0	0	0	0	4.5	5	0	0
50	0	0	0	0	4.5	5.5	0	3
100	0	0	0	0	5	7	0	5
200	0	0	0	0	5.5	10.5	2	8
250	0	0	0	0	5.5	12	3	10
300	0	0	0	0	6	13	3.5	11.5
400	0	0	0	0	6.5	15.5	4	14.5
500	0	0	0	0	7	17.5	5	17.5
600	0	0	0	0	8	19.5	5.5	19.5
750	0	0	0	16.5	9	21.5	6.5	22
1000	0	0	0	23.5	10.5	25.5	8	25
1250	0	0	0	27.5	12	29.5	10	27.5
1500	0	0	0	30	13	36	11.5	29.5

6-meter band ground plane, 45-degree radials, Frequency = 52.0 MHz, Height above ground = 30 feet

Horizontal distance (feet) from any part of the antenna for compliance with occupational/controlled or general population/uncontrolled exposure limits*

Height above ground (feet) where exposure occurs

Power** (watts)	6 feet con.	6 feet unc.	12 feet con.	12 feet unc.	20 feet con.	20 feet unc.	30 feet con.	30 feet unc.
10	0	0	0	0	0	0	4	4.5
25	0	0	0	0	0	0	4.5	5
50	0	0	0	0	0	0	4.5	5.5
100	0	0	0	0	0	0	5	7
200	0	0	0	0	0	0	5.5	10.5
250	0	0	0	0	0	0	5.5	12
300	0	0	0	0	0	0	6	13.5
400	0	0	0	0	0	0	6.5	15.5
500	0	0	0	0	0	0	7	17.5
600	0	0	0	0	0	0	8	19.5
750	0	0	0	0	0	0	9	22
1000	0	0	0	0	0	0	10.5	26
1250	0	0	0	0	0	19	12	30.5
1500	0	0	0	0	0	23.5	13.5	35

* 0 feet indicates that the exposure at the height in the column above or below the antenna is in compliance.
** Power = Average power input to the antenna. See Chapter 5.

2-meter band ground plane, 45-degree radials, Frequency = 146.0 MHz, Height above ground = 6 feet

Horizontal distance (feet) from any part of the antenna for compliance with occupational/controlled or general population/uncontrolled exposure limits*

Height above ground (feet) where exposure occurs

Power** (watts)	6 feet con.	6 feet unc.	12 feet con.	12 feet unc.	20 feet con.	20 feet unc.	30 feet con.	30 feet unc.
10	2	3	0	0	0	0	0	0
25	2.5	4	0	0	0	0	0	0
50	3	6	0	0	0	0	0	0
100	4	9.5	0	0	0	0	0	0
200	5.5	13.5	0	9	0	0	0	0
250	6	14.5	0	10.5	0	0	0	0
300	7	15.5	0	11	0	0	0	0
400	8	17	0	13	0	0	0	0
500	9.5	18	0	22	0	0	0	0
600	10.5	19.5	6	24.5	0	0	0	0
750	12	20.5	7.5	26.5	0	0	0	0
1000	13.5	22.5	9	29.5	0	15	0	0
1250	14.5	24.5	10.5	32	0	17	0	0
1500	15.5	26.5	11	33.5	0	18	0	0

2-meter band ground plane, 45-degree radials, Frequency = 146.0 MHz, Height above ground = 10 feet

Horizontal distance (feet) from any part of the antenna for compliance with occupational/controlled or general population/uncontrolled exposure limits*

Height above ground (feet) where exposure occurs

Power** (watts)	6 feet con.	6 feet unc.	12 feet con.	12 feet unc.	20 feet con.	20 feet unc.	30 feet con.	30 feet unc.
10	0	0	1.5	3	0	0	0	0
25	0	0	2	4.5	0	0	0	0
50	0	0	3	6	0	0	0	0
100	0	0	4	8.5	0	0	0	0
200	0	9	5.5	13	0	0	0	0
250	0	11	6	15	0	0	0	0
300	0	15	7	16	0	0	0	0
400	0	19	8	17.5	0	0	0	0
500	0	20.5	8.5	18.5	0	12.5	0	0
600	0	22	9.5	19.5	0	14	0	0
750	0	23.5	10.5	21.5	0	16	0	0
1000	9	26	13	24.5	0	28	0	0
1250	11	28	15	35	0	30	0	0
1500	15	30	16	39	0	31.5	0	0

2-meter band ground plane, 45-degree radials, Frequency = 146.0 MHz, Height above ground = 20 feet

Horizontal distance (feet) from any part of the antenna for compliance with occupational/controlled or general population/uncontrolled exposure limits*

Height above ground (feet) where exposure occurs

Power** (watts)	6 feet con.	6 feet unc.	12 feet con.	12 feet unc.	20 feet con.	20 feet unc.	30 feet con.	30 feet unc.
10	0	0	0	0	2	3	0	0
25	0	0	0	0	2.5	4	0	0
50	0	0	0	0	3	6	0	0
100	0	0	0	0	3.5	8.5	0	0
200	0	0	0	0	5.5	12	0	0
250	0	0	0	0	6	13.5	0	0
300	0	0	0	0	6.5	14.5	0	0
400	0	0	0	0	7.5	17.5	0	0
500	0	0	0	13	8.5	19.5	0	11.5
600	0	0	0	15	9.5	21	0	14
750	0	0	0	17.5	10.5	22.5	0	20.5
1000	0	0	0	21.5	12	25	0	23
1250	0	15	0	31.5	13.5	34	0	25
1500	0	33.5	0	34	14.5	35.5	0	34.5

* 0 feet indicates that the exposure at the height in the column above or below the antenna is in compliance.
** Power = Average power input to the antenna. See Chapter 5.

2-meter band ground plane, 45-degree radials, Frequency = 146.0 MHz, Height above ground = 30 feet

Horizontal distance (feet) from any part of the antenna for compliance with
occupational/controlled or general population/uncontrolled exposure limits*

Height above ground (feet) where exposure occurs

Power**	6 feet		12 feet		20 feet		30 feet	
(watts)	con.	unc.	con.	unc.	con.	unc.	con.	unc.
10	0	0	0	0	0	0	2	3
25	0	0	0	0	0	0	2.5	4
50	0	0	0	0	0	0	3	6
100	0	0	0	0	0	0	3.5	8.5
200	0	0	0	0	0	0	5.5	12
250	0	0	0	0	0	0	6	13.5
300	0	0	0	0	0	0	6.5	14.5
400	0	0	0	0	0	0	7.5	17
500	0	0	0	0	0	0	8.5	19
600	0	0	0	0	0	0	9.5	20.5
750	0	0	0	0	0	0	10.5	23
1000	0	0	0	0	0	20	12	27.5
1250	0	0	0	0	0	24	13.5	29.5
1500	0	0	0	0	0	32.5	14.5	31.5

222-MHz band ground plane, 45-degree radials, Frequency = 222.0 MHz, Height above ground = 6 feet

Horizontal distance (feet) from any part of the antenna for compliance with
occupational/controlled or general population/uncontrolled exposure limits*

Height above ground (feet) where exposure occurs

Power**	6 feet		12 feet		20 feet		30 feet	
(watts)	con.	unc.	con.	unc.	con.	unc.	con.	unc.
10	2	3	0	0	0	0	0	0
25	2	4.5	0	0	0	0	0	0
50	3	6.5	0	0	0	0	0	0
100	4	8.5	0	0	0	0	0	0
200	5.5	11.5	0	7.5	0	0	0	0
250	6.5	16	0	13.5	0	0	0	0
300	7	18	0	15	0	0	0	0
400	8	20	0	16.5	0	0	0	0
500	8.5	21.5	0	17.5	0	0	0	0
600	9	23	0	19	0	0	0	0
750	10	24.5	0	19.5	0	0	0	0
1000	11.5	26.5	7.5	20.5	0	24.5	0	0
1250	16	28.5	13.5	34	0	27	0	0
1500	18	30.5	15	37.5	0	29	0	0

222-MHz band ground plane, 45-degree radials, Frequency = 222.0 MHz, Height above ground = 10 feet

Horizontal distance (feet) from any part of the antenna for compliance with
occupational/controlled or general population/uncontrolled exposure limits*

Height above ground (feet) where exposure occurs

Power**	6 feet		12 feet		20 feet		30 feet	
(watts)	con.	unc.	con.	unc.	con.	unc.	con.	unc.
10	0	0	1.5	2.5	0	0	0	0
25	0	0	2	4.5	0	0	0	0
50	0	0	2.5	6.5	0	0	0	0
100	0	0	4	8.5	0	0	0	0
200	0	11	5.5	13	0	0	0	0
250	0	12	6.5	14.5	0	0	0	0
300	0	13	7	15.5	0	0	0	0
400	0	15	8	16.5	0	0	0	0
500	0	17	8.5	17.5	0	14	0	0
600	0	18.5	9.5	24.5	0	15.5	0	0
750	7	27	10.5	27.5	0	17	0	0
1000	11	30.5	13	29	0	27	0	0
1250	12	33	14.5	30.5	0	28.5	0	0
1500	13	35.5	15.5	32	0	29.5	0	0

* 0 feet indicates that the exposure at the height in the column above or below the antenna is in compliance.
** Power = Average power input to the antenna. See Chapter 5.

222-MHz band ground plane, 45-degree radials, Frequency = 222.0 MHz, Height above ground = 20 feet

Horizontal distance (feet) from any part of the antenna for compliance with occupational/controlled or general population/uncontrolled exposure limits*

Power** (watts)	6 feet		12 feet		20 feet		30 feet	
	con.	unc.	con.	unc.	con.	unc.	con.	unc.
10	0	0	0	0	2	3	0	0
25	0	0	0	0	2	4.5	0	0
50	0	0	0	0	3	6.5	0	0
100	0	0	0	0	4	8.5	0	0
200	0	0	0	0	5.5	12	0	0
250	0	0	0	0	6.5	13.5	0	0
300	0	0	0	0	7	14.5	0	0
400	0	0	0	0	8	17	0	0
500	0	0	0	12.5	8.5	19.5	0	0
600	0	0	0	18	9.5	20.5	0	15.5
750	0	0	0	19.5	10.5	22	0	18
1000	0	21	0	22	12	28.5	0	24.5
1250	0	23	0	31	13.5	30	0	26
1500	0	24.5	0	33	14.5	31	0	34

222-MHz band ground plane, 45-degree radials, Frequency = 222.0 MHz, Height above ground = 30 feet

Horizontal distance (feet) from any part of the antenna for compliance with occupational/controlled or general population/uncontrolled exposure limits*

Power** (watts)	6 feet		12 feet		20 feet		30 feet	
	con.	unc.	con.	unc.	con.	unc.	con.	unc.
10	0	0	0	0	0	0	2	3
25	0	0	0	0	0	0	2	4.5
50	0	0	0	0	0	0	3	6.5
100	0	0	0	0	0	0	4	8.5
200	0	0	0	0	0	0	5.5	12
250	0	0	0	0	0	0	6.5	13.5
300	0	0	0	0	0	0	7	15
400	0	0	0	0	0	0	8	17
500	0	0	0	0	0	0	8.5	19
600	0	0	0	0	0	0	9.5	21
750	0	0	0	0	0	14.5	10.5	23
1000	0	0	0	0	0	22.5	12	27.5
1250	0	0	0	0	0	24	13.5	29.5
1500	0	0	0	0	0	32.5	15	35.5

420-MHz band ground plane, 45-degree radials, Frequency = 420.0 MHz, Height above ground = 10 feet

Horizontal distance (feet) from any part of the antenna for compliance with occupational/controlled or general population/uncontrolled exposure limits*

Power** (watts)	6 feet		12 feet		20 feet		30 feet	
	con.	unc.	con.	unc.	con.	unc.	con.	unc.
10	0	0	0	0	0	0	0	0
25	0	0	0	3	0	0	0	0
50	0	0	0	5	0	0	0	0
100	0	0	2.5	7.5	0	0	0	0
200	0	8.5	4.5	10	0	0	0	0
250	0	9	5	12	0	0	0	0
300	0	10.5	5.5	12.5	0	0	0	0
400	0	15	6.5	16	0	0	0	0
500	0	15.5	7.5	17	0	0	0	0
600	0	17	8	17.5	0	0	0	0
750	0	17.5	9	18.5	0	15	0	0
1000	8.5	26	10	24.5	0	21.5	0	0
1250	9	27.5	12	25	0	23	0	41
1500	10.5	28.5	12.5	33	0	23.5	0	41

* 0 feet indicates that the exposure at the height in the column above or below the antenna is in compliance.
** Power = Average power input to the antenna. See Chapter 5.

420-MHz band ground plane, 45-degree radials, Frequency = 420.0 MHz, Height above ground = 20 feet

Horizontal distance (feet) from any part of the antenna for compliance with
occupational/controlled or general population/uncontrolled exposure limits*

Height above ground (feet) where exposure occurs

Power**	6 feet		12 feet		20 feet		30 feet	
(watts)	con.	unc.	con.	unc.	con.	unc.	con.	unc.
10	0	0	0	0	1.5	3	0	0
25	0	0	0	0	2	4	0	0
50	0	0	0	0	3	5.5	0	0
100	0	0	0	0	3.5	7.5	0	0
200	0	0	0	0	5	10.5	0	0
250	0	0	0	0	5.5	11.5	0	0
300	0	0	0	0	6	12.5	0	0
400	0	0	0	0	7	14.5	0	0
500	0	0	0	0	7.5	16.5	0	0
600	0	0	0	10.5	8.5	17.5	0	0
750	0	0	0	33	9	20	0	0
1000	0	27	0	33	10.5	24	0	16.5
1250	0	27	0	33	11.5	41	0	20.5
1500	0	27	0	33	12.5	41	0	24.5

420-MHz band ground plane, 45-degree radials, Frequency = 420.0 MHz, Height above ground = 30 feet

Horizontal distance (feet) from any part of the antenna for compliance with
occupational/controlled or general population/uncontrolled exposure limits*

Height above ground (feet) where exposure occurs

Power**	6 feet		12 feet		20 feet		30 feet	
(watts)	con.	unc.	con.	unc.	con.	unc.	con.	unc.
10	0	0	0	0	0	0	1.5	3
25	0	0	0	0	0	0	2	4
50	0	0	0	0	0	0	3	5.5
100	0	0	0	0	0	0	3.5	7.5
200	0	0	0	0	0	0	5	10.5
250	0	0	0	0	0	0	5.5	11.5
300	0	0	0	0	0	0	6	12.5
400	0	0	0	0	0	0	7	14.5
500	0	0	0	0	0	0	7.5	16
600	0	0	0	0	0	0	8	17.5
750	0	0	0	0	0	0	9	19.5
1000	0	0	0	0	0	14.5	10.5	23
1250	0	37	0	0	0	19	11.5	26.5
1500	0	37	0	0	0	23.5	12.5	27.5

902-MHz band ground plane, 45-degree radials, Frequency = 902.0 MHz, Height above ground = 10 feet

Horizontal distance (feet) from any part of the antenna for compliance with
occupational/controlled or general population/uncontrolled exposure limits*

Height above ground (feet) where exposure occurs

Power**	6 feet		12 feet		20 feet		30 feet	
(watts)	con.	unc.	con.	unc.	con.	unc.	con.	unc.
10	0	0	0	0	0	0	0	0
25	0	0	0	0	0	0	0	0
50	0	0	0	0	0	0	0	0
100	0	0	0	4.5	0	0	0	0
200	0	0	0	7	0	0	0	0
250	0	0	0	7.5	0	0	0	0
300	0	0	0	8.5	0	0	0	0
400	0	7	4	9.5	0	0	0	0
500	0	9.5	4.5	11	0	0	0	0
600	0	12	5	13	0	0	0	0
750	0	12.5	6	13.5	0	0	0	0
1000	0	15.5	7	17.5	0	0	0	0
1250	0	20	7.5	18	0	0	0	0
1500	0	20.5	8.5	20.5	0	0	0	0

* 0 feet indicates that the exposure at the height in the column above or below the antenna is in compliance.
** Power = Average power input to the antenna. See Chapter 5.

902-MHz band ground plane, 45-degree radials, Frequency = 902.0 MHz, Height above ground = 20 feet

Horizontal distance (feet) from any part of the antenna for compliance with
occupational/controlled or general population/uncontrolled exposure limits*

Height above ground (feet) where exposure occurs

Power**	6 feet		12 feet		20 feet		30 feet	
(watts)	con.	unc.	con.	unc.	con.	unc.	con.	unc.
10	0	0	0	0	1.5	2	0	0
25	0	0	0	0	1.5	3	0	0
50	0	0	0	0	2	4	0	0
100	0	0	0	0	3	5.5	0	0
200	0	0	0	0	3.5	7.5	0	0
250	0	0	0	0	4	8	0	0
300	0	0	0	0	4.5	9	0	0
400	0	0	0	0	5	10	0	0
500	0	0	0	0	5.5	11	0	0
600	0	0	0	0	6	12.5	0	0
750	0	0	0	0	6.5	13.5	0	0
1000	0	0	0	0	7.5	15.5	0	0
1250	0	0	0	11	8	17.5	0	0
1500	0	0	0	15.5	9	20	0	0

1.2 GHz band ground plane, 45-degree radials, Frequency = 1240.0 MHz, Height above ground = 10 feet

Horizontal distance (feet) from any part of the antenna for compliance with
occupational/controlled or general population/uncontrolled exposure limits*

Height above ground (feet) where exposure occurs

Power**	6 feet		12 feet		20 feet		30 feet	
(watts)	con.	unc.	con.	unc.	con.	unc.	con.	unc.
10	0	0	0	0	0	0	0	0
25	0	0	0	0	0	0	0	0
50	0	0	0	0	0	0	0	0
100	0	0	0	3.5	0	0	0	0
200	0	0	0	5.5	0	0	0	0
250	0	0	0	6.5	0	0	0	0
300	0	0	0	7	0	0	0	0
400	0	0	0	8.5	0	0	0	0
500	0	7.5	3.5	9.5	0	0	0	0
600	0	9	4	10.5	0	0	0	0
750	0	11	5	12	0	0	0	0
1000	0	13.5	5.5	14	0	0	0	0
1250	0	14	6.5	15.5	0	0	0	0
1500	0	16	7	17.5	0	0	0	0

* 0 feet indicates that the exposure at the height in the column above or below the antenna is in compliance.
** Power = Average power input to the antenna. See Chapter 5.

2-ELEMENT YAGI ANTENNAS

These tables represent 2-element Yagi monoband antennas. The tables show the horizontal compliance distance at compliance heights of 6 feet, 12 feet, 20 feet and at the height of the antenna, in the direction the antenna is pointing. The distances were calculated from the physical center of the antenna. See Chapter 5 for information on how to use the pattern of the antenna to estimate compliance distance in other directions.

These tables are generally a conservative estimate for shortened or loaded 2-element Yagis. They also provide an approximate estimate for other directive arrays using two elements.

30-meter band horizontal, 2-element Yagi, Frequency = 10.15 MHz, Height above ground = 20 feet

Horizontal distance (feet) from any part of the antenna for compliance with occupational/controlled or general population/uncontrolled exposure limits*

Height above ground (feet) where exposure occurs

Power**	6 feet		12 feet		20 feet		30 feet	
(watts)	con.	unc.	con.	unc.	con.	unc.	con.	unc.
10	0	0	0	0	8	9	0	0
25	0	0	0	0	8.5	9.5	0	0
50	0	0	0	0	9	10.5	0	0
100	0	0	0	0	9.5	11.5	0	0
200	0	0	0	0	10	13	0	0
250	0	0	0	0	10.5	13.5	0	0
300	0	0	0	0	10.5	14	0	0
400	0	0	0	10.5	11	15	0	0
500	0	0	0	12.5	11.5	16	0	0
600	0	0	0	14	12	17	0	0
750	0	0	0	15.5	12.5	18	0	11
1000	0	13	0	18	13	19.5	0	14
1250	0	16.5	0	19.5	13.5	20.5	0	16.5
1500	0	19	0	21	14	22	0	18

30-meter band horizontal, 2-element Yagi, Frequency = 10.15 MHz, Height above ground = 30 feet

Horizontal distance (feet) from any part of the antenna for compliance with occupational/controlled or general population/uncontrolled exposure limits*

Height above ground (feet) where exposure occurs

Power**	6 feet		12 feet		20 feet		30 feet	
(watts)	con.	unc.	con.	unc.	con.	unc.	con.	unc.
10	0	0	0	0	0	0	8	8.5
25	0	0	0	0	0	0	8.5	9.5
50	0	0	0	0	0	0	8.5	10
100	0	0	0	0	0	0	9	11
200	0	0	0	0	0	0	10	12.5
250	0	0	0	0	0	0	10	13
300	0	0	0	0	0	0	10.5	13.5
400	0	0	0	0	0	0	11	14.5
500	0	0	0	0	0	0	11	15.5
600	0	0	0	0	0	0	11.5	16
750	0	0	0	0	0	0	12	17
1000	0	0	0	0	0	12	12.5	18.5
1250	0	0	0	0	0	14.5	13	20
1500	0	0	0	0	0	16.5	13.5	21.5

* 0 feet indicates that the exposure at the height in the column above or below the antenna is in compliance.
** Power = Average power input to the antenna. See Chapter 5.

40-meter band horizontal, 2-element Yagi, Frequency = 7.3 MHz, Height above ground = 20 feet

Horizontal distance (feet) from any part of the antenna for compliance with
occupational/controlled or general population/uncontrolled exposure limits*

Height above ground (feet) where exposure occurs

Power**	6 feet		12 feet		20 feet		30 feet	
(watts)	con.	unc.	con.	unc.	con.	unc.	con.	unc.
10	0	0	0	0	10.5	11	0	0
25	0	0	0	0	10.5	11.5	0	0
50	0	0	0	0	11	12	0	0
100	0	0	0	0	11.5	13	0	0
200	0	0	0	0	12	14.5	0	0
250	0	0	0	0	12	14.5	0	0
300	0	0	0	0	12.5	15	0	0
400	0	0	0	0	12.5	16	0	0
500	0	0	0	0	13	16.5	0	0
600	0	0	0	12	13.5	17	0	0
750	0	0	0	14.5	13.5	18	0	0
1000	0	0	0	16.5	14.5	19	0	0
1250	0	0	0	18.5	14.5	20	0	0
1500	0	14	0	19.5	15	21	0	13.5

40-meter band horizontal, 2-element Yagi, Frequency = 7.3 MHz, Height above ground = 30 feet

Horizontal distance (feet) from any part of the antenna for compliance with
occupational/controlled or general population/uncontrolled exposure limits*

Height above ground (feet) where exposure occurs

Power**	6 feet		12 feet		20 feet		30 feet	
(watts)	con.	unc.	con.	unc.	con.	unc.	con.	unc.
10	0	0	0	0	0	0	10.5	11
25	0	0	0	0	0	0	10.5	11.5
50	0	0	0	0	0	0	11	12
100	0	0	0	0	0	0	11.5	13
200	0	0	0	0	0	0	12	14
250	0	0	0	0	0	0	12	14.5
300	0	0	0	0	0	0	12	14.5
400	0	0	0	0	0	0	12.5	15.5
500	0	0	0	0	0	0	13	16
600	0	0	0	0	0	0	13	16.5
750	0	0	0	0	0	0	13.5	17.5
1000	0	0	0	0	0	0	14	18.5
1250	0	0	0	0	0	12	14.5	19.5
1500	0	0	0	0	0	14.5	14.5	20

* 0 feet indicates that the exposure at the height in the column above or below the antenna is in compliance.
** Power = Average power input to the antenna. See Chapter 5.

3-ELEMENT YAGI ANTENNAS

These tables represent 3-element Yagi monoband antennas. The tables show the horizontal compliance distance at compliance heights of 6 feet, 12 feet, 20 feet and at the height of the antenna, in the direction the antenna is pointing. The distances were calculated from the physical center of the antenna. See Chapter 5 for information on how to use the pattern of the antenna to estimate compliance distance in other directions.

These tables are generally a conservative estimate for shortened or loaded 3-element Yagis ("tribanders" or similar). They also provide an approximate estimate for other directive arrays using three elements.

40-meter band horizontal, 3-element Yagi, Frequency = 7.3 MHz, Height above ground = 20 feet

Horizontal distance (feet) from any part of the antenna for compliance with occupational/controlled or general population/uncontrolled exposure limits*

Height above ground (feet) where exposure occurs

Power** (watts)	6 feet con.	6 feet unc.	12 feet con.	12 feet unc.	20 feet con.	20 feet unc.	30 feet con.	30 feet unc.
10	0	0	0	0	24.5	24.5	0	0
25	0	0	0	0	24.5	25	0	0
50	0	0	0	0	24.5	25.5	0	0
100	0	0	0	0	25	26	0	0
200	0	0	0	0	25.5	27	0	0
250	0	0	0	0	25.5	27.5	0	0
300	0	0	0	0	25.5	27.5	0	0
400	0	0	0	0	26	28	0	0
500	0	0	0	0	26	28.5	0	0
600	0	0	0	0	26.5	29	0	0
750	0	0	0	3.5	26.5	29.5	0	0
1000	0	0	0	25	27	30.5	0	0
1250	0	0	0	28	27.5	31	0	0
1500	0	0	0	29.5	27.5	32	0	9.5

40-meter band horizontal, 3-element Yagi, Frequency = 7.3 MHz, Height above ground = 30 feet

Horizontal distance (feet) from any part of the antenna for compliance with occupational/controlled or general population/uncontrolled exposure limits*

Height above ground (feet) where exposure occurs

Power** (watts)	6 feet con.	6 feet unc.	12 feet con.	12 feet unc.	20 feet con.	20 feet unc.	30 feet con.	30 feet unc.
10	0	0	0	0	0	0	24.5	25
25	0	0	0	0	0	0	24.5	25
50	0	0	0	0	0	0	25	25.5
100	0	0	0	0	0	0	25	26.5
200	0	0	0	0	0	0	25.5	27
250	0	0	0	0	0	0	25.5	27.5
300	0	0	0	0	0	0	26	28
400	0	0	0	0	0	0	26	28.5
500	0	0	0	0	0	0	26.5	29
600	0	0	0	0	0	0	26.5	29.5
750	0	0	0	0	0	0	27	30
1000	0	0	0	0	0	0	27	31
1250	0	0	0	0	0	7.5	27.5	31.5
1500	0	0	0	0	0	25.5	28	32.5

* 0 feet indicates that the exposure at the height in the column above or below the antenna is in compliance.
** Power = Average power input to the antenna. See Chapter 5.

30-meter band horizontal, 3-element Yagi, Frequency = 10.15 MHz, Height above ground = 20 feet

Horizontal distance (feet) from any part of the antenna for compliance with occupational/controlled or general population/uncontrolled exposure limits*

| Power** | 6 feet | | 12 feet | | 20 feet | | 30 feet | |
(watts)	con.	unc.	con.	unc.	con.	unc.	con.	unc.
10	0	0	0	0	17.5	18	0	0
25	0	0	0	0	18	18.5	0	0
50	0	0	0	0	18	19.5	0	0
100	0	0	0	0	18.5	20	0	0
200	0	0	0	0	19	21.5	0	0
250	0	0	0	0	19.5	22	0	0
300	0	0	0	0	19.5	22.5	0	0
400	0	0	0	6	20	23	0	0
500	0	0	0	19	20	23.5	0	0
600	0	0	0	21	20.5	24.5	0	8.5
750	0	0	0	22.5	21	25	0	14.5
1000	0	0	0	25	21.5	26.5	0	18.5
1250	0	21.5	0	27	22	28	0	21
1500	0	25.5	0	28.5	22.5	29	0	23

30-meter band horizontal, 3-element Yagi, Frequency = 10.15 MHz, Height above ground = 30 feet

Horizontal distance (feet) from any part of the antenna for compliance with occupational/controlled or general population/uncontrolled exposure limits*

| Power** | 6 feet | | 12 feet | | 20 feet | | 30 feet | |
(watts)	con.	unc.	con.	unc.	con.	unc.	con.	unc.
10	0	0	0	0	0	0	17.5	18
25	0	0	0	0	0	0	18	18.5
50	0	0	0	0	0	0	18	19.5
100	0	0	0	0	0	0	18.5	20
200	0	0	0	0	0	0	19	21
250	0	0	0	0	0	0	19.5	21.5
300	0	0	0	0	0	0	19.5	22
400	0	0	0	0	0	0	20	23
500	0	0	0	0	0	0	20	23.5
600	0	0	0	0	0	8.5	20.5	24
750	0	0	0	0	0	16.5	20.5	25
1000	0	0	0	0	0	20.5	21	26
1250	0	0	0	0	0	22.5	21.5	27.5
1500	0	0	0	0	0	24.5	22	28.5

20-meter band horizontal, 3-element Yagi, Frequency = 14.35 MHz, Height above ground = 20 feet

Horizontal distance (feet) from any part of the antenna for compliance with occupational/controlled or general population/uncontrolled exposure limits*

| Power** | 6 feet | | 12 feet | | 20 feet | | 30 feet | |
(watts)	con.	unc.	con.	unc.	con.	unc.	con.	unc.
10	0	0	0	0	14.5	15	0	0
25	0	0	0	0	15	16	0	0
50	0	0	0	0	15	16.5	0	0
100	0	0	0	0	15.5	17.5	0	0
200	0	0	0	0	16.5	19.5	0	0
250	0	0	0	11.5	16.5	20	0	0
300	0	0	0	15	17	20.5	0	0
400	0	0	0	18	17.5	22	0	12.5
500	0	0	0	20	17.5	23	0	16
600	0	0	0	21.5	18	24.5	0	18.5
750	0	0	0	23.5	18.5	26	0	22
1000	0	23.5	0	26.5	19.5	29	0	26.5
1250	0	27	11.5	29	20	31.5	0	30.5
1500	0	29.5	15	31.5	20.5	34.5	0	34.5

* 0 feet indicates that the exposure at the height in the column above or below the antenna is in compliance.
** Power = Average power input to the antenna. See Chapter 5.

20-meter band horizontal, 3-element Yagi, Frequency = 14.35 MHz, Height above ground = 30 feet

Horizontal distance (feet) from any part of the antenna for compliance with occupational/controlled or general population/uncontrolled exposure limits*

Height above ground (feet) where exposure occurs

Power**	6 feet		12 feet		20 feet		30 feet	
(watts)	con.	unc.	con.	unc.	con.	unc.	con.	unc.
10	0	0	0	0	0	0	14.5	15
25	0	0	0	0	0	0	15	16
50	0	0	0	0	0	0	15	16.5
100	0	0	0	0	0	0	15.5	17.5
200	0	0	0	0	0	0	16	19
250	0	0	0	0	0	0	16.5	19.5
300	0	0	0	0	0	0	16.5	20
400	0	0	0	0	0	10	17	21.5
500	0	0	0	0	0	13.5	17.5	22.5
600	0	0	0	0	0	16	18	24
750	0	0	0	0	0	18.5	18	26
1000	0	0	0	0	0	22.5	19	29.5
1250	0	0	0	0	0	26	19.5	33
1500	0	0	0	0	0	29.5	20	36

17-meter band horizontal, 3-element Yagi, Frequency = 18.168 MHz, Height above ground = 20 feet

Horizontal distance (feet) from any part of the antenna for compliance with occupational/controlled or general population/uncontrolled exposure limits*

Height above ground (feet) where exposure occurs

Power**	6 feet		12 feet		20 feet		30 feet	
(watts)	con.	unc.	con.	unc.	con.	unc.	con.	unc.
10	0	0	0	0	8	9	0	0
25	0	0	0	0	8.5	10	0	0
50	0	0	0	0	9	11.5	0	0
100	0	0	0	0	10	13	0	0
200	0	0	0	11.5	11	16	0	0
250	0	0	0	13	11.5	17.5	0	10.5
300	0	0	0	15	11.5	19	0	13.5
400	0	0	0	18	12.5	22	0	18
500	0	14.5	0	20.5	13	24.5	0	21.5
600	0	18.5	6	22.5	14	27	0	25
750	0	22.5	9	26	14.5	30.5	0	29.5
1000	0	26.5	11.5	30	16	36	0	36
1250	0	30	13	33.5	17.5	40.5	10.5	41.5
1500	0	32.5	15	37	19	44	13.5	47

17-meter band horizontal, 3-element Yagi, Frequency = 18.168 MHz, Height above ground = 30 feet

Horizontal distance (feet) from any part of the antenna for compliance with occupational/controlled or general population/uncontrolled exposure limits*

Height above ground (feet) where exposure occurs

Power**	6 feet		12 feet		20 feet		30 feet	
(watts)	con.	unc.	con.	unc.	con.	unc.	con.	unc.
10	0	0	0	0	0	0	8	9
25	0	0	0	0	0	0	8.5	10
50	0	0	0	0	0	0	9	11
100	0	0	0	0	0	0	10	13
200	0	0	0	0	0	0	11	16.5
250	0	0	0	0	0	7.5	11	18
300	0	0	0	0	0	10	11.5	19.5
400	0	0	0	0	0	14	12.5	22.5
500	0	0	0	0	0	18	13	25
600	0	0	0	0	0	21	14	27.5
750	0	0	0	0	0	26.5	15	31
1000	0	0	0	0	0	36.5	16.5	37
1250	0	0	0	27.5	7.5	43	18	43.5
1500	0	25.5	0	34	10	48	19.5	51

* 0 feet indicates that the exposure at the height in the column above or below the antenna is in compliance.
** Power = Average power input to the antenna. See Chapter 5.

17-meter band horizontal, 3-element Yagi, Frequency = 18.168 MHz, Height above ground = 40 feet

Horizontal distance (feet) from any part of the antenna for compliance with
occupational/controlled or general population/uncontrolled exposure limits*

Height above ground (feet) where exposure occurs

Power**	6 feet		12 feet		20 feet		40 feet	
(watts)	con.	unc.	con.	unc.	con.	unc.	con.	unc.
10	0	0	0	0	0	0	8	9
25	0	0	0	0	0	0	8.5	10.5
50	0	0	0	0	0	0	9	11.5
100	0	0	0	0	0	0	10	13.5
200	0	0	0	0	0	0	11	16.5
250	0	0	0	0	0	0	11.5	18
300	0	0	0	0	0	0	12	19.5
400	0	0	0	0	0	0	12.5	22
500	0	0	0	0	0	0	13.5	24.5
600	0	0	0	0	0	0	14	26.5
750	0	0	0	0	0	0	15	29
1000	0	0	0	0	0	0	16.5	33.5
1250	0	0	0	0	0	0	18	37
1500	0	0	0	0	0	39	19.5	40.5

15-meter band horizontal, 3-element Yagi, Frequency = 21.45 MHz, Height above ground = 20 feet

Horizontal distance (feet) from any part of the antenna for compliance with
occupational/controlled or general population/uncontrolled exposure limits*

Height above ground (feet) where exposure occurs

Power**	6 feet		12 feet		20 feet		30 feet	
(watts)	con.	unc.	con.	unc.	con.	unc.	con.	unc.
10	0	0	0	0	10	11	0	0
25	0	0	0	0	10.5	12	0	0
50	0	0	0	0	11	13	0	0
100	0	0	0	0	11.5	15	0	0
200	0	0	0	13	12.5	19.5	0	0
250	0	0	0	15.5	13	22	0	13
300	0	0	0	18	13.5	24	0	17.5
400	0	0	0	22.5	14.5	28.5	0	24
500	0	19	0	26	15	33	0	29.5
600	0	23	6.5	29	16	37	0	34
750	0	27	9.5	33	17.5	41.5	0	40
1000	0	32	13	38.5	19.5	48	0	50
1250	0	35.5	15.5	42.5	22	53	13	57.5
1500	0	38.5	18	46	24	57	17.5	63.5

15-meter band horizontal, 3-element Yagi, Frequency = 21.45 MHz, Height above ground = 30 feet

Horizontal distance (feet) from any part of the antenna for compliance with
occupational/controlled or general population/uncontrolled exposure limits*

Height above ground (feet) where exposure occurs

Power**	6 feet		12 feet		20 feet		30 feet	
(watts)	con.	unc.	con.	unc.	con.	unc.	con.	unc.
10	0	0	0	0	0	0	10	11
25	0	0	0	0	0	0	10.5	12
50	0	0	0	0	0	0	11	13
100	0	0	0	0	0	0	11.5	15
200	0	0	0	0	0	5.5	12.5	19.5
250	0	0	0	0	0	10.5	13	21.5
300	0	0	0	0	0	14.5	13.5	23.5
400	0	0	0	0	0	21	14.5	27
500	0	0	0	0	0	26	15	30
600	0	0	0	0	0	34	16	33.5
750	0	0	0	28	0	42	17.5	38
1000	0	0	0	38.5	5.5	51.5	19.5	47
1250	0	30.5	0	45	10.5	58.5	21.5	59.5
1500	0	37	0	49.5	14.5	64.5	23.5	69

* 0 feet indicates that the exposure at the height in the column above or below the antenna is in compliance.
** Power = Average power input to the antenna. See Chapter 5.

15-meter band horizontal, 3-element Yagi, Frequency = 21.45 MHz, Height above ground = 40 feet

Horizontal distance (feet) from any part of the antenna for compliance with
occupational/controlled or general population/uncontrolled exposure limits*

Height above ground (feet) where exposure occurs

Power**	6 feet		12 feet		20 feet		40 feet	
(watts)	con.	unc.	con.	unc.	con.	unc.	con.	unc.
10	0	0	0	0	0	0	10	11
25	0	0	0	0	0	0	10.5	12
50	0	0	0	0	0	0	11	13
100	0	0	0	0	0	0	12	15.5
200	0	0	0	0	0	0	13	19.5
250	0	0	0	0	0	0	13	21
300	0	0	0	0	0	0	13.5	23
400	0	0	0	0	0	0	14.5	26
500	0	0	0	0	0	0	15.5	29
600	0	0	0	0	0	0	16	31.5
750	0	0	0	0	0	0	17.5	34.5
1000	0	0	0	0	0	22.5	19.5	39.5
1250	0	0	0	0	0	50.5	21	43.5
1500	0	0	0	43	0	61.5	23	47

12-meter band horizontal, 3-element Yagi, Frequency = 24.99 MHz, Height above ground = 20 feet

Horizontal distance (feet) from any part of the antenna for compliance with
occupational/controlled or general population/uncontrolled exposure limits*

Height above ground (feet) where exposure occurs

Power**	6 feet		12 feet		20 feet		30 feet	
(watts)	con.	unc.	con.	unc.	con.	unc.	con.	unc.
10	0	0	0	0	6	6.5	0	0
25	0	0	0	0	6	8	0	0
50	0	0	0	0	6.5	9.5	0	0
100	0	0	0	0	7.5	13	0	0
200	0	0	0	13	9	19.5	0	8
250	0	0	0	18.5	9.5	22.5	0	13.5
300	0	0	0	22	10.5	25.5	0	17.5
400	0	18	0	27.5	11.5	32	0	24
500	0	22	0	31	13	37	0	30
600	0	25	5	34	14.5	41.5	0	35.5
750	0	29	8.5	38	16.5	46.5	0	45
1000	0	33.5	13	43	19.5	53	8	56
1250	0	37	18.5	47	22.5	58.5	13.5	64
1500	0	40	22	50.5	25.5	63	17.5	70

12-meter band horizontal, 3-element Yagi, Frequency = 24.99 MHz, Height above ground = 30 feet

Horizontal distance (feet) from any part of the antenna for compliance with
occupational/controlled or general population/uncontrolled exposure limits*

Height above ground (feet) where exposure occurs

Power**	6 feet		12 feet		20 feet		30 feet	
(watts)	con.	unc.	con.	unc.	con.	unc.	con.	unc.
10	0	0	0	0	0	0	6	7
25	0	0	0	0	0	0	6.5	8
50	0	0	0	0	0	0	7	10
100	0	0	0	0	0	0	7.5	13
200	0	0	0	0	0	12	9	18
250	0	0	0	0	0	15.5	10	20
300	0	0	0	0	0	18.5	10.5	22
400	0	0	0	0	0	23	12	25
500	0	0	0	25.5	0	27.5	13	28
600	0	0	0	32	0	35.5	14	30.5
750	0	0	0	38.5	0	45.5	15.5	34.5
1000	0	28	0	46	12	56.5	18	42
1250	0	35.5	0	51.5	15.5	64.5	20	52.5
1500	0	41	0	56.5	18.5	70.5	22	70.5

* 0 feet indicates that the exposure at the height in the column above or below the antenna is in compliance.
** Power = Average power input to the antenna. See Chapter 5.

12-meter band horizontal, 3-element Yagi, Frequency = 24.99 MHz, Height above ground = 40 feet

Horizontal distance (feet) from any part of the antenna for compliance with occupational/controlled or general population/uncontrolled exposure limits*

Height above ground (feet) where exposure occurs

Power**	6 feet		12 feet		20 feet		40 feet	
(watts)	con.	unc.	con.	unc.	con.	unc.	con.	unc.
10	0	0	0	0	0	0	6	7
25	0	0	0	0	0	0	6.5	8
50	0	0	0	0	0	0	7	9.5
100	0	0	0	0	0	0	7.5	13
200	0	0	0	0	0	0	9	18.5
250	0	0	0	0	0	0	9.5	20.5
300	0	0	0	0	0	0	10.5	22.5
400	0	0	0	0	0	0	11.5	26.5
500	0	0	0	0	0	0	13	30.5
600	0	0	0	0	0	16	14	33
750	0	0	0	0	0	24	16	36.5
1000	0	0	0	38.5	0	33	18.5	40.5
1250	0	0	0	49	0	55.5	20.5	44
1500	0	0	0	55.5	0	67.5	22.5	47.5

12-meter band horizontal, 3-element Yagi, Frequency = 24.99 MHz, Height above ground = 50 feet

Horizontal distance (feet) from any part of the antenna for compliance with occupational/controlled or general population/uncontrolled exposure limits*

Height above ground (feet) where exposure occurs

Power**	6 feet		12 feet		20 feet		50 feet	
(watts)	con.	unc.	con.	unc.	con.	unc.	con.	unc.
10	0	0	0	0	0	0	6	7
25	0	0	0	0	0	0	6.5	8
50	0	0	0	0	0	0	7	10
100	0	0	0	0	0	0	7.5	13
200	0	0	0	0	0	0	9	18
250	0	0	0	0	0	0	10	20
300	0	0	0	0	0	0	10.5	22
400	0	0	0	0	0	0	12	25.5
500	0	0	0	0	0	0	13	28.5
600	0	0	0	0	0	0	14	31.5
750	0	0	0	0	0	0	15.5	35.5
1000	0	0	0	0	0	0	18	43.5
1250	0	0	0	0	0	23.5	20	53.5
1500	0	0	0	43	0	33	22	60

10-meter band horizontal, 3-element Yagi, Frequency = 29.7 MHz, Antenna height = 30 feet

Horizontal distance (feet) from any part of the antenna for compliance with occupational/controlled or general population/uncontrolled exposure limits*

Height above ground (feet) where exposure occurs

Power**	6 feet		12 feet		20 feet		30 feet	
(watts)	con.	unc.	con.	unc.	con.	unc.	con.	unc.
10	0	0	0	0	0	0	8	9
25	0	0	0	0	0	0	8.5	11
50	0	0	0	0	0	0	9	13.5
100	0	0	0	0	0	0	10.5	18.5
200	0	0	0	0	0	21.5	12.5	25
250	0	0	0	0	0	25	13.5	27.5
300	0	0	0	0	0	28.5	14.5	30
400	0	0	0	39	0	35	16.5	34
500	0	0	0	47	0	48	18.5	37.5
600	0	0	0	52.5	0	59.5	20	40.5
750	0	36	0	59	16.5	70.5	22	45.5
1000	0	46.5	0	67	21.5	82.5	25	61.5
1250	0	53	0	73.5	25	91.5	27.5	95.5
1500	0	58.5	0	79	28.5	99	30	108

* 0 feet indicates that the exposure at the height in the column above or below the antenna is in compliance.
** Power = Average power input to the antenna. See Chapter 5.

10-meter band horizontal, 3-element Yagi, Frequency = 29.7 MHz, Antenna height = 40 feet

Horizontal distance (feet) from any part of the antenna for compliance with occupational/controlled or general population/uncontrolled exposure limits*

Height above ground (feet) where exposure occurs

Power**	6 feet		12 feet		20 feet		40 feet	
(watts)	con.	unc.	con.	unc.	con.	unc.	con.	unc.
10	0	0	0	0	0	0	8	9
25	0	0	0	0	0	0	8.5	11
50	0	0	0	0	0	0	9	13.5
100	0	0	0	0	0	0	10.5	18.5
200	0	0	0	0	0	0	12.5	26
250	0	0	0	0	0	0	13.5	29.5
300	0	0	0	0	0	0	14.5	32.5
400	0	0	0	0	0	0	16.5	40
500	0	0	0	0	0	28.5	18.5	45.5
600	0	0	0	0	0	34.5	20	49
750	0	0	0	55	0	41.5	22.5	53
1000	0	0	0	68	0	75.5	26	58
1250	0	43.5	0	77	0	92	29.5	61
1500	0	54.5	0	84	0	103	32.5	64.5

10-meter band horizontal, 3-element Yagi, Frequency = 29.7 MHz, H, Antenna height = 50 feet

Horizontal distance (feet) from any part of the antenna for compliance with occupational/controlled or general population/uncontrolled exposure limits*

Height above ground (feet) where exposure occurs

Power**	6 feet		12 feet		20 feet		50 feet	
(watts)	con.	unc.	con.	unc.	con.	unc.	con.	unc.
10	0	0	0	0	0	0	8	9
25	0	0	0	0	0	0	8.5	11
50	0	0	0	0	0	0	9	13.5
100	0	0	0	0	0	0	10.5	18.5
200	0	0	0	0	0	0	12.5	25.5
250	0	0	0	0	0	0	13.5	28.5
300	0	0	0	0	0	0	14.5	31
400	0	0	0	0	0	0	16.5	35.5
500	0	0	0	0	0	0	18.5	39
600	0	0	0	0	0	0	20	42.5
750	0	0	0	0	0	0	22.5	47.5
1000	0	0	0	0	0	37	25.5	57.5
1250	0	0	0	71.5	0	48	28.5	78
1500	0	0	0	82.5	0	84	31	85.5

10-meter band horizontal, 3-element Yagi, Frequency = 29.7 MHz, H, Antenna height = 60 feet

Horizontal distance (feet) from any part of the antenna for compliance with occupational/controlled or general population/uncontrolled exposure limits*

Height above ground (feet) where exposure occurs

Power**	6 feet		12 feet		20 feet		60 feet	
(watts)	con.	unc.	con.	unc.	con.	unc.	con.	unc.
10	0	0	0	0	0	0	8	9
25	0	0	0	0	0	0	8.5	11
50	0	0	0	0	0	0	9	13.5
100	0	0	0	0	0	0	10.5	18.5
200	0	0	0	0	0	0	12.5	25.5
250	0	0	0	0	0	0	13.5	28.5
300	0	0	0	0	0	0	14.5	31.5
400	0	0	0	0	0	0	16.5	36.5
500	0	0	0	0	0	0	18.5	42.5
600	0	0	0	0	0	0	20	47.5
750	0	0	0	0	0	0	22.5	53
1000	0	0	0	0	0	0	25.5	58.5
1250	0	0	0	0	0	0	28.5	62.5
1500	0	0	0	65.5	0	0	31.5	66.5

* 0 feet indicates that the exposure at the height in the column above or below the antenna is in compliance.
** Power = Average power input to the antenna. See Chapter 5.

6-meter band horizontal, 3-element Yagi, Frequency = 50.5 MHz, Height above ground = 30 feet

Horizontal distance (feet) from any part of the antenna for compliance with
occupational/controlled or general population/uncontrolled exposure limits*

Power**	6 feet		12 feet		20 feet		30 feet	
(watts)	con.	unc.	con.	unc.	con.	unc.	con.	unc.
10	0	0	0	0	0	0	4	6
25	0	0	0	0	0	0	5	8.5
50	0	0	0	0	0	0	6	12
100	0	0	0	0	0	0	7.5	17.5
200	0	0	0	0	0	23.5	11	23.5
250	0	0	0	0	0	28	12	25.5
300	0	0	0	0	0	30.5	13.5	28
400	0	31.5	0	24.5	0	34	15.5	33.5
500	0	40	0	30	0	36.5	17.5	48
600	0	45.5	0	53	11	38.5	18.5	52.5
750	0	52	0	68.5	14.5	41.5	20.5	56.5
1000	0	59.5	0	81.5	23.5	46	23.5	60.5
1250	0	65	0	90.5	28	49.5	25.5	63
1500	0	70	0	97.5	30.5	104	28	65

6-meter band horizontal, 3-element Yagi, Frequency = 50.5 MHz, Height above ground = 40 feet

Horizontal distance (feet) from any part of the antenna for compliance with
occupational/controlled or general population/uncontrolled exposure limits*

Power**	6 feet		12 feet		20 feet		40 feet	
(watts)	con.	unc.	con.	unc.	con.	unc.	con.	unc.
10	0	0	0	0	0	0	4	6
25	0	0	0	0	0	0	5	8.5
50	0	0	0	0	0	0	6	12
100	0	0	0	0	0	0	7.5	17
200	0	0	0	0	0	0	11	23.5
250	0	0	0	0	0	0	12	26
300	0	0	0	0	0	0	13.5	29
400	0	0	0	0	0	0	15.5	35
500	0	0	0	0	0	37	17	44
600	0	0	0	0	0	42	18.5	47
750	0	44.5	0	28	0	46	20.5	50
1000	0	58	0	39.5	0	50	23.5	53
1250	0	66.5	0	87.5	0	53.5	26	56
1500	0	73	0	99.5	0	57	29	58.5

6-meter band horizontal, 3-element Yagi, Frequency = 50.5 MHz, Height above ground = 50 feet

Horizontal distance (feet) from any part of the antenna for compliance with
occupational/controlled or general population/uncontrolled exposure limits*

Power**	6 feet		12 feet		20 feet		50 feet	
(watts)	con.	unc.	con.	unc.	con.	unc.	con.	unc.
10	0	0	0	0	0	0	4	6
25	0	0	0	0	0	0	5	8.5
50	0	0	0	0	0	0	6	12
100	0	0	0	0	0	0	7.5	17
200	0	0	0	0	0	0	11	23.5
250	0	0	0	0	0	0	12	26.5
300	0	0	0	0	0	0	13	29
400	0	0	0	0	0	0	15.5	36
500	0	0	0	0	0	0	17	42
600	0	0	0	0	0	0	18.5	45
750	0	0	0	0	0	0	20.5	47.5
1000	0	0	0	0	0	52.5	23.5	51.5
1250	0	59.5	0	0	0	58	26.5	55.5
1500	0	70	0	43.5	0	61	29	82.5

* 0 feet indicates that the exposure at the height in the column above or below the antenna is in compliance.
** Power = Average power input to the antenna. See Chapter 5.

5-, 6- AND 8-ELEMENT YAGI ANTENNAS

These tables represent 5-, 6- and 8-element Yagi monoband antennas. The tables show the horizontal compliance distance at compliance heights of 6 feet, 12 feet, 20 feet and at the height of the antenna, in the direction the antenna is pointing. The distances were calculated from the physical center of the antenna.

See Chapter 5 for information on how to use the pattern of the antenna to estimate compliance distance in other directions.

The tables are generally are a conservative estimate for shortened or loaded directional antennas with the same number of elements.

20-meter band horizontal, 5-element Yagi, Frequency = 14.35 MHz, Height above ground = 30 feet

Horizontal distance (feet) from any part of the antenna for compliance with occupational/controlled or general population/uncontrolled exposure limits*

Power** (watts)	Height above ground (feet) where exposure occurs							
	6 feet		12 feet		20 feet		30 feet	
	con.	unc.	con.	unc.	con.	unc.	con.	unc.
10	0	0	0	0	0	0	24.5	25
25	0	0	0	0	0	0	24.5	25
50	0	0	0	0	0	0	25	25.5
100	0	0	0	0	0	0	25	26.5
200	0	0	0	0	0	0	25.5	27
250	0	0	0	0	0	0	25.5	27.5
300	0	0	0	0	0	0	26	28
400	0	0	0	0	0	0	26	28.5
500	0	0	0	0	0	0	26.5	29.5
600	0	0	0	0	0	0	26.5	30
750	0	0	0	0	0	5.5	27	31
1000	0	0	0	0	0	23	27	32.5
1250	0	0	0	0	0	26.5	27.5	34
1500	0	0	0	0	0	28.5	28	36

17-meter band horizontal, 5-element Yagi, Frequency = 18.168 MHz, Height above ground = 30 feet

Horizontal distance (feet) from any part of the antenna for compliance with occupational/controlled or general population/uncontrolled exposure limits*

Power** (watts)	Height above ground (feet) where exposure occurs							
	6 feet		12 feet		20 feet		30 feet	
	con.	unc.	con.	unc.	con.	unc.	con.	unc.
10	0	0	0	0	0	0	15.5	16.5
25	0	0	0	0	0	0	16	17
50	0	0	0	0	0	0	16.5	18
100	0	0	0	0	0	0	17	19.5
200	0	0	0	0	0	0	18	22.5
250	0	0	0	0	0	12.5	18	23.5
300	0	0	0	0	0	15.5	18.5	25
400	0	0	0	0	0	19	19	28.5
500	0	0	0	0	0	23	19.5	31.5
600	0	0	0	0	0	26.5	20	34
750	0	0	0	0	0	31.5	21	38.5
1000	0	0	0	0	0	42	22.5	45.5
1250	0	0	0	28.5	12.5	50	23.5	54
1500	0	0	0	37.5	15.5	55.5	25	61.5

* 0 feet indicates that the exposure at the height in the column above or below the antenna is in compliance.
** Power = Average power input to the antenna. See Chapter 5.

17-meter band horizontal, 5-element Yagi, Frequency = 18.168 MHz, Height above ground = 40 feet

Horizontal distance (feet) from any part of the antenna for compliance with occupational/controlled or general population/uncontrolled exposure limits*

Height above ground (feet) where exposure occurs

Power** (watts)	6 feet con.	6 feet unc.	12 feet con.	12 feet unc.	20 feet con.	20 feet unc.	40 feet con.	40 feet unc.
10	0	0	0	0	0	0	15.5	16.5
25	0	0	0	0	0	0	16	17
50	0	0	0	0	0	0	16.5	18
100	0	0	0	0	0	0	17	19.5
200	0	0	0	0	0	0	18	22.5
250	0	0	0	0	0	0	18	23.5
300	0	0	0	0	0	0	18.5	25
400	0	0	0	0	0	0	19	28
500	0	0	0	0	0	0	19.5	30.5
600	0	0	0	0	0	0	20	33
750	0	0	0	0	0	0	21	36.5
1000	0	0	0	0	0	0	22.5	42
1250	0	0	0	0	0	0	23.5	47
1500	0	0	0	0	0	45.5	25	51.5

15-meter band horizontal, 5-element Yagi, Frequency = 21.45 MHz, Height above ground = 30 feet

Horizontal distance (feet) from any part of the antenna for compliance with occupational/controlled or general population/uncontrolled exposure limits*

Height above ground (feet) where exposure occurs

Power** (watts)	6 feet con.	6 feet unc.	12 feet con.	12 feet unc.	20 feet con.	20 feet unc.	30 feet con.	30 feet unc.
10	0	0	0	0	0	0	18.5	19.5
25	0	0	0	0	0	0	19	20
50	0	0	0	0	0	0	19.5	21
100	0	0	0	0	0	0	20	23
200	0	0	0	0	0	13.5	20.5	27
250	0	0	0	0	0	19	21	29.5
300	0	0	0	0	0	22	21.5	31.5
400	0	0	0	0	0	28	22	36
500	0	0	0	0	0	33	23	40
600	0	0	0	0	0	39	23.5	44
750	0	0	0	0	0	49	25	50
1000	0	0	0	38	13.5	60.5	27	63
1250	0	0	0	50	19	68.5	29.5	75
1500	0	37.5	0	56.5	22	75	31.5	84

15-meter band horizontal, 5-element Yagi, Frequency = 21.45 MHz, Height above ground = 40 feet

Horizontal distance (feet) from any part of the antenna for compliance with occupational/controlled or general population/uncontrolled exposure limits*

Height above ground (feet) where exposure occurs

Power** (watts)	6 feet con.	6 feet unc.	12 feet con.	12 feet unc.	20 feet con.	20 feet unc.	40 feet con.	40 feet unc.
10	0	0	0	0	0	0	18.5	19.5
25	0	0	0	0	0	0	19	20
50	0	0	0	0	0	0	19.5	21
100	0	0	0	0	0	0	20	22.5
200	0	0	0	0	0	0	20.5	26.5
250	0	0	0	0	0	0	21	29
300	0	0	0	0	0	0	21.5	31
400	0	0	0	0	0	0	22	35
500	0	0	0	0	0	0	22.5	38.5
600	0	0	0	0	0	0	23.5	41.5
750	0	0	0	0	0	0	24.5	46
1000	0	0	0	0	0	0	26.5	52.5
1250	0	0	0	0	0	63	29	58.5
1500	0	0	0	0	0	74.5	31	65

* 0 feet indicates that the exposure at the height in the column above or below the antenna is in compliance.
** Power = Average power input to the antenna. See Chapter 5.

12-meter band horizontal, 5-element Yagi, Frequency = 24.99 MHz, Height above ground = 30 feet

Horizontal distance (feet) from any part of the antenna for compliance with occupational/controlled or general population/uncontrolled exposure limits*

Height above ground (feet) where exposure occurs

Power**	6 feet		12 feet		20 feet		30 feet	
(watts)	con.	unc.	con.	unc.	con.	unc.	con.	unc.
10	0	0	0	0	0	0	11	12
25	0	0	0	0	0	0	11.5	13
50	0	0	0	0	0	0	12	15
100	0	0	0	0	0	0	13	18.5
200	0	0	0	0	0	17.5	14	24.5
250	0	0	0	0	0	22	15	27.5
300	0	0	0	0	0	26	15.5	30
400	0	0	0	0	0	32.5	17	34.5
500	0	0	0	30.5	0	43	18.5	38.5
600	0	0	0	40.5	0	51.5	20	43
750	0	0	0	48	9	60.5	21.5	50.5
1000	0	36	0	56.5	17.5	71.5	24.5	70.5
1250	0	45	0	62.5	22	80	27.5	85.5
1500	0	50.5	0	68	26	86.5	30	96

12-meter band horizontal, 5-element Yagi, Frequency = 24.99 MHz, Height above ground = 40 feet

Horizontal distance (feet) from any part of the antenna for compliance with occupational/controlled or general population/uncontrolled exposure limits*

Height above ground (feet) where exposure occurs

Power**	6 feet		12 feet		20 feet		40 feet	
(watts)	con.	unc.	con.	unc.	con.	unc.	con.	unc.
10	0	0	0	0	0	0	11	12
25	0	0	0	0	0	0	11.5	13
50	0	0	0	0	0	0	12	15
100	0	0	0	0	0	0	12.5	18.5
200	0	0	0	0	0	0	14	25
250	0	0	0	0	0	0	15	27.5
300	0	0	0	0	0	0	15.5	30
400	0	0	0	0	0	0	17	34.5
500	0	0	0	0	0	0	18.5	38
600	0	0	0	0	0	0	20	41.5
750	0	0	0	0	0	32.5	22	45.5
1000	0	0	0	47.5	0	62.5	25	51
1250	0	0	0	60	0	78	27.5	55.5
1500	0	0	0	68	0	88	30	59.5

12-meter band horizontal, 5-element Yagi, Frequency = 24.99 MHz, Height above ground = 50 feet

Horizontal distance (feet) from any part of the antenna for compliance with occupational/controlled or general population/uncontrolled exposure limits*

Height above ground (feet) where exposure occurs

Power**	6 feet		12 feet		20 feet		50 feet	
(watts)	con.	unc.	con.	unc.	con.	unc.	con.	unc.
10	0	0	0	0	0	0	11	12
25	0	0	0	0	0	0	11.5	13
50	0	0	0	0	0	0	12	15
100	0	0	0	0	0	0	13	18.5
200	0	0	0	0	0	0	14	25
250	0	0	0	0	0	0	15	27.5
300	0	0	0	0	0	0	15.5	30
400	0	0	0	0	0	0	17	34.5
500	0	0	0	0	0	0	18.5	39
600	0	0	0	0	0	0	20	42.5
750	0	0	0	0	0	0	22	49
1000	0	0	0	0	0	0	25	59.5
1250	0	0	0	0	0	0	27.5	66.5
1500	0	0	0	0	0	41	30	71.5

* 0 feet indicates that the exposure at the height in the column above or below the antenna is in compliance.
** Power = Average power input to the antenna. See Chapter 5.

10-meter band horizontal, 5-element Yagi, Frequency = 29.7 MHz, Height above ground = 30 feet

Horizontal distance (feet) from any part of the antenna for compliance with occupational/controlled or general population/uncontrolled exposure limits*

Height above ground (feet) where exposure occurs

Power**	6 feet		12 feet		20 feet		30 feet	
(watts)	con.	unc.	con.	unc.	con.	unc.	con.	unc.
10	0	0	0	0	0	0	14	15
25	0	0	0	0	0	0	14.5	16.5
50	0	0	0	0	0	0	15	19
100	0	0	0	0	0	0	16	24
200	0	0	0	0	0	27.5	18	33
250	0	0	0	0	0	33.5	19	36.5
300	0	0	0	0	0	38.5	20	40
400	0	0	0	0	0	53	22	46.5
500	0	0	0	51.5	0	66	24	54.5
600	0	0	0	58.5	10.5	74.5	26	62
750	0	0	0	66.5	16.5	84.5	29	84.5
1000	0	52	0	76	27.5	97	33	106.5
1250	0	60	0	83	33.5	106	36.5	120.5
1500	0	66	0	89	38.5	114	40	131.5

10-meter band horizontal, 5-element Yagi, Frequency = 29.7 MHz, Height above ground = 40 feet

Horizontal distance (feet) from any part of the antenna for compliance with occupational/controlled or general population/uncontrolled exposure limits*

Height above ground (feet) where exposure occurs

Power**	6 feet		12 feet		20 feet		40 feet	
(watts)	con.	unc.	con.	unc.	con.	unc.	con.	unc.
10	0	0	0	0	0	0	14	15
25	0	0	0	0	0	0	14.5	16.5
50	0	0	0	0	0	0	15	19
100	0	0	0	0	0	0	16	24.5
200	0	0	0	0	0	0	18	33.5
250	0	0	0	0	0	0	19	37
300	0	0	0	0	0	0	20	40.5
400	0	0	0	0	0	0	22.5	46.5
500	0	0	0	0	0	0	24.5	51.5
600	0	0	0	0	0	43	26.5	55.5
750	0	0	0	0	0	74.5	29.5	60.5
1000	0	0	0	75	0	98	33.5	67.5
1250	0	0	0	86	0	111.5	37	73.5
1500	0	0	0	94	0	122	40.5	79.5

10-meter band horizontal, 5-element Yagi, Frequency = 29.7 MHz, Height above ground = 50 feet

Horizontal distance (feet) from any part of the antenna for compliance with occupational/controlled or general population/uncontrolled exposure limits*

Height above ground (feet) where exposure occurs

Power**	6 feet		12 feet		20 feet		50 feet	
(watts)	con.	unc.	con.	unc.	con.	unc.	con.	unc.
10	0	0	0	0	0	0	14	15
25	0	0	0	0	0	0	14.5	16.5
50	0	0	0	0	0	0	15	19
100	0	0	0	0	0	0	16	24
200	0	0	0	0	0	0	18	33.5
250	0	0	0	0	0	0	19	37
300	0	0	0	0	0	0	20	40.5
400	0	0	0	0	0	0	22	47
500	0	0	0	0	0	0	24	52.5
600	0	0	0	0	0	0	26.5	58
750	0	0	0	0	0	0	29	66.5
1000	0	0	0	0	0	0	33.5	79.5
1250	0	0	0	74.5	0	102.5	37	87
1500	0	0	0	91	0	121	40.5	92.5

* 0 feet indicates that the exposure at the height in the column above or below the antenna is in compliance.
** Power = Average power input to the antenna. See Chapter 5.

10-meter band horizontal, 6-element Yagi, Frequency = 29.7 MHz, Height above ground = 30 feet

Horizontal distance (feet) from any part of the antenna for compliance with
occupational/controlled or general population/uncontrolled exposure limits*

Height above ground (feet) where exposure occurs

Power**	6 feet		12 feet		20 feet		30 feet	
(watts)	con.	unc.	con.	unc.	con.	unc.	con.	unc.
10	0	0	0	0	0	0	18.5	19.5
25	0	0	0	0	0	0	19	20.5
50	0	0	0	0	0	0	19.5	22.5
100	0	0	0	0	0	0	20	27
200	0	0	0	0	0	30	21.5	36.5
250	0	0	0	0	0	36.5	22.5	40.5
300	0	0	0	0	0	42	23	44.5
400	0	0	0	0	0	57.5	25	51.5
500	0	0	0	53.5	0	70	27	60
600	0	0	0	61	0	79	29	70.5
750	0	0	0	69	18.5	89	32	93.5
1000	0	53.5	0	79	30	101.5	36.5	114
1250	0	62.5	0	86.5	36.5	111	40.5	127.5
1500	0	69	0	93	42	119	44	138.5

10-meter band horizontal, 6-element Yagi, Frequency = 29.7 MHz, Height above ground = 40 feet

Horizontal distance (feet) from any part of the antenna for compliance with
occupational/controlled or general population/uncontrolled exposure limits*

Height above ground (feet) where exposure occurs

Power**	6 feet		12 feet		20 feet		40 feet	
(watts)	con.	unc.	con.	unc.	con.	unc.	con.	unc.
10	0	0	0	0	0	0	18.5	19.5
25	0	0	0	0	0	0	19	20.5
50	0	0	0	0	0	0	19.5	22.5
100	0	0	0	0	0	0	20.5	27.5
200	0	0	0	0	0	0	22	36
250	0	0	0	0	0	0	22.5	40
300	0	0	0	0	0	0	23.5	43.5
400	0	0	0	0	0	0	25.5	50
500	0	0	0	0	0	0	27.5	55.5
600	0	0	0	0	0	49.5	29.5	59.5
750	0	0	0	0	0	83.5	32	65
1000	0	0	0	79	0	105	36	73
1250	0	0	0	90.5	0	118.5	40	80.5
1500	0	0	0	99	0	129	43.5	88.5

10-meter band horizontal, 6-element Yagi, Frequency = 29.7 MHz, Height above ground = 50 feet

Horizontal distance (feet) from any part of the antenna for compliance with
occupational/controlled or general population/uncontrolled exposure limits*

Height above ground (feet) where exposure occurs

Power**	6 feet		12 feet		20 feet		50 feet	
(watts)	con.	unc.	con.	unc.	con.	unc.	con.	unc.
10	0	0	0	0	0	0	18.5	19.5
25	0	0	0	0	0	0	19	20.5
50	0	0	0	0	0	0	19.5	22.5
100	0	0	0	0	0	0	20	27.5
200	0	0	0	0	0	0	21.5	36.5
250	0	0	0	0	0	0	22.5	40.5
300	0	0	0	0	0	0	23	44.5
400	0	0	0	0	0	0	25	50.5
500	0	0	0	0	0	0	27	56.5
600	0	0	0	0	0	0	29.5	61.5
750	0	0	0	0	0	0	32	70
1000	0	0	0	0	0	0	36.5	82.5
1250	0	0	0	75	0	113	40.5	90.5
1500	0	0	0	95.5	0	129.5	44.5	95.5

* 0 feet indicates that the exposure at the height in the column above or below the antenna is in compliance.
** Power = Average power input to the antenna. See Chapter 5.

6-meter band horizontal, 5-element Yagi, Frequency = 50.5 MHz, Height above ground = 30 feet

Horizontal distance (feet) from any part of the antenna for compliance with occupational/controlled or general population/uncontrolled exposure limits*

Power** (watts)	6 feet con.	6 feet unc.	12 feet con.	12 feet unc.	20 feet con.	20 feet unc.	30 feet con.	30 feet unc.
10	0	0	0	0	0	0	7	8.5
25	0	0	0	0	0	0	7.5	11.5
50	0	0	0	0	0	0	8.5	16
100	0	0	0	0	0	0	10.5	22
200	0	0	0	0	0	29	14.5	30.5
250	0	0	0	0	0	33	16	34
300	0	0	0	0	0	35.5	17.5	38.5
400	0	0	0	30	0	39.5	20	51.5
500	0	47	0	65	0	43	22	56
600	0	53.5	0	75.5	14	46	24	59.5
750	0	60.5	0	86	20	50	26.5	62.5
1000	0	69	0	98	29	103.5	30.5	66.5
1250	0	75.5	0	107.5	33	123.5	34	69.5
1500	0	81.5	0	115	35.5	136.5	38.5	72

6-meter band horizontal, 5-element Yagi, Frequency = 50.5 MHz, Height above ground = 40 feet

Horizontal distance (feet) from any part of the antenna for compliance with occupational/controlled or general population/uncontrolled exposure limits*

Power** (watts)	6 feet con.	6 feet unc.	12 feet con.	12 feet unc.	20 feet con.	20 feet unc.	40 feet con.	40 feet unc.
10	0	0	0	0	0	0	7	8.5
25	0	0	0	0	0	0	7.5	11.5
50	0	0	0	0	0	0	8.5	16
100	0	0	0	0	0	0	10.5	22
200	0	0	0	0	0	0	14.5	30.5
250	0	0	0	0	0	0	16	34.5
300	0	0	0	0	0	0	17.5	39
400	0	0	0	0	0	0	20	47
500	0	0	0	0	0	42.5	22	51
600	0	0	0	0	0	47	24	53.5
750	0	47.5	0	0	0	51	26.5	56.5
1000	0	68	0	97	0	56.5	30.5	62
1250	0	77.5	0	111	0	62	34.5	101.5
1500	0	85	0	122	0	66	39	108

6-meter band horizontal, 5-element Yagi, Frequency = 50.5 MHz, Height above ground = 50 feet

Horizontal distance (feet) from any part of the antenna for compliance with occupational/controlled or general population/uncontrolled exposure limits*

Power** (watts)	6 feet con.	6 feet unc.	12 feet con.	12 feet unc.	20 feet con.	20 feet unc.	50 feet con.	50 feet unc.
10	0	0	0	0	0	0	7	8.5
25	0	0	0	0	0	0	7.5	11.5
50	0	0	0	0	0	0	8.5	16
100	0	0	0	0	0	0	10.5	22
200	0	0	0	0	0	0	14.5	31
250	0	0	0	0	0	0	16	34.5
300	0	0	0	0	0	0	17.5	38.5
400	0	0	0	0	0	0	20	45.5
500	0	0	0	0	0	0	22	49
600	0	0	0	0	0	0	24	52
750	0	0	0	0	0	0	27	56.5
1000	0	0	0	0	0	57.5	31	78.5
1250	0	67.5	0	0	0	63	34.5	87.5
1500	0	81.5	0	117.5	0	66.5	38.5	91

* 0 feet indicates that the exposure at the height in the column above or below the antenna is in compliance.
** Power = Average power input to the antenna. See Chapter 5.

6-meter band horizontal, 8-element Yagi, Frequency = 50.5 MHz, Height above ground = 30 feet

Horizontal distance (feet) from any part of the antenna for compliance with occupational/controlled or general population/uncontrolled exposure limits*

Height above ground (feet) where exposure occurs

Power** (watts)	6 feet		12 feet		20 feet		30 feet	
	con.	unc.	con.	unc.	con.	unc.	con.	unc.
10	0	0	0	0	0	0	17.5	18.5
25	0	0	0	0	0	0	18	20
50	0	0	0	0	0	0	18.5	24.5
100	0	0	0	0	0	0	19.5	33
200	0	0	0	0	0	34	23	42.5
250	0	0	0	0	0	42.5	24.5	46.5
300	0	0	0	0	0	47.5	26.5	50.5
400	0	0	0	68	0	55.5	30	59.5
500	0	0	0	86.5	0	63.5	33	65.5
600	0	0	0	96.5	0	101.5	35.5	71.5
750	0	67.5	0	107.5	0	126.5	38.5	74.5
1000	0	81.5	0	121	34	148	42.5	82
1250	0	91	0	131.5	42.5	163	46.5	171.5
1500	0	98.5	0	140.5	47.5	175	50.5	193.5

6-meter band horizontal, 8-element Yagi, Frequency = 50.5 MHz, Height above ground = 40 feet

Horizontal distance (feet) from any part of the antenna for compliance with occupational/controlled or general population/uncontrolled exposure limits*

Height above ground (feet) where exposure occurs

Power** (watts)	6 feet		12 feet		20 feet		40 feet	
	con.	unc.	con.	unc.	con.	unc.	con.	unc.
10	0	0	0	0	0	0	17.5	18.5
25	0	0	0	0	0	0	18	20
50	0	0	0	0	0	0	18.5	24.5
100	0	0	0	0	0	0	19.5	32
200	0	0	0	0	0	0	23	42.5
250	0	0	0	0	0	0	24.5	47
300	0	0	0	0	0	0	26	52.5
400	0	0	0	0	0	0	29.5	64
500	0	0	0	0	0	0	32	70
600	0	0	0	0	0	0	34.5	75.5
750	0	0	0	102	0	63	37.5	85.5
1000	0	0	0	126	0	77	42.5	106
1250	0	84	0	140.5	0	167	47	114
1500	0	97.5	0	152	0	185.5	52.5	119.5

6-meter band horizontal, 8-element Yagi, Frequency = 50.5 MHz, Height above ground = 50 feet

Horizontal distance (feet) from any part of the antenna for compliance with occupational/controlled or general population/uncontrolled exposure limits*

Height above ground (feet) where exposure occurs

Power** (watts)	6 feet		12 feet		20 feet		50 feet	
	con.	unc.	con.	unc.	con.	unc.	con.	unc.
10	0	0	0	0	0	0	17.5	18.5
25	0	0	0	0	0	0	18	20
50	0	0	0	0	0	0	18.5	24.5
100	0	0	0	0	0	0	19.5	32
200	0	0	0	0	0	0	23	43
250	0	0	0	0	0	0	24.5	48.5
300	0	0	0	0	0	0	26	56
400	0	0	0	0	0	0	29	63.5
500	0	0	0	0	0	0	32	68
600	0	0	0	0	0	0	34.5	72.5
750	0	0	0	0	0	0	38	79.5
1000	0	0	0	0	0	0	43	94.5
1250	0	0	0	137	0	0	48.5	103.5
1500	0	0	0	154.5	0	81.5	56	111.5

* 0 feet indicates that the exposure at the height in the column above or below the antenna is in compliance.
** Power = Average power input to the antenna. See Chapter 5.

2-meter band horizontal, 5-element Yagi, Frequency = 146.0 MHz, Height above ground = 20 feet

Horizontal distance (feet) from any part of the antenna for compliance with
occupational/controlled or general population/uncontrolled exposure limits*

Height above ground (feet) where exposure occurs

Power** (watts)	6 feet		12 feet		20 feet		30 feet	
	con.	unc.	con.	unc.	con.	unc.	con.	unc.
10	0	0	0	0	3.5	7.5	0	0
25	0	0	0	0	5.5	11	0	0
50	0	0	0	0	7.5	15.5	0	0
100	0	0	0	15.5	10	23	0	0
200	0	19	0	25	14	28	0	26
250	0	21	0	43.5	15.5	41	0	27.5
300	0	25	0	46.5	16	43	0	38.5
400	0	63.5	0	50	18.5	45.5	0	42.5
500	0	74	15.5	52	23	46.5	0	43.5
600	0	81	22	53.5	25	76	0	65
750	0	89	23.5	55	26.5	81.5	16	68
1000	19	99.5	25	57	28	86	26	70.5
1250	21	107.5	43.5	121.5	41	89	27.5	113.5
1500	25	114	46.5	139	43	91	38.5	121.5

2-meter band horizontal, 5-element Yagi, Frequency = 146.0 MHz, Height above ground = 30 feet

Horizontal distance (feet) from any part of the antenna for compliance with
occupational/controlled or general population/uncontrolled exposure limits*

Height above ground (feet) where exposure occurs

Power** (watts)	6 feet		12 feet		20 feet		30 feet	
	con.	unc.	con.	unc.	con.	unc.	con.	unc.
10	0	0	0	0	0	0	3.5	7.5
25	0	0	0	0	0	0	5.5	11.5
50	0	0	0	0	0	0	7.5	15
100	0	0	0	0	0	0	10	22
200	0	0	0	0	0	26	14	31.5
250	0	0	0	0	0	27.5	15	34
300	0	0	0	0	0	38.5	16	35
400	0	0	0	34.5	0	42.5	19.5	47
500	0	0	0	36.5	0	43.5	22	48.5
600	0	29.5	0	67.5	0	65	23	50
750	0	35.5	0	72	18	68	24.5	68.5
1000	0	103.5	0	76.5	26	70.5	31.5	71
1250	0	117	0	79	27.5	113.5	34	103
1500	0	127	0	81	38.5	121.5	35	106

2-meter band horizontal, 5-element Yagi, Frequency = 146.0 MHz, Height above ground = 40 feet

Horizontal distance (feet) from any part of the antenna for compliance with
occupational/controlled or general population/uncontrolled exposure limits*

Height above ground (feet) where exposure occurs

Power** (watts)	6 feet		12 feet		20 feet		40 feet	
	con.	unc.	con.	unc.	con.	unc.	con.	unc.
10	0	0	0	0	0	0	3.5	7.5
25	0	0	0	0	0	0	5.5	11.5
50	0	0	0	0	0	0	7.5	15
100	0	0	0	0	0	0	10	21.5
200	0	0	0	0	0	0	14	30
250	0	0	0	0	0	0	15	32
300	0	0	0	0	0	0	16.5	34.5
400	0	0	0	0	0	0	19.5	41.5
500	0	0	0	0	0	35.5	21.5	43.5
600	0	0	0	0	0	37	22.5	54
750	0	0	0	45	0	57	25	56
1000	0	35	0	87	0	85.5	30	71.5
1250	0	40.5	0	94.5	0	90.5	32	92.5
1500	0	118	0	98.5	0	93	34.5	95

* 0 feet indicates that the exposure at the height in the column above or below the antenna is in compliance.
** Power = Average power input to the antenna. See Chapter 5.

2-meter band horizontal, 5-element Yagi, Frequency = 146.0 MHz, Height above ground = 50 feet

Horizontal distance (feet) from any part of the antenna for compliance with
occupational/controlled or general population/uncontrolled exposure limits*

Height above ground (feet) where exposure occurs

Power**	6 feet		12 feet		20 feet		50 feet	
(watts)	con.	unc.	con.	unc.	con.	unc.	con.	unc.
10	0	0	0	0	0	0	3.5	7.5
25	0	0	0	0	0	0	5.5	11.5
50	0	0	0	0	0	0	7.5	15
100	0	0	0	0	0	0	10	21
200	0	0	0	0	0	0	13.5	29.5
250	0	0	0	0	0	0	15	32
300	0	0	0	0	0	0	16.5	37
400	0	0	0	0	0	0	19.5	40
500	0	0	0	0	0	0	21	48
600	0	0	0	0	0	0	22.5	50
750	0	0	0	0	0	0	26	60.5
1000	0	0	0	0	0	45.5	29.5	74
1250	0	0	0	56	0	71.5	32	77
1500	0	0	0	59	0	73.5	37	94.5

2-meter band horizontal, 5-element Yagi, Frequency = 146.0 MHz, Height above ground = 60 feet

Horizontal distance (feet) from any part of the antenna for compliance with
occupational/controlled or general population/uncontrolled exposure limits*

Height above ground (feet) where exposure occurs

Power**	6 feet		12 feet		20 feet		60 feet	
(watts)	con.	unc.	con.	unc.	con.	unc.	con.	unc.
10	0	0	0	0	0	0	3.5	7.5
25	0	0	0	0	0	0	5.5	11.5
50	0	0	0	0	0	0	7.5	15
100	0	0	0	0	0	0	10	21
200	0	0	0	0	0	0	13.5	29
250	0	0	0	0	0	0	15	32.5
300	0	0	0	0	0	0	16.5	36.5
400	0	0	0	0	0	0	19	40
500	0	0	0	0	0	0	21	46.5
600	0	0	0	0	0	0	22.5	48.5
750	0	0	0	0	0	0	26	57
1000	0	0	0	0	0	0	29	68.5
1250	0	0	0	0	0	0	32.5	81
1500	0	0	0	0	0	53	36.5	83.5

* 0 feet indicates that the exposure at the height in the column above or below the antenna is in compliance.
** Power = Average power input to the antenna. See Chapter 5.

OTHER YAGI ANTENNAS

These tables represent large Yagi monoband antennas. The tables show the horizontal compliance distance at compliance heights of 6 feet, 12 feet, 20 feet and at the height of the antenna, in the direction the antenna is pointing. The distances were calculated from the physical center of the antenna. See Chapter 5 for information on how to use the pattern of the antenna to estimate compliance distance in other directions.

These tables are generally an approximate estimate for directive antennas with similar gain

---

2-meter band horizontal, 13-element Yagi, Frequency = 146.0 MHz, Height above ground = 30 feet

Horizontal distance (feet) from any part of the antenna for compliance with occupational/controlled or general population/uncontrolled exposure limits*

Height above ground (feet) where exposure occurs

Power** (watts)	6 feet con.	6 feet unc.	12 feet con.	12 feet unc.	20 feet con.	20 feet unc.	30 feet con.	30 feet unc.
10	0	0	0	0	0	0	11	15
25	0	0	0	0	0	0	12.5	21.5
50	0	0	0	0	0	0	15	29
100	0	0	0	0	0	0	19.5	40
200	0	0	0	0	0	50.5	26	56.5
250	0	0	0	0	0	54	29	59.5
300	0	0	0	0	0	58	31.5	63
400	0	0	0	45.5	0	73.5	35	78.5
500	0	122.5	0	77.5	0	112.5	40	83
600	0	135.5	0	81	0	123.5	42.5	106
750	0	149.5	0	84.5	35.5	130	47	110.5
1000	0	167.5	0	208	50.5	136.5	56.5	181.5
1250	0	181.5	0	234.5	54	140.5	59.5	192
1500	0	193	0	254	58	143.5	63	198

2-meter band horizontal, 13-element Yagi, Frequency = 146.0 MHz, Height above ground = 40 feet

Horizontal distance (feet) from any part of the antenna for compliance with occupational/controlled or general population/uncontrolled exposure limits*

Height above ground (feet) where exposure occurs

Power** (watts)	6 feet con.	6 feet unc.	12 feet con.	12 feet unc.	20 feet con.	20 feet unc.	40 feet con.	40 feet unc.
10	0	0	0	0	0	0	11	15
25	0	0	0	0	0	0	12.5	21.5
50	0	0	0	0	0	0	15	29
100	0	0	0	0	0	0	19.5	39.5
200	0	0	0	0	0	0	26	54.5
250	0	0	0	0	0	0	29	62
300	0	0	0	0	0	0	31	65
400	0	0	0	0	0	0	36	80
500	0	0	0	0	0	66.5	39.5	83.5
600	0	0	0	0	0	70.5	42.5	98
750	0	142.5	0	0	0	95	48.5	105.5
1000	0	174.5	0	104	0	164	54.5	131.5
1250	0	194	0	109	0	172.5	62	182
1500	0	208.5	0	113	0	178	65	192

---

* 0 feet indicates that the exposure at the height in the column above or below the antenna is in compliance.
** Power = Average power input to the antenna. See Chapter 5.

2-meter band horizontal, 13-element Yagi, Frequency = 146.0 MHz, Height above ground = 50 feet

Horizontal distance (feet) from any part of the antenna for compliance with occupational/controlled or general population/uncontrolled exposure limits*

| Power** | 6 feet | | 12 feet | | 20 feet | | 50 feet | |
| | Height above ground (feet) where exposure occurs | | | | | | | |
(watts)	con.	unc.	con.	unc.	con.	unc.	con.	unc.
10	0	0	0	0	0	0	11	15
25	0	0	0	0	0	0	12.5	21.5
50	0	0	0	0	0	0	15	29
100	0	0	0	0	0	0	19.5	39
200	0	0	0	0	0	0	26	55.5
250	0	0	0	0	0	0	29	60.5
300	0	0	0	0	0	0	31	67.5
400	0	0	0	0	0	0	36	75.5
500	0	0	0	0	0	0	39	86
600	0	0	0	0	0	0	43.5	89.5
750	0	0	0	0	0	0	48	106.5
1000	0	0	0	0	0	0	55.5	126.5
1250	0	189.5	0	0	0	117.5	60.5	133
1500	0	211	0	127	0	203	67.5	157.5

2-meter band horizontal, 17-element Yagi, Frequency = 146.0 MHz, Height above ground = 30 feet

Horizontal distance (feet) from any part of the antenna for compliance with

occupational/controlled or general population/uncontrolled exposure limits*

| Power** | 6 feet | | 12 feet | | 20 feet | | 30 feet | |
| | Height above ground (feet) where exposure occurs | | | | | | | |
(watts)	con.	unc.	con.	unc.	con.	unc.	con.	unc.
10	0	0	0	0	0	0	16	19.5
25	0	0	0	0	0	0	17	26
50	0	0	0	0	0	0	19.5	34.5
100	0	0	0	0	0	0	24	48
200	0	0	0	0	0	55.5	31	63.5
250	0	0	0	0	0	60.5	34.5	78.5
300	0	0	0	0	0	78	37	83.5
400	0	117	0	0	0	85.5	43	88.5
500	0	138	0	81	0	122.5	48	103.5
600	0	151	0	86	0	129.5	50.5	113.5
750	0	166	0	209.5	43.5	135.5	57	176
1000	0	185	0	245	55.5	141.5	63.5	192.5
1250	0	200	0	269	60.5	145.5	78.5	200.5
1500	0	212.5	0	288.5	78	303.5	83.5	206

2-meter band horizontal, 17-element Yagi, Frequency = 146.0 MHz, Height above ground = 40 feet

Horizontal distance (feet) from any part of the antenna for compliance with occupational/controlled or general population/uncontrolled exposure limits*

| Power** | 6 feet | | 12 feet | | 20 feet | | 40 feet | |
| | Height above ground (feet) where exposure occurs | | | | | | | |
(watts)	con.	unc.	con.	unc.	con.	unc.	con.	unc.
10	0	0	0	0	0	0	16	19.5
25	0	0	0	0	0	0	17	26
50	0	0	0	0	0	0	19.5	34.5
100	0	0	0	0	0	0	24	47
200	0	0	0	0	0	0	31.5	67
250	0	0	0	0	0	0	34.5	72
300	0	0	0	0	0	0	37	80.5
400	0	0	0	0	0	0	42.5	88
500	0	0	0	0	0	71	47	108
600	0	0	0	0	0	76	52	111.5
750	0	164	0	0	0	152.5	56	117
1000	0	195	0	104.5	0	171.5	67	143
1250	0	215.5	0	269	0	179	72	192
1500	0	231.5	0	302	0	184	80.5	198

* 0 feet indicates that the exposure at the height in the column above or below the antenna is in compliance.
** Power = Average power input to the antenna. See Chapter 5.

2-meter band horizontal, 17-element Yagi, Frequency = 146.0 MHz, Height above ground = 50 feet

Horizontal distance (feet) from any part of the antenna for compliance with occupational/controlled or general population/uncontrolled exposure limits*

Height above ground (feet) where exposure occurs

Power**	6 feet		12 feet		20 feet		50 feet	
(watts)	con.	unc.	con.	unc.	con.	unc.	con.	unc.
10	0	0	0	0	0	0	16	19.5
25	0	0	0	0	0	0	17	26
50	0	0	0	0	0	0	19.5	34.5
100	0	0	0	0	0	0	24	46.5
200	0	0	0	0	0	0	31.5	64.5
250	0	0	0	0	0	0	34.5	74.5
300	0	0	0	0	0	0	37	78
400	0	0	0	0	0	0	42.5	91.5
500	0	0	0	0	0	0	46.5	105
600	0	0	0	0	0	0	51.5	110
750	0	0	0	0	0	0	56.5	133
1000	0	0	0	0	0	0	64.5	139
1250	0	215	0	0	0	203	74.5	166.5
1500	0	239	0	0	0	213.5	78	173

222-MHz band horizontal, 22-element Yagi, Frequency = 222.0 MHz, Height above ground = 30 feet

Horizontal distance (feet) from any part of the antenna for compliance with occupational/controlled or general population/uncontrolled exposure limits*

Height above ground (feet) where exposure occurs

Power**	6 feet		12 feet		20 feet		30 feet	
(watts)	con.	unc.	con.	unc.	con.	unc.	con.	unc.
10	0	0	0	0	0	0	14.5	20.5
25	0	0	0	0	0	0	17	29
50	0	0	0	0	0	0	20.5	38.5
100	0	0	0	0	0	0	26	53.5
200	0	0	0	0	0	65	35.5	76
250	0	0	0	0	0	84	38.5	80.5
300	0	0	0	0	0	88	43	97
400	0	0	0	77.5	0	93.5	50	101.5
500	0	179	0	121.5	0	118	53.5	123.5
600	0	198.5	0	126.5	0	185.5	57.5	128
750	0	219.5	0	131.5	52.5	198.5	65.5	171
1000	0	245.5	0	275.5	65	209	76	178
1250	0	266	0	331	84	215.5	80.5	289.5
1500	0	282.5	0	363	88	220	97	300.5

222-MHz band horizontal, 22-element Yagi, Frequency = 222.0 MHz, Height above ground = 40 feet

Horizontal distance (feet) from any part of the antenna for compliance with occupational/controlled or general population/uncontrolled exposure limits*

Height above ground (feet) where exposure occurs

Power**	6 feet		12 feet		20 feet		40 feet	
(watts)	con.	unc.	con.	unc.	con.	unc.	con.	unc.
10	0	0	0	0	0	0	14.5	20.5
25	0	0	0	0	0	0	17	29
50	0	0	0	0	0	0	20.5	39.5
100	0	0	0	0	0	0	26.5	53.5
200	0	0	0	0	0	0	35	73
250	0	0	0	0	0	0	39.5	83.5
300	0	0	0	0	0	0	42	89
400	0	0	0	0	0	0	48	110.5
500	0	0	0	0	0	0	53.5	115
600	0	0	0	0	0	112	58	137
750	0	197.5	0	150.5	0	146	64.5	141
1000	0	256	0	165	0	247	73	169
1250	0	285	0	171.5	0	263.5	83.5	212
1500	0	308	0	176	0	272.5	89.5	293.5

* 0 feet indicates that the exposure at the height in the column above or below the antenna is in compliance.
** Power = Average power input to the antenna. See Chapter 5.

222-MHz band horizontal, 22-element Yagi, Frequency = 222.0 MHz, Height above ground = 50 feet

Horizontal distance (feet) from any part of the antenna for compliance with
occupational/controlled or general population/uncontrolled exposure limits*

Height above ground (feet) where exposure occurs

Power**	6 feet		12 feet		20 feet		50 feet	
(watts)	con.	unc.	con.	unc.	con.	unc.	con.	unc.
10	0	0	0	0	0	0	14.5	20.5
25	0	0	0	0	0	0	17	29
50	0	0	0	0	0	0	20.5	39
100	0	0	0	0	0	0	26.5	54
200	0	0	0	0	0	0	35.5	73.5
250	0	0	0	0	0	0	39	82.5
300	0	0	0	0	0	0	42	90
400	0	0	0	0	0	0	48.5	103
500	0	0	0	0	0	0	54	116.5
600	0	0	0	0	0	0	58.5	130
750	0	0	0	0	0	0	64	150.5
1000	0	0	0	0	0	0	73.5	177
1250	0	274.5	0	193.5	0	187	82.5	182
1500	0	312	0	203.5	0	306	90	210.5

420-MHz band horizontal, 22-element Yagi, Frequency = 432 MHz, Height above ground = 30 feet

Horizontal distance (feet) from any part of the antenna for compliance with
occupational/controlled or general population/uncontrolled exposure limits*

Height above ground (feet) where exposure occurs

Power**	6 feet		12 feet		20 feet		30 feet	
(watts)	con.	unc.	con.	unc.	con.	unc.	con.	unc.
10	0	0	0	0	0	0	9	15.5
25	0	0	0	0	0	0	12	23
50	0	0	0	0	0	0	15.5	31.5
100	0	0	0	0	0	0	21	43.5
200	0	0	0	0	0	0	28.5	64
250	0	0	0	0	0	54	31.5	71
300	0	0	0	0	0	69	34.5	74
400	0	0	0	0	0	83	39	92
500	0	0	0	0	0	98	43.5	105.5
600	0	0	0	126	0	100.5	48	108
750	0	117	0	133.5	0	151.5	53	124
1000	0	123.5	0	221.5	0	157	64	175
1250	0	127.5	0	232.5	54	222	71	180
1500	0	130.5	0	239.5	69	226.5	74	182.5

420-MHz band horizontal, 22-element Yagi, Frequency = 432 MHz, Height above ground = 40 feet

Horizontal distance (feet) from any part of the antenna for compliance with
occupational/controlled or general population/uncontrolled exposure limits*

Height above ground (feet) where exposure occurs

Power**	6 feet		12 feet		20 feet		40 feet	
(watts)	con.	unc.	con.	unc.	con.	unc.	con.	unc.
10	0	0	0	0	0	0	9	15.5
25	0	0	0	0	0	0	12	23
50	0	0	0	0	0	0	15.5	31.5
100	0	0	0	0	0	0	21	44.5
200	0	0	0	0	0	0	28.5	62.5
250	0	0	0	0	0	0	31.5	69
300	0	0	0	0	0	0	34.5	75.5
400	0	0	0	0	0	0	39	89.5
500	0	0	0	0	0	0	44.5	98
600	0	0	0	0	0	0	48.5	107
750	0	0	0	0	0	92.5	54	130
1000	0	0	0	0	0	111.5	62.5	146.5
1250	0	152	0	176	0	202.5	69	163.5
1500	0	159	0	180	0	207	75.5	184.5

* 0 feet indicates that the exposure at the height in the column above or below the antenna is in compliance.
** Power = Average power input to the antenna. See Chapter 5.

420-MHz band horizontal, 22-element Yagi, Frequency = 432 MHz, Height above ground = 50 feet

Horizontal distance (feet) from any part of the antenna for compliance with
occupational/controlled or general population/uncontrolled exposure limits*

Height above ground (feet) where exposure occurs

Power**	6 feet		12 feet		20 feet		50 feet	
(watts)	con.	unc.	con.	unc.	con.	unc.	con.	unc.
10	0	0	0	0	0	0	9	15.5
25	0	0	0	0	0	0	12	23
50	0	0	0	0	0	0	15.5	31.5
100	0	0	0	0	0	0	21	44.5
200	0	0	0	0	0	0	28.5	61.5
250	0	0	0	0	0	0	31.5	70
300	0	0	0	0	0	0	34.5	76
400	0	0	0	0	0	0	39.5	86.5
500	0	0	0	0	0	0	44.5	99
600	0	0	0	0	0	0	49	107
750	0	0	0	0	0	0	54	122
1000	0	0	0	0	0	0	61.5	143
1250	0	0	0	0	0	0	70	168.5
1500	0	0	0	0	0	114.5	76	172

902-MHz band horizontal, 22-element Yagi, Frequency = 915 MHz, Height above ground = 30 feet

Horizontal distance (feet) from any part of the antenna for compliance with
occupational/controlled or general population/uncontrolled exposure limits*

Height above ground (feet) where exposure occurs

Power**	6 feet		12 feet		20 feet		30 feet	
(watts)	con.	unc.	con.	unc.	con.	unc.	con.	unc.
10	0	0	0	0	0	0	5.5	10.5
25	0	0	0	0	0	0	8	15.5
50	0	0	0	0	0	0	10.5	22
100	0	0	0	0	0	0	14	30.5
200	0	0	0	0	0	0	19.5	43
250	0	0	0	0	0	0	22	48
300	0	0	0	0	0	0	24	53
400	0	0	0	0	0	0	27.5	60
500	0	0	0	0	0	54.5	30.5	67.5
600	0	0	0	0	0	62.5	33	75
750	0	0	0	0	0	79.5	36.5	85
1000	0	0	0	0	0	95	43	103.5
1250	0	134.5	0	78	0	105	48	112.5
1500	0	140	0	150	0	129.5	53	121.5

902-MHz band horizontal, 22-element Yagi, Frequency = 915 MHz, Height above ground = 40 feet

Horizontal distance (feet) from any part of the antenna for compliance with
occupational/controlled or general population/uncontrolled exposure limits*

Height above ground (feet) where exposure occurs

Power**	6 feet		12 feet		20 feet		40 feet	
(watts)	con.	unc.	con.	unc.	con.	unc.	con.	unc.
10	0	0	0	0	0	0	5.5	10.5
25	0	0	0	0	0	0	8	16
50	0	0	0	0	0	0	10.5	22
100	0	0	0	0	0	0	14	30.5
200	0	0	0	0	0	0	19.5	42.5
250	0	0	0	0	0	0	22	47.5
300	0	0	0	0	0	0	24	52
400	0	0	0	0	0	0	27	61
500	0	0	0	0	0	0	30.5	66.5
600	0	0	0	0	0	0	33	75
750	0	0	0	0	0	0	36.5	82.5
1000	0	0	0	0	0	0	42.5	94
1250	0	0	0	0	0	0	47.5	112
1500	0	0	0	0	0	89.5	52	124.5

* 0 feet indicates that the exposure at the height in the column above or below the antenna is in compliance.
** Power = Average power input to the antenna. See Chapter 5.

1.2 GHz band horizontal, 22-element Yagi, Frequency = 1296 MHz, Height above ground = 30 feet

Horizontal distance (feet) from any part of the antenna for compliance with
occupational/controlled or general population/uncontrolled exposure limits*

Height above ground (feet) where exposure occurs

Power**	6 feet		12 feet		20 feet		30 feet	
(watts)	con.	unc.	con.	unc.	con.	unc.	con.	unc.
10	0	0	0	0	0	0	4.5	9
25	0	0	0	0	0	0	6.5	13.5
50	0	0	0	0	0	0	9	18.5
100	0	0	0	0	0	0	12	26
200	0	0	0	0	0	0	16.5	36.5
250	0	0	0	0	0	0	18.5	40
300	0	0	0	0	0	0	20	43.5
400	0	0	0	0	0	0	23	50.5
500	0	0	0	0	0	0	26	56.5
600	0	0	0	0	0	0	28.5	64
750	0	0	0	0	0	58	31.5	70
1000	0	0	0	0	0	76.5	36.5	82.5
1250	0	0	0	0	0	87	40	95.5
1500	0	0	0	0	0	98.5	43.5	105.5

VHF VERTICAL COLLINEAR ANTENNAS

These tables represent collinear arrays consisting of 2 half-wave vertical antennas stacked about a wavelength apart and 4 half-wave vertical antennas stacked about a wavelength apart. This would be typical of the antennas used by Amateur Radio repeaters. The height shown is for the lowest part of the antenna.

2-meter band vertical collinear array, 2 half wave dipoles, Frequency = 146.0 MHz, Height above ground = 10 feet

Horizontal distance (feet) from any part of the antenna for compliance with
occupational/controlled or general population/uncontrolled exposure limits*

Height above ground (feet) where exposure occurs

Power**	6 feet		12 feet		20 feet		30 feet	
(watts)	con.	unc.	con.	unc.	con.	unc.	con.	unc.
10	0	0	1	2	1	2	0	0
25	0	0	1.5	3	1.5	2.5	0	0
50	0	0	2	4.5	2	3.5	0	0
100	0	0	3	7	2.5	5	0	0
200	0	0	4	14.5	3.5	6.5	0	0
250	0	0	4.5	16.5	3.5	7	0	0
300	0	0	5	18	4	7.5	0	0
400	0	7	6	21.5	4.5	8.5	0	0
500	0	8	7	26.5	5	9	0	0
600	0	8.5	8	30.5	5.5	13.5	0	0
750	0	9.5	10	33.5	6	29	0	0
1000	0	10.5	14.5	38	6.5	32	0	12
1250	0	11	16.5	43	7	36	0	14
1500	0	12	18	48.5	7.5	41	0	15

* 0 feet indicates that the exposure at the height in the column above or below the antenna is in compliance.
** Power = Average power input to the antenna. See Chapter 5.

2-meter band vertical collinear array, 2 half wave dipoles, Frequency = 146.0 MHz, Height above ground = 20 feet

Horizontal distance (feet) from any part of the antenna for compliance with
occupational/controlled or general population/uncontrolled exposure limits*

Height above ground (feet) where exposure occurs

Power** (watts)	6 feet con.	6 feet unc.	12 feet con.	12 feet unc.	20 feet con.	20 feet unc.	30 feet con.	30 feet unc.
10	0	0	0	0	1.5	2	1	2
25	0	0	0	0	1.5	2.5	1.5	2.5
50	0	0	0	0	2	3.5	2	3.5
100	0	0	0	0	2.5	5	2.5	5
200	0	0	0	0	3	6.5	3.5	6.5
250	0	0	0	0	3.5	7	3.5	7
300	0	0	0	0	4	7.5	4	7.5
400	0	0	0	0	4.5	9.5	4.5	8.5
500	0	0	0	0	5	20.5	5	9
600	0	0	0	0	5	23.5	5.5	22
750	0	0	0	0	5.5	32	6	25
1000	0	0	0	0	6.5	35.5	6.5	36
1250	0	0	0	11.5	7	38.5	7	38

2-meter band vertical collinear array, 2 half wave dipoles, Frequency = 146.0 MHz, Height above ground = 30 feet

Horizontal distance (feet) from any part of the antenna for compliance with
occupational/controlled or general population/uncontrolled exposure limits*

Height above ground (feet) where exposure occurs

Power** (watts)	6 feet con.	6 feet unc.	12 feet con.	12 feet unc.	20 feet con.	20 feet unc.	30 feet con.	30 feet unc.
10	0	0	0	0	0	0	1.5	2
25	0	0	0	0	0	0	1.5	2.5
50	0	0	0	0	0	0	2	3.5
100	0	0	0	0	0	0	2.5	5
200	0	0	0	0	0	0	3	6.5
250	0	0	0	0	0	0	3.5	7
300	0	0	0	0	0	0	4	7.5
400	0	0	0	0	0	0	4.5	9.5
500	0	0	0	0	0	0	5	20.5
600	0	0	0	0	0	0	5	26.5
750	0	0	0	0	0	0	5.5	29
1000	0	0	0	0	0	0	6.5	38.5
1250	0	0	0	0	0	0	7	41.5
1500	0	0	0	0	0	10.5	7.5	43.5

2-meter band vertical collinear array, 4 half wave dipoles, Frequency = 146.0 MHz, Height above ground = 10 feet

Horizontal distance (feet) from any part of the antenna for compliance with
occupational/controlled or general population/uncontrolled exposure limits*

Height above ground (feet) where exposure occurs

Power** (watts)	6 feet con.	6 feet unc.	12 feet con.	12 feet unc.	20 feet con.	20 feet unc.	30 feet con.	30 feet unc.
10	0	0	1	1.5	1	1.5	1.5	2
25	0	0	1	2	1	2	1.5	2.5
50	0	0	1.5	3.5	1.5	3	2	3.5
100	0	0	2	5	2	4.5	2	5
200	0	0	3	8.5	3	6.5	3	7.5
250	0	0	3.5	10.5	3	7.5	3.5	8.5
300	0	0	3.5	12	3.5	8	3.5	9
400	0	0	4.5	13	4	9.5	4.5	9.5
500	0	0	5	14.5	4.5	27	5	10.5
600	0	5	5.5	16	5	30	5.5	11.5
750	0	6.5	6.5	17.5	6	39	6	19
1000	0	7.5	8.5	25	6.5	46.5	7.5	22.5
1250	0	8.5	10.5	27	7.5	56.5	8.5	28.5
1500	0	8.5	12	28	8	62	9	30

* 0 feet indicates that the exposure at the height in the column above or below the antenna is in compliance.
** Power = Average power input to the antenna. See Chapter 5.

2-meter band vertical collinear array, 4 half wave dipoles, Frequency = 146.0 MHz, Height above ground = 20 feet

Horizontal distance (feet) from any part of the antenna for compliance with
occupational/controlled or general population/uncontrolled exposure limits*

Height above ground (feet) where exposure occurs

Power** (watts)	6 feet con.	unc.	12 feet con.	unc.	20 feet con.	unc.	30 feet con.	unc.
10	0	0	0	0	1	1.5	1	1.5
25	0	0	0	0	1.5	2	1	2
50	0	0	0	0	1.5	2.5	1.5	3
100	0	0	0	0	2	3.5	2	4.5
200	0	0	0	0	2.5	4.5	2.5	6.5
250	0	0	0	0	2.5	5	3	7.5
300	0	0	0	0	3	5.5	3.5	8
400	0	0	0	0	3	6.5	4	9
500	0	0	0	0	3.5	10	4.5	23
600	0	0	0	0	3.5	12.5	5	34.5
750	0	0	0	0	4	14.5	6	38
1000	0	0	0	0	4.5	17.5	6.5	49
1250	0	0	0	0	5	29	7.5	58
1500	0	0	0	8	5.5	30.5	8	61.5

2-meter band vertical collinear array, 4 half wave dipoles, Frequency = 146.0 MHz, Height above ground = 30 feet

Horizontal distance (feet) from any part of the antenna for compliance with
occupational/controlled or general population/uncontrolled exposure limits*

Height above ground (feet) where exposure occurs

Power** (watts)	6 feet con.	unc.	12 feet con.	unc.	20 feet con.	unc.	30 feet con.	unc.
10	0	0	0	0	0	0	1	1.5
25	0	0	0	0	0	0	1.5	2
50	0	0	0	0	0	0	1.5	2.5
100	0	0	0	0	0	0	2	3.5
200	0	0	0	0	0	0	2.5	4.5
250	0	0	0	0	0	0	2.5	5
300	0	0	0	0	0	0	3	5.5
400	0	0	0	0	0	0	3	6.5
500	0	0	0	0	0	0	3.5	9.5
600	0	0	0	0	0	0	3.5	12.5
750	0	0	0	0	0	0	4	14.5
1000	0	0	0	0	0	0	4.5	16.5
1250	0	0	0	0	0	0	5	27
1500	0	0	0	0	0	0	5.5	28.5

* 0 feet indicates that the exposure at the height in the column above or below the antenna is in compliance.
** Power = Average power input to the antenna. See Chapter 5.

VHF YAGI ARRAYS

The number of combinations of Yagi arrays, frequencies and heights are endless. This set of tables represents the Yagi array featured in Table 12 of the FCC tables that follow this section. The array is modeled at a height of 30 feet and 60 feet, pointed at the horizon and pointed 45 degrees up in the air.

2-meter band Yagi array, 8 17-element horizontal Yagis at horizon, Frequency = 146.0 MHz
Array center height above ground = 30 feet

Horizontal distance (feet) from any part of the antenna for compliance with occupational/controlled or general population/uncontrolled exposure limits*

Power** (watts)	Height above ground (feet) where exposure occurs							
	6 feet		12 feet		20 feet		30 feet	
	con.	unc.	con.	unc.	con.	unc.	con.	unc.
10	0	0	0	0	0	0	0	0
25	0	0	0	0	0	0	0	0
50	0	0	0	0	0	0	0	44
100	0	0	0	0	0	0	0	79
200	0	0	0	0	0	78.5	17	115
250	0	0	0	0	0	106.5	44	131
300	0	0	0	0	0	137	60	141
400	0	173	0	249	0	155	66.5	198
500	0	200	0	285	0	163.5	79	217
600	0	218	0	311	0	169.5	84.5	227
750	0	239.5	0	340	0	388	100	236
1000	0	267.5	0	378	78.5	453	115	246
1250	0	288.5	0	408	106.5	498	131	252.5
1500	0	306.5	0	433	137	533	141	576

2-meter band Yagi array, 8 17-element horizontal Yagis at horizon, Frequency = 146.0 MHz
Lower array height above ground = 60 feet

Horizontal distance (feet) from any part of the antenna for compliance with occupational/controlled or general population/uncontrolled exposure limits*

Power** (watts)	Height above ground (feet) where exposure occurs							
	6 feet		12 feet		20 feet		30 feet	
	con.	unc.	con.	unc.	con.	unc.	con.	unc.
10	0	0	0	0	0	0	0	0
25	0	0	0	0	0	0	0	0
50	0	0	0	0	0	0	0	0
100	0	0	0	0	0	0	0	0
200	0	0	0	0	0	0	0	0
250	0	0	0	0	0	0	0	0
300	0	0	0	0	0	0	0	0
400	0	0	0	0	0	0	0	0
500	0	0	0	0	0	0	0	0
600	0	0	0	0	0	0	0	0
750	0	0	0	0	0	0	0	0
1000	0	0	0	0	0	0	0	0
1250	0	0	0	0	0	0	0	222.5
1500	0	0	0	0	0	0	0	264

* 0 feet indicates that the exposure at the height in the column above or below the antenna is in compliance.
** Power = Average power input to the antenna. See Chapter 5.

2-meter band Yagi array, 8 17-element horizontal Yagis at 45 degr, Frequency = 146.0 MHz, Array center height above ground = 30 feet

Horizontal distance (feet) from any part of the antenna for compliance with occupational/controlled or general population/uncontrolled exposure limits*

Power**	6 feet		12 feet		20 feet		30 feet	
(watts)	con.	unc.	con.	unc.	con.	unc.	con.	unc.
10	0	0	0	0	0	0	0	0
25	0	0	0	0	0	0	0	0
50	0	0	0	0	0	0	0	8.5
100	0	0	0	0	0	0	0	11.5
200	0	0	0	0	0	0	8	13.5
250	0	0	0	0	0	1	8.5	14
300	0	0	0	0	0	1.5	9	14.5
400	0	0	0	0	0	2	11	15.5
500	0	0	0	0	0	2	11.5	16
600	0	0	0	0	0	2	12	16.5
750	0	0	0	0	0	2.5	12.5	21.5
1000	0	0	0	16.5	0	3.5	13.5	22.5
1250	0	0	0	17.5	1	4	14	23.5
1500	0	36	0	25.5	1.5	26.5	14.5	30.5

* 0 feet indicates that the exposure under the antenna is in compliance.
** Power = Average power input to the antenna. See Chapter 5.

FCC TABLES

The following tables (Tables 5-17 from Supplement B) give estimated distances to meet RF power-density MPE limits in the main beam of typical amateur station antennas. These tables were supplied by Professor Wayne Overbeck of California State University, Fullerton, CA, Mr. Kai Siwiak, P.E., KE4PT and by the FCC. These tables are based on the far-field equations discussed in Bulletin 65. They provide values that assume a surface reflection as an estimate of the ground reflection and other factors that surround most antenna installations.

These tables were featured in Supplement B with all distances in meters. The distances have been converted to feet for this printing.

Table 5
Three-element "triband" Yagi assuming surface (ground) reflection

Distance (feet) from any part of the antenna for compliance with either occupational/controlled or general population/uncontrolled exposure limits

Power (watts)	14 MHz, 6.5 dBi		21 MHz, 7 dBi		28 MHz, 8 dBi	
	con.	unc.	con.	unc.	con.	unc.
100	4.6	10.2	7.2	16.4	11.2	24.6
500	10.2	23	16.4	36.7	24.6	54.8
1,000	14.8	32.8	23.3	51.8	34.8	77.8
1,500	18	40	28.5	63.6	42.7	95.1

Table 6
Omnidirectional HF quarter-wave vertical or ground plane antenna (estimated gain 1dBi) assumes surface (ground) reflection

Distance (feet) from any part of the antenna for compliance with either occupational/controlled or general population/uncontrolled exposure limits

Transmitter power watts	3.5 MHz		7 MHz		14 MHz		21 MHz		28 MHz	
	con.	unc.	con.	unc.	con.	unc.	con.	unc.	con.	unc.
100	.7	1.3	1.3	2.6	2.6	5.6	3.6	8.2	4.9	10.8
500	1.3	3	2.6	6.2	5.6	12.1	8.2	18.4	10.8	24.6
1000	2	4.3	3.9	8.9	7.9	17.4	11.5	25.9	15.4	34.8
1500	2.3	5.2	4.6	10.5	9.5	21.3	14.1	31.8	19	42.3

Table 7
Horizontal, half-wave dipole wire antenna (estimated gain 2 dBi) assuming surface (ground) reflection

Distance (feet) from any part of the antenna for compliance with either occupational/controlled or general population/uncontrolled exposure limits .

Transmitter power watts	3.5 MHz		7 MHz		14 MHz		21 MHz		28 MHz	
	con.	unc.	con.	unc.	con.	unc.	con.	unc.	con.	unc.
100	.7	1.6	1.3	3	3	6.2	4.3	9.2	5.6	12.1
500	1.6	3.3	3	6.9	6.2	13.8	9.2	20.7	12.5	27.6
1000	2.3	4.9	4.3	9.5	8.5	19.4	13.1	29.2	17.4	38.7
1500	2.6	5.9	5.2	11.8	10.8	23.6	16.1	35.8	21.3	47.6

Table 8
VHF 1/4 wave ground plane or mobile whip antenna at 146 MHz (estimated gain 1 dBi) assuming surface (ground) reflection

Transmitter power (watts)	Distance (feet) to comply with occupational/ controlled exposure limit	Distance (feet) to comply with gen. population/uncontrolled exposure limit
10	1.6	3.6
50	3.6	8.2
150	6.6	14.4

Table 9
UHF 5/8 wave ground plane or whip antenna at 446 MHz (estimated gain 4 dBi); main beam exposure, assumes surface (ground) reflection

Transmitter power (watts)	Distance (feet) to comply with occupational/ controlled exposure limit	Distance (feet) to comply with gen. population/uncontrolled exposure limit
10	2	4.3
50	4.3	9.5
150	7.5	16.7

Table 10
Seventeen (17) element Yagi on five-wavelength boom designed for weak-signal communications on 144 MHz (estimated gain 16.8 dBi); main beam exposure assuming surface (ground) reflection

Transmitter power (watts)	Distance (feet) to comply with occupational/ controlled exposure limit	Distance (feet) to comply with gen. population/uncontrolled exposure limit
10	10.2	23
100	32.5	72.5
500	72.5	160.8
1500	125.3	280.5

Table 11
Seventeen (17) element Yagi on five-wavelength boom designed for weak-signal communications on 144 MHz (estimated gain 16.8 dBi); main beam exposure; this table does not assume ground reflection and can only be used if the antenna is pointed significantly above the horizon

Transmitter power (watts)	Distance (feet) to comply with occupational/ controlled exposure limit	Distance (feet) to comply with gen. population/uncontrolled exposure limit
10	6.6	14.4
100	20	45.6
500	45.6	101
1500	78.4	175.5

Table 12
Eight 17-element Yagis with five-wavelength booms designed for "moonbounce" communications on 144 MHz (estimated gain 24 dBi); main beam exposure, assumes surface (ground) reflection

Transmitter power (watts)	Distance (feet) to comply with occupational/ controlled exposure limit	Distance (feet) to comply with gen. population/uncontrolled exposure limit
150	90.9	203.4
500	166	370.7
1500	287.4	643

Table 13
Eight 17-element Yagis with five-wavelength booms designed for "moonbounce" communications on 144 MHz (estimated gain 24 dBi); main beam exposure; this table does not assume ground reflection and can only be used if the antenna is pointed significantly above the horizon

Transmitter power (watts)	Distance (feet) to comply with occupational/ controlled exposure limit	Distance (feet) to comply with gen. population/uncontrolled exposure limit
150	57.1	127
500	103.3	232
1500	180.1	399.9

Table 14
HF discone antenna (estimated gain 2 dBi); main beam exposure, assumes surface (ground) reflection

	Distance (feet) from any part of the antenna for compliance with either occupational/controlled or general population/uncontrolled exposure limits							
	3.5 MHz		7 MHz		14 MHz		28 MHz	
Transmitter power (watts)	con.	unc.	con.	unc.	con.	unc.	con.	unc
50	.3	1	1	2	1.6	3.9	3.6	7.9
100	.7	1.3	1.3	2.6	2.3	5.6	4.9	10.8
250	1	2.3	2	4.3	3.9	8.5	7.9	17.4
500	1.3	3	2.6	6.2	5.6	12.1	10.8	24.6

Table 15
VHF/UHF Discone antenna (estimated gain 2 dBi) main beam exposure, assumes surface (ground) reflection

	Distance (feet) from any part of the antenna for compliance with either occupational/controlled or general population/uncontrolled exposure limits							
	50 MHz		144 MHz		220 MHz		440 MHz	
Transmitter power (watts)	con.	unc.	con.	unc.	con.	unc.	con.	unc.
50	4.3	9.2	4.3	9.2	4.3	9.2	3.3	7.5
100	5.9	13.1	5.9	13.1	5.9	13.1	4.9	10.8
250	9.2	21	9.2	21	9.2	21	7.5	17.1
500	13.1	29.5	13.1	29.5	13.1	29.5	10.8	24

Table 16
Quarter-wave half-sloper antenna (estimated gain 6.7 dBi); main beam exposure, assumes surface reflection

	Distance (feet) from any part of the antenna for compliance with either occupational/controlled or general population/uncontrolled exposure limits							
	7 MHz		14 MHz		21 MHz		28 MHz	
Transmitter power (watts)	con.	unc.	con.	unc.	con.	unc.	con.	unc.
100	2.3	5.2	4.6	10.5	7.2	15.7	9.5	21
500	5.2	11.8	10.5	23.6	16.1	35.1	21	46.9
1000	7.5	16.4	14.8	33.5	22.6	49.9	29.9	66.3
1500	9.2	20.3	18.4	41	27.6	61	36.4	81.4

Table 17
(Submitted by Kai Siwiak, P.E., KE4PT) One meter diameter HF loop, 150 W, assumes surface (ground) reflection

Frequency (MHz)	Distance (feet) from loop center for compliance with either occupational/controlled or general population/uncontrolled exposure limits	
	con.	unc.
7	6.6	9.2
10	6.9	9.2
14	6.9	10.5
18	7.5	11.5
21	7.5	12.1
24	7.9	12.8
28	7.9	13.8

FCC Regulations Relating To RF Exposure

For your reference, here is the actual text of the FCC Regulations.

Hams tend to think of Part 97 of Title 47 of the Code of Federal Regulations when it comes to rules governing the Amateur Radio Service. However, all radio services also are governed by some of the provisions of Part 1 and Part 2.

The complete set of FCC regulations occupies 5 volumes. The entire set is updated yearly and can be ordered from the US Government Printing Office, Superintendent of Documents, P.O. Box 371954, Pittsburgh, PA 15250-7954. Phone: 202-512-1800, Web Site: http://www.gpo.gov. You may also find it in your nearest Government Document Depository Library; check with your local library.

For those who have access to the Internet, the ARRL RIB (Regu-

latory Information Branch) maintains the current versions of Part 97 on the following sites–World Wide Web (http://www.arrl.org/field/regulations/news/part97/); FTP (oak.oakland.edu); ARRL's automated e-mail information server (info@arrl.org) and ARRL's landline BBS (860-594-0306). RIB also maintains current versions on CompuServe's HamNet Forum and the Ham Radio Section of America On Line. All of these versions are updated whenever a rules change takes effect.

This appendix contains the parts of Title 47 of the Code of Federal Regulations, Parts 1, 2, and 97, that were amended to reflect the new RF exposure requirements.

PART 1— PRACTICE AND PROCEDURE

1. The authority citation for part 1 continues to read as follows:

AUTHORITY: 47 U.S.C. 151, 154, 303 and 309(j), unless otherwise noted, and Section 704 of the Telecommunications Act of 1996.

2. Section 1.1307 is amended by revising paragraphs (b)(1), (b)(2), (b)(3) and (b)(4) and by adding paragraph (b)(5) to read as follows:

§1.1307 Actions which may have a significant environmental effect, for which Environmental Assessments (EAs) must be prepared.

* * * * *

(b) * * *

(1) The appropriate exposure limits in §1.1310 and §2.1093 are generally applicable to all facilities, operations and transmitters regulated by the Commission. However, a determination of compliance with the exposure limits in §1.1310 or §2.1093 (routine environmental evaluation), and preparation of an EA if the limits are exceeded, is necessary only for facilities, operations and transmitters that fall into the categories listed in Table 1, or those specified in paragraph (b)(2) of this section. All other facilities, operations and transmitters

are categorically excluded from making such studies or preparing an EA, except as indicated in paragraphs (c) and (d) of this section. For purposes of Table 1, "building-mounted antennas" means antennas mounted in or on a building structure that is occupied as a workplace or residence. The term "power" in column 2 of Table 1 refers to total operating power of the transmitting operation in question in terms of effective radiated power (ERP), equivalent isotropically radiated power (EIRP), or peak envelope power (PEP), as defined in § 2.1 of this chapter. For the case of the Cellular Radiotelephone Service, subpart H of part 22 of this chapter; the Personal Communications Service, part 24 of this chapter and the Specialized Mobile Radio Service, part 90 of this chapter, the phrase "total power of all channels" in column 2 of Table 1 means the sum of the ERP or EIRP of all co located simultaneously operating transmitters owned and operated by a single licensee. When applying the criteria of Table 1, radiation in all directions should be considered. For the case of transmitting facilities using sectorized transmitting antennas, applicants and licensees should apply the criteria to all transmitting channels in a given sector, noting that for a highly directional antenna there is relatively little contribution to ERP or EIRP summation for other directions.

(2) Mobile and portable transmitting devices that operate in the Cellular Radiotele-

phone Service, the Personal Communications Services (PCS), the Satellite Communications Services, the General Wireless Communications Service, the Wireless Communications Service, the Maritime Services (ship earth stations only) and the Specialized Mobile Radio Service authorized under subpart H of part 22, part 24, part 25, part 26, part 27, part 80, and part 90 of this chapter are subject to routine environmental evaluation for RF exposure prior to equipment authorization or use, as specified in §§2.1091 and 2.1093 of this chapter. Unlicensed PCS, unlicensed NII and millimeter wave devices are also subject to routine environmental evaluation for RF exposure prior to equipment authorization or use, as specified in §§15.253(f), 15.255(g), and 15.319(i) and 15.407(f) of this chapter. All other mobile, portable, and unlicensed transmitting devices are categorically excluded from routine environmental evaluation for RF exposure under §§2.1091 and 2.1093 of this chapter except as specified in paragraphs (c) and (d) of this section.

(3) In general, when the guidelines specified in §1.1310 are exceeded in an accessible area due to the emissions from multiple fixed transmitters, actions necessary to bring the area into compliance are the shared responsibility of all licensees whose transmitters produce, at the area in question, power density levels that exceed 5% of the power density exposure limit applicable to their particular transmitter

TABLE 1: TRANSMITTERS, FACILITIES AND OPERATIONS SUBJECT TO ROUTINE ENVIRONMENTAL EVALUATION

SERVICE (TITLE 47 CFR RULE PART)	EVALUATION REQUIRED IF:
Experimental Radio Services (part 5)	power > 100 W ERP (164 W EIRP)
Multipoint Distribution Service (subpart K of part 21)	<u>non-building-mounted antennas</u>: height above ground level to lowest point of antenna < 10 m <u>and</u> power > 1640 W EIR <u>building-mounted antennas</u>: power > 1640 W EIRP
Paging and Radiotelephone Service (subpart E of part 22)	<u>non-building-mounted antennas</u>: height above ground level to lowest point of antenna < 10 m <u>and</u> power > 1000 W ERP (1640 W EIRP) <u>building-mounted antennas</u>: power > 1000 W ERP (1640 W EIRP)
Cellular Radiotelephone Service (subpart H of part 22)	<u>non-building-mounted antennas</u>: height above ground level to lowest point of antenna < 10 m <u>and</u> total power of all channels > 1000 W ERP (1640 W EIRP) <u>building-mounted antennas</u>: total power of all channels > 1000 W ERP (1640 W EIRP)
Personal Communications Services (part 24)	(1) Narrowband PCS (subpart D): <u>non-building-mounted antennas</u>: height above ground level to lowest point of antenna < 10 m <u>and</u> total power of all channels > 1000 W ERP (1640 W EIRP) <u>building-mounted antennas</u>: total power of all channels > 1000 W ERP (1640 W EIRP) (2) Broadband PCS (subpart E): <u>non-building-mounted antennas</u>: height above ground level to lowest point of antenna < 10 m <u>and</u> total power of all channels > 2000 W ERP (3280 W EIRP) <u>building-mounted antennas</u>: total power of all channels > 2000 W ERP (3280 W EIRP)
Satellite Communications (part 25)	all included
General Wireless Communications Service (part 26)	total power of all channels > 1640 W EIRP
Wireless Communications Service (part 27)	total power of all channels > 1640 W EIRP
Radio Broadcast Services (part 73)	all included
Experimental, auxiliary, and special broadcast and other program distributional services (part 74)	subparts A, G, L: power > 100 W ERP subpart I: <u>non-building-mounted antennas</u>: height above ground level to lowest point of antenna < 10 m <u>and</u> power > 1640 W EIRP <u>building-mounted antennas</u>: power > 1640 W EIRP
Stations in the Maritime Services (part 80)	ship earth stations only
Private Land Mobile Radio Services Paging Operations (part 90)	<u>non-building-mounted antennas</u>: height above ground level to lowest point of antenna < 10 m <u>and</u> power > 1000 W ERP (1640 W EIRP) <u>building-mounted antennas</u>: power > 1000 W ERP (1640 W EIRP)
Private Land Mobile Radio Services Specialized Mobile Radio (part 90)	<u>non-building-mounted antennas</u>: height above ground level to lowest point of antenna < 10 m <u>and</u> total power of all channels > 1000 W ERP (1640 W EIRP) <u>building-mounted antennas</u>: total power of all channels > 1000 W ERP (1640 W EIRP)
Amateur Radio Service (part 97)	**transmitter output power > levels specified in § 97.13(c)(1) of this chapter**
Local Multipoint Distribution Service (subpart L of part 101)	<u>non-building-mounted antennas</u>: height above ground level to lowest point of antenna < 10 m <u>and</u> power > 1640 W EIRP <u>building-mounted antennas</u>: power > 1640 W EIRP LMDS licensees are required to attach a label to subscriber transceiver antennas that: (1) provides adequate notice regarding potential radiofrequency safety hazards, <u>e.g.,</u> information regarding the safe minimum separation distance required between users and transceiver antennas; and (2) references the applicable FCC-adopted limits for radiofrequency exposure specified in §1.1310 of this chapter.

[1]Emphasis added.—Ed.

or field strength levels that, when squared, exceed 5% of the square of the electric or magnetic field strength limit applicable to their particular transmitter. Owners of transmitter sites are expected to allow applicants and licensees to take reasonable steps to comply with the requirements contained in §1.1307(b) and, where feasible, should encourage co-location of transmitters and common solutions for controlling access to areas where the RF exposure limits contained in §1.1310 might be exceeded.

(i) Applicants for proposed (not otherwise excluded) transmitters, facilities or modifications that would cause non-compliance with the limits specified in §1.1310 at an accessible area previously in compliance must submit an EA if emissions from the applicant's transmitter or facility would result, at the area in question, in a power density that exceeds 5% of the power density exposure limit applicable to that transmitter or facility or in a field strength that, when squared, exceeds 5% of the square of the electric or magnetic field strength limit applicable to that transmitter or facility.

(ii) Renewal applicants whose (not otherwise excluded) transmitters or facilities contribute to the field strength or power density at an accessible area not in compliance with the limits specified in §1.1310 must submit an EA if emissions from the applicant's transmitter or facility results, at the area in question, in a power density that exceeds 5% of the power density exposure limit applicable to that transmitter or facility or in a field strength that, when squared, exceeds 5% of the square of the electric or magnetic field strength limit applicable to that transmitter of facility.

(4) Transition Provisions. Applications filed with the Commission prior to October 15, 1997 (or January 1, 1998, for the Amateur Radio Service only), for construction permits, licenses to transmit or renewals thereof, modifications in existing facilities or other authorizations or renewals thereof require the preparation of an Environmental Assessment if the particular facility, operation or transmitter would cause human exposure to levels of radiofrequency radiation that are in excess of the requirements contained in paragraphs (b)(4)(i)—(4)(iii) of this section. In accordance with section 1.1312, if no new application or Commission action is required far a license to construct a new facility or physically modify an existing facility, e.g., geographic area licensees, and construction begins on or after October 15, 1997, the licensee will be required to prepare an Environmental Assessment if construction or modification of the

facility would not comply with the provisions of paragraph (b)(1) of this section. These transition provisions do not apply to applications for equipment authorization or use of mobile, portable and unlicensed devices specified in paragraph (2) of this section.

(5) Existing transmitting facilities, devices and operations: All existing transmitting facilities, operations and devices regulated by the Commission must be in compliance with the requirements of paragraphs (1) - (3) of this section by September 1, 2000, or, if not in compliance, file an Environmental Assessment as specified in 47 CFR §1.1311.

§1.1308 Consideration of environmental assessments (EAs); findings of no significant impact.

(a) Applicants shall prepare EAs for actions that may have a significant environmental impact (see §1.1307). An EA is described in detail in §1.1311 of this part of the Commission rules.

(b) The EA is a document which shall explain the environmental consequences of the proposal and set forth sufficient analysis for the Bureau or the Commission to reach a determination that the proposal will or will not have a significant environmental effect. To assist in making that determination, the Bureau or the Commission may request further information from the applicant, interested persons, and agencies and authorities which have jurisdiction by law or which have relevant expertise.

*****

(c) If the Bureau or the Commission determines, based on an independent review of the EA and any applicable mandatory consultation requirements imposed upon federal agencies (see note above), that the proposal will have a significant environmental impact upon the quality of the human environment, it will so inform the applicant. The applicant will then have an opportunity to amend its application so as to reduce, minimize, or eliminate environmental problems. See §1.1309. If the environmental problem is not eliminated, the Bureau will publish in the Federal Register a Notice of Intent (see §1.1314) that EISs will be prepared (see §1.1315 and 1.1317), or

(d) If the Bureau or Commission determines, based on an independent review of the EA, and any mandatory consultation requirements imposed upon federal agencies (see the note to paragraph (b) of this section), that the proposal would not have a significant impact, it will make a finding of no significant impact. Thereafter, the application will be processed without further

documentation of environmental effect. Pursuant to CEQ regulations, see 40 CFR 1501.4 and 1501.6, the applicant must provide the community notice of the Commission's finding of no significant impact.

§1.1309 Application amendments.

Applicants are permitted to amend their applications to reduce, minimize or eliminate potential environmental problems. As a routine matter, an applicant will be permitted to amend its application within thirty (30) days after the Commission or the Bureau informs the applicant that the proposal will have a significant impact upon the quality of the human environment (see §1.1308(c)). The period of thirty (30) days may be extended upon a showing of good cause.

§1.1310 Radiofrequency radiation exposure limits.

The criteria listed in table 1 shall be used to evaluate the environmental impact of human exposure to radiofrequency (RF) radiation as specified in §1.1307(b), except in the case of portable devices which shall be evaluated according to the provisions of §2.1093 of this chapter. Further information on evaluating compliance with these limits can be found in the FCC's OST/OET Bulletin Number 65, "Evaluating Compliance with FCC-Specified Guidelines for Human Exposure to Radiofrequency Radiation."

NOTE TO INTRODUCTORY PARAGRAPH: These limits are generally based on recommended exposure guidelines published by the National Council on Radiation Protection and Measurements (NCRP) in "Biological Effects and Exposure Criteria for Radiofrequency Electromagnetic Fields," NCRP Report No. 86, Sections 17.4.1, 17.4.1.1, 17.4.2 and 17.4.3. Copyright NCRP, 1986, Bethesda, Maryland 20814. In the frequency range from 100 MHz to 1500 MHz, exposure limits for field strength and power density are also generally based on guidelines recommended by the American National Standards Institute (ANSI) in Section 4.1 of "IEEE Standard for Safety Levels with Respect to Human Exposure to Radio Frequency Electromagnetic Fields, 3 kHz to 300 GHz," ANSI/IEEE C95.1-1992, Copyright 1992 by the Institute of Electrical and Electronics Engineers, Inc., New York, New York 10017.

§1.1311 Environmental information to be included in the environmental assessment (EA).

(a) The applicant shall submit an EA with each application that is subject to environmental processing (see §1.1307).

The EA shall contain the following information:

(1) For antenna towers and satellite earth stations, a description of the facilities as well as supporting structures and appurtenances, and a description of the site as well as the surrounding area and uses. If high intensity white lighting is proposed or utilized within a residential area, the EA must also address the impact of this lighting upon the residents.

(2) A statement as to the zoning classification of the site, and communications with, or proceedings before and determinations (if any) made by zoning, planning, environmental or other local, state or federal authorities on matters relating to environmental effect.

(3) A statement as to whether construction of the facilities has been a source of controversy on environmental grounds in the local community.

(4) A discussion of environmental and other considerations which led to the selection of the particular site and, if relevant, the particular facility; the nature and extent of any unavoidable adverse environmental effects, and any alternative sites or facilities which have been or might reasonably be considered.

(5) Any other information that may be requested by the Bureau or Commission.

(6) If endangered or threatened species or their critical habitats may be affected, the applicant's analysis must utilize the best scientific and commercial data available, see 50 CFR 402.14(c).

(b) The information submitted in the EA shall be factual (not argumentative or conclusory) and concise with sufficient detail to explain the environmental consequences and to enable the Commission or Bureau, after an independent review of the EA, to reach a determination concerning the proposal's environmental impact, if any. The EA shall deal specifically with any feature of the site which has special environmental significance (e.g., wilderness areas, wildlife preserves, natural migration paths for birds and other wildlife, and sites of historic, architectural, or archeological value). In the case of historically significant sites, it shall specify the effect of the facilities on any district, site, building, structure or object listed, or eligible for listing, in the National structure or object listed, or eligible for listing, in the National Register of Historic Places. It shall also detail any substantial change in the character of the land utilized (e.g., deforestation, water diversion, wetland fill, or other extensive change of surface features). In the case of wilderness areas, wildlife preserves, or other like areas, the statement shall discuss the effect of any continuing pattern of human intrusion into the area (e.g., necessitated by the operation and maintenance of the facilities).

(c) The EA shall also be accompanied with evidence of site approval which has been obtained from local or federal land use authorities.

(d) To the extent that such information is submitted in another part of the application, it need not be duplicated in the EA, but adequate cross-reference to such information shall be supplied.

(e) An EA need not be submitted to the Commission if another agency of the Federal Government has assumed responsibility for determining whether the facilities in question will have a significant effect on the quality of the human environment and, if it will, for invoking the environmental impact statement process.

§1.1312 Facilities for which no pre-construction authorization is required.

(a) In the case of facilities for which no Commission authorization prior to construction is required by the Commission's rules and regulations the licensee or applicant shall initially ascertain whether the proposed facility may have a significant environmental impact as defined in § 1.1307 of this part or is categorically excluded from environmental processing under §1.1306 of this part.

(b) If a facility covered by paragraph (a) of this section may have a significant environmental impact, the information required by §1.1311 of this part shall be submitted by the licensee or applicant and ruled on by the Commission, and environmental processing (if invoked) shall be completed, see §1.1308 of this part, prior to the initiation of construction of the facility.

(c) If a facility covered by paragraph (a) of this section is categorically excluded from environmental processing, the licensee or applicant may proceed with construction and operation of the facility in accordance with the applicable licensing rules and procedures.

(d) If, following the initiation of construction under this section, the licensee or applicant discovers that the proposed facility may have a significant environmental effect, it shall immediately cease construction which may have that effect, and submit the information required by §1.1311 of this part. The Commission shall rule on that submission and complete further environmental processing (if invoked), §1.1308 of this part, before such construction is resumed.

(e) Paragraphs (a) through (d) of this section shall not apply to the construction of mobile stations.

Table A.1
Maximum Permissible Exposure (MPE) Limits

Frequency Range (MHz)	Controlled Exposure (6-Minute Average)			Uncontrolled Exposure (30-Minute Average)		
	Electric Field Strength (V/m)	Magnetic Field Strength (A/m)	Power Density (mW/cm^2)	Electric Field Strength (V/m)	Magnetic Field Strength (A/m)	Power Density (mW/cm^2)
0.3-3.0	614	1.63	(100)*			
3.0-30	1842/f	4.89/f	(900/f^2)*			
0.3-1.34				614	1.63	(100)*
1.34-30				824/f	2.19/f	(180/f^2)*
30-300	61.4	0.163	1.0	27.5	0.073	0.2
300-1500	—	—	f/300	—	—	f/1500
1,500-100,000	—	—	5	—	—	1.0

f = frequency, in MHz.
* = Plane-wave equivalent power density

Note 1: Occupational/controlled limits apply in situations in which persons are exposed as a consequence of their employment provided those persons are fully aware of the potential for exposure and can exercise control over their exposure. Limits for occupational/controlled exposure also apply in situations when an individual is transient through a location where occupational/controlled limits apply provided he or she is made aware of the potential for exposure.

Note 2: General population/uncontrolled exposures apply in situations in which the general public may be exposed, or in which persons that are exposed as a consequence of their employment may not be fully aware of the potential for exposure or cannot exercise control over their exposure.

PART 2 — FREQUENCY ALLOCATIONS AND RADIO TREATY MATTERS; GENERAL RULES AND REGULATIONS

1. The authority citation for part 2 continues to read as follows:

AUTHORITY: Sec. 4, 302, 303 and 307 of the Communications Act of 1934, as amended, 47 U.S.C. Sections 154, 302, 303 and 307, unless otherwise noted.

2. Section 2.1091 is amended by revising the section caption, by revising paragraphs (b), (c) and (d)(3) and by adding a new paragraph (d)(4) to read as follows:

§2.1091 Radiofrequency radiation exposure evaluation: mobile devices.

(b) For purposes of this section, a mobile device is defined as a transmitting device designed to be used in other than fixed locations and to generally be used in such a way that a separation distance of at least 20 centimeters is normally maintained between the transmitter's radiating structure(s) and the body of the user or nearby persons. In this context, the term "fixed location" means that the device is physically secured at one location and is not able to be easily moved to another location. Transmitting devices designed to be used by consumers or workers that can be easily re-located, such as wireless devices associated with a personal computer, are considered to be mobile devices if they meet the 20 centimeter separation requirement.

(c) Mobile devices that operate in the Cellular Radiotelephone Service, the Personal Communications Services, the Satellite Communications Services, the General Wireless Communications Service, the Wireless Communications Service, the Maritime Services and the Specialized Mobile Radio Service authorized under subpart H of part 22 of this chapter, part 24 of this chapter, part 25 of this chapter, part 26 of this chapter, part 27 of this chapter, part 80 of this chapter (ship earth stations devices only) and part 90 of this chapter are subject to routine environmental evaluation for RF exposure prior to equipment authorization or use if they operate at frequencies of 1.5 GHz or below and their effective radiated power (ERP) is 1.5 watts or more, or if they operate at frequencies above 1.5 GHz and their ERP is 3 watts or more. Unlicensed personal communications service devices, unlicensed millimeter wave devices and unlicensed NII devices authorized under §15.253, §15.255, and subparts D and E of part 15 of this chapter are also subject to routine environmental evaluation for RF exposure prior to equipment authorization or use if their ERP is 3 watts or more or if they meet the definition of a portable device as specified in §2.1093 (b) requiring evaluation under the provisions of that section. All other mobile and unlicensed transmitting devices are categorically excluded from routine environmental evaluation for RF exposure prior to equipment authorization or use, except as specified in §§1.1307(c) and 1.1307(d) of this chapter. Applications for equipment authorization of mobile and unlicensed transmitting devices subject to routine environmental evaluation must contain a statement confirming compliance with the limits specified in paragraph (d) of this section as part of their application. Technical information showing the basis for this statement must be submitted to the Commission upon request.

(d) * * *

(3) If appropriate, compliance with exposure guidelines for devices in this section can be accomplished by the use of warning labels and by providing users with information concerning minimum separation distances from transmitting structures and proper installation of antennas.

(4) In some cases, e.g., modular or desktop transmitters, the potential conditions of use of a device may not allow easy classification of that device as either mobile or portable (also see 47 CFR 2.1093). In such cases, applicants are responsible for determining minimum distances for compliance for the intended use and installation of the device based on evaluation of either specific absorption rate (SAR), field strength or power density, whichever is most appropriate.

3. Section 2.1093 is amended by revising paragraphs (b), (c) and (d) to read as follows:

§2.1093 Radiofrequency radiation exposure evaluation: portable devices.

* * * * *

(b) For purposes of this section, a portable device is defined as a transmitting device designed to be used so that the radiating structure(s) of the device is/are within 20 centimeters of the body of the user.

(c) Portable devices that operate in the Cellular Radiotelephone Service, the Personal Communications Services, the Satellite Communications services, the

General Wireless Communications Service, the Wireless Communications Service, the Maritime Services and the Specialized Mobile Radio Service authorized under subpart H of part 22 of this chapter, part 24 of this chapter, part 25 of this chapter, part 26 of this chapter, part 27 of this chapter, part 80 of this chapter (ship earth station devices only), part 90 of this chapter, and portable unlicensed personal communication service, unlicensed NII devices and millimeter wave devices authorized under §15.253, §15.255 or subparts D and E of part 15 of this chapter are subject to routine environmental evaluation for RF exposure prior to equipment authorization or use. All other portable transmitting devices are categorically excluded from routine environmental evaluation for RF exposure prior to equipment authorization or use, except as specified in §§1.1307(c) and 1.1307(d) of this chapter. Applications for equipment authorization of portable transmitting devices subject to routine environmental evaluation must contain a statement confirming compliance with the limits specified in paragraph (d) of this section as part of their application. Technical information showing the basis for this statement must be submitted to the Commission upon request.

(d) The limits to be used for evaluation are based generally on criteria published by the American National Standards Institute (ANSI) for localized specific absorption rate ("SAR") in Section 4.2 of "IEEE Standard for Safety Levels with Respect to Human Exposure to Radio Frequency Electromagnetic Fields, 3 kHz to 300 GHz," ANSI/IEEE C95.1-1992, Copyright 1992 by the Institute of Electrical and Electronics Engineers, Inc., New York, New York 10017. These criteria for SAR evaluation are similar to those recommended by the National Council on Radiation Protection and Measurements (NCRP) in "Biological Effects and Exposure Criteria for Radiofrequency Electromagnetic Fields," NCRP Report No. 86, Section 17.4.5. Copyright NCRP, 1986, Bethesda, Maryland 20814. SAR is a measure of the rate of energy absorption due to exposure to an RF transmitting source. SAR values have been related to threshold levels for potential biological hazards. The criteria to be used are specified in paragraphs (d)(1) and (d)(2) of this section and shall apply for portable devices transmitting in the frequency range from 100 kHz to 6 GHz. Portable devices that transmit at frequencies above 6 GHz are to be evaluated in terms of the MPE limits specified in §1.1310 of this chapter.

Measurements and calculations to demonstrate compliance with MPE field strength or power density limits for devices operating above 6 GHz should be made at a minimum distance of 5 cm from the radiating source.

* * * * *

PART 97 — AMATEUR RADIO SERVICE

1. The authority citation for part 97 continues to read as follows:

AUTHORITY: 48 Stat. 1066, 1082, as amended; 47 U.S.C. §§154, 303. Interpret or apply 48 Stat. 1064-1068, 1081-1105, as amended; 47 U.S.C. §§151-155, 301-609, unless otherwise noted.

2. Section 97.13 is amended by revising paragraph (c) and adding paragraphs (c)(1) and (c)(2) to read as follows:

Wavelength Band	Evaluation Required if Power* (watts) Exceeds:
MF	
160 m	500
HF	
80 m	500
75 m	500
40 m	500
30 m	425
20 m	225
17 m	125
15 m	100
12 m	75
10 m	50
VHF (all bands)	50
UHF	
70 cm	70
33 cm	150
23 cm	200
13 cm	250
SHF (all bands)	250
EHF (all bands)	250
Repeater stations (all bands)	non-building-mounted antennas: height above ground level to lowest point of antenna < 10 m and power > 500 W ERP building-mounted antennas: power > 500 W ERP

* Power = PEP input to antenna except, for repeater stations only, power exclusion is based on ERP (effective radiated power).

§ 97.13 Restrictions on station location.

* * * * *

(c) Before causing or allowing an amateur station to transmit from any place where the operation of the station could cause human exposure to RF electromagnetic field levels in excess of those allowed under §1.1310 of this chapter, the licensee is required to take certain actions.

(1) The licensee must perform the routine RF environmental evaluation prescribed by §1.1307(b) of this chapter, if the power of the licensee's station exceeds the limits given in the following table:

(2) If the routine environmental evaluation indicates that the RF electromagnetic fields could exceed the limits contained in §1.1310 of this chapter in accessible areas, the licensee must take action to prevent human exposure to such RF electromagnetic fields. Further information on evaluating compliance with these limits can be found in the FCC's OET Bulletin 65, "Evaluating Compliance with FCC-Specified Guidelines for Human Exposure to Radio Frequency Electromagnetic Fields."

§97.503 Element standards.

* * * * *

(b) * * *

(1) Element 2: 35 questions concerning the privileges of a Novice Class operator license. The minimum passing score is 26 questions answered correctly.

(2) Element 3(A): 30 questions concerning the privileges of a Technician Class operator license. The minimum passing score is 22 questions answered correctly.

(3) Element 3(B): 30 questions concerning the privileges of a General Class operator license. The minimum passing score is 22 questions answered correctly.

* * * * *

(c) * * *

Topics:	Element:				
	2	3(A)	3(B)	4(A)	4(B)

* * * * *

(10) Radiofrequency environmental safety practices at an amateur station					
	5	5	5	0	0

FCC Reports And Orders

This condensation contains the parts of these FCC documents pertinent to Amateur Radio. ARRL comments are added to the text in italics. ARRL has reproduced the FCC documents as they were released, including a few non-critical typographical errors.

ARRL's Condensation of ET Docket 93-62:

Report and Order, FCC 96-326, August 1, 1996

First Memorandum Opinion and Order, FCC 96-487, December 23, 1996

Second Memorandum Opinion and Order and Notice of Proposed Rulemaking, FCC 97-303, August 25, 1997

ARRL's Condensation of ET Docket 93-62
First Memorandum Opinion and Order (FCC 96-326)
August 1, 1996

Before the
FEDERAL COMMUNICATIONS COMMISSION
Washington, D.C. 20554

FCC 96-326

In the Matter of   )
)
Guidelines for Evaluating the Environmental   ) ET Docket No. 93-62
Effects of Radiofrequency Radiation   )
)

REPORT AND ORDER

Adopted: August 1, 1996   ; Released: August 1, 1996

By the Commission: Commissioners Quello and Chong issuing a joint statement.

TABLE OF CONTENTS

This condensation does not contain all items listed in this table of contents.–Ed.

I. INTRODUCTION

1. By this action, we are amending our rules to adopt new guidelines and methods for evaluating the environmental effects of radiofrequency (RF) radiation from FCC-regulated transmitters. We are adopting Maximum Permissible Exposure (MPE) limits for electric and magnetic field strength and power density for transmitters operating at frequencies from 300 kHz to 100 GHz.[1] We are also adopting limits for localized ("partial body") absorption that will apply to certain portable transmitting devices.[2] We believe that the guidelines we are adopting will protect the public and workers from exposure to potentially harmful RF fields.

2. In reaching our decision on the adoption of new RF exposure guidelines we have carefully considered the large number of comments submitted in this proceeding, and particularly those submitted by the U.S. Environmental Protection Agency (EPA), the Food and Drug Administration (FDA) and other federal health and safety agencies. The new guidelines we are adopting are based substantially on the recommendations of those agencies, and we believe that these guidelines represent a consensus view of the federal agencies responsible for matters relating to the public safety and health.[3]

3. The MPE limits adopted herein are based on exposure criteria quantified in terms of specific absorption rate (SAR), a measure of the rate of RF energy absorption. The basis for these limits, as well as the basis for the 1982 ANSI limits that the Commission previously specified in our rules, is an SAR limit of 4 watts per kilogram. The new MPE limits are derived by incorporating safety factors that lead, in some cases, to limits that are more conservative than the limits specified by ANSI in 1982. The more conservative limits do not arise from a fundamental change in the RF safety criteria for SAR, but from a precautionary desire for more rigor in the derivation of factors which allow limits for MPE to be derived from SAR limits.

II. BACKGROUND

5. The National Environmental Policy Act of 1969 (NEPA) requires agencies of the Federal Government to evaluate the effects of their actions on the quality of the human environment.[4] To meet its responsibilities under NEPA, the Commission has adopted requirements for evaluating the environmental impact of its actions.[5] One of several environmental factors addressed by these requirements is human exposure to RF energy emitted by FCC-regulated transmitters and facilities.

6. In 1985, the Commission adopted a 1982 ANSI standard for use in evaluating the effects of RF radiation on the environment, noting that the ANSI standard was widely accepted and was technically and scientifically supportable.[6] Since then the Commission has used this standard as its processing guideline for determining the potential environmental impact of RF emissions. The rules now require applicants for certain facilities to prepare an Environmental Assessment (EA) if the transmitter or facility under consideration could expose the general public or workers to levels of RF radiation that are in excess of the 1982 ANSI guidelines.[7] Examples of facilities that could potentially cause exposures in excess of these guidelines because of their relatively high operating power include radio and television broadcast stations and satellite uplink facilities. The rules also address other related matters, such as the evaluation of sites with multiple transmitters.

7. The Commission has "categorically excluded" many low-power, intermittent, or normally inaccessible RF transmitters and facilities from routine evaluation for RF radiation exposure based on calculations and measurement data indicating that they would not cause exposures in excess of the guidelines under normal and routine conditions of use.[8] Examples of currently excluded transmitters include land mobile, cellular and amateur radio stations.

8. In 1992, ANSI adopted a new standard for RF exposure, designated ANSI/IEEE C95.1-1992 to replace its 1982 standard.[9] This new standard contains a number of significant differences from the 1982 ANSI standard. In some respects, the 1992 ANSI/IEEE standard is more restrictive in the amount of environmental RF exposure permitted, although for some situations recommended MPE levels are similar to the 1982 limits. The 1992 ANSI/IEEE standard also extends the frequency range under consideration to cover frequencies from 3 kHz to 300 GHz.[10] The 1992 ANSI/IEEE standard specifies two tiers of exposure criteria, one tier for "controlled environments" (usually involving workers) and another, more stringent tier, for "uncontrolled environments" (usually involving the general public). The 1982 ANSI standard specified only one set of exposure limits, regardless of whether the individual exposed was a worker or a member of the general public. The 1992 ANSI/IEEE standard also, for the first time, includes specific restrictions on currents induced in the human body by RF fields. These restrictions apply to both "induced" currents and "contact" currents related to shock and burn hazards.

9. The 1992 ANSI standard is generally more stringent in the evaluation of low-power devices, such as hand-held radios and cellular telephones, than the 1982 standard. That is, the 1982 ANSI standard permitted exclusion from compliance with the MPE limits if the localized specific absorption rate (SAR) of a low-power device could be shown to be 8 watts/kilogram (8 W/kg) or less, or if the input power of the radiating device at frequencies between 300 kHz and 1 GHz was 7 watts or less.[11] The 1992 guidelines reduce the allowable localized SAR level for devices operating in "uncontrolled" environments by a factor of five to 1.6 W/kg, while maintaining the 8 W/kg limit for "controlled" environments. Further, the exclusion thresholds based on operating power are significantly reduced for devices that operate in uncontrolled environments and for devices that operate above 450 MHz in controlled environments. The 1992 ANSI/IEEE standard also prohibits the application of the power exclusion to hand-held devices where the radiating structure is maintained less than 2.5 centimeters (cm) from the body of the user.

10. On April 8, 1993, we issued the Notice in this proceeding to consider amending and updating the guidelines and methods used by the Commission for evaluating the environmental effects of RF radiation.[12] In the Notice, we proposed to base our RF safety regulations on the ANSI/IEEE C95.1-1992 standard instead of the 1982 ANSI standard. The major issues addressed in the Notice were: 1) the selection of the appropriate RF exposure standard; 2) use of the 1992 ANSI/IEEE definitions for "controlled" and "uncontrolled" environments to determine application of exposure criteria; 3) implementation of new limits on induced and contact currents; 4) evaluation of low-power devices such as cellular telephones; 5) categorical exclusions from environmental evaluation for certain transmitters; 6) compliance and measurement issues; and 7) administrative procedures and effective dates for implementation.

III. DISCUSSION

A. New RF Exposure Guidelines

12. In the Notice, we noted that the 1992 ANSI/IEEE standard reflects recent scientific studies of the biological effects of RF radiation and that use of this standard would thus ensure that FCC-regulated facilities comply with the latest safety guidelines for RF exposure.[13] We also noted that other RF exposure criteria are available, such as those of the National Council on Radiation Protection and Measurements (NCRP) and those of the International Radiation Protection Association (IRPA).[14] We requested comment on whether the differences between these other guidelines and the 1992 ANSI/IEEE guidelines are significant, and

whether it would be appropriate to adopt limits for RF exposure that differ from those in the 1992 ANSI/IEEE guidelines.

13. The comments filed in this proceeding have focused primarily on the 1992 ANSI/IEEE and NCRP exposure criteria. In many ways, these two sets of exposure guidelines are similar. Both organizations identify the same threshold level at which harmful biological effects may occur, and the MPE limits recommended for electric and magnetic field strength and power density in both documents are based on this threshold level.[15] Both the 1992 ANSI/IEEE and NCRP guidelines also are frequency dependent, based on knowledge of how whole-body-averaged human exposure is a function of the frequency of the RF signal. Further, both ANSI/IEEE and NCRP recommend two exposure tiers, one for "controlled environments" (ANSI/IEEE) or "occupational exposure" (NCRP) and another, more stringent, tier for "uncontrolled environments" (ANSI/IEEE) or "general population" exposure (NCRP). Tables 1, 2 and 3 in Appendix B show the MPE limits for the 1982 ANSI, 1992 ANSI/IEEE and NCRP exposure criteria, respectively.

14. The two sets of guidelines, however, do differ in some respects. The NCRP MPE limits are generally more stringent than the ANSI/IEEE limits for magnetic field strength at frequencies below 3 MHz and for power density at frequencies above 1500 MHz.[16] The NCRP guidelines also include a unique provision (that we are not adopting here) that reduces the exposure limit for workers with respect to certain forms of modulated RF carrier frequencies.[17] The NCRP guidelines specify that the general population MPE limits at higher frequencies are to be averaged over longer periods of time than those recommended by the ANSI/IEEE guidelines.[18] The NCRP, unlike ANSI/IEEE, only specifies MPE limits for frequencies up to 100 GHz. With respect to evaluating low-power devices, although both ANSI/IEEE and NCRP generally recommend the same localized SAR limits, ANSI/IEEE also includes an exclusion clause based on radiated power that is not a part of the NCRP guidelines. Although the ANSI/IEEE and NCRP guidelines differ at higher and lower frequencies, at frequencies used by the majority of FCC licensees the MPE limits are essentially the same regardless of whether ANSI/IEEE or NCRP guidelines are used. Therefore, the overall impact on most of our licensees from our adoption of new guidelines should not be significantly different regardless of which limits we choose.

15. Several federal agencies filed comments in this proceeding expressing varying viewpoints on whether we should adopt the ANSI/IEEE guidelines or some alternative. Within the Federal Government, the EPA is generally responsible for investigating and making recommendations with regard to environmental issues. In its comments, the EPA states that the new ANSI/IEEE guidelines are a significant revision of the 1982 ANSI guidelines and notes that certain aspects of the new guidelines are improvements with regard to protection.[19] However, the EPA submits that some of the provisions of the new ANSI/IEEE guidelines are not acceptable. For example, EPA does not support the relaxation in MPE limits for power density at higher microwave frequencies, and it opposes the application of the same exposure limits to both controlled and uncontrolled environments for frequencies from 15 GHz to 300 GHz. The EPA states that the ANSI/IEEE exposure limits for these frequencies are not sufficiently protective for public exposure. The EPA also argues that the terms controlled and uncontrolled environments used in the ANSI/IEEE guidelines are not well defined and are not directly applicable to any specific population group.

16. The EPA recommends that we adopt the NCRP's recommended MPE limits along with sections of the 1992 ANSI/IEEE guidelines dealing with localized exposure and induced and contact body currents. In terms of MPEs for power density and field strength, the EPA argues that the NCRP guidelines would better protect the public from potential long term effects of RF exposure at higher microwave frequencies where the two sets of guidelines differ. The EPA maintains that, "[t]here are no substantive differences in the literature base supporting 1986 NCRP and 1992 ANSI/IEEE except for the literature on RF shocks and burns." In addition, the EPA notes that NCRP is chartered by the U.S. Congress to develop radiation protection recommendations.

17. The EPA generally supports the use of the ANSI/IEEE limits for dealing with induced and contact currents to protect against shock and burn hazards. EPA states that those guidelines are not included in the NCRP exposure criteria, and they are a result of research and knowledge acquired since development of the NCRP recommendations. The EPA also supports the FCC proposal to use ANSI/IEEE SAR limits that apply to low-power devices such as cellular telephones (see discussion below). These values are similar to those recommended by the NCRP.

18. The FDA has general jurisdiction for protecting the public from potentially harmful radiation from consumer and industrial devices and in that capacity is expert in RF exposures that would result from consumer or industrial use of hand-held devices such as cellular telephones.[20] The FDA generally supports our proposed use of the 1992 ANSI/IEEE guidelines, although it does express some reservations. It states that these guidelines will provide a greater level of protection to the general public, and it particularly supports use of the values for SAR that would apply to consumer and industrial devices. As discussed below, however, the FDA expresses significant concern about the radiated power exclusion clause included in the ANSI/IEEE standard that would apply to some hand-held devices.[21]

19. The National Institute for Occupational Safety and Health (NIOSH), an agency of the U. S. Department of Health and Human Services, is responsible for performing research and analysis with respect to worker safety and health. In its comments, NIOSH expresses general support for our efforts to update our RF exposure regulations and notes that the MPE limits defined in the 1992 ANSI/IEEE guidelines are similar to those contained in recommendations of the NCRP and the International Radiation Protection Association.[22] NIOSH states that we should take a more conservative approach when uncertainty exists with respect to applying certain features of the exposure guidelines. In particular, NIOSH agrees with the EPA that it would be more appropriate to use the MPE limits recommended by NCRP guidelines at higher frequencies. NIOSH also supports the use of the ANSI/IEEE limits on induced RF currents.

20. The Occupational Safety and Health Administration (OSHA) has jurisdiction over Federal regulations dealing with worker safety and health. In its comments, OSHA generally endorses our proposal to update our RF exposure guidelines by adopting the new ANSI/IEEE guidelines.[23] OSHA also urges us to require applicants to implement a written RF exposure protection program which appropriately addresses traditional safety and health program elements including training, medical monitoring, protective procedures and engineering controls, signs, hazard assessments, employee involvement, and designated responsibilities for program implementation. It notes that the exposure limits in the ANSI/IEEE guidelines may be useful in determining when specific elements of an RF safety program should be implemented. However, OSHA objects to the two categories of exposure environments contained in the new ANSI/IEEE standard, finding it unacceptable that employees may be subjected to a higher level of risk than the general public simply because they "are aware of the potential for exposure as a concomitant of employment." Rather, OSHA proposes that we adopt the uncontrolled environment criteria as an "action limit" which determines

when an RF protection program will be required. That is, under OSHA's proposal, persons who are exposed in excess of the limits specified for uncontrolled environments would be protected by a program designed to mitigate any potential increase in risk.

21. The majority of industry comments favor adoption of the 1992 ANSI/IEEE guidelines. For example, American Personal Communications (APC), American Telephone and Telegraph Company (AT&T), Electromagnetic Energy Policy Alliance (EEPA), Ericsson Corporation (Ericsson), McCaw Cellular Communications, Inc. (McCaw), National Association of Broadcasters (NAB), Telecommunications Industry Association (TIA), and others urge that we adopt the new ANSI/IEEE guidelines, arguing that they represent the most up-to-date standard available. Telocator (now the Personal Communications Industry Association, PCIA) agrees that the ANSI/IEEE standard is the most recent and comprehensive RF exposure guideline, noting that an international committee of over 120 scientists and engineers was involved in its drafting. However, Telocator submits that the actual impact of the ANSI/IEEE, NCRP or IRPA standards would be about the same on Personal Communications Service (PCS) operations, since all three standards are based on the same specific absorption rates, and the power densities each provides for the PCS band are essentially the same.[24]

26. The IEEE's Standards Coordinating Committee 28 (IEEE/SCC28), which developed the ANSI/IEEE guidelines, took issue with several of the points made by the EPA. IEEE/SCC28 states that the new guidelines and the NCRP recommendations are actually quite similar, with the exception of the MPEs at higher microwave frequencies. In addition, it points out that both the ANSI/IEEE and the NCRP guidelines are based on the use of SAR as the fundamental dosimetric parameter, the same criterion for biological effect (behavioral disruption), and the same safety factors to define the two tiers of exposure.[25]

27. In comments filed late in this proceeding, Dr. Arthur W. Guy, former Chairman of both ANSI/IEEE and NCRP committees on RF exposure expresses his view that, "it would be a mistake for the FCC to adopt the older 1986 NCRP standard at this time considering the fact that newer and more advanced standards have been developed since the publication of the NCRP standard."[26] Similar views are expressed in letters submitted to the Commission by Dr. Eleanor Adair and Dr. C.K. Chou, both of whom have been involved in ANSI/IEEE and NCRP RF committees.[27] All of these individuals urge that we adopt the ANSI/IEEE standard instead of the NCRP exposure criteria.

28. Decision. Although most commenting parties generally support our proposal to adopt the 1992 ANSI/IEEE guidelines, some of the Federal agencies filing comments in this proceeding, principally those with responsibility for oversight regarding health and safety issues, object to the use of certain aspects of these guidelines. In the past, the Commission has stressed repeatedly that it is not a health and safety agency and would defer to the judgment of these expert agencies with respect to determining appropriate levels of safe exposure to RF energy.[28] We continue to believe that we must place special emphasis on the recommendations and comments of Federal health and safety agencies because of their expertise and their responsibilities with regard to health and safety matters. Accordingly, as recommended by the EPA, we are adopting exposure limits for field strength and power density based on those recommended by the NCRP for frequencies from 300 kHz to 100 GHz (see Appendix C). As noted previously, over a wide frequency range these limits are also based on those recommended in the ANSI/IEEE 1992 standard.[29] We believe that the exposure criteria we are adopting will protect workers and the general public from potentially harmful RF emissions due to FCC-regulated transmitters.

29. We recognize that the NCRP guidelines do not address exposure at frequencies below 300 kHz or above 100 GHz, as do the ANSI/IEEE guidelines. However, the FCC-regulated transmitters of concern operate at frequencies between 300 kHz and 100 GHz. Therefore, we see no need at this time to adopt guidelines for frequencies outside of the range of the NCRP recommendations.

30. We appreciate the concerns raised by NAB with respect to NCRP guidelines for low-frequency magnetic-field exposure, and we recognize that the NCRP guidelines may be conservative for frequencies below 100 MHz. However, compliance with these limits would appear to be an issue only in occupational situations, e.g., in the immediate vicinity of an AM broadcast transmitter; and, there is nothing in the record to indicate that significant problems exist with respect to compliance with these magnetic field limits in the workplace.

31. We also recognize the merit of arguments as to whether, at the higher microwave frequencies, incorporating different time-averaging values, such as those specified by the ANSI/IEEE guidelines may be desirable. As discussed by JC&A, IEEE/SCC28 and others, the level of energy density allowed by the 1992 ANSI/IEEE guidelines can actually be more restrictive at higher frequencies than the NCRP guidelines when time-averaging is considered. For frequencies above 3 GHz (uncontrolled) and 15 GHz (controlled) the ANSI/IEEE time-averaging values are less than those of NCRP, and they continue to decrease at higher frequencies. Because of the lengthier NCRP averaging times at these frequencies, very short exposures at very high power densities might comply with NCRP limits as long as they are followed by insignificant exposures for the duration of the time-averaging interval. In that sense, ANSI/IEEE could be viewed as affording a greater degree of protection from skin burning at the higher microwave frequencies. However, we are not aware of any practical situations involving FCC-regulated transmitting facilities where such exposures are likely to occur. Of far greater significance, we believe, is the case of a consumer-product without any identifiable usage pattern, where continuous exposure would have to be assumed and time-averaging would not be relevant.

32. We agree with those commenters who maintain that there is insufficient evidence to give special consideration to modulation effects at this time. Since we have no specific indication of exposure hazards related to modulation caused by FCC-regulated transmitters, we believe it would be premature at this time to adopt the NCRP modulation criteria.

33. We believe that OSHA's suggestion that we use the uncontrolled exposure tier of the ANSI/IEEE standard as the basis for an "action limit" for establishment of an RF safety program is beyond the scope of our jurisdiction. Our NEPA responsibilities do not appear to encompass the issuance of specific rules on workplace practices and procedures. If such a policy were to be instituted by the Federal Government it would seem more appropriate for OSHA itself to promulgate this type of rule.

34. Both the IEEE and the NCRP have committees that are working on revisions of their respective exposure guidelines. We encourage these organizations and other similar groups developing exposure criteria to work together, along with the relevant federal agencies, to develop consistent, harmonized guidelines that will address the concerns and issues raised in this proceeding. We will consider amending our rules at any appropriate time if these groups conclude that such action is desirable.

B. Definitions of Controlled and Uncontrolled Environments

35. The 1992 ANSI/IEEE guidelines specify two sets of exposure limits based on the "environment" in which the exposure takes place.[30] These environments are classified as either "con-

trolled" or "uncontrolled." Controlled environments are defined as locations where "there is exposure that may be incurred by persons who are aware of the potential for exposure as a concomitant of employment, by other cognizant persons, or as the incidental result of transient passage through areas where analysis shows the exposure levels may be above [the exposure and induced current levels permitted for uncontrolled environment but not those permitted for controlled environments]." Uncontrolled environments are defined as "locations where there is the exposure of individuals who have no knowledge or control of their exposure. The exposures may occur in living quarters or workplaces where there are no expectations that the exposure levels may exceed [the exposure and induced current levels permitted for uncontrolled environments]." The NCRP designates exposure limits in terms of "occupational" and "general population" exposure. However, the NCRP report does not provide specific definitions of these terms.

36. In the Notice, we requested comment on the criteria to be used in determining which set of exposure limits would apply to the various situations that would be subject to environmental analysis and whether the definitions of controlled and uncontrolled environments used in the ANSI/IEEE guidelines were practical and supportable for the Commission's purposes. We stated that because matters of possible health and safety are involved, a conservative approach would be appropriate. Accordingly, we proposed to provide that where there is any question of possible exposure of the general public, the more stringent guidelines for uncontrolled environments would apply. We also specifically stated that the guidelines for uncontrolled environments would apply to any transmitter or facility located in a residential area where proximity to the transmitter is unrestricted. On the other hand, we indicated that controlled environment limits would apply to situations where exposure is incidental and transitory or where exposure is incurred when individuals are aware of the exposure potential.

37. Most parties support the use of a two-tier RF exposure standard and the ANSI/IEEE definitions for "controlled environment" and "uncontrolled environment." In general, these parties support applying the ANSI/IEEE definition for uncontrolled environment to those transmitters and facilities in residential areas or locations with unrestricted access. They suggest that the controlled environment should apply to incidental and transitory exposure and in areas where people are aware of potential exposure through warning signs and instructions. The Land Mobile Communication Council (LMCC), NAB, and others propose that the distinction between the two environments be based on the context of the equipment's use and types of communication operations being performed. They argue that the controlled standards should be applied when the equipment is used in a commercial or business setting where the operator is "knowledgeable" in the use of his/her equipment. They state that the uncontrolled standard should apply to the general public where the user or party exposed is not considered "knowledgeable" about the transmitting device and the use of those devices is incidental or personal in nature.[31]

40. The EPA opposes use of the terms controlled and uncontrolled environments and recommends that we define exposure environments using the traditional terms of "occupational" and "general population or public" contained in the NCRP guidelines. EPA contends that its own 1984 report on the biological effects of RF radiation and the NCRP have concluded that the general population has groups of individuals particularly susceptible to heat, including the elderly, infants, pregnant women and others.[32] EPA argues that the ANSI/IEEE terms are not directly applicable to any population group and are not well defined.[33] OSHA and

NIOSH do not oppose the use of the ANSI/IEEE definitions but raise questions about their application. OSHA, for example, states that employees should not be subjected to a higher level of risk as a condition of their employment just because they are made aware of the potential for exposure.[34] NIOSH states that where there is any question about exposure category, the more conservative uncontrolled criteria should be applied.[35]

41. The American Radio Relay League, Inc. (ARRL) also opposes use of the ANSI/IEEE definitions, arguing that under these definitions amateur operations would unjustly be categorized as operating in an uncontrolled environment. It suggests that there is no reason to require amateurs to meet the high safety factor below the threshold for adverse health effects that is the basis for the uncontrolled MPE limits. The ARRL indicates that the controlled environment MPE limits "should be safe for all."[36]

42. <u>Decision</u>. We find it appropriate to use the terms "occupational" and "general population" contained in the NCRP report. We note, however, that the NCRP report does not provide explicit definitions of these terms, and we agree with the commenting parties that we need to define these terms more completely and clearly to minimize any ambiguity in the application of the exposure limits. We believe that the ANSI/IEEE definitions for controlled and uncontrolled environments can be used as a basis for applying our use of the two exposure tiers we are adopting, while at the same time accomplishing the intent of the NCRP criteria to protect workers and the public.

43. Accordingly, "occupational/controlled" exposure, as used by the Commission, will apply to situations in which persons are exposed as a consequence of their employment and in which those persons who are exposed have been made fully aware of the potential for exposure and can exercise control over his or her exposure. Occupational/controlled exposure will also apply where exposure is of a transient nature as a result of incidental passage through a location where exposure levels may be above general population/uncontrolled limits (see below), as long as the exposed person has been made fully aware of the potential for exposure and can exercise control over his or her exposure by leaving the area or by some other appropriate means. We will apply the occupational/controlled exposure limits to amateur radio operators and members of their immediate household, as discussed later (see para. 162, infra).

44. "General population/uncontrolled" exposure, as used by the Commission, will apply to situations in which the general public may be exposed or in which persons who are exposed as a consequence of their employment may not be made fully aware of the potential for exposure or can not exercise control over their exposure. Therefore, members of the general public always fall under this category when exposure is not employment-related, as in the case of residents in an area near a broadcast tower. We believe that these definitions will clarify the ambiguities pointed out by many of the commenting parties and will thus ensure that the appropriate level of protection is applied in each situation. We do not agree with those parties that support applying the general population or uncontrolled limits to all situations. This approach would impose significant and unnecessary economic and technical burdens for which adequate justification has not been presented.

45. For purposes of these definitions, awareness of the potential for RF exposure can be provided through specific training as part of an RF safety program. Warning signs and labels can also be used to establish such awareness as long as they provide information, in a prominent manner, on risk of potential exposure and instructions on methods to minimize such exposure risk.[37] However, warning labels placed on low-power consumer devices such as cellular telephones will not be considered sufficient to achieve

the awareness necessary to qualify these devices as operating in a controlled environment. We plan to provide further instructions on the application of these definitions in an upcoming revision of OST Bulletin No. 65 concerning compliance with RF exposure guidelines.[38]

Note: All of the above information about amateurs and controlled environments remains correct, but in the August, 1997 release of the revised OET Bulletin 65, the FCC clarified the status of the household of Amateur Radio operators by adding:

"In complying with the Commission's Report and Order, amateur operators should follow a policy of systematic avoidance of excessive RF exposure. The Commission has said that it will continue to rely upon amateur operators, in constructing and operating their stations, to take steps to ensure that their stations comply with the MPE limits for both occupational/controlled and general public/uncontrolled situations, as appropriate. In that regard, amateur radio operators and members of their immediate household are considered to be in a 'controlled environment' and are subject to the occupational/controlled MPE limits. Neighbors who are not members of an amateur operator's household are considered to be members of the general public, since they cannot reasonably be expected to exercise control over their exposure. In those cases general population/uncontrolled exposure MPE limits will apply.

"In order to qualify for use of the occupational/controlled exposure criteria, appropriate restrictions on access to high RF field areas must be maintained and educational instruction in RF safety must be provided to individuals who are members of the amateur operator's household. Persons who are not members of the amateur operator's household but who are present temporarily on an amateur operator's property may also be considered to fall under the occupational/controlled designation provided that appropriate information is provided them about RF exposure potential if transmitters are in operation and such persons are exposed in excess of the general population/uncontrolled limits." —Ed.

Evaluation of Low-Power Devices

46. The 1992 ANSI/IEEE guidelines permit low-power devices designed to be used in the immediate vicinity of the body, such as portable and hand-held radios and telephones, to be excluded from compliance with the prescribed limits for field strength and power density provided that such devices comply with specific SAR limits or that the radiated power of the device is below a certain level.[39] "Low-power" devices include mobile transmitters such as automobile and marine radio transceivers, and hand-held portable devices such as cellular telephones and "walkie-talkie" type radios. These low-power exclusions would eliminate the need for making MPE field strength measurements in areas extremely near to the transmitting device where they may not be an appropriate measure of actual energy absorption. For low-power devices in controlled environments, SAR levels must be less than 0.4 W/kg as averaged over the whole-body, and the spatial peak SAR must be less than 8 W/kg as averaged over any 1 gram of tissue at frequencies between 100 kHz and 6 GHz. The corresponding limits for devices operated in uncontrolled environments are 0.08 W/kg for whole-body average exposure and 1.6 W/kg for spatial peak SAR. These SAR limits are also essentially the same as those recommended by the NCRP for occupational and general population exposure, respectively.[40]

47. With regard to exclusions based on radiated power, the ANSI/IEEE guidelines permit an exclusion in controlled environments if the radiated power of a device is 7 watts or less at frequencies between 100 kHz and 450 MHz. At frequencies between 450 and 1500 MHz, the radiated power is limited to 7(450/f) watts, where f is the frequency in MHz. In uncontrolled environments, the guidelines permit exclusion if the radiated power is 1.4 watts or less for frequencies between 100 kHz and 450 MHz and 1.4(450/f) watts for frequencies between 450 and 1500 MHz. The ANSI/IEEE guidelines also state that exclusions based on radiated power do not apply when the "radiating structure" of the device is within 2.5 cm of the body. The NCRP guidelines do not provide exclusions based on radiated power.

48. In the Notice, we proposed to adopt the ANSI/IEEE SAR exclusion for low-power devices for both controlled and uncontrolled environments, depending on the actual environment in which the device would be used. We also proposed to adopt the radiated power exclusion, but only for those low-power devices that meet the more conservative guidelines for uncontrolled environments. We also requested comment on whether proof of compliance should be required to be submitted as part of the equipment authorization process, and, if so, the form such a showing should take.

62. Decision. Most commenting parties, including Federal health and safety agencies, support the use of the ANSI/IEEE SAR limits for localized (partial body) exposure for evaluating low-power devices designed to be used in the immediate vicinity of the body. As mentioned above, the SAR limits specified by the ANSI/IEEE guidelines for devices used in controlled and uncontrolled environments are essentially the same as those recommended by NCRP for occupational and general population exposure, respectively. Therefore, in view of the consensus and the scientific support in the record, we are adopting SAR limits for the determination of safe exposure from low-power devices designed to be used in the immediate vicinity of the body based upon the 1992 ANSI/IEEE guidelines. We will apply the MPE limits we are adopting to certain mobile and unlicensed devices that, although not normally used within the immediate vicinity of the body, can use higher power and may be relatively close to the body of the user and to nearby persons. Examples of the latter are cellular "bag phones."

63. The SAR limits we are adopting will generally apply to portable devices submitted for Commission authorization that are designed to be used with any part of the radiating structure of the device in direct contact with the body of the user or within 20 cm of the body of the user under normal conditions of use. For example, this definition would apply to hand-held cellular telephones. We believe that a threshold of 20 cm is appropriate, since the ANSI/IEEE standard specifies 20 cm as the minimum separation distance where reliable MPE measurements can be made.[41] At these closer distances, we believe an SAR determination is a more appropriate measure of exposure.

64. In addition to SAR limits for portable devices, exposure criteria in terms of the MPE limits will apply to certain mobile and unlicensed devices that would normally be used with radiating structures maintained 20 cm or more from the body of the user. Examples include transportable cellular telephones ("bag" phones), cellular telephones and other radio devices that use vehicle-mounted antennas and certain other transportable transmitting devices. For these types of transmitters, evaluation of compliance with MPE limits rather than SAR limits is more appropriate because of the greater separation distance between radiator and user.

65. We will require routine SAR evaluation, either by laboratory measurement techniques or by computational modeling, prior to equipment authorization or use for the following categories of portable devices: (1) portable telephones or portable telephone devices to be used in the Cellular Radiotelephone Service under Part 22 Subpart H or to be used in the Private Land Mobile Radio Services for certain "covered" SMR systems under Part 90 of our

rules;[42] (2) portable devices to be used for PCS under Part 24 of our rules; (3) mobile devices to be used for earth-satellite communication under Part 25 and Part 80 of our rules; and (4) portable unlicensed PCS and portable unlicensed millimeter wave devices authorized under Part 15 of our rules. In all cases the term "portable" means that the telephone or device is intended for use within 20 cm of the body of the user as defined previously. The applicable SAR limit will normally be the 1.6 W/kg as recommended by ANSI/IEEE for uncontrolled environments, such as those typical for consumer use. However, devices intended solely for use in the workplace may be considered under the less restrictive occupational/controlled environment category.

66. We also will require routine evaluation prior to equipment authorization or use for the following mobile transmitters if the effective radiated power (ERP) of the station, in its normal configuration, will be 1.5 watts or greater[43]: (1) mobile radio telephones to be used in the Cellular Radiotelephone Service authorized under Part 22 Subpart H or in the Private Land Mobile Radio Services for covered SMR systems under Part 90 of our rules; (2) mobile devices to be used for PCS under Part 24 of our rules; and (3) mobile devices to be used for earth-satellite communication as authorized under Part 25 and Part 80 of our rules. For purposes of this rule, "mobile devices" means devices for which radiating structures would normally be maintained at least 20 cm from the body of the user or of nearby persons. We will also continue to require routine evaluation of unlicensed PCS and unlicensed millimeter wave devices authorized under Sections 15.253, 15.255, and Part 15 Subpart D of our rules unless these devices are portable devices, as defined above.[44] The general population/uncontrolled MPE limits will apply to such mobile and unlicensed devices. Mobile transmitters intended solely for use in the workplace may be considered under the less restrictive occupational/controlled environment category. We recognize that it may not be possible for the manufacturer of the mobile or unlicensed transmitter to ensure that persons will not be located in areas in which the MPE limits could be exceeded. Accordingly, manufacturers may address such concerns by the use of warning labels and instructional material provided to users and installers that advises as to minimum separation distances required between users and radiating antennas to meet the appropriate limits.

67. Although our exposure criteria will apply to portable and mobile devices in general, at this time routine evaluation for compliance will not be required of devices such as "push-to-talk" portable radios and "push to talk" mobile radios used in taxicabs, business, police and fire vehicles and used by amateur radio operators. These transmitting devices will be excluded from routine evaluation because their duty factors (percentage of time during use when the device is transmitting) are generally low and, for mobile radios, because the antennas are normally mounted on the body of a vehicle which provide some shielding and separation from the user. This significantly reduces the likelihood of human exposure in excess of the RF safety guidelines due to emissions from these transmitters. Duty factors associated with transmitting devices that are not "push-to-talk," such as transportable cellular telephones ("bag" phones) or cellular telephones that use vehicle-mounted antennas, would be generally higher, and we will require that these devices be subject to routine evaluation for compliance with general population/uncontrolled MPE limits. Although we are not requiring routine evaluation of all portable and mobile devices, under Sections 1.1307(c) and 1.1307(d) of the FCC's Rules, 47 CFR 1.1307(c) and (d), the Commission reserves the right to require evaluation for environmental significance of any device (in this case with respect to SAR or compliance with MPE limits).

Note: This paragraph tells us that amateur mobile operation is *generally categorically excluded from evaluation. However, high power, high operating duty cycles or antenna placement can result in excessive RF exposure to the occupants of the vehicle. The language of 1.1307(c) and 1.1307(d) could require that these types of installations be evaluated.—Ed.*

D. Categorical Exclusions

75. Our existing environmental rules regarding RF radiation exposure delineate particular categories of existing and proposed transmitting facilities for which licensees and applicants are required to conduct an initial environmental evaluation and prepare Environmental Assessments if their environmental evaluation indicates that their facilities exceed or will exceed the specified RF exposure guidelines. See 47 CFR §1.1307(b)(Note 1). As for transmitting facilities not specifically delineated under Section 1.1307(b)(Note 1), the Commission had determined, based on calculations, measurement data and other information, that such transmitters offered little potential for causing exposures in excess of the guidelines,[45] and thus "categorically excluded" those transmitters from the initial environmental evaluation requirement. Categorical exclusions from routine environmental evaluation are allowed under NEPA when actions are judged individually and cumulatively to have no significant potential for effect on the human environment. See 47 CFR §1.1306(a); see also, Notice at para. 5, ET Docket No. 93-62, 8 FCC Rcd 2849 (1993). However, the Commission, under §1.1307(c) and (d), retains the authority to request that a licensee or an applicant conduct an environmental evaluation and, if appropriate, file environmental information pertaining to an otherwise categorically excluded application if it is determined that in that particular case there is a possibility for significant environmental impact. All transmitting facilities and devices regulated by this Commission are expected to be in compliance with the RF radiation exposure guidelines, and, if not, to file an Environmental Assessment for review under our NEPA procedures.

76. Examples of currently excluded transmitters are those used for land mobile, cellular radio and fixed microwave communications. In the Notice, we noted that some existing categorical exclusions may not be consistent with the more stringent provisions of the 1992 ANSI/IEEE guidelines or may not warrant automatic categorical exclusions because of new data or other information on exposure potential. We, therefore, requested comment, information and analysis relating to the existing categorical exclusions.

83. Several parties address continuation of the categorical exclusion for the amateur radio service. The ARRL and the ARRL-Bioeffects Committee support prudent avoidance[46] and state that most of the amateur radio users do not possess the requisite equipment, technical skills, and/or financial resources to conduct an environmental analysis.[47] Both the ARRL and the ARRL Bio-Effects Committee submit that we could raise an amateur radio applicant's awareness concerning RF energy by placing relevant questions on the amateur license examination.[48] On the other hand, Dr. Wayne Overbeck and the Amateur Radio Health Group comment that it would be inappropriate for this Commission to exempt the amateur service automatically from all requirements for compliance with radiation safety guidelines.[49] Overbeck and the Amateur Radio Health Group state that education is not enough and suggest that we create a version of OST Bulletin No. 65 for radio amateur operations. They state this bulletin could supplement Part 97 rules and be used by amateurs to certify compliance with the RF exposure safety guidelines.[50]

86. Decision. We continue to believe that it is desirable and appropriate to categorically exclude from routine evaluation those transmitting facilities that offer little or no potential for exposure in excess of the specified guidelines. Requiring routine

environmental evaluation of such facilities would place an unnecessary burden on licensees. However, we believe that some alteration of our previous categorical exclusion policy is necessary. Several commenters have submitted technical documentation indicating the power levels and distances at which transmitting sources in various services will comply with the exposure guidelines.[51] Our staff has evaluated this material and has performed analyses of its own. Based on these studies, we now believe that in certain cases we should no longer exempt entire services from demonstrating compliance. Examples include high-power paging and cellular telephone sites on relatively short towers or rooftops where access may not be restricted. There is also evidence that certain amateur radio facilities have the potential for exceeding our new limits.

87. Our current rules require that environmental evaluation for RF exposure be performed for facilities and operations authorized under Parts 5 (Experimental Radio Services); 15 (millimeter wave and unlicensed PCS devices); 21, Subpart K, (Multipoint Distribution Service); 24 (Personal Communications Service); 25 (Satellite Communications); 73 (Radio Broadcast Services); 74, Subparts A, G, I, and L (Experimental, Auxiliary, and Special Broadcast and other Program Distributional Services) and 80 (ship earth stations in the Maritime Services).[52] We believe it is appropriate to continue to subject these facilities and operations to routine environmental evaluation with certain modifications. With respect to transmitting facilities not in these categories, there are certain cases where we no longer believe that an automatic categorical exclusion is justified, and we will require evaluation of some transmitting facilities that were previously excluded. This expansion of the list of transmitting facilities subject to routine evaluation would be necessary regardless of whether our MPE guidelines are based on 1992 ANSI/IEEE or NCRP recommendations.

88. It is important to emphasize, however, that even if a transmitting source or facility is not automatically excluded from routine evaluation, no further environmental processing is required once it has been determined that exposures are within the guidelines, as specified in Part 1 of our rules. There are various ways to accomplish compliance, including restrictions on access, implementation of appropriate work procedures for personnel, incorporation of RF shielding, mounting of appropriate warning signs, control of time of exposure and reduction of power during periods when personnel or the public are present. The revised edition of the FCC's OST Bulletin 65 will include a detailed discussion of this topic.

89. Our new policy on categorical exclusion is designed to bring consistency in the way that we decide what transmitters or facilities warrant an automatic exemption from evaluation. This policy is based on our own calculations and analyses, along with information and data acquired in the record of this proceeding and from other sources. We believe that some transmitting facilities, regardless of service, may offer the potential for causing exposures in excess of MPE limits because of such factors as their relatively high operating power, location or relative accessibility. We believe that it is more reasonable to base our exclusions on such variables since they apply generally to all transmitting facilities. In that regard, our new exclusion policy will also eliminate the requirement for routine evaluation of some relatively low-powered transmitters in some of the services for which routine evaluation was previously required such as certain broadcast services.

90. Routine environmental evaluation for RF exposure will only be required for transmitters, facilities or operations that are included in the categories listed in Table 1 of the new rule Section 1.1307(b)(1) that we are adopting, as shown in Appendix C. This includes some, but not necessarily all, transmitters, facilities or

operations that are authorized under the following Parts of our rules: 5, 15, 21 Subpart K, 22 Subpart E, 22 Subpart H, 24, 25, 73, 74 (Subparts A, G, I, and L), 80, 90, and 97. Within a specific service category, conditions are listed to determine which transmitters will be subject to evaluation. These conditions are generally based on one or more of the following variables: (1) operating power, (2) location, or (3) height above ground. In the case of Part 15 devices, only devices that transmit on millimeter wave frequencies and unlicensed PCS devices are covered, as noted in Table 1. Transmitters and facilities not included in these categories will continue to be categorically excluded from routine evaluation.[53] Such transmitting facilities generally pose little or no risk for causing exposures in excess of the guidelines. Our new policy will provide a clear, "bright line" standard for categorical exclusions that is administratively easy to apply and affords adequate protection from harmful RF exposure.

E. Compliance Evaluation, Measurement Procedures and Transition Provisions

94. In the Notice, we requested comment on issues related to the procedures to be used for demonstrating compliance with exposure guidelines and also on issues concerning quantitative measurement of RF fields and exposure. We recognized that compliance with new guidelines could impose new and significant burdens on some licensees and equipment manufacturers and stated that we would seek to minimize this impact wherever possible. With respect to measurements, we proposed that the procedures established by ANSI/IEEE C95.3-1992 would be appropriate for determining compliance with the new RF exposure guidelines.[54] We further proposed to continue the requirement that facilities and operations subject to environmental evaluation provide environmental information at the time of application for a construction permit, license renewal, or other Commission authorization.[55] We requested comment on whether we should require more complete documentation or evidence from applicants who claim compliance with environmental RF guidelines and what form that documentation should take. Finally, we requested comments, opinions, data and other information concerning devices that are commercially available for measuring electromagnetic fields and currents.

112. Decision. We believe that the rules we are adopting should provide a reasonable transition period for applicants and stations to come into compliance with the new requirements. After considering the comments and the impact of these new requirements, we conclude that the new RF guidelines will apply to station applications filed after January 1, 1997, as described in Appendix C, Section 1.1307(b)(4). [Note: This date has changed! See the note after Paragraph 119. — Ed.] During the period between the effective date of the rules we are adopting and January 1, 1997, our existing RF guidelines will continue to apply to station applications. We recognize that this relatively short transition period may cause some difficulties for certain applicants. Accordingly, for a period of one year from the date this Order is adopted, we will allow our Bureaus to address under delegated authority the specific needs of individual parties that make a good cause showing that they require additional time to meet the new RF guidelines. Such relief could come through waivers of our rules or through other similar actions.

113. The new guidelines for SAR and MPE will apply immediately to non-excluded applications for equipment authorization for portable, mobile, and unlicensed devices as described in Appendix C, Section 1.1307(b)(2). We see no need to delay implementation of the new guidelines for these devices. As previously discussed, information on techniques and procedures for SAR evaluation is already available from several references including

ANSI/IEEE C95.3-1992. There are several acceptable techniques for SAR evaluation, including numerical analytic techniques such as the FDTD procedure discussed earlier, and we do not believe it is practical or necessary at this time for us to institute a certification program for laboratories that perform such services. In fact, as noted previously, we already require SAR evaluation from manufacturers of PCS and portable unlicensed devices, and we have already granted authorizations based on SAR data submitted to us. In addition, certification programs for hand-held devices such as cellular telephones are being developed by other organizations.[56] Similarly, for mobile devices, typical exposure levels can be determined by the use of simple calculational methods and equations such as those described in the current edition of the FCC's OST Bulletin 65.

114. We appreciate the desires of many commenting parties that we delay the effective date for implementation of the new RF exposure guidelines. We recognize that applicants may need to undertake significant analysis and study in order to comply with the new guidelines. Detailed information on evaluating compliance, in the form of a revised version of OST Bulletin No. 65, would provide significant assistance to those attempting to comply with these new guidelines. Therefore, it is our intent to issue in the near future a draft revised OST Bulletin 65. We plan to solicit comments on the draft from individuals and organizations who are active and knowledgeable in this area. This was the same approach that the Commission took in developing the original version of OST Bulletin No. 65.

116. We find that the record generally supports our proposal to endorse the measurement procedures and techniques contained in the ANSI/IEEE C95.3-1992 document for use in evaluating RF exposure potential. In addition, we note that the NCRP has recently published NCRP Report No. 119, which contains practical guidelines and information for performing field measurements in broadcast and other environments, and we also endorse its use.[57] If, in the future, questions arise as to measurement procedures or instrumentation issues, we intend to rely on the above documents. We may also consult expert bodies such as the appropriate NCRP or IEEE committees and other groups, organizations and agencies, as appropriate. Any decisions regarding such issues will be addressed in official Commission notices, proceedings or bulletins, or in response to individual inquiries.

118. With respect to grandfathering previously-authorized portable, mobile and unlicensed devices, we recognize that it would be impractical to require re-authorization of these devices. Furthermore, we believe that most existing devices already comply with the limits that we are adopting. Therefore, we will generally not require re-authorization or testing of previously approved devices solely to demonstrate compliance with our new RF guidelines. If we have reason to believe that a previously authorized device may cause exposures in excess of the guidelines, we may request environmental information and require that the device be re-authorized based on compliance with the guidelines.[58]

119. With respect to previously-licensed stations, we note that we expect our licensees to comply with our RF radiation environmental rules as applicable to them. See, e.g., 47 CFR §§1.1307, 1.1311, and 1.1312. The environmental processing requirements contained in these rules ensure that, at the time of licensing and authorization, transmitting facilities are operating within the applicable RF radiation limits. Once a license is granted, we expect our licensees to continue to operate their facilities in compliance with these limits.

Note: Paragraphs 118 and 119 above are partially superseded by the provisions of Paragraph 113 and 114 of the Second MO&O (Memorandum Opinion and Order). While no stations have been grandfathered, the transition period for existing stations has been extended to September 1, 2000.

The implementation date for these rules has been changed a number of times throughout the reconsideration period. Although the requirements were actually effective immediately when the Report and Order was released in 1996, the FCC established a "transition period" that gave certain stations time to complete any required station evaluation and comply with the rules. During this transition period, the provisions of the old RF-exposure rules apply.

As originally described in paragraph 112 of the Report and Order, the implementation date was January 1, 1997. ARRL immediately petitioned the FCC, asking for a more reasonable amount of time for amateurs to comply with these rules. In paragraph 10 of the First MO&O, the FCC extended the transition period for the Amateur Radio Service to January 1, 1998. This was reiterated in paragraph 4 of the Second MO&O.

The Second MO&O, however, added an additional provision that gave existing stations, including amateur, a date certain of September 1, 2000 as a transition period. The FCC has clarified, however, that new stations, or stations that renew or modify their licenses after January 1, 1998, must be in compliance as certified on the 610 application.

Amateur stations, however, are not subject to prior approval before authorization. Existing stations that make changes to their station that could affect the RF exposure from that station must evaluate those changes before the new configuration is put into use. See 47 CFR §1.1312.—Ed.

IV. ADDITIONAL ISSUES

A. Induced and Contact Current Compliance

130. As discussed in the Notice, the new ANSI/IEEE guidelines contain recommendations regarding maximum permissible limits for induced and contact currents that result from RF exposure. The previous 1982 ANSI guidelines did not address this issue. The ANSI/IEEE recommendations require exposure evaluation over the frequency range from 3 kHz to 100 MHz for RF currents induced in the human body as well as for RF contact currents that can result in shock and burn hazards. We recognize that this new provision has raised many issues relative to interpretation and implementation, and we requested comment on whether we should adopt these requirements.

145. The EPA recommends that we "consider including limits for induced and contact RF currents for the frequency range of 300 kHz to 100 MHz to protect against shock and burn . . . ."[59] This recommendation was in addition to EPA's support for our selection of the NCRP guidelines for field strength and power density that are somewhat different than those of ANSI/IEEE (see earlier discussion). EPA states that it agrees that the ANSI/IEEE induced current limits are useful and should also be implemented.

147. Decision. Most comments, including those of federal health and safety agencies, generally support the use of ANSI/IEEE limits for induced and contact currents as a means of controlling potentially harmful exposure to RF fields. However, in view of the continuing questions and difficulties relating to evaluation of induced and contact currents, especially with regard to measurements, we are not adopting the exposure guidelines for induced and contact currents at this time. Until these questions are satisfactorily resolved, we see no practical way to require compliance with these limits. We see merit in the suggestion of NAB and others that it may be possible to determine compliance with the induced current limits using the magnitude of the electric field strength. However, at this time we do not believe there is sufficient documentation in the record to support the accuracy and reliability of this method. Although we are not adopting limits for

induced and contact currents in this proceeding, we recognize the desirability for limits to be adopted in the future, particularly if more accurate measuring instruments become available. Accordingly, we will continue to monitor the issues raised in this proceeding with respect to induced and contact currents, and we may revisit this issue and issue a specific proposal for controlling such exposures.

B. Amateur Radio

152. Amateur stations present an unusual case with respect to compliance with RF exposure guidelines. First, over 700,000 amateur stations in the United States are authorized by our rules to transmit from any place where the Commission regulates the service, as well as on the high seas. The Commission does not pre-approve individual amateur station transmitting facilities and no additional application is made for permission to relocate an amateur station or to add additional stations at the same or other locations. Second, the granting of a license is solely conditional upon the applicant passing an examination demonstrating that the examinee possesses the operational and technical qualifications required to perform properly the duties of an amateur operator under our rules. Third, amateur stations vary greatly. Amateur stations are located in dwellings, in air, surface and space craft, and carried on the person. Many of these stations transmit from residential or other areas where individuals may be in close proximity to an RF radiator. In addition, amateur station transmissions are made intermittently and may involve as many as 1,300 different emission types — each with a distinctive on-off duty cycle. Finally, most amateur stations engage only in two-way communications. Thus, even when in operation, the station is usually transmitting but half of the time. There are many variables, therefore, to be considered in determining whether an amateur station complies with guidelines for environmental RF radiation.

153. Measurements made during a Commission/EPA study of several typical amateur stations in 1990 indicated that there may be some situations where excessive exposures could occur.[60] Further, among amateur operators themselves there appears to be varying degrees of knowledge concerning the potential hazards of RF radiation. At least one prominent amateur radio publication has a comprehensive section dealing with potential RF hazards at amateur stations.[61]

154. Comments on continuing to exempt amateur stations from demonstrating compliance are divided. The ARRL opposes inclusion, and claims that most amateur operators adopt the philosophy of prudent avoidance, that is, they avoid unnecessary exposure to electromagnetic radiation as a common-sense response to potential — but not yet proven — health hazards. The ARRL also states that its publications, which include sections on RF safety, urge amateur operators to practice prudent avoidance wherever possible and are sufficient to keep the amateur community informed of the hazards of RF radiation. The ARRL and the ARRL Bio-Effects Committee support "prudent avoidance" and state that most amateur operators do not possess the requisite equipment, technical skills, and/or financial resources to conduct an environmental analysis if the categorical exclusion for Part 97 were eliminated.

155. The ARRL argues that amateur stations, because of their intermittent operation, low duty cycles, and relatively low power levels, rarely exceed the 1992 ANSI/IEEE standard. Further, the ARRL suggests that the risk of exceeding those levels would only be relevant for a licensee and his or her family. The ARRL maintains that in this experimental service it is better to rely on education and testing of licensees than on submission of a complex environmental assessment which would not be valid for long in most cases since much amateur station transmitting equipment,

especially antennas, is constructed and designed by the licensee and often changes. Therefore, the ARRL argues that amateur service licensees should not be subjected to routine environmental processing.

156. The ARRL states that if the Commission applied these rules to the amateur radio service, it then must facilitate the installation of amateur station antennas in configurations that will permit compliance with the RF exposure guidelines by issuing a more comprehensive preemption statement with respect to amateur station antennas than now exists, and must completely preempt the judicial enforcement of restrictive covenants which result in amateurs installing station antennas indoors or at locations on a horizontal plane with human occupants of residences. Indeed, the ARRL continues, such an order is overdue anyway; but the combination of adoption of a strict RF exposure standard and continuation of a hands-off attitude with respect to antenna covenants is tantamount to a license revocation, as it would preclude the operation of any amateur station subject to both restrictions.

157. The ARRL Bio-Effects Committee claims that amateur operators normally would be exempted from environmental review requirements, since most engage in operations that would not cause the ANSI/IEEE guidelines to be exceeded. However, it notes, a 100 watt VHF "vehicular installation" may produce higher fields inside the vehicle than the ANSI/IEEE standard would allow. Furthermore, hand-held transceivers, facilities employing indoor antennas, and facilities engaging in specialized activities such as "moonbounce" communication, may produce significant localized fields near the antenna.

158. Further, the ARRL Bio-Effects Committee notes that a comprehensive environmental review would be too burdensome both for the amateur operators and the Commission staff. It therefore recommends that a tabular chart showing the calculated field intensities at various distances from antennas having directive patterns, driven by transmitters of various power output levels common in the amateur service be added to Part 97. The ARRL Bio-Effects Committee also recommends inserting questions about electromagnetic radiation safety in each amateur operator license examination and requiring certification on the license application that the applicant has read the Commission guidelines, understands them, and agrees to comply. Under this scheme, the ARRL Bio-Effects Committee argues, amateur operators would follow the policy of "prudent avoidance" that the ARRL publications now advocate.

159. Professor Wayne Overbeck, filing comments as an individual, believes that few amateur operators are aware of the electromagnetic radiation levels present near their own amateur stations and that rather than being excluded from our requirements, the amateur service should be subject to the standard for "uncontrolled environments" through language added to Part 97. Professor Overbeck points out that vast numbers of amateurs are neither members of the ARRL nor subscribers to any amateur service magazines and consequently these educational sources are not sufficient to ensure adherence to our guidelines. Because actual measurements would be financially prohibitive for most amateur operators, Professor Overbeck recommends that we promulgate a rule requiring amateur operators to adopt operating and antenna-placement practices calculated to meet the exposure limits and that they be required to certify on their application forms that they have read and will adhere to the guidelines for antenna placement. Finally, Professor Overbeck suggests that we promulgate an amateur service version of OST Bulletin No. 65 that would include charts and tables showing required separation distances between antennas and inhabited areas for various power levels. He also suggests that amateurs be tested on this topic as part of operator license examinations.

160. <u>Decision</u>. The Commission expects all its licensees to comply with the RF guidelines specified in our rules, or, if not, to file an Environmental Assessment for review under our NEPA procedures. After a thorough review of the comments and the results of an FCC/EPA measurement study,[62] we conclude that, although it appears to be relatively small, there is a potential for amateur stations to cause exposures to RF radiation in excess of these guidelines. Amateur stations can transmit with up to 1500 watts peak envelope power on frequencies in specified bands from 1,800 kHz to over 300 GHz. Certain of the emission types permitted have high duty cycles, for example frequency or phase shifted digital signals. Amateur stations are not subject generally to restrictions on antenna gain, antenna placement and other relevant exposure variables. Even though situations where exposures are excessive may be relatively uncommon and even though most amateur stations transmit for short periods of time at power levels considerably lower than the maximum allowed, the possibility of human exposure to RF radiation in excess of the guidelines cannot be disregarded. Therefore, a blanket exemption for all amateur stations does not appear to be justified, and we will apply our new guidelines to amateur stations. We will rely upon amateur licensees to demonstrate their knowledge of our guidelines through examinations. We will also rely on amateur licensees to evaluate their own stations if they transmit using more than 50 watts of output power (*Note: this threshold level has changed! See the note below. — Ed.*) Applicants for new licenses and renewals also will be required to demonstrate that they have read and that they understand our applicable rules regarding RF exposure.

Note: In response to an ARRL petition, the FCC has significantly relaxed the 50-watt threshold on most amateur bands. The 50-watt threshold still applies to amateur frequencies between 28 and 225 MHz , but on other bands, the threshold has essentially been changed to match the way the maximum-permissible exposure varies with frequency in the rules. See Chapter 4 of this book, FCC Regulations; Part 97 (reprinted in Appendix A of this book) and paragraph 57 in the Second MO&O included later in this appendix.

The FCC has also added an additional categorical exclusion for most amateur repeaters. Repeaters that use 500 W ERP or less generally do not need to be evaluated. Those that use more than 500 W ERP need to be evaluated if their antenna is mounted on a building or rooftop or is ground mounted and has any part of the antenna system located less than 10 meters (32.8 feet) above ground.—Ed.

161. We find it to be the duty of the licensee of an amateur station to prevent the station from transmitting from any place where the operation of the station could cause human exposure to levels of RF radiation that are in excess of the limits we are adopting. We concur with the ARRL that amateur operators should follow a policy of prudent avoidance of excessive RF exposure. We will continue to rely upon amateur operators, in constructing and operating their stations, to take steps to ensure that their stations comply with the MPE limits for both occupational/controlled and general public/uncontrolled environments. In this regard, we recognize and agree with the ARRL's position that the occupational/controlled limits generally can be considered adequate for situations involving amateur stations considering the most commonly used power levels, intermittent operation and frequencies involved. We recognize that operation in the amateur radio service presents certain unique conditions. Nonetheless, we are concerned that amateur radio operations are likely to be located in residential neighborhoods and may expose persons to RF fields in excess of the MPE guidelines. We will consider amateur radio operators and members of their immediate household to be in a "controlled environment" and will apply the occupational/con-

trolled MPE limits to those situations. Neighbors who are not members of an amateur operator's household, are considered to be members of the general public, however, since they cannot reasonably be expected to exercise control over their exposure. In those cases general population/uncontrolled exposure MPE limits will apply.

Note: In Bulletin 65 Supplement B (Chapter 7 of this book), the FCC has clarified that the immediate household of amateur operators and their guests need to be given instruction about RF exposure in order to be considered as being in a controlled RF environment.—Ed.

162. We believe that the burden for action to assure compliance with RF exposure limits should fall on the relatively few licensees who operate stations that can potentially cause individuals, knowingly or unknowingly, to be exposed to RF energy in excess of these guidelines. We want the licensees of such stations to provide adequately for RF safety. We do not believe, however, that a detailed EA or other routine environmental filing is practical or necessary. To make the complex determination of possible excessive exposure as simple as possible, we are specifying a threshold limit for transmitter power that will apply regardless of frequency used. Below 50 watts transmitter power, the licensee will not be required to take any action, unless requested by Commission staff pursuant to Section 1.1307(c) or 1.1307(d) of our rules. Above this power threshold, the licensee must perform a routine evaluation to predict if the RF radiation could be in excess of that allowed by the criteria listed in §1.1310. If so, the licensee must take action to prevent such an occurrence. The action could be in the form of altering operating patterns, relocating the antenna, revising the station's technical parameters such as frequency, power or emission type or combinations of these and other remedies. To assist with routine evaluation of exposure levels in accordance with the guidelines, we encourage the amateur community to develop and disseminate information in the form of tables, charts and computer analytical tools that relate such variables as operating patterns, emission types, frequencies, power and distance from antennas. We also intend to provide straightforward methods for amateur operators to determine potential exposure levels. This information could be included in our updated version of OST Bulletin No. 65, or we may follow the suggestion to develop a separate bulletin tailored for the amateur service community. As a result of the adoption of a transition period, which was discussed earlier, the new guidelines will apply to amateur stations beginning January 1, 1997. This should provide sufficient time for the amateur community and the Commission staff to prepare the necessary information to help amateur operators comply with these requirements.

Note: The 50-watt threshold has changed. See the editorial note following paragraph 160 earlier in this document.— Ed.

163. As suggested by the ARRL, the ARRL Bio-Effects Committee and Professor Overbeck, we are amending our rules to require the operator license examination question pools to include questions concerning RF safety at amateur stations. We are requiring an additional five questions on RF safety within each of three written examination elements. We also are adopting ARRL's proposal that amateur operators should be required to certify, as part of their license application process, that they have read and understand our bulletins and the relevant FCC rules.[63] We will rely on our Wireless Telecommunications Bureau to develop suitable methods for obtaining this certification.

C. Federal Preemption
Note: The Second MO&O also included a Notice of Proposed Rulemaking (NPRM) to extend federal preemption of RF-exposure matters to cover Commercial Mobile Radio Service Trans-

mitting Facilities. At press time, the ARRL is considering this NPRM and how it can continue to ask the FCC for preemption for the Amateur Radio Service.— Ed.

164. In the past, parties have requested that the Commission preempt state and local authority over RF exposure matters.[64] To date the Commission has declined to preempt on health and safety matters. However, the Commission has noted that should non-Federal RF radiation standards be adopted that adversely affect a licensee's ability to engage in Commission-authorized activities, the Commission would consider reconsidering whether Federal action is necessary.[65]

165. In the <u>Notice</u>, we did not discuss Federal preemption of state and local regulations regarding RF radiation exposure. However, many commenters request that we address this matter by establishing Federal preemption of state and local regulations concerning RF radiation exposure.[66] Two Petitions for Rule Making have been filed in this docket requesting a Further Notice of Proposed Rule Making to address the preemption of non-Federal government regulations concerning RF radiation hazards.[67] The Village of Wilmette, Illinois, and Ergotec Assocation, Inc, in late-filed reply comments, oppose federal preemption of local RF exposure regulations.

166. <u>Decision</u>. In the past the Commission has hesitated to intrude on the ability of states and localities to make regulations affecting health and safety. Many of the comments indicate that a patchwork of divergent local and State regulations could pose a burden on interstate communications. However, since these comments were filed, Congress has passed the Telecommunications Act of 1996, Pub. L. No. 104-104, 110 Stat. 56 (1996). Section 704 of the Telecommunications Act amends the Communications Act by providing for federal preemption of state and local regulation of personal wireless service facilities on the basis of RF environmental effects.[68] The Telecommunications Act also provides for resolution of conflicts related to the regulation of RF emissions by the courts or by petition to the Commission.[69] Accordingly, we are amending §1.1307 of our rules to incorporate the provisions of Section 704 of the Telecommunications Act.

167. The Telecommunications Act does not preempt state or local regulations relating to RF emissions of broadcast facilities or other facilities that do not fall within the definition of "personal wireless services."[70] It would appear from the comments that a few such regulations have been imposed, generally as a result of health and safety concerns. At this point, it does not appear that the number of instances of state and local regulation of RF emissions in non-personal wireless services situations is large enough to justify considering whether or not they should be preempted. We have traditionally been reluctant to preempt state or local regulations enacted to promote <u>bona fide</u> health and safety objectives. We have no reason to believe that the instances cited in the comments were motivated by anything but <u>bona fide</u> concerns.

168. We believe that the regulations that we are adopting herein represent the best scientific thought and are sufficient to protect the public health. Once states and localities have had an opportunity to review and analyze the guidelines we are adopting, we expect they will agree that no further state or local regulation is warranted. Should our expectations prove to be misplaced and should FCC licensees encounter a pattern of state or local activities which constitute an obstacle to the scheme of federal control of radio facilities set forth in the Communications Act, they should present us with such evidence as well as their view of the legal basis which could justify FCC preemption of state and local ordinances. At this time, however, we deny the petitions from the EEA and from Hammett and Edison, as well as the comments from several parties, requesting a broad-based preemption policy to cover all transmitting sources.

V. CONCLUSION

169. To protect public health with respect to RF radiation from FCC-regulated transmitters, and to fulfill our responsibilities under NEPA, we are updating our guidelines for evaluating the environmental impact of RF emissions. We believe that the guidelines we are adopting will be of benefit both to the public and to the telecommunications industry. They will provide assurance that recent scientific knowledge is taken into account regarding future decisions on approval of FCC-authorized facilities and equipment.

VI. ORDERING CLAUSES

170. Section 704(b) of the Telecommunications Act of 1996 requires that we prescribe and make effective these new rules by August 6, 1996. Accordingly, we find that good cause exists, pursuant to 5 U.S.C. Sec. 553(d)(3), to make these rules effective upon publication in the Federal Register rather than to follow the normal practice of making them effective 30 days after publication in the Federal Register.[71] Completion of this rulemaking has required an extensive amount of work to resolve extremely complex issues. In addition, coordination with the various affected federal agencies through to the Interdepartment Radio Advisory Committee has consumed more time than anticipated. The time required to review the comments, decide on the best possible guidelines and coordinate that decision with other federal agencies has made it impossible to delay the effective date for 30 days and still meet the Congressionally imposed deadline. Thus, we have no alternative but to make these rules effective immediately. We note that the <u>Notice</u> in this proceeding was first issued in 1993. In addition, we note that the Telecommunications Act of 1996, containing a deadline for implementation, was enacted in early February of this year. Therefore, most parties to this proceeding have had considerable notice of the likely actions we would be taking, and they should have had sufficient opportunity to prepare for the implementation of new guidelines pursuant to the implementation schedule set forth above.

171. Accordingly, pursuant to the authority contained in Sections 4(i), 7(a), 303(c), 303(f), 303(g), 303(r) and 332(c)(7) of the Communications Act of 1934, as amended, 47 U.S.C. Sections 154(i), 157(a), 303(c), 303(f), 303(g), 303(r) and 332(c)(7), IT IS ORDERED, that effective August 6, 1996, Parts 1, 2, 15, 24, and 97 of the Commission's Rules and Regulations, 47 CFR Parts 1, 2, 15, 24, and 97, ARE AMENDED as specified in Appendix C.

172. IT IS FURTHER ORDERED, that the respective petitions of the Electromagnetic Energy Association, Hammett and Edison, Inc., and Ken Hollady ARE DENIED.

VII. PROCEDURAL MATTERS

173. For further information concerning this rulemaking, contact the Commission's radiofrequency safety program at (202) 418-2464. Address: Office of Engineering and Technology, Federal Communications Commission, Washington, DC 20554. Internet email address: rfsafety@fcc.gov.

APPENDIX A

FINAL REGULATORY FLEXIBILITY ANALYSIS

IV. SUMMARY OF PROJECTED REPORTING, RECORDKEEPING AND OTHER COMPLIANCE REQUIREMENTS:

Applicants that are subject to the new RF radiation guidelines (i.e., not categorically excluded), are required to make a statement on any application filed with the Commission indicating that they comply with the RF radiation limits. Technical information supporting that statement must be retained by the applicant, and provided to the Commission upon request. In some cases, the applicant will be able to determine compliance by making calculations or reading applicable literature, including OST Bulletin No. 65. In other cases, detailed measurements of the transmitting facility may be necessary. In addition, steps to control access to the facility, such as warning signs or fences, may be required. Manufacturers of radio transmitting equipment will, as indicated above, need to make MPE and/or SAR measurements that will need to form part of the manufacturer's records for equipment authorization.

Reporting

Reporting requirements are limited to certain classes of applicants and licensees for which the potential for human exposure to RF emissions is the greatest. Most applicants and licensees are categorically excluded from routinely evaluating their facilities, operations or transmitters for compliance with the new RF exposure guidelines. The National Environmental Policy Act (NEPA), upon which our rules are based, allows "categorical exclusion" of large classes of actions that generally do not provide an opportunity for causing significant environmental impact, such as would result from human exposure to RF emissions in excess of the guidelines. In this case, the "actions" excluded are the granting of Commission applications and authorizations. Therefore, we are categorically excluding many applications submitted to the Commission from routine evaluation for compliance with the RF guidelines. This exclusion significantly limits burden on our regulatees, including many small businesses. The category exclusions apply to all radio services except those listed in section IV above and the radio amateur service. This means, for example, that all land mobile and public safety two-way systems are categorically excluded.

Applicants in services that are not categorically excluded may also be categorically excluded from determining compliance based on antenna location or station power. Applicants who are not categorically excluded are required to make a statement on certain application forms filed with the Commission indicating whether they comply with our environmental rules. This action by a licensee or applicant is the primary reporting requirement. In addition, supporting information (such as measurement data, site drawings, and calculations) may be requested, in certain cases, to justify the statement made on a Commission form.

Recordkeeping

The Commission has no specific recordkeeping requirements related to compliance with the RF exposure guidelines. This has not changed from the rules previously in place regarding compliance with RF exposure guidelines. The Commission does reserve the right to request information supporting the answer an applicant gives on a form. Such information would normally be technical in nature and could involve a report of calculations performed or measurements made to determine compliance. Therefore, many applicants and licensees may keep information related to their compliance on file in some form for their own records. The Commission provides applicants with guidance on performing calculations or measurements through its OST Bulletin No. 65, which is being updated to reflect the new guidelines. In many cases, an applicant or licensee can easily use this bulletin to determine compliance through the use of charts, figures and tables. This largely eliminates the need for keeping a detailed analytic report in many cases. Manufacturers of equipment who are required to evaluate portable or mobile devices would likely have to perform more detailed analysis and keep on file a specific technical report for review by the Commission if requested. Also, in a few cases involving multiple transmitters at large antenna farms detailed measurement studies may be necessary. Reports of such studies would be retained by an applicant to provide evidence of compliance if required.

Other Compliance Requirements

As was true for the previous rules, there are no specific compliance requirements, as such. Under the Commission's NEPA rules, applicants and licensees are required to submit an Environmental Assessment (EA) if they do not comply with our RF exposure guidelines (47 CFR §1.1311). An EA is a detailed accounting of the consequences created by a specific action that may have a significant environmental impact, in this case a Commission authorization of a transmitter or facility that exceeds the RF guidelines. An EA would be evaluated by the Commission to determine whether the authorization should be granted in view of the environmental impact. In reality, this leads to a de facto compliance requirement, since most applicants and licensees who are not categorically excluded (see above) undertake measures to ensure compliance before submitting an application in order to avoid the preparation of a costly and time-consuming EA. For this reason EAs are rarely filed with the Commission. This has not changed from the existing rules. As for determining compliance, as mentioned above, the Commission provides applicants with specific guidance in the form of a technical bulletin. This bulletin is designed to minimize the effort and burden required by an applicant to determine compliance with the guidelines prior to submitting an application. Many options are available for ensuring compliance, including restricting access to an area where high RF levels exist, using warning signs or fences to provide notice of potential RF exposure, use or protective shielding or warning devices, reduction of power when people are in high RF areas and, in the case of portable and mobile devices, designing devices to minimize RF absorption in the body of the user.

Skills Needed to Meet Requirements

If a station is not categorically excluded, then the licensee or applicant must make a determination of whether the station will comply with the RF radiation limits. This study can be done by calculation or measurement, depending upon the situation. The calculations can be done in many cases by a radio technician or engineer familiar with radio propagation. If measurements are necessary, then a radio technician or engineer will also be required.

The applicant must indicate on its application that it meets the NEPA requirements and, therefore, does not exceed the RF radiation limits. The is usually done by checking a box on a form, which can be done by a clerical person.

NOTES

[1]Specifically, we are adopting limits for field strength and power density that are generally based on Sections 17.4.1 and 17.4.2, and the time-averaging provisions recommended in Sections 17.4.1.1 and 17.4.3, of "Biological Effects and Exposure Criteria for Radiofrequency Electromagnetic Fields," NCRP Report No. 86 (1986), National Council on Radiation Protection and Measurements (NCRP). With the exception of the limits on exposure to power density above 1500 MHz and the limits for exposure to lower frequency magnetic fields, these MPE limits are also generally based on the guidelines contained in the RF safety standard developed by the Institute of Electrical and Electronic Engineers, Inc. (IEEE) and adopted by the American National Standards Institute (ANSI). See Section 4.1 of ANSI/IEEE C95.1-1992, "Safety Levels with Respect to Human Exposure to Radio Frequency Electromagnetic Fields, 3 kHz to 300 GHz."

[2]These guidelines are based on those recommended by ANSI/IEEE and NCRP. See Sections 4.2.1 and 4.2.2 of ANSI/IEEE C95.1-1992 and Section 17.4.5 of NCRP Report No. 86.

[3]For example, see letter from Carol M. Browner, Administrator, U.S. Environmental Protection Agency, to Reed. E. Hundt, Chairman, FCC, dated July 25, 1996; and letter from Elizabeth D. Jacobson, Ph.D., Deputy Director for Science, Center for Devices and Radiological Health, Food and Drug Administration, to Richard M. Smith, Chief, Office of Engineering and Technology, FCC, dated July 17, 1996. Both letters have been placed into the docket record as ex parte filings in this proceeding.

[4]National Environmental Policy Act of 1969, 42 U.S.C. Section 4321, et seq.

[5]See 47 CFR §1.1301, et seq.

[6]See Report and Order, GEN Docket No. 79-144, 100 FCC 2d 543 (1985); Memorandum Opinion and Order, 58 RR 2d 1128 (1985); see also ANSI C95.1-1982, "American National Standard Safety Levels with Respect to Human Exposure to Radio Frequency Electromagnetic Fields, 300 kHz to 100 GHz," ANSI, New York, NY.

[7]47 CFR Section 1.1307(b).

[8]See Second Report and Order, GEN Docket No. 79-144, 2 FCC Rcd 2064 (1987); Erratum, 2 FCC Rcd 2526 (1987). Facilities that are otherwise categorically excluded from RF environmental evaluation may still be required, on a case-by-case basis, to undergo evaluation pursuant to the provisions of 47 CFR §1.1307(c) and (d). The Council on Environmental Quality, which has oversight responsibility with regard to NEPA, permits Federal agencies to categorically exclude certain actions from routine environmental processing when the potential for individual or cumulative environmental impact is judged to be negligible. See 40 CFR §§ 1507, 1508.4; see also Regulations for Implementing the Procedural Provisions of NEPA, 43 Fed. Reg. 55,978 (1978).

[9]ANSI/IEEE C95.1-1992, "Safety Levels with Respect to Human Exposure to Radio Frequency Electromagnetic Fields, 3 kHz to 300 GHz."

[10]The 1982 ANSI guidelines cover the frequency range 300 kHz to 100 GHz.

[11]Specific absorption rate is a measure of the rate of energy absorption by the body. SAR levels are specified for both whole-body exposure and for partial-body or localized exposure (generally specified in terms of spatial peak values), such as might occur to the head of the user of a hand-held radiotelephone.

[12]See Notice of Proposed Rule Making, ET Docket No. 93-62, 8 FCC Rcd 2849 (1993); see also 8 FCC Rcd 5528 (1993), 9 FCC Rcd 985 (1993), 9 FCC Rcd 317 (1994), 9 FCC Rcd 989 (1994) extending the comment deadlines.

[13]The ANSI/IEEE standard was developed by the IEEE Standards Coordinating Committee 28 on Non-Ionizing Radiation Hazards (IEEE SCC28) and subsequently adopted by the IEEE Standards Board and the American National Standards Institute.

[14]Notice at para. 23. The NCRP is a non-profit corporation chartered by Congress to develop information and recommendations concerning radiation protection. NCRP consists of the members and participants who serve on its various scientific committees. Several government agencies and non-government organizations have established relationships with NCRP as "Collaborating Organizations." The FCC is one of these Collaborating Organizations.

[15]Both the ANSI/IEEE and NCRP exposure criteria are based on a determination that potentially harmful biological effects can occur at an SAR level of 4 W/kg as averaged over the whole-body. Appropriate safety factors were then added to arrive at limits for both whole-body exposure (0.4 W/kg for "controlled" or "occupational" exposure and 0.08 W/kg for "uncontrolled" or "general population" exposure, respectively) and for partial-body (localized SAR), such as might occur in the head of the user of a hand-held cellular telephone.

[16]For example, in uncontrolled environments the 1992 ANSI/IEEE guidelines recommend a safe power density level of 1 mW/cm^2 at 1500 MHz increasing to a maximum of 10 mW/cm^2 at 15 GHz to 300 GHz, a significant change from the 1982 ANSI standard. The NCRP guidelines specify a fixed level of 1 mW/cm^2 for exposure of the general public at frequencies above 1500 MHz. NCRP limits for magnetic field exposure are also generally more stringent for frequencies below 100 MHz.

[17]This provision recommends that the stricter public exposure limits apply where workers are exposed to electromagnetic fields with carrier frequencies that are modulated at a depth of 50 percent or greater at frequencies between 3 and 100 hertz. See NCRP, supra, Section 17.4.7.

[18]For measuring MPE levels, the NCRP guidelines use an averaging time of 6 minutes for occupational exposure and 30 minutes for public exposure. For frequencies above 15 GHz, the ANSI/IEEE guidelines reduce this averaging time in a manner that is inversely proportional to the frequency raised to the 1.2 power.

[19]EPA Comments at 1.

[20]See 21 CFR §1000 et seq.

[21]FDA Comments at 1.

[22]NIOSH Comments at 1.

[23]OSHA Reply Comments at 1.

[24]Telocator Comments at 3.

[25]IEEE/SCC28 Reply Comments at 1-7.

[26]See "Reply Comments of Arthur W. Guy, Ph.D.," March 9, 1996, and letter of A. W. Guy to Reed E. Hundt, Chairman, FCC, dated March 14, 1996. Both placed in the record of this proceeding as ex parte filings.

[27]See, letter from Eleanor R. Adair, Ph.D., to Reed E. Hundt, Chairman, FCC, dated March 14, 1996, and letter from C. K. Chou, Ph.D., to Thomas P. Stanley, Chief Engineer, FCC, dated March 20, 1996.

[28]See, e.g., Report and Order, GEN Docket 79-144, at para. 26 note 6 supra. See also, letter from Mark S. Fowler, Chairman, FCC, to Anne M. Burford, Administrator, EPA, February 22, 1983; letter from Dennis R. Patrick, Chairman, FCC, to Lee M. Thomas, Administrator, EPA, November 29, 1988; and letter from Thomas P. Stanley, Chief Engineer, FCC, to Ken Sexton, Director, Office of Health Research, Office of Research and Development, EPA, October 24, 1990.

[29]See note 1, supra.

[30]The 1982 ANSI guidelines contain a single level of MPE limits and do not differentiate based on environment.

[31]LMCC Comments at 4. NAB Comments at 2.

[32]EPA Comments at 3.

[33]EPA Comments at 3-4.

[34]OSHA Reply Comments at 1-2.

[35]NIOSH Comments at 2.

[36]ARRL Comments at 11-12.

[37]For example, a sign warning of RF exposure risk and indicating that individuals should not remain in the area for more than a certain period of time could be acceptable.

[38]"Evaluating Compliance with FCC-Specified guidelines for Human Exposure to Radiofrequency Radiation," OST Bulletin No. 65, October 1985. OST Bulletin No. 65 will be renamed OET Bulletin No. 65 when it is released.

[39]See Notice at para. 14 ("Low-Power Devices/Exclusions"). The ANSI/IEEE low-power exclusions are based on consideration of either SAR or a device's radiated power ("radiated power exclusion"). See also ANSI/IEEE C95.1-1992, clauses 4.2.1.1 and 4.2.2.1.

[40]See NCRP Report No. 86, Section 17.4.5. The NCRP guidelines specify that the criterion for general-population, localized exposure "should allow no more than one-fifth the levels of SAR allowed for occupational exposures [8 W/kg]," i.e., 1.6 W/kg as also recommended by ANSI/IEEE. However, the NCRP also notes that exposure of individuals in the general population who use "radio emitters" such as hand-held transceivers is permitted, "as a per-

sonal decision by the individual, provided that the devices are designed and used as designed so that the exposure of the individual does not exceed the occupational guidelines [8 W/kg] and provided that . . . . the individual does not expose other persons above the population guidelines."

[41]Although ANSI/IEEE does not explicitly state a rule for determining when SAR measurements are preferable to MPE measurements, we believe that the 20 cm distance is appropriate based on Sec. 4.3 (3) of ANSI/IEEE C95.1-1992.

[42]"Covered SMR" systems include two classes of SMR licensees: geographic area SMR licensees in the 800 MHz and 900 MHz SMR bands that offer real-time, two-way switched voice service that is interconnected with the public switched network; and Incumbent Wide Area SMR licensees, defined in Section 20.3 as "licensees who have obtained extended implementation authorizations in the 800 MHz or 900 MHz service, either by waiver or under Section 90.629 of these rules, and who offer real-time, two-way voice service that is interconnected with the public switched network."

[43]The effective radiated power (ERP) limit of 1.5 watts was determined by calculating the ERP that could result in the most restrictive power density limit for general public/uncontrolled exposure at the relevant frequencies of the devices to be evaluated at a distance of 20 cm from the radiating structure. For 800-900 MHz transmitting devices this limit is in the range of 0.5-0.6 mW/cm^2.

[44]These devices are already subject to such requirements, as specified in Sections 15.253(f), 15.255(g), and 15.319(i) of our existing rules.

[45]Second Report and Order, GEN Docket No. 79-144, id.; Erratum, 2 FCC Rcd 2526 (1987).

[46]ARRL Comments at 17, ARRL Bio-Effects Committee Comments at 4.

[47]ARRL Comments at 14.

[48]ARRL Comments at 16, ARRL Bio-Effects Committee Comments at 5.

[49]Overbeck Comments at 2, Overbeck and Amateur Radio Health Group Reply Comments at 11.

[50]Overbeck and Amateur Radio Health Group Reply Comments at 13.

[51]See comments of JC&A, AFCCE, Motorola, MSTV/NBC, and NAB.

[52]See 47 CFR Parts 5, 15 (§15.253, §15.255, and Subpart D), 21 (Subpart K), 24, 25, 73, 74 (Subparts A, G, I, and L) and 80 (ship earth stations).

[53]However, as noted previously, Sections 1.1307(c) and (d) of our rules allow that, even though a transmitter may be categorically excluded, the Commission may still require environmental evaluation on a case-by-case basis.

[54]"Recommended Practice for the Measurement of Potentially Hazardous Electromagnetic Fields –RF and Microwave." ANSI/IEEE C95.3-1992. See Notice at para. 28.

[55]As addressed above, we also requested comment on whether proof of compliance for low-power devices should be submitted as part of the equipment authorization process.

[56]See para. 70, supra.

[57]"A Practical Guide to the Determination of Human Exposure to Radiofrequency Fields," Report No. 119. Copyright 1993, NCRP. Copies may be purchased from NCRP Publications, 7910 Woodmont Ave., Suite 800, Bethesda, MD 20814. Telephone: (800) 229-2652.

[58]47 CFR §1.1307(c) and (d).

[59]EPA Comments at 2.

[60]"Measurements of Environmental Electromagnetic Fields at Amateur Radio Stations," Report No. FCC/OET ASD-9601 (February 1996). Copies can be ordered through the National Technical Information Service (NTIS) at (800) 553-6847. NTIS Order No. PB 96-145016. [See Appendix C—Ed.]

[61]See The ARRL Radio Amateur Handbook For Radio Amateurs. Copyright ARRL, Newington, CT.

[62]See, note 194, supra.

[63]ARRL Comments at 17. ARRL Bio-Effects Committee Comments at 5.

[64]See, 5 FCC Rcd 486 (1990).

[65]See, GEN Dkt 79-144, Report and Order, 100 FCC 2d at 558.

[66]See, for example, comments of MSTV/NBC, McCaw, PacTel, Hammet & Edison, Joint Broadcasters, Celpage, Ericsson, AMSC, the New Jersey Broadcasters Association, and ARRL.

[67]See Electromagnetic Energy Association (formerly EEPA), Petition for Further Notice of Proposed Rulemaking and Hammett & Edison Comments requesting that it serve as a Petition for Rule Making concerning the preemption of state and local RF regulations.

[68]Telecommunications Act of 1996, Section 704. Facilities Siting: Radio Frequency Emission Standards. Sec. 704 (a) (7) (B) (iv). This section states that: "No State or local government or instrumentality thereof may regulate the placement, construction, and modification of personal wireless service facilities on the basis of the environmental effects of radio frequency emissions to the extent that such facilities comply with the Commission's regulations concerning such emissions."

[69]Telecommunications Act of 1996, Section 704 (a) (7) (B) (v). This section states that, "Any person adversely affected by any final action or failure to act by a State or local government or any instrumentality thereof that is inconsistent with this subparagraph may, within 30 days after such action or failure to act, commence an action in any court of competent jurisdiction. The court shall hear and decide such action on an expedited basis. Any person adversely affected by an act or failure to act by a State or local government or any instrumentality thereof that is inconsistent with clause (iv) may petition the Commission for relief."

[70]Section 704 (a) (C) (i) of the Act defines "personal wireless services" to mean "commercial mobile services, unlicensed wireless services, and common carrier wireless exchange access services."

[71]See note 4, supra. Unlike other sections of that Act, see, e.g., Secs. 251(d)(d)(1), which directs us to "complete" action, and Sec. 254(a)(2), which directs us to "promulgate" rules, Sec. 704 requires that the RF exposure guidelines be made effective within the prescribed 180 day time period.

ARRL's Condensation of ET Docket 93-62
First Memorandum Opinion and Order (FCC 96-487)
December 23, 1996

Before the
FEDERAL COMMUNICATIONS COMMISSION
Washington, D.C. 20554

FCC 96-487

In the Matter of	)
	)
Guidelines for Evaluating the Environmental	) ET Docket No. 93-62
Effects of Radiofrequency Radiation	)
	)

FIRST MEMORANDUM OPINION AND ORDER

Adopted: December 23, 1996 ; Released: December 24, 1996

By the Commission:

I. INTRODUCTION

1. By this action, we are amending our rules to extend the transition period for applicants and station licensees to determine compliance with our new requirements for evaluating the environmental effects of radiofrequency (RF) electromagnetic fields from FCC-regulated transmitters. For most radio services, we are extending the transition period by eight months, until September 1, 1997. For the Amateur Radio Service, we are extending the transition period for amateur radio operators until January 1, 1998. We also are allowing changes to amateur radio operator license examinations to be made as the examinations are routinely revised between now and July 1, 1998. We believe that these extensions are necessary so that our applicants and licensees will have adequate time to understand the new requirements and ensure that their facilities are in compliance with them.

II. BACKGROUND

2. The National Environmental Policy Act of 1969 (NEPA) requires agencies of the Federal Government to evaluate the effects of their actions on the quality of the human environment.[1] To meet its responsibilities under NEPA, the Commission has adopted requirements for evaluating the environmental impact of its actions.[2] One of several environmental factors addressed by these requirements is human exposure to RF energy emitted by FCC-regulated transmitters and facilities.

3. In 1985, the Commission adopted rules for evaluating the environmental effects of RF electromagnetic fields produced by FCC-regulated transmitters.[3] On August 1, 1996, we adopted the Report and Order in this proceeding which amended those rules by providing for the use of new guidelines and methods.[4] In our Report and Order, we provided a transition period for applicants and stations to come into compliance with the new requirements. After considering the comments and the impact of the new requirements, we concluded that the new requirements would apply to station applications filed after January 1, 1997, as described in the amended 47 CFR §1.1307(b)(4).[5] Recognizing that this relatively short transition period might cause some difficulties for certain applicants, we gave our Bureaus delegated authority for one year to address, through the granting of waivers or similar actions, the specific needs of individual parties that make a good cause showing that they require additional time to comply with the new guidelines.[6] Seventeen petitions for reconsideration and/or clarification, as well as a motion for extension of the effective date, were filed in response to the Report and Order. A list of those organizations and individuals filing petitions, as well as those filing oppositions and replies to the petitions, can be found in Appendix B. Several technical and legal issues have been raised in the petitions. This First Memorandum Opinion and Order addresses those petitions and comments requesting extension of the transition provisions contained in the Report and Order. We intend to address the other issues in a separate action in the very near future.

III. DISCUSSION

4. The American Radio Relay League, Inc. (ARRL), Ameritech Mobile Communications, Inc., AT&T Wireless Services, Inc., BellSouth Corporation, Paging Network, Inc. (PageNet), the Personal Communications Industry Association (PCIA), and U S WEST, Inc., ask in their petitions that we extend the transition period beyond January 1, 1997, arguing that the existing transition period does not allow adequate time for affected parties to achieve compliance with the new requirements.[7] In responding to the petitions and comments, Arch Communications Group, Inc., the Cellular Telecommunications Industry Association (CTIA), the National Association of Broadcasters (NAB) and PageMart II, Inc. (PageMart) also support extending the transition period.[8] Opposition to the proposals to extend the transition period was filed by the Ad-hoc Association of Parties Concerned About the Federal Communications Commission's Radiofrequency Health and Safety Rules (Ad-hoc Association), the Brooklyn Green Party, the Cellular Phone Taskforce, Alan Golden, and Dawn Mason.[9] These parties generally argue that an extension could result in adverse public health risks and would allow the continued proliferation of facilities that do not comply with the new requirements.

5. Ameritech, AT&T, CTIA, NAB, PageMart, PageNet, PCIA and U S WEST request that we extend the end of the transition period to at least one year after a revised OST Bulletin 65 is issued.[10] BellSouth requests that the transition period be extended to six months after release of the revised bulletin.[11] In a Motion filed after the deadline for formal Oppositions and Replies, the ARRL also requests that we extend the transition

period for general compliance by licensed amateur radio operators to January 1, 1998.[12] These petitioners urge us to provide ample time for affected parties to consider the information contained in this bulletin and to conduct the necessary measurements and calculations needed to ensure compliance. PCIA states that applicants may need to undertake significant analysis and study of the revised bulletin in order to comply with the new regulations. PageNet states that if an extension is not granted, compliance will be "rendered impossible," requiring the filing of waiver requests by virtually every carrier in the country.[13] This view is echoed by AT&T, CTIA, PCIA, and U S WEST.

6. The ARRL also requests that we provide a reasonable transition period for compliance with the requirements adopted in the Report and Order regarding amateur operator license examinations and question pools.[14] The ARRL says that it would be impossible for the thousands of volunteer examiners to comply with those requirements, which went into effect immediately, absent a transition period. Our new rules require that at least five questions on the examinations for Elements 2, 3(A), and 3(B) must be related to "radiofrequency environmental safety practices at an amateur station."[15] The new rules also require that the total number of questions on the examinations for Elements 2, 3(A), and 3(B) be increased.[16] Based on the new rules, the question pool for Element 2 examinations would need to include 350 questions, while the question pools for Element 3(A) and 3(B) examinations would need to include 300 questions.[17] The practical problem, according to the ARRL, is not the number of questions in each question pool related to RF environmental safety practices, but rather that the examinations now in circulation do not contain the requisite total number of questions, and the present Element 3(A) and 3(B) question pools, slated for revision in the near term, do not contain at least 300 questions. The ARRL requests that we delay the implementation date for increasing the number of questions on the Element 2 and 3(A) examinations to July 1, 1997, and that we delay the implementation date for increasing the number of questions on the Element 3(B) examinations to July 1, 1998, in order to coincide with the current schedule for the routine revision of the question pools.[18]

Note: The implementation date for these rules has been changed a number of times throughout the reconsideration period. Although the requirements were actually effective immediately when the Report and Order was released in 1996, the FCC established a "transition period" that gave certain stations time to complete any required station evaluation and comply with the rules. During this transition period, the provisions of the old RF-exposure rules apply.

*As originally described in paragraph 112 of the Report and Order, the implementation date was January 1, 1997. ARRL immediately petitioned the FCC, asking for a more reasonable amount of time for amateurs to comply with these rules. In paragraph 10 of the First MO&O, the FCC extended the transition period for the Amateur Radio Service to **January 1, 1998**. This was reiterated in paragraph 4 of the Second MO&O.*

*The Second MO&O, however, added an additional provision that gave existing stations, including amateur, a date certain of **September 1, 2000** as a transition period. The FCC has clarified, however, that new stations, or stations that renew or modify their licenses after January 1, 1998, must be in compliance as certified on the 610 application.*

Amateur stations, however, are not subject to prior approval before authorization. Existing stations that make changes to their station that could affect the RF exposure from that station must evaluate those changes before the new configuration is put into use. See 47 CFR §1.1312.—Ed.

7. Decision. We are extending the transition period so that the new RF guidelines will apply to station applications filed after September 1, 1997, as described in Appendix A, Section 1.1307(b)(4). When we adopted the Report and Order, we anticipated that it might cause difficulties for certain applicants to have to determine compliance with the new RF guidelines by January 1, 1997. Accordingly, we gave delegated authority to our Bureaus to extend this transition period on a case-by-case basis. Based on the petitions and comments we have now received, it is clear that most station applicants will need additional time to determine that they comply with the new requirements. An extension of the transition period would eliminate the need for the filing and granting of individual waiver requests, and would allow time for our applicants and licensees to review the results of the decisions we will be taking in the near future to address the other issues raised in the petitions. It would also allow applicants to review the revised Bulletin 65 and to make the necessary measurements or calculations to determine that they are in compliance.

8. While we concur with petitioners who request that we extend the transition period, we believe that it would be unnecessary, in most circumstances, to extend the transition period for a full year or more after a revised Bulletin 65 is issued. At the same time, we do not concur with petitioners who suggest that granting any extension of the transition period will have significant adverse effects on public health. Accordingly, we are extending the transition period for station applications until September 1, 1997.

9. We are also extending the transition period to January 1, 1998, for amateur operators to come into compliance with the new requirements. We see merit in the arguments expressed by the ARRL that, due to the uniqueness of the Amateur Radio Service, additional time is need to ensure compliance. In particular, we note that amateur stations can use a wide variety of equipment and antennas, and this can make it very difficult to determine whether excessive RF electromagnetic fields may be produced by individual stations. Furthermore, all amateur radio stations in the past had been categorically exempt from these regulations, and many amateur operators may not be familiar with the new requirements and may need additional time to determine how to perform correctly a routine environmental evaluation. This extended transition period for amateur operators will have the advantage of allowing our staff ample time to work with the amateur radio community to refine and issue a special supplement to Bulletin 65 for the specific use of amateur operators.[19] With respect to amateur operator license examination requirements, we agree with the arguments raised by the ARRL. The volunteers recently released revised versions of two of the pools which contain the required questions.[20] Teachers and publishers are currently incorporating the new material into training manuals and courses for use by those preparing to take the examinations starting July 1, 1997. Work is also underway to similarly revise the third and final question pool for use starting July 1, 1998. We are, therefore, staying the enforcement of the new examination provisions adopted in the Report and Order in the amended 47 CFR §97.503(b) to July 1, 1997, with respect to Element 2 and 3(A) examinations and to July 1, 1998, with respect to Element 3(B) examinations. Recognizing that a relatively short transition period might cause some difficulties for certain applicants, we are delegating authority, as we did in the Report and Order, to our Bureaus until July 1, 1998, to address the specific needs of individual parties that make a good cause showing that they require additional time to meet the new guidelines. Such relief could come through waivers of our rules or through other similar actions.

IV. ORDERING CLAUSES

10. The rules we are adopting temporarily relieve existing restrictions. Pursuant to 5 U.S.C. §§553(d)(1) and 553(d)(3), we find that good cause exists to make these rules effective immediately rather than to follow the normal practice of making them effective 30 days after publication in the Federal Register. This will permit all parties filing applications during the next 30 days to take advantage of the extension of the transition periods. Accordingly, pursuant to the authority contained in Sections 4(i), 7(a), 303(c), 303(f), 303(g), 303(r) and 332(c)(7) of the Communications Act of 1934, as amended, 47 U.S.C. Sections 154(i), 157(a), 303(c), 303(f), 303(g), 303(r) and 332(c)(7), IT IS ORDERED THAT, effective upon adoption, Part 1 of the Commission's Rules and Regulations, 47 CFR Part 1, IS AMENDED as specified in Appendix A.

11. IT IS FURTHER ORDERED THAT, to the extent discussed above and as reflected in the new rules contained in Appendix A, certain aspects of the various petitions and motions filed in this proceeding ARE GRANTED. IT IS ALSO ORDERED THAT motions filed by the Ad-hoc Association to accept a late-filed petition for reconsideration, by the Ad-hoc Association to accept a late filed reply to an opposition to a petition for reconsideration, and by the Cellular Taskforce to accept a late-filed opposition to petition for reconsideration and clarification ARE GRANTED. Because the decisions we are taking in this proceeding relate specifically to important public health issues, we believe that it is in the public interest to consider these late-filed documents along with all of the other timely petitions and comments in this proceeding. IT IS ALSO ORDERED THAT enforcement of the amendments to 47 CFR §§97.503(b)(1) and 97.503(b)(2) adopted in the <u>Report and Order</u> ARE STAYED until July 1, 1997, and enforcement of the amendments to 47 CFR §97.503(b)(3) IS STAYED until July 1, 1998.

V. PROCEDURAL MATTERS

12. For further information concerning this rule making, contact the Commission's radiofrequency safety program at (202) 418-2464. Address: Office of Engineering and Technology, Federal Communications Commission, Washington, D.C. 20554. Internet e-mail address: rfsafety@fcc.gov.

FEDERAL COMMUNICATIONS COMMISSION

William F. Caton
Acting Secretary

APPENDIX C

FINAL REGULATORY FLEXIBILITY ANALYSIS

IV. DESCRIPTION AND ESTIMATE OF THE SMALL ENTITIES SUBJECT TO THE RULES:

The rules being adopted in this <u>First Memorandum Opinion and Order</u> apply to the following eleven industry categories and services. The RFA generally defines the term "small business" as having the same meaning as the term "small business concern" under the Small Business Act, 15 U.S.C. §632. Based on that statutory provision, we will consider a small business concern one which (1) is independently owned and operated; (2) is not dominant in its field of operation; and (3) satisfies any additional criteria established by the Small Business Administration (SBA). The RFA SBREFA provisions also apply to non-profit organizations and to governmental organizations. Since the Regulatory Flexibility Act amendments were not in effect until the record in this proceeding was closed, the Commission was unable to request information regarding the number of small business within each of these services or the number of small business that would be affected by this action. We have, however, made estimates based on our knowledge about applications that have been submitted in the past. To the extent that a government entity may be a licensee or an applicant, the impact on those entities is included in the estimates for small businesses below.

Under the new rules adopted in the <u>Report and Order</u>, many radio services are categorically excluded from having to determine compliance with the new RF exposure limits. This exclusion is based on a determination that there is little potential for these services causing exposures in excess of the limits. Within the following services that are not categorically excluded in their entirety, many transmitting facilities are categorically excluded based on antenna location and power. These categorical exclusions significantly reduce the burden associated with these rules, and may reduce the impact of these rules on small businesses. Furthermore, the extension of the transition periods contained in this <u>First Memorandum Opinion and Order</u> will reduce the impact on applicants, particularly small businesses, by allowing them adequate time to understand the new requirements and ensure that their facilities are in compliance with them in a orderly and reasonable manner.

K. Amateur Radio Service Volunteer Examiner Coordinator (VECs)

In our original FRFA, we did not analyze the possible impact and burden on Amateur Radio Service (ARS) VECs. The ARRL has commented that our original FRFA is flawed because it fails to address the impact of the rules on small business entities such as itself and one other VEC.[21] The Commission has not developed a definition for a small business or small organization that is applicable for VECs. The RFA defines the term "small organization" as meaning "any not-for-profit enterprise which is independently owned and operated and is not dominant in its field..."[22] Our rules do not specify the nature of the entity that may act as a VEC.[23] However, all of the sixteen VEC organizations would appear to meet the RFA definition for small organization. Consequently, we have now analyzed the burden associated with this action on VECs.

The VECs coordinate the activities of the VEs who prepare and administer the Commission's amateur operator license examination system. The administering VEs prepare written examinations using questions drawn from common question pools.[24] The VEs also prepare the questions for the question pools which are maintained by the VECs. The questions in the pools are updated and revised periodically. In the Report and Order, we required that new examination questions on RF safety be added to the examinations. That requirement was made effective immediately. In response to the Report and Order, the ARRL filed a petition requesting that we allow the examinations to be modified according to the VECs' normal revision schedule. We are adopting such an implementation plan into this First Memorandum Opinion and Order. As a result, the VECs can proceed with their normal schedule for soliciting questions from the VEs and revising the question pools. The VECs, therefore, will have a minimum burden in meeting the new requirements.

* * * * *

In this First Memorandum Opinion and Order, we have also taken the following additional steps to reduce the burden on small businesses and organizations:

1. We extended the transition period for station applicants to come into compliance with the new requirements. This will give licensees, and applicants for new stations many of which may be small businesses, more time to learn the nature of the new requirements, make studies to determine whether they comply, and take steps to come into compliance if necessary.

2. We decided to permit the required changes in the ARS examinations to be made as the examinations are being routinely revised. This ensures that a minimal burden is put on the small organizations acting as VECs.

Report to Congress: The Commission shall send a copy of this Final Regulatory Flexibility Analysis, along with this Report and Order, in a report to Congress pursuant to the Small Business Regulatory Enforcement Fairness Act of 1996, 5 U.S.C. §801(a)(1)(A). A copy of this FRFA will also be published in the Federal Register.

NOTES

[1] National Environmental Policy Act of 1969, 42 U.S.C. Section 4321, et seq.

[2] See 47 CFR §1.1201, et seq.

[3] See Report and Order, GEN Docket No. 79-144, 100 FCC 2d 543 (1985); Memorandum Opinion and Order, 58 RR 2d 1128 (1985).

[4] See Report and Order, ET Docket 93-62, released August 1, 1996, FCC 96-326, 61 FR 41006 (August 7, 1996).

[5] We applied the new guidelines immediately to applications for equipment authorization for portable, mobile, and unlicensed devices, as described in the amended 47 CFR §1.1307(b)(2). In addition, we amended our rules to require, immediately, that amateur operator license examination questions pools include questions concerning RF safety at amateur stations. See Report and Order at para. 163 and the amended 47 CFR §97.503.

[6] See Report and Order at para. 112.

[7] ARRL Petition at 16, Ameritech Petition at 4-6, AT&T Petition at 1-2, BellSouth Petition at 5, PageNet Petition at 5, PCIA Petition at 11, U S WEST Petition at 3-5.

[8] Arch Comments at 3, CTIA Comments at 1, NAB Comments at 3, PageMart Reply at 3.

[9] Ad-hoc Association Late-Filed Reply Comments, filed October 28, 1996, at 1-2; Cell Phone Taskforce Opposition at 2 and Reply at 6; Brooklyn Green Party Reply at 1; Alan Golden Reply at 2; Dawn Mason Reply at 2.

[10] Ameritech Petition at 4-6, AT&T Petition at 1-2, CTIA Comments at 1, NAB Comments at 3, PageMart Reply Comments at 3, PageNet Petition at 5, PCIA Petition at 10-14, and U S WEST Petition at 3-5. OST Bulletin No. 65, "Evaluating Compliance with FCC-Specified Guidelines for Human Exposure to Radiofrequency Radiation", was last updated in October, 1985. In the Report and Order, we indicated that it was our intention to issue an updated OST Bulletin 65 (which would be renamed OET Bulletin 65) in the near future. See Report and Order at para 114. The revised Bulletin 65 will contain detailed information on evaluating compliance and would provide significant assistance to those attempting to comply with the new guidelines. Our staff prepared a draft revision of Bulletin 65 and, in mid-October, solicited comments on the draft from individuals and organizations who are active and knowledgeable in this area. Numerous comments have been received and our staff is finalizing the revised bulletin based on the comments. We intend to release a revised Bulletin 65 shortly after we address the other outstanding issues in this proceeding.

[11] BellSouth at 5.

[12] ARRL "Motion for Extension of Effective Date of Rules," filed on November 7, 1996, at 1-6.

[13] PageNet at 5.

[14] See "Emergency Motion for Extension of Effective Date of Rules" filed on August 12, 1996, by the ARRL.

[15] See amended 47 CFR §97.503(c)(10).

[16] See amended 47 CFR §97.503(b).

[17] See 47 CFR §97.523, which requires that each question pool contain at least 10 times the number of questions required for a single examination.

[18] The ARRL points out that we failed to consider, in our Regulatory Flexibility Act analysis contained in the Report and Order, the impact that the application of the new rules would have on small business entities, such as the ARRL and other Volunteer Examiner Coordinators (VECs). We have analyzed the burden associated with the administration of the new examination requirements, and are adopting a Revised Final Regulatory Flexibility Analysis, as shown in Appendix C.

[19] The revised Bulletin 65 will contain sufficient information for amateur operators to begin to evaluate whether their stations comply with our requirements. The supplement, as currently envisioned, would contain additional information on specific amateur antennas and configurations that could make these evaluations more simple.

[20] See Release of Question Pools into Public Domain, from Question Pool Committee, National Conference of Volunteer Examiner Coordinators, to volunteer-examiner coordinators, volunteer examiners, publishers, amateur radio media and other interested parties, dated December 1. 1996.

[21] The ARRL/VEC and the W5YI-VEC are components of organizations that publish materials marketed to persons for the purpose of preparing for passing the examinations required for the grant of an amateur operator license. This publishing activity, however, is separate from their VEC activity.

[22] 5 U.S.C. §601(4)

[23] Our rules, however, require that a VEC be an organization that has entered into a written agreement with the FCC to coordinate the examinations for amateur operator licenses. The examinations are prepared and administered by tens of thousands of amateur operators who serve as VEs. The VEC organization must exist for the purpose of furthering the amateur service, be capable of serving as a VEC in at least one of the thirteen VEC regions, agree to coordinate the examinations, agree to assure that every examinee is registered without regard to race, sex, religion, national origin or membership in any amateur service organization, and cooperate in maintaining the question pools for the VEs. See 47 CFR §§97.521 and 97.523, which outline the qualifications for VECs and question pools.

[24] See 47 CFR §97.507, which outlines the requirements for preparing examinations for an amateur operator license.

ARRL's Condensation of ET Docket 93-62
Second Memorandum Opinion and Order and
Notice of Proposed Rulemaking (FCC 97-303)
August 25, 1997

Note: The implementation date for these rules has been changed a number of times throughout the reconsideration period. Although the requirements were actually effective immediately when the Report and Order was released in 1996, the FCC established a "transition period" that gave certain stations time to complete any required station evaluation and comply with the rules. During this transition period, the provisions of the old RF-exposure rules apply.

*As originally described in paragraph 112 of the Report and Order, the implementation date was **January 1, 1997**. ARRL immediately petitioned the FCC, asking for a more reasonable amount of time for amateurs to comply with these rules. In paragraph 10 of the First MO&O, the FCC extended the transition period for the Amateur Radio Service to **January 1, 1998**. This was reiterated in paragraph 4 of the Second MO&O.*

*The Second MO&O, however, added an additional provision that gave existing stations, including amateur, a date certain of **September 1, 2000** as a transition period. The FCC has clarified, however, that new stations, or stations that renew or modify their licenses after **January 1, 1998**, must be in compliance as certified on the 610 application.*

Amateur stations, however, are not subject to prior approval before authorization. Existing stations that make changes to their station that could affect the RF exposure from that station must evaluate those changes before the new configuration is put into use. See 47 CFR §1.1312.—Ed.

Before the
FEDERAL COMMUNICATIONS COMMISSION   FCC 97-303
Washington, D.C. 20554

In the Matter of	)
	)
Procedures for Reviewing Requests for	) WT Docket No. 97-197
Relief From State and Local Regulations	)
Pursuant to Section 332(c)(7)(B)(v) of the	)
Communications Act of 1934	)
	)
Guidelines for Evaluating the Environmental	) ET Docket No. 93-62
Effects of Radiofrequency Radiation	)
	)
Petition for Rulemaking of the Cellular	)
Telecommunications Industry Association	) RM-8577
Concerning Amendment of the Commission's	)
Rules to Preempt State and Local Regulation	)
of Commercial Mobile Radio Service	)
Transmitting Facilities	)

SECOND MEMORANDUM OPINION AND ORDER
AND
NOTICE OF PROPOSED RULEMAKING

Adopted: August 25, 1997   Released: August 25, 1997
Comment Date (WT Docket No. 97-197): **October 9, 1997**
Reply Comment Date (WT Docket No. 97-197): **October 24, 1997**

TABLE OF CONTENTS

This condensation does not contain all items listed in this table of contents.–Ed.

I. INTRODUCTION

1. By this action, we are adopting a *Second Memorandum Opinion and Order* in ET Docket No. 93-62, responding to petitions and amending certain aspects of our guidelines for evaluating the environmental effects of radiofrequency (RF) emissions produced by FCC-regulated transmitters. We are also adopting a *Notice of Proposed Rulemaking* in WT Docket No. 97-197, opening a new proceeding to establish procedures for filing and reviewing requests for relief from state or local regulations based directly or indirectly on the environmental effects of RF emissions.

II. SECOND MEMORANDUM OPINION AND ORDER

A. Introduction and Executive Summary

2. In this *Second Memorandum Opinion and Order*, we are amending our rules to refine and clarify the decisions adopted in the *Report and Order* in ET Docket No. 93-62 regarding the use of new guidelines and methods in the evaluation of the environmental effects of RF electromagnetic fields or emissions produced by FCC-regulated transmitters. This *Second Memorandum Opinion and Order* responds to petitions for reconsideration and/or clarification filed in this proceeding. In reaching our decisions, we have considered carefully the petitions and comments that were received in this proceeding. We believe our decisions provide a proper balance between the need to protect the public and workers from exposure to potentially harmful RF electromagnetic fields and the requirement that industry be allowed to provide telecommunications services to the public in the most efficient and practical manner possible. Specifically, we are: 1) affirming the RF exposure limits that were previously adopted; 2) modifying in a few areas our policy that categorically excludes certain transmitters from routine environmental evaluation; 3) revising and clarifying our guidelines regarding RF emissions involving multiple transmitter operating at one site; and 4) modifying our rules to extend the initial transition period to October 15, 1997, and to require that all existing facilities be brought into compliance with our new guidelines within three years (by September 1, 2000). We are also adopting a number of minor changes and clarifications.

Note: The transition date for the Amateur Radio Service is January 1, 1998. — Ed.

3. In the *Report and Order*, the Commission adopted limits for Maximum Permissible Exposure (MPE) and localized, partial-body exposure of humans based on criteria published by the National Council on Radiation Protection and Measurements (NCRP) and by the American National Standards Institute/Institute of Electrical and Electronics Engineers, Inc. (ANSI/IEEE). The *Report and Order* also modified the Commission's policy on categorical exclusions that exempts many radio services and transmitters from routine environmental evaluation for RF exposure. In accordance with Section 704 of the Telecommunications Act of 1996, the *Report and Order* followed Congressional direction with respect to completion of the docket in this proceeding. The new rules became effective immediately; however, a transition period (originally to January 1, 1997) was provided for implementation of the new requirements for transmitters other than portable and mobile devices.

4. Several technical and legal issues were raised in the petitions. A *First Memorandum Opinion and Order*, adopted on December 23, 1996, addressed comments in those petitions requesting extension of the transition provisions of the *Report and Order* and extended the transition period to September 1, 1997 (January 1, 1998 for the Amateur Radio Service, only). This *Second Memorandum Opinion and Order* addresses the other issues raised in the petitions, including whether we should: (1) reconsider the RF exposure limits originally adopted; (2) reconsider our policy on categorical exclusion of certain transmitters from routine evaluation for compliance with our guidelines; (3) modify our policy with respect to evaluation of RF exposure at multiple transmitter sites; (4) revise our policy with respect to routine evaluation for SMR transmitters; and (5) broaden our authority to preempt state and local regulations concerning RF exposure.

5. Some petitioners ask that we reconsider our previous decision not to adopt ANSI/IEEE C95.1-1992 in its entirety. Several other petitioners claim that the limits we adopted were not protective enough. The staff believes that no new and compelling justifications have been provided that would warrant a modification of the limits adopted in the *Report and Order*. Those limits were crafted to address concerns about ANSI/IEEE C95.1-1992 that had been raised by several agencies of the Federal Government with responsibility for health and safety. Furthermore, all of these agencies have written letters to the Commission supporting our new guidelines. We believe that the limits adopted in the *Report and Order* provide a proper balance between the need to protect the public and workers from exposure to excessive RF electromagnetic fields and the need to allow communications services to readily address growing marketplace demands.

6. The Commission's environmental rules identify particular categories of existing or proposed transmitters or facilities for which licensees and applicants are required to conduct routine environmental evaluations to determine whether these transmitters or facilities comply with our RF guidelines. Other transmitting facilities are categorically excluded from these rules because we have judged them to offer little potential for causing exposures in excess of the applicable guidelines. In the *Report and Order*, we revised our rules related to this policy of categorical exclusion based on our own calculations and analyses of the implications of the new limits, along with information and data acquired during the proceeding. Whereas previously we had categorically excluded entire service categories, such as paging and cellular transmitters, the *Report and Order* concluded that some transmitting facilities, regardless of service, may offer the potential for causing exposures in excess of MPE limits.

7. Several petitioners ask that we return to our earlier policy of categorical exclusion for entire services. However, these petitioners present no new evidence that would lead us to change our basic premise for categorical exclusion. We continue to believe that it is desirable and appropriate to categorically exclude from routine environmental evaluation only those transmitting facilities that offer little or no potential for exposure in excess of our limits. However, some transmitting facilities, regardless of service, offer the potential for causing exposures in excess of MPE limits because of such factors as their relatively high operating power, location or relative accessibility, and these facilities should not be categorically excluded from routine evaluation.

8. Except in a few limited areas, we do not believe it is appropriate to modify the categorical exclusion policies adopted in the *Report and Order*. We are modifying our policy related to unlicensed millimeter-wave devices that do not meet the definition of a portable device and unlicensed and licensed PCS and other mobile devices operating above 1.5 GHz. Secondly, we are revising the 50-watt threshold for routine evaluation of amateur radio stations so that it reflects the manner in which the RF exposure limits change in the different amateur frequency bands. We are also revising categorical exclusions currently

based on the height of the antenna radiation center above ground so that they are based on the height of the lowest portion of the antenna above ground. In addition to these areas, we are revising our policy on categorical exclusions for SMR transmitters so that all SMR operations are covered, and we are changing our definition of "rooftop" so that antennas that are mounted on the sides of buildings or otherwise don't fit the previous definition will be considered, if appropriate.

9. Several petitioners argue that our policy regarding evaluation at sites with multiple FCC-regulated transmitters is overly burdensome. Our rules state that when the RF exposure limits are exceeded in an accessible area due to the RF fields of multiple fixed transmitters, actions necessary to bring the area into compliance are the shared responsibility of all licensees whose transmitters produce power densities in excess of 1% of the exposure limit applicable to their transmitter. After considering the various arguments, we conclude that the 1% level should be changed. We concur that a 1% level is difficult to measure or calculate. We believe that a 5% threshold represents a more reasonable and supportable compromise, by offering relief to relatively low-powered site occupants who do not contribute significantly to areas of non-compliance and, at the same time, by providing for the appropriate allocation of responsibility among major site emitters.

10. Some petitioners request that the Commission broaden its preemptive authority beyond the category of "personal wireless services" authorized in the Telecommunications Act of 1996. Based upon the current record in this proceeding, we find that there is insufficient evidence at this time to warrant our preempting state and local actions that are based on concerns over RF emissions for services other than those defined by Congress as "personal wireless services." However, additional issues concerning preemption of state and local regulations involving advanced television facilities have been raised in a Petition for Further Rulemaking filed by the National Association of Broadcasters which will be considered in a separate proceeding.

11. Several additional petitions were received in response to our earlier *First Memorandum Opinion and Order* extending the transition period for fixed stations and transmitters. Some petitioners request that we end the transition period immediately because of the potential for large scale exposure of the public to harmful RF emissions. Others argue that additional time is needed to consider the Commission's response to earlier petitions relating to OET Bulletin 65 on RF compliance. This bulletin will be released simultaneously with this Order. In order to provide applicants and licensees with sufficient time to review the final version of the bulletin, we will extend the initial transition period to October 15, 1997. The transition period for the Amateur Radio Service, only, will remain the same, and will end on January 1, 1998.

12. Finally, we are revising our rules to require that existing sites and transmitters come into compliance with the new guidelines as of a date certain. Accordingly, we will require all existing facilities, operations and devices to comply with the new FCC RF guidelines no later than September 1, 2000.

B. Background

13. The National Environmental Policy Act of 1969 (NEPA) requires agencies of the Federal Government to evaluate the effects of their actions on the quality of the human environment.[1] To meet its responsibilities under NEPA, the Commission has adopted requirements for evaluating the environmental impact of its actions.[2] One of several environmental factors addressed by these requirements is human exposure to RF energy emitted by FCC-regulated transmitters and facilities.

14. The Commission's environmental processing rules, 47 C.F.R. §§1.1301-1.1319, generally require an applicant to perform the necessary analysis (e.g., calculations and/or measurements) to ascertain whether a particular transmitting facility or device complies with the Commission's adopted RF exposure guidelines set forth in Section 1.1307(b), in effect at the time the applicant files for an initial construction permit, license, or renewal or modification of an existing license. If on the basis of the applicant's analysis the applicant determines that the facility complies (or will comply) with the Commission's adopted RF guidelines, the applicant certifies compliance as part of its application. If, on the other hand, the applicant determines that operation of the facility or device will not comply with the RF guidelines, the applicant is required to prepare an Environmental Assessment, and undergo environmental review by Commission staff unless the applicant amends its application so as to comply with the Commission's adopted RF guidelines. See 47 C.F.R. §§1.1311; see also 47 C.F.R. §§1.1308, 1.1309, 1.1314-1.1317.

15. If no pre-construction Commission authorization is required (as is the case for PCS and cellular licenses, for example, where the Commission authorizes blanket licenses that are not site-specific), Section 1.1312 of the Commission's environmental processing rules requires that the licensee conduct the appropriate calculations and determine whether the facility will comply with the Commission's adopted RF guidelines in effect at that time (i.e., at the pre-construction, not the initial application, stage) prior to the commencement of construction, rather than prior to licensing under the Commission's general environmental processing scheme. The processing requirements remain the same — if the calculations indicate compliance with the RF guidelines, the licensee may proceed with construction; if the calculations indicate non-compliance, the licensee will either modify its proposal to ensure compliance or submit an Environmental Assessment and undergo Commission environmental review prior to construction. The only difference lies in the timing: environmental calculations must take place prior to construction rather than prior to the applicable licensing.

16. Finally, it should be noted that if the facility or device has been categorically excluded from environmental processing requirements with respect to the RF exposure guidelines based on the Commission's prior determination that the operation of such facility or device, individually or cumulatively, will not exceed the Commission's adopted RF exposure limits, the applicant or licensee is exempt from the requirement of performing any calculations and/or measurements to determine whether there is compliance; the Commission presumes that the operation of a categorically excluded facility or equipment is in compliance.

17. In 1985, the Commission adopted a 1982 American National Standards Institute (ANSI) standard for use in evaluating the effects of RF electromagnetic fields on the environment, noting that the ANSI standard was widely accepted and was technically and scientifically supportable.[3] In 1992, ANSI adopted a new standard for RF exposure, designated ANSI/IEEE C95.1-1992, to replace its 1982 standard.[4] This new standard contained a number of significant differences from the 1982 ANSI standard and, in some respects, was more restrictive in the amount of environmental RF exposure permitted. On April 8, 1993, the Commission issued the *Notice of Proposed Rule Making* (*Notice*) in this proceeding to consider amending and updating the guidelines and methods used by the Commission for evaluating the environmental effects of RF electromagnetic fields.[5] In the *Notice*, we proposed to base our regulations on the ANSI/IEEE C95.1-1992 standard instead of the 1982 ANSI stan-

dard. More than 100 parties, including telecommunications organizations, other Federal Government agencies, state and local authorities, and individuals, submitted comments in response to the *Notice*.

18. On August 1, 1996, we adopted the *Report and Order* in this proceeding amending our rules to provide for the use of new guidelines and methods in the evaluation of the environmental effects of RF electromagnetic fields produced by FCC-regulated transmitters.[6] Seventeen petitions for reconsideration and/or clarification were filed in response to the *Report and Order*. A list of those organizations and individuals filing petitions, as well as those filing oppositions and replies to the petitions, can be found in Appendix B. Several technical and legal issues have been raised in the petitions. In the *First Memorandum Opinion and Order* in this proceeding, we addressed those petitions, motions, and comments that requested extensions of the transition periods adopted in the *Report and Order*.[7] This *Second Memorandum Opinion and Order* addresses the other issues that were raised in the petitions and comments.

C. Discussion

1. RF Exposure Limits

19. In the *Notice* in this proceeding, we proposed to base our RF exposure guidelines on limits for RF exposure contained in the ANSI/IEEE C95.1-1992 standard. However, comments filed in this proceeding from federal health and safety agencies, notably the U.S. Environmental Protection Agency (EPA) and the U.S. Food and Drug Administration (FDA), raised questions about certain aspects of those limits and recommended against the adoption of the entire ANSI/IEEE C95.1-1992 standard. After careful consideration of those views as well as the views of those commenters who opposed the federal agencies' views, we decided to adopt guidelines and limits that are generally based on elements of the exposure criteria recommended by the National Council on Radiation Protection and Measurements (NCRP) as well as those contained in the ANSI/IEEE C95.1-1992 standard.[8]

22. The DOD and the Hewlett-Packard Company (HP) state that our decision to adopt RF exposure limits that differ significantly from those initially proposed in the *Notice*, without issuing a second Notice of Proposed Rule Making allowing comment, does not appear to conform to Section 553(b) of the Administrative Procedure Act (APA).[9] The American Radio Relay League, Inc. (ARRL) also claims that we violated provisions of the APA in adopting the *Report and Order*.[10] DOD says that our decision was made in "an unnecessarily closed and narrow-focused" process, and denied interested parties with safety and health responsibilities, such as DOD, an opportunity to evaluate a draft decision and present comments. DOD also alleges that our decision did not receive adequate coordination with all federal agencies or departments having responsibility for RF safety and health. The ARRL argues that our *Notice* in this proceeding was faulty in that it failed to identify the nature of the rules to be adopted and did not adequately apprise radio amateurs of the obligations that would be placed on them in the *Report and Order*.

29. Decision. We reaffirm our decision to adopt exposure limits for field strength and power density based on recommendations contained in NCRP Report No. 86 and ANSI/IEEE C95.1-1992. We continue to believe that these RF exposure limits provide a proper balance between the need to protect the public and workers from exposure to excessive RF electromagnetic fields and the need to allow communications services to readily address growing marketplace demands.

30. We appreciate the views of some petitioners that we should

have adopted all provisions of the ANSI/IEEE C95.1-1992 standard. However, as discussed in our *Report and Order*, although most commenting parties generally supported our proposal to adopt the ANSI/IEEE C95.1-1992 standard, certain agencies of the Federal Government with oversight responsibilities for safety and health objected to the use of certain aspects of this standard.[11] In the past, the Commission has stressed repeatedly that it is not a health and safety agency and would give great weight to the judgment of these expert agencies with respect to determining appropriate levels of safe exposure to RF electromagnetic fields.[12] The guidelines and rules we adopted in the *Report and Order* addressed the concerns raised by the health and safety agencies and, at the same time, contained limits that over a wide frequency range are based on those recommended in the ANSI/IEEE C95.1-1992 standard.

31. As for claims that our guidelines are not protective enough, we reiterate that these guidelines are based on recommendations of expert organizations and federal agencies with responsibilities for health and safety. It would be impracticable for us to independently evaluate the significance of studies purporting to show biological effects, determine if such effects constitute a safety hazard, and then adopt stricter standards that those advocated by federal health and safety agencies. This is especially true for such controversial issues as non-thermal effects and whether certain individuals might be "hypersensitive" or "electrosensitive."

32. Concerning objections that our guidelines are not scientifically-based or technically sound, we note that our guidelines are based on recommendations of both the ANSI/IEEE C95.1-1992 standard and the NCRP exposure criteria. Both of these organizations are internationally recognized for their expertise in this area, and there is little evidence to support a claim that these guidelines are not based on science. In fact, both the ANSI/IEEE and NCRP guidelines are based on the same threshold for potentially hazardous whole-body exposure.[13] We recognize that ongoing research in a number of areas may ultimately result in changes in the fundamental understandings upon which ANSI/IEEE C95.1-1992 and the NCRP Report No. 86 are based. Both the IEEE and the NCRP have committees that are working on revisions of their respective exposure guidelines. As indicated in the *Report and Order*, we encourage these organizations and other similar groups developing exposure criteria to work together, along with the relevant federal agencies, to develop consistent, harmonized guidelines that will address the concerns and issues raised in this proceeding. We will, of course, consider amending our rules at any appropriate time if these groups conclude that such action is desirable.

33. Regarding the criticism from the Ad-hoc Association over our failure to adopt the NCRP's clause related to carrier modulation, we reiterate our previous conclusion that there is insufficient evidence to give special consideration to modulation effects.[14] Since we have no specific indication of exposure hazards related to modulation caused by FCC-regulated transmitters, and since at this time no new proof of such hazards has been presented by petitioners, we continue to believe that it would be premature to adopt the NCRP modulation criteria. However, we will evaluate and consider any new evidence relating to modulation effects this is submitted to us in the future.

35. As noted previously, DOD, HP and the ARRL allege that we did not comply with provisions of the APA in adopting guidelines different than those originally proposed. However, we point out that our *Notice* incorporated a prominent discussion and request for comment on whether we should adopt alternative guidelines from those that were the principal focus of our proposal.[15] This discussion specifically mentioned the MPE

limits recommended by the NCRP which, along with ANSI/ IEEE C95.1-1992, formed the basis for the limits we adopted in the *Report and Order*. Similarly, we indicated in the *Notice* that our categorical exclusions, such as previously applied to all amateur radio stations, would be reviewed in light of the new guidelines.[16] We believe that the final rules that were adopted were a "logical outgrowth" of that proposed in the *Notice*. *See American Water Works Ass'n. v. EPA*, 40 F. 3d 1266, 1274 (D.C. Cir. 1994). The Courts have generally ruled that "[A] final rule may properly differ from a proposed rule ... when the record evidence warrants the change." *See United Steelworkers of America v. Marshall*, 647 F.2d 1189, 1221 (D.C. Cir.), *cert. denied*, 453 U.S. 913 (1980). A final rule is not a logical outgrowth of a proposed rule generally "when the changes are so major that the original notice did not adequately frame the subject for discussion." *Connecticut Light and Power Co. v. Nuclear Regulatory Commission*, 673 F.2d 525, 533 (D.C. Cir.), *cert. denied*, 459 U.S. 835 (1982). Given that the *Notice* raised the issues of whether an alternative guideline such as that recommended by NCRP should be adopted and whether the categorical exclusions should be changed, as well as the substantial discussion of the issues in the comments in this proceeding, we conclude that the notice and comment provisions of the APA were followed and that a further Notice on these issues is unnecessary.

39. In summary, in considering the arguments raised with respect to the RF exposure limits adopted in the *Report and Order*, we place special emphasis on the recommendations and comments of federal health and safety agencies because of their expertise and responsibilities with regard to health and safety matters. In the *Report and Order*, we adopted RF exposure limits that addressed specific safety concerns raised by these agencies about the limits we had originally proposed to adopt. We do not believe that the petitioners and commenters have provided reasonable alternatives that similarly would adequately address these safety concerns. Accordingly, we conclude that the RF exposure limits adopted in the *Report and Order* are appropriate because they address those concerns and, at the same time, allow applicants and licensees to meet the growing marketplace demand for communications services.

2. Categorical Exclusions

40. Our rules identify particular categories of existing and proposed transmitting facilities for which licensees and applicants are required to conduct an initial, routine environmental evaluation to determine whether these transmitting facilities comply with our RF guidelines.[17] *See* 47 CFR §1.1307(b)(1). Our rules also identify certain types of mobile and portable transmitting devices that are subject to routine environmental evaluation prior to equipment authorization. *See* 47 CFR §§2.1091(c) and 2.1093(c). As for transmitting facilities and devices not specifically identified under 47 CFR §§1.1307(b)(1), 2.1091(c) or 2.1093(c), we have determined, based on calculations, measurement data, and other information, that such transmitting facilities offer little potential for causing exposure in excess of the applicable guidelines, and thus have "categorically excluded" those transmitters from the initial, routine environmental evaluation requirement.[18]

41. In the *Report and Order*, we revised our RF exposure rules to require routine evaluation of certain transmitting facilities that were previously categorically excluded from performing routine evaluation. These revisions were based on our own calculations and analyses of the implications of the new limits, along with information and data acquired in the record of this proceeding and from other sources. We attempted to bring con-

sistency to the categorical exclusions, by adopting power, antenna height, and transmitter site criteria that would apply across similar services.

45. Decision. After considering the arguments raised by the petitioners, we generally are maintaining the categorical exclusions adopted in the *Report and Order*, except with respect to modifying Table 1 of Section 1.1310 regarding unlicensed PCS and millimeter wave devices, and categorical exclusions based on the height of the antenna "radiation center" above ground level, as discussed below (and with respect to amateur radio stations, as discussed later). We continue to believe that it is desirable and appropriate to categorically exclude from routine evaluation only those transmitting facilities that we have reason to believe offer little or no potential for exposure in excess of our limits. We believe that our revised categorical exclusions meet this objective.

3. Amateur Radio Service (ARS)

53. Historically, all licensees and applicants in the ARS have been categorically excluded from performing routine environmental evaluations for compliance with our RF exposure guidelines. In the *Report and Order*, however, we concluded that there was a potential for amateur stations to cause RF exposure that would exceed our new limits. Accordingly, we decided to require amateur station licensees to: 1) conduct a routine environmental evaluation if they transmit using more than 50 watts; 2) take action to prevent human exposure to excessive RF electromagnetic fields if the routine environmental evaluation indicates that our limits could be exceeded; 3) demonstrate their knowledge of our guidelines through examinations; and 4) indicate in their applications for new licenses and renewals that they have read and understand our rules for limiting RF exposure.[19]

54. In its petition, the ARRL claims that the 50-watt threshold we adopted in the *Report and Order*, above which amateur radio operators must evaluate their stations, is arbitrary and inappropriate.[20] The ARRL points out that this threshold does not consider important factors, such as frequency, antenna height, antenna gain, emission mode, or duty cycle. The ARRL also notes that many other radio services, including some with higher duty cycles, are categorically excluded from performing routine evaluations even though they may operate with similar or higher power. The ARRL requests that the 50-watt threshold be modified to incorporate power levels contained in its petition, which vary by frequency, or else be increased to at least 150 watts transmitter power output if all parts of the antenna are located at least 10 meters from any area of uncontrolled exposure.

55. Alan Dixon, an amateur radio operator, maintains that it is burdensome and unnecessary for amateur radio operators to perform routine environmental evaluations and, when necessary, EAs.[21] Mr. Dixon states that the amateur radio community utilizes long-established customs of limiting duration of transmissions, using minimal power levels and establishing antenna installations which maximize propagation while inherently limiting unintended exposures. He believes that amateur operators should continue their traditional self-policing, free of "rigid overly-specific RF radiation parameters," given the "utter lack of evidence of detrimental effects thereby."

56. Decision. In the *Report and Order*, we noted that amateur stations can transmit with up to 1,500 watts peak envelope power on a wide range of frequency bands from 1,800 kHz to over 300 GHz. We also noted that amateur stations are not subject generally to restrictions on antenna gain, antenna placement, duty cycles, and other relevant exposure variables and, as a result, the possibility of human exposure to RF electromagnetic fields in excess of the guidelines could not be completely disre-

garded. Therefore, we came to the conclusion that a categorical exclusion for all amateur stations is not justified. We continue to believe that is the case. However, we now conclude that a uniform 50-watt categorical exclusion threshold, as adopted in the *Report and Order*, would cause many amateur station licensees to perform unnecessary routine environmental evaluations.

57. The ARRL is correct that our MPE limits are frequency dependent. Because amateur stations are permitted to transmit in frequency bands covering a wide range of frequencies, the MPE limits that might apply to any particular amateur station operation can vary dramatically.[22] The ARRL argues, quite correctly, that by applying a single power threshold above which a routine environmental evaluation must be performed, the variations that occur in the RF exposure limit as the station transmitter frequency changes are disregarded. The ARRL proposes, in its petition, that we scale the power threshold to match the RF exposure limit. We believe that this proposal makes sense for frequency bands above 10 MHz. However, on frequency bands below 10 MHz, persons are more likely to be located in the "near-field" of the amateur station antenna, where the field strength can vary dramatically in a very short distance.[23] In addition, a simple scaling of the power threshold to match the RF exposure limit below 10 MHz would result in extremely high-powered operations being permitted without any routine environmental evaluation. We believe that a flat 500-watt power threshold below 10 MHz is necessary to ensure that these high-powered amateur stations do not cause human exposure to excessive RF electromagnetic fields. Accordingly, we are adopting the ARRL's proposal by specifying a transmitter power threshold for each individual ARS frequency band. As indicated in the table shown in 47 CFR §97.13(c) of the revised rules, the power threshold for transmissions in the frequency bands below 10 MHz is 500 watts. We have also established this threshold for amateur repeater stations, which are normally located high above ground level and often at commercial sites, and we will base exclusions for these antennas on factors similar to those for paging and cellular antennas, as shown in the revised table, since their operation is similar. For frequency bands above 10 MHz, the power threshold varies from 50 watts to 450 watts. We believe the revised power thresholds for the ARS will eliminate burdensome and unnecessary requirements for most radio amateurs, and thus address the overall concerns raised by the ARRL and Mr. Dixon. These new thresholds, as well as some clarifying language we have added to 47 CFR §97.13(c), also help protect the public from excessive exposure to RF electromagnetic fields produced by ARS stations by requiring that their licensees perform routine environmental evaluations and take appropriate actions if they operate their station in a manner that could cause human exposure to RF electromagnetic fields above that permitted under our guidelines.

4. Compliance at Multiple Transmitter Sites

58. In our *Report and Order*, we generally retained our policies regarding the environmental evaluation of RF electromagnetic fields at sites with multiple FCC-regulated transmitters.[24] Our existing rules state that, when the RF exposure limits are exceeded in an accessible area due to the RF electromagnetic fields produced by multiple fixed transmitters, actions necessary to bring the area into compliance are the shared responsibility of all licensees whose transmitters produce fields at the non-complying area in excess of 1% of the exposure limits applicable to their transmitter.[25] The rules also state that applicants for proposed (not otherwise excluded) transmitters, facilities, or modifications that would cause non-compliance with our limits at an accessible area previously in compliance are responsible for submitting an EA if the emissions from the applicant's transmitter or facility would result in a field strength or power density at the non-complying area in excess of 1% of the exposure limit applicable to that transmitter or facility.[26] In the case of renewal applicants, a similar requirement applies — renewal applicants whose (not otherwise excluded) transmitters or facilities contribute field levels in excess of 1% of the applicable exposure limit at an accessible area must submit an EA if the area in question is not in compliance with the applicable RF guidelines.[27]

59. Several petitioners and commenters believe that the 1% level used as our threshold for determining responsibility at a non-complying area is too low. Arch, AT&T Wireless Services (AT&T), BellSouth Corporation (BellSouth), PageNet and PCIA all support raising this threshold from 1% to 10%.[28] However, this proposal is opposed by the Cellular Taskforce and others, who advocate increased regulation and scrutiny at multiple emitter sites.[29]

66. <u>Decision</u>. For the reasons set forth below, we are amending our rules to raise the responsibility threshold, above which licensees at multiple transmitter locations must share responsibility for addressing RF exposure non-compliance problems, from 1% to 5%. We believe that a 5% responsibility threshold will offer relief to relatively low-powered site occupants who do not contribute significantly to the non-compliance and, at the same time, provide for the appropriate allocation of responsibility among major site emitters. Similarly, we are raising the filing threshold that determines whether an applicant must file an EA if the applicant contributes to field levels at an area of non-compliance. We are raising the present threshold of 1% to 5%. Therefore, if an applicant's contribution to the area of non-compliance exceeds 5%, the applicant must file an EA. We are also modifying the language used in our rules somewhat to better explain what is required at multiple-user sites.

67. Our policy with respect to multiple transmitter sites was adopted several years ago and has essentially remained unchanged. The 1% responsibility and filing thresholds have not been seriously questioned until now. These new questions undoubtedly reflect the fact that we have now removed the categorical exclusions for a number of different transmitting facilities, and this has resulted in the necessity for evaluating many more multiple-transmitter situations than was the case previously. Many petitioners give valid reasons for modifying the 1% thresholds. First and foremost, we believe, is the issue of accuracy of determination of field contributions, either through measurements or calculations. BellSouth makes a good point when it notes the difficulties of making accurate determinations to the 1% level. We also see merit in the arguments that a threshold of 1% is too encompassing, particularly in light of the potential that an applicant or licensee could be required to undergo an unnecessary and expensive evaluation and that such a requirement could actually discourage co-location. However, we believe that changing the threshold to 10% goes too far in the other direction, and could lead to the creation of areas of non-compliance. It could also result in some transmitter operators escaping their responsibilities for compliance at multiple transmitter sites.

68. For example, consider the case of a multiple-transmitter site where most of the antennas are paging antennas operating at ERPs of 1000 W or greater. Often such sites involve numerous, densely packed antennas, especially in urban areas. At some points during the day, due to high traffic, most of the antennas may be transmitting almost simultaneously. If there is a compliance problem at such a site, many or most of the antennas may be contributing to the area of non-compliance but not necessarily at the 10% level. Calculations can be used to demonstrate

that non-complying areas are more likely to be the result of the contributions of several of these antennas, rather than just one or two. For this reason, it is important not to establish an exclusion threshold that is too high. On the other hand, as noted before, upon reconsideration, we agree that a level of 1% is unreasonable considering the problems of measurement and prediction accuracy and also the potential for unnecessary impact on small contributors. We believe that a 5% threshold represents a reasonable and supportable compromise, and are amending 47 CFR §1.1307(b)(3) accordingly.

69. We agree with Ameritech and AirTouch, and others, that further guidance is needed on how to address multiple transmitter situations. In general, we intend that our rules, along with the guidance given in a revised FCC bulletin on evaluating compliance, OET Bulletin 65, will be sufficiently clear and complete so that licensees can readily determine their compliance with our RF exposure requirements.[30] In adopting this *Second Memorandum Opinion and Order*, we are attempting to address those areas where parties have indicated that confusion may exist. We recognize, however, that additional questions are likely to arise over time, especially with regard to particular multiple-transmitter situations. We direct staff to work with the industry to address such questions that may arise, both through the revision of Bulletin 65 and in response to inquiries regarding specific situations.

70. The key trigger with respect to our RF exposure rules is the existence of an accessible area where RF field levels will exceed our MPE limits. As delineated in 47 CFR §1.1307(b)(3) as amended by this *Second Memorandum Opinion and Order*, responsibility is to be shared among those transmitter facilities contributing above the 5% threshold at a non-complying area. Since such situations can arise according to a variety of criteria, including transmitter power, antenna height, frequency and associated RF exposure limit, location of fencing to restrict access, etc., we can see no easy way to define a "site" or to specify some arbitrary radius around antennas at which compliance must be evaluated. However, we believe that it will not be difficult for most applicants to determine areas which are accessible. Applicants should be able to calculate, based on frequency, power, and antenna configuration, the distance from their transmitting antenna where their signal produces field levels equal to, or greater than, 5% of the relevant RF exposure limit. Applicants are then responsible for evaluating compliance in any accessible areas within this distance from their transmitting antenna.

71. In evaluating compliance in accessible areas, applicants are expected to make a good-faith effort to consider RF emissions from other nearby transmitters. However, we do not believe it is realistic, practical, or necessary for applicants to consider extremely weak signals that are not likely to present a significant risk for exposure in excess of our limits. Accordingly, applicants need only consider those RF emissions produced by nearby transmitting facilities that exceed 5% of their relevant RF exposure limit.[31] The percentages of the relevant RF exposure limits produced by each station are added, to determine whether the limits are (or would be) exceeded as a result of the RF emissions from the multiple transmitter facilities.[32] If the limits are exceeded, then the applicant and the other responsible parties must address the problem (or the applicant can file an EA).

73. We appreciate the arguments raised by the petitioners who advocate that site owners (rather than individual licensees) be responsible for determining and ensuring compliance with our RF exposure requirements. However, in an earlier decision regarding the streamlining of our antenna structure clearance procedure, we determined that responsibilities pertaining to RF electromagnetic fields properly belonged with our licensees and applicants, rather than with site owners. We agree with the concerns raised by Holly Fournier and Mary Beth Freeman that many site owners may not have the capability or understanding to make sure that transmitter facilities on their property are in compliance. Finally, since the area in which a licensee is responsible for addressing non-compliance problems (i.e., the contour within which the station's power density exceeds 5% of the relevant RF exposure limit) can extend for several meters from the transmitting antenna itself, it is conceivable that the accessible areas where our RF exposure limits are exceeded may involve multiple site owners or transmitting antennas located at other sites, making it difficult for a single site owner to ensure compliance.[33]

74. Nevertheless, we recognize that a site owner has significant control over applicants' and licensees' abilities to comply with our RF exposure requirements. For example, a site owner can determine whether a licensee will be permitted to erect a fence to limit public access in areas where the uncontrolled RF exposure limits may be exceeded. For sites where there are multiple licensees, the site owner also may be able to encourage the licensees to cooperate to find a common solution to problems caused by multiple transmitters. In addition, site owners may be able to take steps that would allow co-location of transmitting facilities. We believe that such co-location is highly desirable — it can reduce the number of locations at which the potential for RF exposure must be evaluated, and it can facilitate the ability of applicants to get through the state and local zoning approval processes. Accordingly, we urge site owners to allow applicants and licensees to take reasonable steps to comply with our RF exposure requirements and, where feasible, encourage co-location of transmitters and common solutions for controlling access to areas where our RF exposure limits might be exceeded.

75. In response to the questions posed by Ameritech, PCIA, and U S WEST regarding how the responsibility for compliance is to be shared at multiple transmitter sites, we do not intend to specify detailed instructions on how to allocate responsibility. One logical suggestion would be to assign compliance costs according to the percentage contributions at the non-complying area(s) for situations involving no change in transmitter facilities.[34] An alternative would be, as suggested by PCIA, to require an applicant for a new facility to resolve the problem. Section 1.1307(b)(3)(i) of our new rules states that it is the responsibility of a new applicant to submit an EA if their transmitter will create a non-complying situation at a location previously in compliance. However, we recognize that some particular circumstances may dictate different solutions. Accordingly, we encourage our licensees and applicants to work in a cooperative manner to address these problems. We note that, at most broadcast antenna farms, cooperative agreements have been developed to ensure compliance with applicable RF exposure guidelines. We see no reason why such agreements also cannot be used at other antenna sites. In response to the concern raised by Ameritech, we encourage any applicant or licensee to notify the appropriate Commission licensing bureau if the operator of a co-located transmitter will not cooperate in addressing a non-compliance problem. This has occurred in the past with respect to broadcast sites, and our staff, as needed, has encouraged the non-cooperating licensee to assist in correcting the problem when appropriate. Similarly, we encourage applicants to notify our licensing bureaus if they believe that existing licensees are not allowing them reasonable access to a site, or are attempting to place unreasonable financial burdens on them. In this regard, we emphasize that if a transmitter at a multiple-transmitter site

is approved under one set of guidelines but, later, another transmitter locates at the site and, as is required, operates under new exposure criteria, then the new criteria must be used to evaluate the entire site.

76. We are amending 47 CFR §1.1307(b)(1), as requested by PCIA, to clarify the meaning of the phrase "total power of all channels" in Table 1. PCIA is correct that the term "facility" used in this context refers to the co-located transmitters owned and operated by a single carrier and is not intended to apply to all other transmitters that may be co-located at an antenna farm or on a rooftop for purposes of exclusion from routine evaluation.

77. Finally, in reviewing the issues raised in the various petitions, we have found that the rules adopted in the *Report and Order* are imprecise with respect to how to calculate the 5% threshold of responsibility for addressing non-compliance situations. Our rules specify RF exposure limits in terms of electric field strength, magnetic field strength, and power density.[35] It is the square of the field strength or power density that is most relevant in determining the potential effect of RF emissions on the human body.[36] Therefore, we are modifying our rules to make it clear that the 5% threshold applies to the power density limit or to the square of electric or magnetic field strength limit.

5. Preemption of State and Local RF Regulations

88. Decision. Based upon the current record in this proceeding, we find that there is insufficient evidence at this time to warrant our preempting state and local actions that are based on concerns over RF emissions for services other than those defined by Congress as "personal wireless services."[37] We note that on May 30, 1997, the National Association of Broadcasters (NAB) and the Association of Maximum Service Television (MSTV) (jointly NAB/MSTV) filed a Petition for Further Notice of Proposed Rulemaking, urging preemption of certain state and local government restrictions on the siting of broadcast transmission facilities, based on petitioner's claims that unreasonable state and local regulations have frustrated the siting of broadcast facilities and could impede the Commission's scheduled conversion to the new digital television service. The NAB/MSTV petition, which raises additional preemption issues for broadcasting, will be addressed in a subsequent Commission action.

89. Concerning Ameritech's proposal that the Commission preempt state and local regulations concerning the *operation* of facilities based on RF-emission considerations, we agree with Ameritech that Congress did not intend to prevent the Commission from preempting state and local regulations concerning the operations of facilities simply by deleting the term "operation" from the final version of Section 332(c)(7). On the contrary, Congress made it clear, in the *Conference Report*, that enactment of Section 332(c)(7) of the Communications Act was not meant to affect the Commission's general authority to regulate the operation of radio facilities.[38] We find that the alternative reading is illogical and would render the statute useless and produce absurd results which Congress could not have intended. Therefore, we will continue to consider requests for relief of state and local government actions that prescribe or restrict the operation of personal wireless facilities pursuant to the authority granted to the Commission by Congress in Section 332(c)(7).

90. Regarding Ameritech's argument that the Commission should specify a federal rule of liability for torts related to RF emissions, we believe that such action is beyond the scope of this proceeding and we question whether such an action, which would preempt too broad a scope of legal actions, would otherwise be appropriate. Therefore, we cannot grant Ameritech's request.

7. Development of a Revised Version of OST Bulletin 65

100. Since 1985, the Commission has made available a technical publication designed for use by Commission licensees and applicants as an aid in evaluating compliance with our RF exposure guidelines. As mentioned previously, we are now updating this publication, OST Bulletin 65, to reflect our adoption of new guidelines.

101. Some of the petitioners and commenters express opinions and offer suggestions about our procedures for developing this document and for allowing review of the revised draft. Ameritech maintains that we should ensure that "all affected parties" are given an opportunity to participate in the formulation of the bulletin.[39] Ameritech points out that we will likely receive the most useful comments from those industry representatives who are faced with concrete compliance responsibilities and who may have a greater incentive to focus on the practical impact of the new guidelines. The EEA urges us to establish an "open consultative" process for revising and issuing any bulletins that are aimed at implementation of the new guidelines.[40] PageNet notes that the forthcoming bulletin is needed to clarify the new RF rules as issued in the *Report and Order*.[41] PCIA proposes that the revised Bulletin 65 be subject to public notice and comment procedures, arguing that this could highlight areas where guidance is needed by industry.[42]

102. Decision. It should be emphasized that the guidance provided in Bulletin 65 is not binding and cannot be construed as a substantive rule; rather the Bulletin merely provides information and interpretations that may be used in complying with our RF exposure guidelines. Other methods of determining compliance are acceptable so long as they are based on generally accepted scientific methods. In the introduction of the existing bulletin, we indicate that: 1) the bulletin is not designed to establish mandatory procedures; 2) the bulletin is meant to provide guidance and assistance in evaluating compliance; and 3) other methods and procedures for evaluating compliance may be acceptable if based on sound engineering practice.

103. In September, 1996, a draft of a revised Bulletin 65 was sent to approximately fifty outside reviewers for comment and suggestions. The reviewers included a broad spectrum of technical experts and representatives from government, industry and academia, and many of these individuals are affiliated with telecommunications entities regulated by the Commission. Many comments were received by late October. Our staff has reviewed these comments and incorporated many of them into the final bulletin. Any additional review would needlessly delay the release of this important document. Therefore, we will not grant requests made by PCIA and others for a more extensive period of public comment. We will, however, take under consideration the comments of PageNet and others regarding areas that need to be addressed in the bulletin. In addition, Bulletin 65 may be revised periodically based upon feedback and questions from industry and the public.

8. Miscellaneous Clarifications and Corrections

104. Since issuing our *Report and Order* in this proceeding, we have identified a few corrections and clarifications that need to be made to rule sections that were amended. We are hereby making these changes (see Appendix A) to our rules as follows:

(1) Paragraph (b)(1) of 47 CFR §1.1307 is modified to make it clear that both our MPE limits contained in 47 CFR §1.1310 and our SAR limits contained in 47 CFR §2.1093 generally apply, as appropriate, to all facilities, operations, and transmitters regulated by the Commission. The rule adopted in the *Report and Order* only made this specific statement with respect to MPE limits. This was an oversight, and a modification is being

made here to prevent possible confusion.

(2) Table 1 in paragraph (b)(1) of 47 CFR §1.1307 is modified to insert the words "ERP" that were inadvertently omitted from column 2 in the section of the table referencing evaluation criteria for Personal Communications Services in Part 24.

(3) We are amending our rules to make it clear that our categorical exclusions apply to transmitters mounted on the sides of buildings as well as those mounted on building roofs. Therefore, we are replacing the term "rooftop" with the term "building-mounted" in our rules for purposes of defining categorical exclusion. We believe that this change will remove possible confusion in the existing rules and will avoid potential situations where persons could be exposed to RF emissions in excess of our guidelines.

(4) Minor language changes have been made to the entry in Table 1 of Appendix A for Local Multipoint Distribution Service (LMDS) requirements (subpart L of part 101) to clearly reference the FCC adopted RF exposure limits in 47 CFR §1.1310.

(5) Paragraph (b)(4) of 47 CFR §1.1307 is modified to correct a typographical error.

(6) Paragraph (b) of 47 CFR §2.1091, which applies to mobile devices, excluded devices intended to be used in "fixed locations." However, the term "fixed locations" was not defined. There was a possibility that some parties might incorrectly assume that certain consumer devices, such as wireless transmitters attached to a computer, are not covered by this paragraph. Accordingly, a definition for "fixed location" has now been added. Language has also been added to this paragraph, and to paragraph (b) of 47 CFR §2.1093, to clarify our definitions of these devices and to make it clear that radiating "antenna" is intended to mean the "radiating structure" or structures of a mobile, unlicensed or portable device. We have also deleted the words "unlicensed devices" from the caption for Section 2.1091 to avoid confusion, since unlicensed devices can also be evaluated under 47 CFR §2.1093, if they are classified as a "portable" device.

(7) A new paragraph (d)(4) is added to 47 CFR §2.1091 to cover special cases where devices may not be easily classified as either mobile or portable. Examples would be modular or desktop transmitters. The wording in paragraph (d)(3) has also been modified to make it clear that warning labels and instructional materials may be used to attain compliance, if appropriate, for all devices covered by this rule part.

(8) Paragraph (d) of 47 CFR §2.1093 is modified to reflect the fact that evaluation for RF exposure due to portable devices in terms of specific absorption rate (SAR) is only valid in the frequency range of 100 kHz to 6 GHz and that evaluation of portable devices above 6 GHz should be in terms of compliance with MPE limits for power density. It is further stipulated that measurements or calculations for compliance can be made at a minimum distance of 5 cm from the transmitting source.

(9) The *Report and Order* failed to amend 47 CFR §26.51(d) and 47 CFR §26.52 that deal with RF hazards in the General Wireless Communications Service (GWCS). These sections have been changed to conform to the new guidelines, and a category for GWCS transmitters has been added to Table 1 in Appendix A. In addition, 47 CFR §2.1091 and 47 CFR §2.1093 have been amended to require evaluation of GWCS portable devices and mobile devices operating above 3 watts EIRP. Exclusion levels for non-mobile and non-portable GWCS transmitters have been established as 1640 watts EIRP, in conformance with the exclusion threshold established for the Wireless Communications Service authorized under Part 27 of the Commission's rules. This threshold is based on calculations of reasonable distances from antennas where individuals might be expected to approach an antenna and where exposures would likely exceed the MPE limits.

Since all of the above changes to the rules involve minor or merely technical clarifying amendments, additional public notice and comment on these changes, beyond that given in the original *Notice* are unnecessary pursuant to Section 553(b)(3)(B) of the Administrative Procedure Act.[43]

9. Petitions for Reconsideration of Transition Period Extension

105. The *First Memorandum Opinion and Order (First MO&O)* in this proceeding extended the transition period for implementing the FCC's policies and guidelines for RF compliance.[44] Additional petitions for reconsideration were submitted to the Commission in response to the *First MO&O*, in accordance with Section 1.429 of the Commission's rules [47 CFR §1.429(i)].[45] For various reasons, these petitioners request that we reconsider our decision on extending the transition period.

110. Decision. In our *First MO&O* in this proceeding we stated that we have no evidence that extending the transition period would have a significant adverse effect on public health.[46] We re-state that conclusion. The new RF exposure guidelines are in certain respects more restrictive than those they replace, particularly with respect to exposure of the general public. However, with regard to most of the personal wireless facilities that are the subject of the petitions of the Ad-hoc Association and the Cellular Phone Taskforce, there is ample evidence that most of these facilities will result in levels of exposure of the general public that are many times lower than our new guidelines.

111. As previously discussed in this *Order* and in the original *Report and Order* in this proceeding, we have relied on the advice and comments of the federal health and safety agencies as to what levels of RF exposure are protective of the public health. The Commission does not have the expertise to make independent judgements on such alleged health effects as "electrosensitivity" or other reported effects on human health. This is the responsibility of the federal health and safety agencies and other qualified public health organizations. Therefore, we continue to consider our new guidelines appropriately protective of public health. There is no evidence to suggest that transmitters or facilities that comply with our guidelines will cause adverse health effects. Our guidelines adopt the most conservative aspects of the ANSI/IEEE and the NCRP recommended exposure criteria and have been recommended by all of the relevant health and safety agencies. Moveover, we do not agree with the Ad-hoc Association and the Cellular Phone Taskforce that even a minimal extension of the initial transition period should be denied. We agree with Ameritech, Northeast, Airtouch and AT&T Wireless that a further extension is necessary to allow applicants and licensees sufficient time to analyze the newly revised version of OET Bulletin 65.

112. For these reasons we will agree to a limited further extension of the transition period to October 15, 1997. Since this *Order* and the revised Bulletin 65 will be issued at the same time, this will allow sufficient time for applicants and licensees to review these documents. Copies of this *Order* and the revised Bulletin 65 will be immediately available on the Commission's World Wide Web page (www.fcc.gov). We do not agree that there is a need for a period as long as eight months to one year beyond issuance of the final version of Bulletin 65. Ample time has already been given to applicants and licensees to begin considering compliance issues, and, as noted, a preliminary draft of

Bulletin 65 was made available to many outside reviewers several months ago. Therefore, the petitions of Ameritech and Northeast are partially granted.[47]

10. Treatment of Existing Facilities, Operations and Devices

113. Under the rules adopted in the *Report and Order* in this proceeding, as modified by the *First MO&O,* all applications to the FCC for construction permits, license renewals and requests for station modifications filed after September 1, 1997 are subject to analysis under our new RF exposure guidelines, whereas existing sites are required to come into compliance only at the time of renewal or modification. In our Order today, we extend the initial transition period under Section 1.1307(b)(4) for implementing the new RF exposure guidelines to October 15, 1997, and clarify that all new facilities constructed after that date must comply with the new guidelines, regardless of whether an application is filed with the Commission. Licensees filing applications for new facilities, renewals or modifications are also required to bring their operations into compliance with the new guidelines. We also revise our rules to require existing sites to come into compliance as of a date certain.

114. We are revising our rules because we believe that the health and safety concerns that underlie the adoption of our new guidelines warrant reconsideration of the ways we have applied these requirements in the past. Previously, our rules have been triggered by applications for new facilities, modifications to existing facilities, or renewals of existing licenses. Although this approach is appropriate for most of the broad range of environmental issues our rules were designed to address, we believe that a different approach is warranted in matters of RF exposure. Because of potential public heath and safety concerns, we adopted more conservative RF exposure guidelines based on the recommendations of the relevant federal health and safety agencies, and we will require all new facilities constructed after the effective date of this Order to comply with the new guidelines by a date certain.[48] We also believe this approach is consistent with Congressional intent underlying Section 704 of the Telecommunications Act of 1996, that the Commission's rules in this proceeding "contain adequate, appropriate and necessary levels of protection to the public."[49] We recognize that licensees require a reasonable amount of time to bring existing facilities into compliance due to the variety of different site configurations and settings. Accordingly, we will require all existing facilities to be brought into compliance with the new rules no later than September 1, 2000. If a licensee believes that its facility cannot be brought into compliance, the licensee must file an Environmental Assessment by this date.[50]

APPENDIX C

REVISED FINAL REGULATORY FLEXIBILITY ANALYSIS

Second Memorandum Opinion and Order

III. SUMMARY OF ISSUES RAISED REGARDING THE FINAL REGULATORY FLEXIBILITY ANALYSIS (FRFA) BY THE PETITIONS, MOTIONS, AND COMMENTS IN RESPONSE TO THE *REPORT AND ORDER*:

The American Radio Relay League, Inc., Paging Network, Inc., and the Personal Communications Industry Association raised concerns in their petitions, motions and comments regarding the FRFA that was associated with the *Report and Order*. Those concerns were addressed in the revised FRFA contained in the *First Memorandum Opinion and Order* in this proceeding.

This was addressed in Paragraph 2 of Appendix C of the Second MO&O: 2. We decided to permit the required changes in the ARS examinations to be made as the examinations are being routinely revised. This ensures that a minimal burden is put on the small organizations acting as VECs.

As further explained in Appendix C of the Second MO&O:

Reporting, Recordkeeping, and Other Compliance Requirements: The proposals under consideration in the *NPRM* include the possibility of imposing a new filing requirement for parties seeking relief pursuant to Section 332(c)(7)(B)(v) of the Communications Act. The filing requirement would be used to determine whether to grant relief from the State or local regulation in question. This filing will be in the form of a request for declaratory ruling filed pursuant to Section 1.2 of the Commission's Rules.[51] Only interested parties or those parties demonstrating the requisite standing will be permitted to participate in the proceeding. The *NPRM* also seeks comment on whether to adopt either a simple certification of compliance or more detailed demonstration of compliance that personal wireless service providers will be required to submit to State and local governments as evidence of RF emissions compliance.

We estimate that the average burden on the party seeking relief will be approximately two hours to prepare the request for relief and file it with the Commission. We estimate an equal amount of time for the State or local authority or other interested party (referred to jointly herein as the "respondents) to prepare and file their comments on and/or oppositions to the preemption request. We estimate that 75 percent of both the requesting parties and the respondents (which may include small businesses) will contract out the burden of preparing their filings. We estimate that it will take approximately one hour to coordinate information with those contractors. The remaining 25 percent of parties filing requests and respondents (which may include small businesses) are estimated to employ in-house staff to provide the information. We estimate that parties requesting relief and respondents that contract out the task of preparing their filings will use an attorney or engineer (average $200 per hour) to prepare the information.

We estimate that the average burden on the party required to prepare a simple certification of RF compliance to be less than one hour. We estimate that the average burden on the party required to prepare a more detailed demonstration of RF compliance to be approximately 5 hours. We estimate that 75 percent of these parties (which may include small businesses) will contract out the burden of prepare their filings. We estimate that it will take approximately 1 hour to coordinate information with those contractors. The remaining 25 percent of parties (which may include small businesses) are estimated to employ in-house staff to provide the information. We estimate that parties that contract out the task of preparing their filings will use an engineer (average $200 per hour) to prepare the information.

NOTES

[1] National Environmental Policy Act of 1969, 42 U.S.C. Section 4321, *et seq.*

[2] *See* 47 CFR §1.1301, *et seq.*

[3] *See Report and Order*, GEN Docket No. 79-144, 100 FCC 2d 543 (1985); *Memorandum Opinion and Order*, 58 RR 2d 1128 (1985); *see also* ANSI C95.1-1982, "American National Standard Safety Levels with Respect to Human Exposure to Radio Frequency Electromagnetic Fields, 300 kHz to 100 GHz," ANSI, New York, NY.

[4] ANSI/IEEE C95.1-1992, "Safety Levels with Respect to Human Exposure to Radio Frequency Electromagnetic Fields, 3 kHz to 300 GHz." This standard had been developed by the Institute of Electrical and Electronics Engineers, Inc. (IEEE), in 1991.

[5] *See Notice of Proposed Rule Making*, ET Docket No. 93-62, 8 FCC Rcd 2849 (1993); *see also* 8 FCC Rcd 5528 (1993), 9 FCC Rcd 985 (1993), 9 FCC Rcd 317 (1994), 9 FCC Rcd 989 (1994) extending the comment deadlines.

[6] *See Report and Order*, ET Docket 93-62, released August 1, 1996, FCC 96-326, 11 FCC Rcd 15123 (1997).

[7] *See First Memorandum Opinion and Order*, ET Docket 93-62, released December 24, 1996, FCC 96-487, 11 FCC Rcd 17512 (1997).

[8] *See Report and Order*, ET Docket No. 93-62, *supra.*, at paras. 12-34. We adopted Maximum Permissible Exposure (MPE) limits for electric and magnetic field strength and power density for transmitters operating at frequencies from 300 kHz to 100 GHz that are generally based on Sections 17.4.1 and 17.4.2, and the time-averaging provisions recommended in Sections 17.4.1.1 and 17.4.3, of "Biological Effects and Exposure Criteria for Radiofrequency Electromagnetic Fields," NCRP Report No. 86 (1986). With the exception of the limits on exposure to power density above 1500 MHz and the limits for exposure to lower frequency magnetic fields, these MPE limits are also generally based on the guidelines contained in Section 4.1 of ANSI/IEEE C95.1-1992. We also adopted limits for localized ("partial body") absorption for certain portable transmitting devices based on Sections 4.2.1 and 4.2.2 of ANSI/IEEE C95.1-1992 and Section 17.4.5 of NCRP Report No. 86.

[9] DOD Petition at 2-3, HP Petition at 3.

[10] ARRL Petition at 5-9.

[11] *See Report and Order* at paras. 15-20.

[12] *See, e.g., Report and Order*, GEN Docket 79-144, 100 FCC 2d 543 (1985), at para. 26 note 6 and *Report and Order*, ET Docket 93-62, *supra.*, at para. 28. *See also,* letter from Mark S. Fowler, Chairman, FCC, to Anne M. Burford, Administrator, EPA, February 22, 1983; letter from Dennis R. Patrick, Chairman, FCC, to Lee M. Thomas, Administrator, EPA, November 29, 1988; and letter from Thomas P. Stanley, Chief Engineer, FCC, to Ken Sexton, Director, Office of Health Research, Office of Research and Development, EPA, October 24, 1990.

[13] *See Report and Order* at Note 16.

[14] *See Report and Order* at para. 32.

[15] *See Notice of Proposed Rule Making* at paras. 23-25.

[16] *See Notice of Proposed Rule Making* at 19.

[17] If a routine evaluation is required, and if it is subsequently determined that the transmitting facility cannot be brought into compliance, the applicant or licensee is then required to submit to us a narrative statement known as an Environmental Assessment (EA). An EA describes why the transmitter or facility will not comply with the guidelines, and includes other pertinent information as is specified in our environmental rules. *See* 47 CFR §1.1311. An EA would be considered in determining whether an application should be approved in view of the environmental impact or whether Environmental Impact Statements (EISs) should be prepared as specified in 47 CFR §1.1314. However, EAs are rarely filed since most applicants and licensees who are not categorically excluded undertake measures to ensure compliance before submitting an application.

[18] Categorical exclusions from routine environmental evaluation are allowed under NEPA when actions are judged individually and cumulatively to have no significant potential for effect on the human environment. *See* 47 CFR §1.1306(a); *see also, Notice* at para. 5, ET Docket No. 93-62, 8 FCC Rcd 2849 (1993). However, we retain, under §1.1307(c) and (d), the authority to request that a licensee or an applicant conduct an environmental evaluation and, if appropriate, file environmental information pertaining to an otherwise categorically excluded application if it is determined that in that particular case there is a possibility for significant environmental impact.

[19] *See Report and Order* at para. 160-163. As discussed previously, we also amended our rules to require the amateur radio operator license examination question pools to include questions concerning RF safety at amateur stations, requiring an additional five questions on RF safety within each of three written examination elements.

[20] ARRL Petition at 9-13.

[21] Alan Dixon Petition at 2-4.

[22] For example, at 1,897 kHz (in the 160 meter amateur band) the MPE limit for general population/uncontrolled exposure is 50 mW/cm^2. At 29 MHz (in the 10 meter amateur band) the MPE limit for general population/uncontrolled exposure is about 0.2 mW/cm^2. The authorized frequency bands are contained in 47 CFR §97.301.

[23] The near-field of an antenna generally extends out to a distance of $2L^2/\lambda$ from the antenna, where L is the effective length of the antenna and l is the wavelength of the signal. For a typical amateur station using a half-wave dipole and operating on 10.125 MHz, the near-field would extend out to points approximately 15 meters from the antenna. As frequency decreases below 10 MHz, the size of the near-field increases (provided the effective length of the antenna is maintained). As frequency increases above 10 MHz, the size of the near-field decreases.

[24] Prior to the effective date of the *Report and Order*, these policies were contained in Note 2 to 47 CFR §1.1307(b). The *Report and Order* recodified these policies, essentially unchanged, into 47 CFR §1.1307(b)(3), as amended.

[25] *See* 47 CFR §1.1307(b)(3).

[26] *See* 47 CFR §1.1307(b)(3)(i).

[27] *See* 47 CFR §1.1307(b)(3)(ii).

[28] Arch Comments at 3, AT&T Petition at 6-8, AT&T Comments at 5, BellSouth Petition at 2, PageNet Petition at 5, PCIA Petition at 14-16.

[29] Cellular Taskforce Reply at 6, Holly Fournier and Mary Beth Freeman Reply at 3, Alan Golden Reply at 2, Dawn Mason Reply at 2.

[30] See later discussion in this *Order* of issues related to the OST Bulletin 65, "Evaluating Compliance with FCC-Specified Guidelines for Human Exposure to Radiofrequency Radiation," which was published in October, 1985. This bulletin is being revised to reflect the Commission's newly adopted RF guidelines and procedures. We expect it to be issued shortly after adoption and release of this *Order*.

[31] We note that, if an area of non-compliance is found, it would be these other stations that would share in the responsibility for correcting the problem.

[32] For example, if a TV station produces a power density 50% of its limit, an FM station produces a power density 25% of its limit, and a second FM station produces a power density of 30% of its limit at a particular accessible area, then the RF emissions would cumulatively equal 105% of the composite limit, and the RF exposure limits would be exceeded.

[33] Consider the example of a high-powered broadcast station on the rooftop of a building. On an apartment building across the street there is a rooftop sundeck with several high-powered, high duty-factor, transmitting antennas used for paging that are located on the same rooftop within a few meters of the sundeck. Assume that at several locations on the sundeck the MPE limits for the general population are exceeded due to emissions of both the paging and broadcast transmitters and that all emission levels exceed the 5% threshold for the respective emitters at the accessible non-complying locations on the sundeck. In such a case the responsibility for compliance should belong to not only the paging transmitters, but also to the broadcast station, which is located several meters away from the sundeck. In such a situation a requirement for responsibility that only included the paging transmitters on the same building as

the sundeck would not include a major contributor, the broadcast station. Therefore, if our RF exposure rules were applied only to site owners, a primary contributor might totally escape responsibility for necessary corrective action to ensure compliance, leaving the burden for compliance with the paging licensees. A similar situation could occur on the rooftop of a building located nearby to a high-powered broadcast station, regardless of whether any additional transmitters were located on the building.

[34]For example, when an applicant files for renewal of license at a location that was previously subject to our old RF exposure guidelines.

[35]*See* 47 CFR §1.1310.

[36]Power density is equal to the square of the electric field strength divided by the characteristic impedance of free space (377 ohms). Similarly, power density is equal to the square of the magnetic field strength times the characteristic impedance of free space.

[37]*See* 47 CFR 1.1307(e), as amended.

[38]Ameritech Reply at 2 citing H. Rep. No. 104-458, 94th Cong. 2nd Sess. 208-09 (1996) *Conference Report*.

[39]Ameritech Petition at 7.

[40]EEA Petition at 14.

[41]PageNet Petition at 3.

[42]PCIA Petition at 8-9.

[43]*See* 5 U.S.C. 553(b).

[44]*First Memorandum Opinion and Order,* ET Docket 93-62, adopted December 23, 1996, 11 FCC Rcd 17,512 (1997).

[45]Petitions for Partial Reconsideration were filed by Ameritech Mobile Communications, Inc. (Ameritech) and Northeast Louisiana Telephone Company, Inc. (Northeast). Petitions for Reconsideration were filed by the Ad-hoc Association of Parties Concerned About the Federal Communications Commission's Radiofrequency Health and Safety Rules (Ad-hoc Association) and the Cellular Phone Taskforce.

[46]See *First MO&O* at Paragraph 8.

[47]Since we are taking this action the late petitions recently filed by Ameritech and PCIA requesting immediate deferral of the September 1, 1997, implementation date are moot and are denied. *See "Emergency Request for Immediate Deferral of Transition Date,"* filed August 8, 1997, by the Personal Communications Industry Association, and *"Request for Extension of Compliance Deadline,"* filed August 15, 1997, by Ameritech Mobile Communications, Inc.

[48]In the Notice of Proposed Rule Making in this docket we specifically asked for comment on "how best to treat equipment and facilities that are in use but do not comply with the new guidelines." See Notice of Proposed Rule Making, ET Docket 93-62, at para. 26.

[49]H. R. Rep. No. 204, 104th Cong., 1st Sess. 95 (1995).

[50]*See* 47 CFR Section 1.1308(a).

[51]47 C.F.R. §1.2.

The New FCC Form 610

This new FCC Form 610 must be used after January 1, 1998. It contains an RF Safety Certification that must be signed by the applicant.

The new FCC Form 610 requires all applicants sign an *RF Safety Certification*. The certification reads:

I have READ and WILL COMPLY WITH Section 97.13(c) of the Commission's Rules regarding RADIOFREQUENCY (RF) RADIATION SAFETY and the amateur service section of OST/OET Bulletin Number 65.

Appendix A of this book contains "Section 97.13(c) of the Commission's Rules." Chapter 6 has "the amateur service section of OST/OET Bulletin Number 65" and Chapter 7 the amateur supplement to OET 65. Thus, all you are missing is the actual FCC Form 610.

This appendix contains a copy of this form, along with Forms 610-A and 610-B. If you wish, you can get additional copies by telephone, US mail, the internet or fax.

There are several 610 forms available:

• Form 610—used for new license applications and upgrades, license renewals and address, name and systematic call sign changes. For license renewals, the Form 610 should be submit-

ted between 60 and 90 days before the license expiration date.

• Form 610-A—used by foreign amateurs to apply for a US reciprocal permit.

• Form 610-B—used for applying for a new club station license, or for renewing a club, military recreation or RACES station license.

• Form 610-V—used to obtain a vanity call sign. As of 9/15/97, a $50 fee ($5 per year with 10 years' fees paid up front) is currently required. At the present time, a fee also will be due and payable at each 10-year renewal of a vanity call sign.

• Form 159—to be submitted with a vanity call sign fee remittance

To get the forms by mail, call the FCC Forms Distribution Center. They accept orders at 800-418-3676. The ARRL/VEC also makes available most FCC forms for the amateur service. You can obtain these forms by writing to:

ARRL
225 Main Street
Newington, CT 06111

An SASE is appreciated. Please be sure to indicate the form(s) you want.

FCC forms are available by fax at 202-418-0177. To request a Form 610 for an

individual station (new application or renewal), request Fax Document Number 000610.

Form	Fax Document Number
159	000159
610	000610
610-A	006101
610-B	006102
610-V	006108

You can also get copies of these forms on the Internet. The ARRL internet site (**http://www.arrl.org/fcc/forms.html**) has active links to the FCC forms page. They also are available directly from the FCC. You can pick the form you want at **http://www.fcc.gov/formpage.html**. The direct address for a standard Form 610 is **http://www.fcc.gov/Forms/Form610**.

The ARRL page gives you the choice of downloading a file by the Web or by FTP. Most FCC forms (and documents) are available as .PDF files. You will need Acrobat Reader software, which is free from Adobe Systems, Inc at **http://www.adobe.com**. The FCC offers a copy at **http://www.fcc.gov/pdf_ref. html**.

The following pages are FCC Forms 610, Form 610-A and Form 610-B. You can use them by making a photocopy.

FEDERAL COMMUNICATIONS COMMISSION
1270 FAIRFIELD ROAD
GETTYSBURG, PENNSYLVANIA
17325-7245

APPLICATION FORM 610 FOR
AMATEUR OPERATOR/PRIMARY STATION LICENSE

Approved by OMB
3060-0003
See instructions for
public burden.

SECTION 1 - TO BE COMPLETED BY APPLICANT (See Instructions)

1. Print or type last name — Suffix — First name — Middle initial

2. Date of birth — month — day — year

3. Mailing address (Number and street)

3A. Internet Address

City

State Code — ZIP Code

4. I HEREBY APPLY FOR (make an X in the appropriate box(es)):

4A. ☐ **EXAMINATION** for a new license

4B. ☐ **EXAMINATION** for upgrade of my operator license class

4C. ☐ **CHANGE** my name on my license to my new name in Item 1. My former name was:

_____
(Last name)   (Suffix)   (First name)   (MI)

4D. ☐ **CHANGE** my mailing address on my license to my new address in Item 3

4E. ☐ **CHANGE** my station call sign systematically (See instructions)
Applicant's Initials _____

4F. ☐ **RENEWAL** of my license

5. Unless you are requesting a new license, attach the original or a photocopy of your license to the back of this Form 610 and complete Items 5A and 5B.

5A. Call sign shown on license

5B. Operator class shown on license

6. If you have filed another Form 610 that we have not acted upon, complete Items 6A and 6B.

6A. Purpose of other form

6B. Date filed — month — day — year

WILLFUL FALSE STATEMENTS MADE ON THIS FORM ARE PUNISHABLE BY FINE AND/OR IMPRISONMENT, (U.S. CODE, TITLE 18, SECTION 1001), AND/OR REVOCATION OF ANY STATION LICENSE OR CONSTRUCTION PERMIT (U.S. CODE, TITLE 47, SECTION 312(A)(1)) AND/OR FORFEITURE (U.S. CODE, TITLE 47, SECTION 503).

I certify that:
* all statements and attachments are true, complete, and correct to the best of my knowledge and belief and are made in good faith;
* I am not a representative of a foreign government;
* I waive any claim to the use of any particular frequency regardless of prior use by license or otherwise;
* the station to be licensed will be inaccessible to unauthorized persons;
* the construction of the station would NOT be an action which is likely to have a significant environmental effect (see the Commission's Rules 47 C.F.R. Sections 1.1301-1.1319 and Section 97.13(a);
* I have READ and WILL COMPLY WITH Section 97.13(c) of the Commission's Rules regarding RADIOFREQUENCY (RF) RADIATION SAFETY and the amateur service section of OST/OET Bulletin Number 65.

7. Signature of applicant (Do not print, type, or stamp. Must match name in Item 1.)

✗ _____

( ) _____
Daytime Telephone Number

8. Date signed — month — day — year

SECTION 2 - TO BE COMPLETED BY ALL ADMINISTERING VEs

A. Applicant is qualified for operator license class:

☐ NOVICE   (Elements 1(A), 1(B), or 1(C) and 2)
☐ TECHNICIAN   (Elements 2 and 3(A))
☐ TECHNICIAN PLUS (Elements 1(A), 1(B), or 1(C), 2 and 3(A))
☐ GENERAL   (Elements 1(B) or 1(C), 2, 3(A) and 3(B))
☐ ADVANCED   (Elements 1(B) or 1(C), 2, 3(A), 3(B) and 4(A))
☐ AMATEUR EXTRA   (Elements 1(C), 2, 3(A), 3(B), 4(A) and 4(B))

B. VEC receipt date:

C. Name of Volunteer-Examiner Coordinator (VEC):

D. Date of VEC coordinated examination session:

E. Examination session location:

I CERTIFY THAT I HAVE COMPLIED WITH THE ADMINISTERING VE REQUIREMENTS IN PART 97 OF THE COMMISSION'S RULES AND WITH THE INSTRUCTIONS PROVIDED BY THE COORDINATING VEC AND THE FCC

	VEs station call sign	VEs signature (must match name)	Date signed
1st VEs name (Print First, MI, Last, Suffix)			
2nd VEs name (Print First, MI, Last, Suffix)			
3rd VEs name (Print First, MI, Last, Suffix)			

FCC Form 610 - Page 1
September 1997

C.2   Appendix C

ATTACH ORIGINAL OR A PHOTOCOPY OF YOUR LICENSE HERE:

SECTION 3 - TO BE COMPLETED BY PHYSICIAN

PHYSICIAN'S CERTIFICATION OF DISABILITY
Please see notice below

Print, type, or stamp physician's name: _____

Street address: _____

City, State, ZIP code: _____

Office telephone number: (___) _____

I CERTIFY THAT I have read the Notice to Physician Certifying to a Disability, and that the person named in Item 1 on the reverse is severely handicapped, the duration of which will extend for more than 365 days beyond this date. Because of this severe handicap, this person is unable to pass a 13 or 20 words per minute telegraphy examination. I am licensed to practice in the United States or its Territories as a doctor of medicine (M.D.) or doctor of osteopathy (D.O.). I have considered the accommodations that could be made for this person's disability and have determined that, even with accommodations, this person would be unable to pass a 13 or 20 words per minute telegraphy examination.

WILLFUL FALSE STATEMENT IS PUNISHABLE BY FINE AND IMPRISONMENT (U.S. CODE TITLE 18, SECTION 1001)

➡ _____   _____   _____
PHYSICIAN'S SIGNATURE (DO NOT PRINT, TYPE, OR STAMP)   M.D. or D.O.   DATE SIGNED

PATIENT'S RELEASE
Authorization is hereby given to the physician named above, who participated in my care, to release to the Federal Communications Commission any medical information deemed necessary to process my application for an amateur operator/primary station license.

➡ _____   _____
APPLICANT'S SIGNATURE (DO NOT PRINT, TYPE, OR STAMP)   DATE SIGNED

NOTICE TO PHYSICIAN CERTIFYING TO A DISABILITY

You are being asked by a person who has already passed a 5 words per minute telegraphy examination to certify that, because of a severe handicap, he/she is unable to pass a 13 or 20 words per minute telegraphy examination. If you sign the certification, the person will be exempt from the examination. Before you sign the certification, please consider the following:

THE REASON FOR THE EXAMINATION - Telegraphy is a method of electrical communication that the Amateur Radio Service community strongly desires to preserve. We support their objective by authorizing additional operating privileges to amateur operators who increase their skill to 13 and 20 words per minute. Normally, to attain these levels of skill, intense practice is required. Annually, thousands of amateur operators prove by passing examinations that they have acquired the skill. These examinations are prepared and administered by amateur operators in the local community who volunteer their time and effort.

THE EXAMINATION PROCEDURE - The volunteer examiners (VEs) send a short message in the Morse code. The examinee must decipher a series of audible dots and dashes into 43 different alphabetic, numeric and punctuation characters used in the message. To pass, the examinee must correctly answer questions about the content of the message. Usually, a fill-in-the-blanks format is used. With your certification, they will give the person credit for passing the examination, even though they do not administer it.

MUST A PERSON WITH A HANDICAP SEEK EXEMPTION?

No handicapped person is required to request exemption from the higher speed telegraphy examinations, nor is anyone denied the opportunity to take the examinations because of a handicap. There is available to all otherwise qualified persons, handicapped or not, the Technician Class operator license that does not require passing a telegraphy examination. Because of international regulations, however, any handicapped applicant requesting exemption from the 13 or 20 words per minute examination must have passed the 5 words per minute examination.

ACCOMMODATING A HANDICAPPED PERSON - Many handicapped persons accept and benefit from the personal challenge of passing the examination in spite of their hardships. For handicapped persons without an exemption who have difficulty in proving that they can decipher messages sent in the Morse code, the VEs make exceptionally accommodative arrangements. They will adjust the tone in frequency and volume to suit the examinee. They will administer the examination at a place convenient and comfortable to the examinee, even at bedside. For a deaf person, they will send the dots and dashes to a vibrating surface or flashing light. They will write the examinee's dictation. Where warranted, they will pause in sending the message after each sentence, each phrase, each word, or each character to allow the examinee additional time to absorb and interpret what was sent. They will even allow the examinee to send the message, rather than receive it.

YOUR DECISION - The VEs rely upon you to make the necessary medical determination for them using your professional judgement. You are being asked to decide if the person's handicap is so severe that he/she cannot pass the examination even when the VEs employ their accommodative procedures. The impairment, moreover, will last more than one year. This procedure is not intended to exempt a person who simply wants to avoid expending the effort necessary to acquire greater skill in telegraphy. The person requesting that you sign the certification will give you names and addresses of VEs and other amateur operators in your community who can provide you with more information on this matter.

DETAILED INSTRUCTIONS - If you decide to execute the certification, you should complete and sign the Physician's Certification of Disability on the person's FCC Form 610. You must be an M.D. or D.O. licensed to practice in the United States or its Territories. The person must sign a release permitting disclosure to the FCC of the medical information pertaining to the disability.

FCC Form 610 - Page 2
September 1997

UNITED STATES OF AMERICA
FEDERAL COMMUNICATIONS COMMISSION

1270 FAIRFIELD ROAD
GETTYSBURG, PENNSYLVANIA 17325-7245
TELEPHONE 1-888-225-5322

INSTRUCTIONS FOR APPLICATION FORM 610 FOR AMATEUR OPERATOR/PRIMARY STATION LICENSE

(Do Not Return Instructions With Application Form)

GENERAL INSTRUCTIONS

● Use the attached FCC Form 610 to request:

1. An examination for a new amateur operator/primary station license or for modification of your license to a higher operator class.

2. A modification of your name or mailing address as it appears on your license, or a systematic assignment of a different call sign.

3. A renewal of your license if it is unexpired or if it expired within the two year grace period.

● Do NOT use the attached FCC Form 610 to request:

1. A Reciprocal Permit for Alien Amateur Licensee. Use FCC Form 610-A.

2. A renewal or modification of a club, military recreation, or RACES station license. Use FCC Form 610-B.

3. A vanity call sign. Use FCC Form 610-V.

● ANTENNA HEIGHT: Effective July 1, 1996, the Commission adopted rules which require Antenna Structure owners to apply for a registration number on revised FCC Form 854 whenever proposed construction or alteration to existing antenna structures meets FAA notification criteria. Generally, these are antenna structures that are higher than 60.96 meters (200 feet) above ground level or interfere with the flight path of a nearby airport (refer to FCC Rules, Section 97.15). Additionally, owners of existing antenna structures which previously required FAA notification and were cleared by the FCC prior to July 1, 1996, must register before June 30,1998 in accordance with filing windows prescribed by state. As these structures are registered,

owners are required to provide licensees with a copy of FCC Form 854R and are required to display the Registration Number near the base of the antenna structure. The revised FCC Form 854 may be obtained by calling 1-800-418-FORM (3676).

● If you have not received a response from us within 90 days, write to Federal Communications Commission, 1270 Fairfield Road, Gettysburg, PA 17325-7245. Include a photocopy of your completed FCC Form 610, or the following information:

1. Your name, address, and date of birth;
2. Your station call sign and operator class;
3. The date that you filed FCC Form 610;
4. The purpose of the FCC Form 610 you filed;
5. The name of the coordinating VEC;
6. The location of the test site (city and state) and the date of the examination.

● Every amateur operator should have a current copy of the amateur service rules, Part 97, which may be obtained from private publishers, vendors, or you may order 47 CFR, Part 80 to End from the U. S. Government Printing Office, Washington, DC 20402, phone (202) 512-1800.

● Detach your completed FCC Form 610 from these instructions. Make a photocopy of it for your records. File your completed FCC Form 610 with the VEs if you have marked Box 4A or 4B on the application. If you have marked box(es) 4C through 4F, mail FCC Form 610 WITHOUT A FEE to:

FEDERAL COMMUNICATIONS COMMISSION
1270 FAIRFIELD ROAD
GETTYSBURG PA 17325-7245

INSTRUCTIONS TO EXAMINEE

A. Your examination will be administered at a location and time specified by your administering VEs. You must comply with their instructions. The VEs will observe you throughout the examination. They are responsible for the proper conduct and necessary supervision of the examination. They must immediately terminate the examination if you fail to comply with their instructions.

B. If you hold an unexpired license, or if you hold a license that expired less than two years before the date of the examination session, attach a photocopy of it, or the original, to the application.

C. Give your completed FCC Form 610 to your administering VEs. Show your VEs at least two documents that prove your identity. Show your VEs any of the following documents for which you are claiming element credit:

1. Original document of your unexpired (or expired within the grace period) amateur operator/primary station license;

2. Certificate(s) of Successful Completion of Examination, if issued to you within 365 days of this examination session;

3. Photocopy of FCC Form 610 that was filed indicating that you qualified for a Novice Class operator license within 365 days of this examination session;

4. Original document of your unexpired (or expired less than five years prior to this examination session) FCC Commercial Radiotelegraph Operator's Certificate.

INSTRUCTIONS TO PERSONS WITH SEVERE HANDICAPS

A. If you have passed the 5 words per minute telegraphy examination, but you are unable to pass the 13 or 20 words per minute examination because of a severe handicap that will extend for more than 365 days, the administering VEs will give you credit for passing the 20 words per minute examination if you obtain a Physician's Certification of Disability. You should, however, first attempt to pass the examination under the special accommodative procedures the VEs use for handicapped examinees.

B. Detailed Instructions:

1. Complete Items 1 through 8 on FCC Form 610.

2. Present your physician with your completed FCC Form 610 and the Notice to Physician Certifying to a Disability.

3. Provide the physician with the names and addresses of your administering VEs and other amateur operators in your community who can provide more information on this matter.

4. Ask your physician to complete and sign the Physician's Certification of Disability in Section 3 of FCC Form 610.

5. Sign and date the Patient's Release in Section 3 of FCC Form 610.

6. Follow Instructions to Examinee.

FCC Form 610 Instructions - Page 1
September 1997

INSTRUCTIONS FOR COMPLETING APPLICATION FORM 610

ITEM 1 – Print (or type) your last name and any suffix (Jr., Sr., II, etc.), first name, and middle initial. The name you enter in Item 1 must agree with your signature in Item 8. It must also agree with the name on your existing license unless you request a change in Box 4C.

ITEM 2 – Print numbers for the month, day, and year of your birth. Example: If you were born on September 20, 1944, enter 09-20-44.

ITEM 3 – Print your mailing address. It must be an address where you can receive mail delivered by the United States Postal Service. (Mail delivery may not be available in certain territories.) Print your two-letter state/territory code from the table.

ITEM 3A - Print an Internet Address, if available, where you can receive information from the FCC regarding your application.

Alabama	AL	New Hampshire	NH
Alaska	AK	New Jersey	NJ
Arizona	AZ	New Mexico	NM
Arkansas	AR	New York	NY
California	CA	North Carolina	NC
Colorado	CO	North Dakota	ND
Connecticut	CT	Ohio	OH
Delaware	DE	Oklahoma	OK
District of Columbia	DC	Oregon	OR
Florida	FL	Pennsylvania	PA
Georgia	GA	Rhode Island	RI
Hawaii	HI	South Carolina	SC
Idaho	ID	South Dakota	SD
Illinois	IL	Tennessee	TN
Indiana	IN	Texas	TX
Iowa	IA	Utah	UT
Kansas	KS	Vermont	VT
Kentucky	KY	Virginia	VA
Louisiana	LA	Washington	WA
Maine	ME	West Virginia	WV
Maryland	MD	Wisconsin	WI
Massachusetts	MA	Wyoming	WY
Michigan	MI	American Samoa	AS
Minnesota	MN	Guam	GU
Mississippi	MS	Northern Mariana Is	MP
Missouri	MO	Puerto Rico	PR
Montana	MT	Virgin Islands	VI
Nebraska	NE		
Nevada	NV		

ITEM 4 – Place an "X" in the proper box to apply for:

BOX 4A An EXAMINATION for a new amateur operator/primary station license. See Instructions to Examinee on reverse. You are eligible for an examination for a new license if you do not have one or if your license has expired beyond the two year grace period.

BOX 4B An EXAMINATION to upgrade your license to a higher class. See Instructions to Examinee on reverse.

BOX 4C CHANGE your name as it appears on your license to your new name in Item 1. Print your former name where indicated.

BOX 4D CHANGE your mailing address as it appears on your license to your new address in Item 3.

BOX 4E CHANGE your station call sign. See Fact Sheet PR-5000, Number 206-S, Amateur Station Sequential Call Sign System, latest date of issue, for information on how the call sign will be systematically assigned. After the call sign change is made, your previous call sign cannot be reinstated. **Initial in the space provided.**

BOX 4F RENEWAL of your unexpired license or RENEWAL of your license if it expired within the grace period. The expiration date must be within the two year grace period. Application must be received by the Commission's Gettysburg office prior to the end of the grace period.

ITEM 5 – If your license document was lost or destroyed, attach to your FCC Form 610 a sheet of paper containing your explanation.

ITEM 5A – Print the call sign shown on your license.

ITEM 5B – Print the operator class shown on your license.

ITEM 6 – If you have filed another Form 610 that we have not acted upon, give the purpose of the other form in Box 6A and print the month, day, and year it was filed in Box 6B.

ITEM 7 – Sign your name. Your signature must agree with your name as printed in Item 1. Provide a telephone number where you can be reached during normal daytime business hours.

ITEM 8 – Print the month, day, and year that you sign your application.

Additional Information for Amateurs Completing the New FCC Form 610

The New Form 610 Requires the Applicant Sign an RF Safety Certification.

The new FCC Form 610 requires that all applicants now sign an <u>RF Safety Certification</u>. The certification that applicants must now sign reads: "I have READ and WILL COMPLY with Section 97.13(c) of the Commission's Rules regarding RADIOFREQUENCY (RF) RADIATION SAFETY and the amateur service section of OST/OET Bulletin Number 65." This is all well and good, but how can you sign this statement if you haven't seen these new rules and Bulletin 65 information? Unfortunately, FCC has not provided this additional information in the instructions to the new Form 610!

Recognizing this need, here is the information you will need to read and must comply with. Section 97.13(c) reads:
c. Before causing or allowing an amateur station to transmit from any place where the operation of the station could cause human exposure to RF electromagnetic field levels in excess of those allowed under §1.1310 of this chapter, the licensee is required to take certain actions.

1. The licensee must perform the routine RF environmental evaluation prescribed by §1.1307(b) of this chapter, if the power of the licensee's station exceeds the limits given in the following table:

Wavelength Band & Evaluation Required if Power* (watts) Exceeds		
MF/HF *160m - 40m = 500 watts* *30m = 425 watts* *20m = 225 watts* *17m = 125 watts* *15m = 100 watts*	*12m = 75 watts* *10m = 50 watts* *VHF all bands = 50 watts* *UHF* *70cm = 70 watts* *33cm = 150 watts*	*23cm = 200 watts* *13cm = 250 watts* *SHF all bands = 250 watts* *EHF all bands = 250 watts*

- Repeater stations (all bands) non-building-mounted antennas:
    height above ground level to lowest point of antenna < 10 m and power > 500 W ERP
- Building-mounted antennas: power > 500 W ERP

* Power = PEP input to antenna except, for repeater stations only, power exclusion is based on ERP (effective radiated power).

2. If the routine environmental evaluation indicates that the RF electromagnetic fields could exceed the limits contained in §1.1310 of this chapter in accessible areas, the licensee must take action to prevent human exposure to such RF electromagnetic fields. Further information on evaluating compliance with these limits can be found in the FCC's OET Bulletin 65, "Evaluating Compliance with FCC-Specified Guidelines for Human Exposure to Radio Frequency Electromagnetic Fields."

The Amateur Section of OET Bulletin Number 65:

In the FCC's Report and Order, certain amateur radio installations were made subject to routine evaluation for compliance with the FCC's RF exposure guidelines.[1] Also, amateur licensees will be expected to demonstrate their knowledge of the FCC guidelines through examinations. Applicants for new licenses and renewals also will be required to demonstrate that they have read and that they understand the applicable rules regarding RF exposure. Before causing or allowing an amateur station to transmit from any place where the operation of the station could cause human exposure to RF radiation levels in excess of the FCC guidelines amateur licensees are now required to take certain actions. A routine RF radiation evaluation is required if the transmitter power of the station exceeds the levels shown and specified in 47 CFR § 97.13(c)(1)[2] (see above). Otherwise the operation is categorically excluded from routine RF radiation evaluation, except as a result of a specific motion or petition as specified in Sections 1.1307(c) and (d) of the FCC's Rules, (see discussion in Section 1 of Bulletin 65 for more information).

The Commission's Report and Order instituted a requirement that operator license examination question pools will include questions concerning RF safety at amateur stations. An additional five questions on RF safety will be required within each of three written examination elements (for Novice, Technician and General written exams).

When routine evaluation of an amateur station indicates that exposure to RF fields could be in excess of the exposure limits specified by the FCC (see Bulletin 65, Appendix A {on reverse side}), the licensee must take action to correct the problem and ensure compliance (see Section 4 of Bulletin 65 on controlling exposure). Such actions could be in the form of modifying patterns of operation, relocating antennas, revising a station's technical parameters such as frequency, power or emission type or combinations of these and other remedies.

(over)

Bulletin 65, Appendix A, Table 1 -- LIMITS FOR MAXIMUM PERMISSIBLE EXPOSURE (MPE)

Limits for Occupational/Controlled Exposure
(f = frequency in MHz   *Plane-wave equivalent power density)

Frequency Range (MHz)	Electric Field Strength (E) (V/m)	Magnetic Field Strength (H) (A/m)	Power Density (S)	Averaging Time $\|E\|^2$, $\|H\|^2$ or S (minutes)
0.3-3.0	614	1.63	(100)*	6
3.0-30	1842/f	4.89/f	(900/f²)*	6
30-300	61.4	0.163	1.0	6
300-1500	--	--	f/300	6
1500-100,000	--	--	5	6

Limits for General Population/Uncontrolled Exposure
(f = frequency in MHz   *Plane-wave equivalent power density)

Frequency Range (MHz)	Electric Field Strength (E) (V/m)	Magnetic Field Strength (H) (A/m)	Power Density (S) (mW/cm²)	Averaging Time $\|E\|^2$, $\|H\|^2$ or S (minutes)
0.3-1.34	614	1.63	(100)*	30
1.34-30	824/f	2.19/f	(180/f²)*	30
30-300	27.5	0.073	0.2	30
300-1500	--	--	f/1500	30
1500-100,000	--	--	1.0	30

In complying with the Commission's Report and Order, amateur operators should follow a policy of systematic avoidance of excessive RF exposure. The Commission has said that it will continue to rely upon amateur operators, in constructing and operating their stations, to take steps to ensure that their stations comply with the MPE limits for both occupational/controlled and general public/uncontrolled situations, as appropriate. In that regard, amateur radio operators and members of their immediate household are considered to be in a "controlled environment" and are subject to the occupational/controlled MPE limits. Neighbors who are not members of an amateur operator's household are considered to be members of the general public, since they cannot reasonably be expected to exercise control over their exposure. In those cases general population/uncontrolled exposure MPE limits will apply.

In order to qualify for use of the occupational/controlled exposure criteria, appropriate restrictions on access to high RF field areas must be maintained and educational instruction in RF safety must be provided to individuals who are members of the amateur operator's household. Persons who are not members of the amateur operator's household but who are present temporarily on an amateur operator's property may also be considered to fall under the occupational/controlled designation provided that appropriate information is provided them about RF exposure potential if transmitters are in operation and such persons are exposed in excess of the general population/uncontrolled limits.

Amateur radio facilities represent a special case for determining exposure, since there are many possible antenna types that could be designed and used for amateur stations. However, several relevant points can be made with respect to analyzing amateur radio antennas for potential exposure that should be helpful to amateur operators in performing evaluations.

First of all, the generic equations described in Bulletin 65 can be used for analyzing fields due to almost all antennas, although the resulting estimates for power density may be overly-conservative in some cases. Nonetheless, for general radiators and for aperture antennas, if the user is knowledgeable about antenna gain, frequency, power and other relevant factors, the equations in this section can be used to estimate field strength and power density as described earlier. In addition, other resources are available to amateur radio operators for analyzing fields near their antennas. The ARRL Handbook For Radio Amateurs contains an excellent section on analyzing amateur radio facilities for compliance with RF guidelines. Also, the FCC and the EPA conducted a study of several amateur radio stations in 1990 that provides a great deal of measurement data for many types of antennas commonly used by amateur operators[3] (see the FCC OET Web site at: <http://www.fcc.gov/oet/info/documents/reports/#ASD-9601> see also <http://www.fcc.gov/oet/rfsafety/>).

Amateur radio organizations and licensees are encouraged to develop their own more detailed evaluation models and methods for typical antenna configurations and power/frequency combinations. The FCC has an Amateur Supplement "B" that is available from the FCC's OET Web site at: <http://www.fcc.gov/oet/rfsafety/>. Information on availability of the supplement, as well as other RF-related questions, can be directed to the FCC's "RF Safety Program" at: (202) 418-2464
or Email to: rfsafety@fcc.gov

See also: Sections 1 and 2 of the FCC Regulations; FCC's "Amateur" Supplement B to OET Bulletin 65; the ARRL's publication entitled "RF Exposure and You" (to be available in early 1998); the ARRLWeb at: <http://www.arrl.org/news/rfsafety/>; and our RF Safety article in January 1998 QST (Pages 50-55) for more information.

[footnotes] -

1 See para. 160 of Report and Order, ET Dkt 93-62. See also, 47 CFR § 97.13, as amended.

2 These levels were chosen to roughly parallel the frequency of the MPE limits of Table 1 in Appendix A. These levels were modified from the Commission's original decision establishing a flat 50 W power threshold for routine evaluation of amateur stations (see Second Memorandum Opinion and Order, ET Docket 93-62, FCC 97-303, adopted August 25, 1997).

3 Federal Communications Commission (FCC), "Measurements of Environmental Electromagnetic Fields at Amateur Radio Stations," FCC Report No. FCC/OET ASD-9601, February 1996. FCC, Office of Engineering and Technology (OET), Washington, D.C. 20554. NTIS Order No. PB96-145016. Copies can also be downloaded from OET's Home Page on the World Wide Web at: http://www.fcc.gov/oet/

UNITED STATES OF AMERICA

FEDERAL COMMUNICATIONS COMMISSION
1270 FAIRFIELD ROAD
GETTYSBURG, PA 17325-7245
1-888-225-5322

Approved by OMB
3060-0022
Expires 8/31/98
See reverse for public
burden estimate

APPLICATION FOR PERMIT OF AN ALIEN AMATEUR RADIO LICENSEE TO OPERATE IN THE UNITED STATES

ARE YOU USING THE CORRECT FORM? Use this form to request a RECIPROCAL PERMIT from the FCC to operate your amateur station in areas where the amateur service is regulated by the FCC. You must possess a valid amateur service license issued by the country of which you are a citizen. If you are a citizen of the United States, you are ineligible for a reciprocal permit even if you are also a citizen of another country. If you will be in the United States for an extended period, you are encouraged to obtain an FCC amateur service license. If you hold an FCC-issued amateur service license, you are ineligible to be issued a reciprocal permit.

1. YOUR LAST NAME:	FIRST NAME:	MI:	2. YOUR DATE OF BIRTH:
			— — MONTH DAY YEAR

3. COUNTRY ISSUING YOUR AMATEUR SERVICE LICENSE:	4. YOUR AMATEUR STATION CALL SIGN:

5. EXPIRATION DATE OF YOUR AMATEUR SERVICE LICENSE: — — MONTH DAY YEAR	6. YOUR COUNTRY OF CITIZENSHIP:
	7. INTERNET ADDRESS:

8. YOUR UNITED STATES MAILING ADDRESS:

9. ADDRESS WHERE YOU WANT YOUR RECIPROCAL PERMIT MAILED, IF ISSUED:

10. YOUR MAILING ADDRESS IN YOUR COUNTRY OF CITIZENSHIP:

11. LIST ALL LOCATIONS FROM WHICH YOUR STATION WILL TRANSMIT FOR ANY PERIOD OF 30 DAYS OR MORE WHILE IN THE UNITED STATES. IF IT WILL NOT BE TRANSMITTING AT ANY ONE LOCATION DURING A PERIOD OF 30 DAYS OR MORE, WRITE "NONE":	12. APPROXIMATE DATE OF YOUR STAY IN THE UNITED STATES:
	(BEGINNING) (ENDING) — — — — MONTH DAY YEAR MONTH DAY YEAR

READ CAREFULLY BEFORE SIGNING

I, the above named alien amateur service licensee, request a reciprocal permit for operation of my amateur station in the United States. I understand that, if a permit is granted, the operation of my amateur station must be in accord with:
* the terms and conditions of the agreement on this subject between my Government and the Government of the United States;
* Part 97 of the FCC Rules;
* the terms and conditions of the amateur service license issued to me by my Government, but not to exceed the Amateur Extra Class operator privileges;
* any further conditions attached to the reciprocal permit by the FCC. In addition, I certify that:
* I am NOT a United States citizen;
* I have READ and WILL COMPLY WITH Section 97.13(c) of the Commission's Rules regarding RADIOFREQUENCY (RF) RADIATION SAFETY and the amateur service section in OST/OET Bulletin Number 65;
* I understand that any reciprocal permit issued to me may be modified, suspended, or cancelled by the FCC without advance notice and that all of the information I have submitted on this application is true, complete and correct to the best of my knowledge.

WILLFUL FALSE STATEMENTS MADE ON THIS FORM ARE PUNISHABLE BY FINE AND/OR IMPRISONMENT (US CODE, TITLE 18, SECTION 1001), AND/OR REVOCATION OF ANY STATION LICENSE OR CONSTRUCTION PERMIT (US CODE, TITLE 47, SECTION 312(A)(1)), AND/OR FORFEITURE (US CODE, TITLE 47, SECTION 503).

12. YOUR SIGNATURE:	13. DATE SIGNED:

SEE REVERSE

FCC FORM 610A
September 1997

GENERAL INSTRUCTIONS

A. Print clearly or type. All items must be completed.

B. You should obtain a current copy of Part 97 of the FCC Rules from a supplier or from the Government Printing Office (GPO). For ordering information from the GPO, you may write to them at the United States Government Printing Office, Washington, D.C. 20402, or you may telephone them at (202) 512-1800.

C. Mail your completed application together with a photocopy of your amateur service license to the Federal Communications Commission, 1270 Fairfield Road, Gettysburg, PA 17325-7245. Allow at least 60 days for processing.

D. Should you become a citizen of the United States, you are no longer eligible to obtain a reciprocal permit. If you wish to apply for an FCC amateur service license, you should contact amateur operators who are accredited as volunteer examiners. Should you obtain an FCC-issued amateur service license, it will supersede your reciprocal permit. No examination credit is allowed toward an FCC amateur service license on the basis of holding an amateur service license issued by another country.

INSTRUCTIONS FOR SPECIFIC ITEMS

Item 6: State the name of the country of which you are a citizen. If different than item 3, do not submit this application. You must be a citizen of the country that issued your amateur service license to be eligible for a reciprocal permit.

Item 7: Please provide an Internet address where you can receive mail concerning your application. If contact via the Internet is not possible, leave this item blank.

Item 8: You must provide the FCC with a United States mailing address at which you may be reached or through which your mail will be forwarded during your stay in the United States. The Embassy or Consulate of your country may be able to serve this purpose for you during short visits to the United States.

Item 9: State a complete mailing address where you wish your reciprocal permit to be mailed.

Item 10: State your mailing address in your own country. If you are an alien resident in the United States, and do not have a mailing address in your own country, please state this in item 10.

BE SURE TO READ THE ENTIRE STATEMENT FOLLOWING ITEMS 11 AND 12, SIGN AND DATE THE APPLICATION.

Public burden for this collection of information is estimated to be five minutes per response, including the time for reviewing instructions, searching existing data sources, gathering and maintaining the data needed, and completing and reviewing the collection of information. Send comments regarding this burden estimate or any other aspect of this collection of information, including suggestions for reducing the burden to the Federal Communications Commission, AMD-PERM, Washington, DC 20554, Paperwork Reduction Project (3060-0022) or via the Internet to jboley@fcc.gov. Individuals are not required to respond to a collection of information unless it displays a currently valid OMB control number.

FCC FORM 610A
September 1997

Federal Communications Commission
1270 Fairfield Road
Gettysburg, PA 17325-7245
1-888-225-5322

Approved by OMB
3060-0079
See reverse for information
regarding public burden estimate.

APPLICATION FOR AN AMATEUR CLUB, RACES OR MILITARY RECREATION STATION LICENSE

BE SURE TO READ INSTRUCTIONS ON REVERSE SIDE. PLEASE PRINT OR TYPE.

➪ ATTACH PRESENT CLUB, RACES OR MILITARY RECREATION LICENSE OR PHOTOCOPY THEREOF TO THE BACK OF APPLICATION. IF THE LICENSE HAS BEEN LOST OR DESTROYED, ATTACH EXPLANATION.

1. CLUB, RACES OR MILITARY RECREATION STATION CALL SIGN:	2. NAME OF CLUB STATION TRUSTEE OR LICENSE CUSTODIAN: (Last, Suffix, First, MI)	3. DATE OF BIRTH: ___ — ___ ___ MONTH DAY YEAR

4. CURRENT MAILING ADDRESS: (Number and Street, City, State, ZIP Code)

5. APPLICATION IS FOR (Check one):

☐ NEW   ☐ MODIFICATION   ☐ RENEWAL

6. TRUSTEE'S PRIMARY STATION CALL SIGN:	7. NAME OF CLUB, RACES ORGANIZATION OR MILITARY RECREATION ENTITY:

8. APPLICANT CLASSIFICATION:

☐ CLUB   ☐ MILITARY RECREATION   ☐ RACES

9. INTERNET ADDRESS:

CERTIFICATION

<u>RESPONSIBLE OFFICIAL</u>: I certify that the above named person is the station trustee or license custodian authorized to apply for and hold an amateur radio station license for this organization, society or entity.

WILLFUL FALSE STATEMENTS MADE ON THIS FORM ARE PUNISHABLE BY FINE AND/OR IMPRISONMENT (US CODE, TITLE 18, SECTION 1001), AND/OR REVOCATION OF ANY STATION LICENSE OR CONSTRUCTION PERMIT (US CODE, TITLE 47, SECTION 312(A)(1)), AND/OR FORFEITURE (US CODE, TITLE 47, SECTION 503).

10. SIGNATURE: (Must not agree with Item 2)   (Title or authority to approve)   DATE SIGNED:

<u>APPLICANT</u>: I certify that:
* all statements herein and attachments herewith are true, complete and correct to the best of my knowledge and belief and are made in good faith;
* I am not the representative of a foreign government;
* I waive the claim to the use of any particular frequency regardless of prior use by license or otherwise;
* the station to be licensed will be inaccessible to unauthorized persons;
* the construction of the station would NOT be an action which is likely to have a significant environmental effect. See the Commission's Rules, 47 C.F.R. Sections 1.1301–1.1319 and Section 97.13(a);
* I have READ and WILL COMPLY WITH Section 97.13(c) of the Commission's Rules regarding RADIOFREQUENCY (RF) RADIATION SAFETY and the amateur service section in OST/OET Bulletin Number 65.

WILLFUL FALSE STATEMENTS MADE ON THIS FORM ARE PUNISHABLE BY FINE AND/OR IMPRISONMENT (US CODE, TITLE 18, SECTION 1001), AND/OR REVOCATION OF ANY STATION LICENSE OR CONSTRUCTION PERMIT (US CODE, TITLE 47, SECTION 312(A)(1)), AND/OR FORFEITURE (US CODE, TITLE 47, SECTION 503).

11. APPLICANT'S SIGNATURE: (Must agree with Item 2)   DATE SIGNED: DAYTIME TELEPHONE NUMBER:

FCC 610B September 1997

INSTRUCTIONS

Use this form to apply for a new club or military recreation station license, a renewed or modified club, military recreation, or RACES (Radio Amateur Civil Emergency Service) station license. For an operator/primary station amateur license, use FCC 610. For a vanity call sign request, use FCC 610-V. New RACES station licenses are not available. To obtain a duplicate copy of a lost, mutilated or destroyed license, submit a letter request giving your name, address, station call sign, license expiration date, and state how the original license was lost, mutilated, or destroyed.

Complete this form as follows:
1. Use typewriter or print clearly in ink.
2. Complete all applicable items carefully.
3. Attach present club, RACES or military recreation station license or photocopy thereof to this side of form.
4. Sign and date the application.
5. Clubs, Military Recreation and RACES: Mail application to Federal Communications Commission, 1270 Fairfield Road, Gettysburg, PA 17325-7245.

For general information, you may call the FCC's National Call Center at 1-888-225-5322.

ITEM 2: TO BE COMPLETED BY ALL APPLICANTS
There are three types of applicants:

1. The applicant for a club station license must be the licensed amateur operator of higher than a Novice class who has been designated by the organization or society to be the station trustee.

2. The applicant for a RACES station license must be the civil defense official responsible for coordination of all civil defense activities in the area concerned and who has been designated by the official responsible for the governmental entity agency served by that civil defense organization to be the RACES station license custodian. The applicant need not be an amateur radio operator.

3. The applicant for a military recreation station license must be the person in charge of a station at a land location provided for the recreational use of amateur radio operators under military auspices of the Armed Forces of the United States and who has been designated by the official in charge of the U. S. Government premises where the military recreation station is located to be the military recreation station license custodian. The applicant need not be an amateur radio operator.

ITEM 9: If one is available, enter an Internet address where the Commission can send you mail regarding your application.

RESPONSIBLE OFFICIAL CERTIFICATION FOR ITEM 10

The signature for the club official must be that of an officer of the club. It must not be that of the station trustee. The signature for the RACES official must be that of the official responsible for the governmental entity served by the civil defense organization. It must not be the same as the applicant's. The signature for the military recreation station must be that of the official in charge of the United States Government premises where the military recreation station is located. It must not be the same as the applicant's.

APPLICANT'S CERTIFICATION FOR ITEM 11

The signature for the club station must be that of the club station trustee. It must agree with item 2. The signature for the RACES station applicant must be that of the civil defense official responsible for the coordination of all civil defense activities in the area concerned. It must agree with item 2. The signature for the military recreation station applicant must be that of the person in charge of a station provided for the recreational use of amateur radio operators, under the military auspices of the Armed Forces of the United States. It must agree with item 2.

ANTENNA HEIGHT

Effective July 1, 1996, the Commission adopted rules which require Antenna Structure owners to apply for a registration number on revised FCC Form 854 whenever proposed construction or alterations to existing antenna structures meet FAA notification criteria. Generally, these are antenna structures that are higher than 60.96 meters (200 feet) above ground level or interfere with the flight path of a nearby airport (refer to FCC Rules 97.15). Additionally, owners of existing antenna structures which previously required FAA notification and were cleared by the FCC prior to July 1, 1996 must register before June 30, 1998 in accordance with prescribed filing windows by state. As these structures are registered, owners are required to provide licensees with a copy of FCC Form 854R and are generally required to display the Registration Number near the base of the antenna structure. The revised FCC Form 854 may be obtained from the FCC's Forms Distribution Center by calling toll free (800) 418-FORM(3676); by Fax-On-Demand by calling (202) 418-0177 from the handset of your fax machine; or via the Internet at http://www.fcc.gov/formpage.html.

FCC 610B
September 1997

FCC Information Sources on RF Safety

This appendix is a reprint of the FCC's material answering the most common questions about RF exposure and RF safety.

In order to make available the latest possible information, the FCC distributes its bulletins and other documents by posting them on their Web site. This address is **http://www.fcc.gov**.

The FCC OET (Office of Engineering and Technology) has links to its documents at **http://www.fcc.gov/oet/info/documents/bulletins**. *Most documents are available in WordPerfect 5.1 and PDF (Adobe Portable Document Format).*

Chapter 6 of this book contains a copy of portions of OET Bulletin 65, as available at press time. Any revisions to this material will be posted on this FCC page.

Another available resource is OET Bulletin 56. It is a summary of the FCC position and activities. The FCC describes this document as:

Questions and Answers About the Biological Effects and Potential Hazards of Radiofrequency Radiation

This is an informative bulletin written as a result of increasing interest and concern of the public with respect to this issue. The expanding use of radiofrequency technology has resulted in speculation concerning the alleged "electromagnetic pollution" of the environment and the potential dangers of exposure to non-ionizing radiation. This publication is designed to provide factual information to the public by answering some of the most commonly asked questions.

You can download OET Bulletin 56 from the OET address above.

The FCC has compiled a list of FAQs (Frequently Asked Questions). It may be found at
http://www.fcc.gov/oet/rfsafety/rf-faqs.html

If a friend or neighbor wants a summary of the RF safety situation, this document is an excellent introduction to the topic, and may answer many questions very quickly. The latest version is reprinted in this appendix.

Frequently Asked Questions

(From **http://www.fcc.gov/oet/rfsafety/rf-faqs.html**)

Question: When are the Commission's new exposure limits effective?

Answer: The Commission released the *2nd Memorandum Opinion and Order* on August 25, 1997 extending the effective date of compliance to October 15, 1997 for all services except the Amateur Radio Service. The Amateur Radio Service has until January 1, 1998. The effective date for mobile and portable devices was August 7, 1996. In the interim, all non-excluded services, except the Personal Communications Service, should continue to evaluate compliance base on the ANSI 1982 exposure guidelines. PCS base stations are required to comply with the ANSI/IEEE C95.1-1992 guidelines.

Question: When will the revised OET Bulletin No. 65 (OET Bulletin 65) be available to the public?

Answer: The Commission issued the revised OET Bulletin Number 65 on August 25, 1997. It is currently available for downloading at **http://www.fcc.gov/oet/info/documents/bulletins/#65**. Hard copies may be obtained by calling 202-418-2464.

Question: Has the FCC adopted the new ANSI/IEEE C95.1-1992 guidelines as proposed?

Answer: The FCC is primarily a regulatory agency and is not an expert on matters pertaining to health and safety. The Commission generally followed the recommendations of expert health and safety agencies such as the EPA, FDA, OSHA, NIOSH, and others, to adopt field and power density limits as recommended by the NCRP Report No. 86 and the SAR limits from the ANSI/IEEE C95.1-1992 guidelines.

Question: Is the FCC going to preempt local and state government regulations relating to radiation guidelines and aesthetics?

Answer: Congress has passed the Telecommunications Act of 1996, P.L. 104-104, 110 Stat. 56(1996). Section 704 of the Act amends the Communications Act by providing federal preemption of state and local regulation of personal wireless service facilities on the basis of RF environmental effects. The Telecommunications Act also provides for resolution of conflicts related to the regulation of RF emissions by the courts or by petition to the Commission. Accordingly, we have amended §1.1307 of our rules to incorporate the provisions of Section 704 of the Act. You may contact the Wireless Telecommunications Bureau at (202)418-0600 for further guidance with respect to the Commission's policies implementing the Telecommunications Act of 1996. In addition, the Wireless Telecomunications Bureau (WTB) released an NPRM on August 25, 1997. Please see *2nd Memorandum Opinion and Order and Notice of Proposed Rulemaking* for further information.

Question: Is it safe to use a cellular phone?

Answer: The ANSI/IEEE and NCRP RF safety guidelines recommend that low-power devices such as cellular hand-held phones not cause a localized exposure in excess of specific absorption rate (SAR) of 1.6W/kg. Studies of human head models using cellular phones have generally reported that the SAR levels are below 1.6 W/kg level as averaged over 1 gram of tissue under normal conditions of use. However, some recent studies have reported higher peak levels under "worst-case" conditions that suggest the need for further dosimetric studies.

Question: Is a person that lives in a house located near a tower, which has multiple antennas on it, in any danger from the radiation emitted from it?

Answer: All FCC licensees, even those categorically excluded or below radiated power and height criteria, are expected to be in compliance with the FCC's exposure limits. It is the responsibility of all the licensees with colocated transmitters to ensure that individual contributions of each transmitter do not cumulatively exceed the Commission's limits in an accessible area. Exposure to RF levels below these levels is considered to have no detrimental biological effect by expert standards bodies such as the Institute of Electrical and Electronics Engineers, Inc. (IEEE) or the National Council on Radiation Protection and Measurements (NCRP).

Question: Does the FCC have a database containing the technical parameters and location of each transmitter?

Answer: No, the Commission does not have transmitter-specific databases for all services it regulates. The Commission has limited information for some services such as AM, FM, and TV broadcast stations, but there is no accuracy standard with respect to location. Location is generally specified in degrees/minutes/seconds format, but this is not sufficient to distinguish between colocated transmitters. In some services, licensees are allowed to use additional transmitters or increase power without filing with the Commission. Other services are licensed by geographic area, such that the Commission has no knowledge concerning the actual number or location of transmitters within a given geographic area. For further information on the Commission's existing databases, please contact Donald Draper Campbell via e-mail or call 202-418-2405.

Question: Does the FCC routinely measure RF radiation emitted by the services it regulates?

Answer: No, the FCC does not routinely measure the RF fields associated with these antennas. However, the FCC has done limited studies in which RF fields from antennas used for AM radio, FM radio, TV, cellular telephony, paging, and Amateur Radio were measured. There is an indication that in some accessible areas, particularly roof-top locations, the exposure limits may be exceeded.

These areas are generally not accessible by the general population, but may be accessible to individuals such as HVAC technicians, roofers, or window washing crews. This information was considered as a basis for requiring additional services to routinely perform an environmental evaluation.

Question: Does the FCC regulate RF radiation exposure from TVs or Computer Monitors?

Answer: The Commission does not regulate exposure from televisions or computer monitors. The Food and Drug Administration (FDA) has primary jurisdiction of consumer devices and inquiries should be made to the FDA's Center for Devices and Radiological Health (CDRH), specifically the Office of Compliance at 301-594-4654. Questions with respect to health or biological effects should be addressed to EPA Hotline at 1-800-363-2383.

Question: If I file my application before the October 15, 1997 (January 1, 1998 for the Amateur Radio Service), am I "grandfathered" with respect to complying with the Commission's new limits?

Answer: No, the National Environmental Policy Act of 1969 does not allow "grandfathering." If you renew before October 15, 1997, there is a "cutoff" date of September 1, 2000 at which all of the FCC's licensees must be in compliance. If you are colocated with another transmitter that renews after October 15, 1997 and you contribute more than 5% of the applicable guideline to the area in question, you must also come into compliance with the new guidelines at that time. October 15, 1997/January 1, 1998 marks the end of use of the ANSI C95.1-1982 as the limits used by the Commission to process applications. After the deadlines, any new station, modification, renewal or new facility constructed under a "blanket" license, for non-excluded transmitters, will be processed using the Commission's new limits. Mobile and portable devices for specific services have been processed under the new limits since August 1, 1996. For specific information with respect to the Commission's requirements for mobile and portable devices, please contact David Means at (301) 725-1585.

Question: Is spatial or time averaging allowed with respect to Maximum Permissible Exposure limits for field strength and power density?

Answer: Yes, both the IEEE C95.3-1991 and NCRP Report No. 119 provide information on spatial and time averaging. The NCRP states, "the concept of spatial and time averaging may be appropriate from a thermal standpoint due to the dynamics of the body's thermal regulation characteristics." The premise of spatial averaging is that the human body can regulate the thermal load caused by high localized exposures as long as the total exposure does not exceed the whole body average limit. Similarly, the premise of time averaging is that the human body can regulate a specific thermal load within a given period. Thus, the human body can endure relatively high exposures for short periods of time as long as the average exposure does not exceed the exposure limit.

For more information on this topic please note:

OET Bulletin No. 56: Questions and Answers About the Biological Effects and Potential Hazards of Radiofrequency Radiation.

Any questions regarding this subject matter should be addressed to: The RF Safety Program

Resources

This appendix contains addresses and contact information for the government agencies and professional organizations that have made major contributions in the development of the FCC's RF exposure safety regulations. Also included are publishers and companies that supply related materials, software, and training.

**Compiled by Joe Bottiglieri, AA1GW
ARRL Technical Information Service
Coordinator**

AGENCIES, ORGANIZATIONS, AND PUBLISHERS

American National Standards Institute
  (ANSI)
11 West 42nd St
New York, NY 10036
212-642-4900
Fax: 212-398-0023
E-mail: sleistne@ansi.org
Web: http://web.ansi.org/

American Radio Relay League (ARRL)
225 Main St
Newington, CT 06111-1494
860-594-0200
BBS: 860-594-0306
Fax: 860-594-0259
E-mail: hq@arrl.org
Web: http://www.arrl.org/

Cellular Telecommunications Industry
  Association (CTIA)
1250 Connecticut Ave NW, Suite 200
Washington, DC 20036
202-785-0081
E-mail: wowcom@ctia.org
Web: http://www.wow-com.com/
  consumer/

CQ Communications
76 North Broadway
Hicksville, NY 11801
516-681-2922
Fax: 516-681-2926
E-mail: CQmagazine@aol.com
Web: http://members.aol.com/
  cqmagazine/

Electronic Industries Association (EIA)
2500 Wilson Blvd
Arlington, VA 22201-3834
703-907-7500
E-mail: PublicAffairs@eia.org
Web: http://www.eia.org/

Electromagnetic Energy Association
  (EEA)
(Formerly Electromagnetic Energy
  Policy Alliance, EEPA)
1255 23rd St NW
Washington, DC 20037-1174
202-452-1070
Fax: 202-833-3636
E-mail: eea@elecenergy.com
Web: http://www.elecenergy.inter.net/

Federal Aviation Administration (FAA)
800 Independence Ave
Washington, DC 20591
202-366-4000
Fax: 202-267-5039
Web: http://www.faa.gov/

Federal Communications Commission
  (FCC)
1270 Fairfield Rd
Gettysburg, PA 17325
888-225-5322
717-338-2500
Fax: 717-338-2696
Web: http://www.fcc.gov/

Federal Communications Commission
  Office of Engineering and Technology
  (OET)
1919 M St NW
Washington, DC 20554
202-418-2464
E-mail: rfsafety@fcc.gov
Web: http://www.fcc.gov/oet/rfsafety/

The Institute of Electrical and Electronics
  Engineers (IEEE) Headquarters
345 East 47th St
New York, NY 10017-2394
212-705-7900
Fax: 212-705-7589
E-mail: member.services@ieee.org
Web: http://www.ieee.org/

IEEE Operations Center
445 Hoes Ln
PO Box 1331
Piscataway, NJ 08855-1331
800-678-4333
732-981-0060
Fax: 732-981-9667
E-mail: customer.service@ieee.org
Web: http://www.ieee.org/

International Commission on Non-Ionizing Radiation Protection (ICNIRP)
Scientific Secretary: Dipl.-Ing. R. Matthes
c/o Bundesamt für Strahlenschutz
Institut für Strahlenhygiene
Ingolstädter Landstraße 1
D-85764 Oberschleißheim
Germany
49 (89) 31603 288
Fax: 49 (89) 31603 289
E-mail: rmatthes@bfs.de
Web: http://www.sz.shuttle.de/dm 1001/icnirp.htm

International Radiation Protection Association (IRPA)
PO Box 662
5600 AR Eindhoven, Netherlands
31 (40) 2473355
Fax: 31 (40) 2435020
E-mail: irpa.exof@sbd.tue.nl
Web: http://www.irpa.at/irpa/gen_info.htm

International Transcription Service, Inc (ITS)
1231 20th St NW
Washington, DC 20036
202-857-3800
Fax: 202-857-3805
Web: http://www.itsi.com/

McGraw-Hill
1333 Burr Ridge Parkway
Burr Ridge, IL 60521
630-789-4000
Web: http://www.mhhe.com/

National Association of Broadcasters (NAB)
1771 N St NW
Washington, DC 20036
202-429-5300
Fax: 202-429-5343
Web: http://www.nab.org/

National Council on Radiation Protection and Measurements (NCRP)
7910 Woodmont Ave, Suite 800
Bethesda, MD 20814-3095
301-657-2652
Fax: 301-907-8768
E-mail: ncrp@ncrp.com
Web: http://www.ncrp.com/

National Institute for Occupational Safety and Health (NIOSH)
Hubert H. Humphrey Building
200 Independence Ave SW
Room 715H
Washington, DC 20201
800-356-4674
E-mail: pubstaft@cdc.gov
Web: http://www.cdc.gov/niosh/

National Technical Information Service (NTIS)
5285 Port Royal Rd
Springfield, VA 22161
800-553-6847
703-605-6000
Fax: 703-321-8547
E-mail: info@ntis.fedworld.gov
Web: http://www.ntis.gov/

Occupational Safety and Health Administration, U.S. Department of Labor (OSHA)
200 Constitution Ave NW
Washington, DC 20210
202-219-5000
Fax: 202-219-7312
Web: http://www.osha.gov/

Prentice-Hall
Division of Simon and Schuster
1 Lake St
Upper Saddle River, NJ 07458
201-236-7000
Fax: 201-236-7696
Web: http://www.prenhall.com/

The Society of Broadcast Engineers, Inc (SBE)
8445 Keystone Crossing, Suite 140
Indianapolis, IN 46240
317-253-1640
Fax: 317-253-0418
Web: http://www.sbe.org/

Telecommunications Industry Association (TIA)
2500 Wilson Boulevard, Suite 300
Arlington, VA 22201-3834
703-907-7700
Fax: 703-907-7727
E-mail: tia@tia.eia.org
Web: http://www.tiaonline.org/

US Environmental Protection Agency (EPA)
401 M St SW
Washington, DC 20460
202-260-4111
Web: http://www.epa.gov/

U.S. Food and Drug Administration (FDA)
Center for Devices and Radiological Health (CDRH)
Division of Life Sciences, Office of Science and Technology (OST)
(HFZ-114) 9200 Corporate Blvd
Rockville, MD 20857
301-443-7118
E-mail: rdo@cdrh.fda.gov
Web: http://www.fda.gov/cdrh/

The W5YI Group
PO Box 565101
Dallas, TX 75356
800-669-9594
E-mail: w5yigroup@w5yi.org
Web: http://www.w5yi.org/

SOFTWARE SOURCES

Adobe Systems Inc.
345 Park Ave
San Jose, CA 95110-2704
408-536-6000
Fax: 408-537-6000
Web: http://www.adobe.com/

Source for Adobe *Acrobat.*

Brian Beezley, K6STI
3532 Linda Vista Dr
San Marcos, CA 92069
760-599-4962
E-mail: k6sti@n2.net

Available software includes *Terrain Analyzer, Antenna Optimizer, Yagi Optimizer, NEC/Wires* and *NEC/Yagis.* You can download a scaled down version of *Antenna Optimizer,* based on *MININEC,* from the Web at http://oak.oakland.edu:8080/pub/hamradio/arrl/bbs/programs/

EM Scientific, Inc
2533 North Carson St, Suite 2107
Carson City, NV 89706
702-888-9449
Fax: 702-883-2384
E-mail: 76111.3171@compuserve.com
Web: http://www.emsci.com/

MININEC for Windows software.
Roy Lewellan, W7EL
W7EL Software
PO Box 6658
Beaverton, OR 97007
503-646-2885
Fax: 503-671-9046
E-mail: w7el@teleport.com
Web: ftp://ftp.teleport.com/vendors/w7el/

Available software includes *ELNEC* and *EZNEC* antenna design/analysis software. *ELNEC* is based *MININEC,* but does not have near-field capability. *EZNEC* is based on *NEC2* and can be used to predict the near-field strength.

Gerald Burke
c/o Lawrence Livermore National
   Laboratory
7000 East Ave
PO Box 808, L-153
Livermore, CA 94550
510-422-8414
Fax: 510-423-3144
Email: burke@llnl.gov

Software source for *NEC4.1* ($850). This program is subject to export restrictions under the Export Administration Act of 1979, extended by Executive Order 12730. Foreign requests shall be submitted through embassy channels and HQDA (DAMI-CIT) to:

Commander, USAISC
ATTN: ASIS
Fort Huachuca, AZ 85613-5000

Further distribution by authorized recipients is not permitted.

NEC2 and documentation is available free of charge from the "NEC Home-Unofficial" at http://www.dec.tis.net/~richesop/nec/index.html

Note: *NEC* is not "user friendly" software, and is best suited for experienced antenna modelers.

RF FIELD STRENGTH MEASUREMENT EQUIPMENT

Holaday Industries, Inc
14825 Martin Dr
Eden Praire, MN 55344
612-934-4920
Fax: 612-934-3604
Email: baron006@gold.tc.umn.edu

Narda Microwave
L-3 Communications Corp
435 Moreland Rd
Hauppauge, NY 11788
516-231-1700
Fax: 516-231-1171
Web: http://www.nardamicrowave.com/

Narda also offers non-ionizing radiation survey training courses designed to train professionals in non-ionizing radiation evaluation and management techniques.

RF EXPOSURE AREA WARNING SIGNS

EMED Company, Inc
PO Box 369
Buffalo, NY 14240
800-442-3633
Fax: 800-344-2578
Email: emed@emedco.com
Web: http://www.emedco.com/

National Association of Broadcasters
   (NAB)
1771 N St NW
Washington, DC 20036
202-429-5300
Fax: 202-429-5343
Web: http://www.nab.org/

Richard Tell Associates
8309 Garnet Canyon Ln
Las Vegas, NV 89129-4897
702-645-3338
Fax: 702-645-8842
Web: http://www.radhaz.com/

This company also offers electromagnetic field evaluation consulting services.

ARRL Technical Information Service Radio-Frequency Exposure Package

Rev.: November 26, 1997

This information package was prepared as a membership service by the American Radio Relay League, Inc., Technical Information Service, 225 Main St., Newington, CT 06111 (860) 594-0200. Email: tis@arrl.org (Internet).

Reprinted from: January 1997 *QST*, "The FCC's New RF-Exposure Regulations"
October 1997 *QST*, "What's New About the FCC's New RF-Exposure Regulations?"
January 1998 *QST*, "FCC RF-Exposure Regulations-the Station Evaluation"
RF Exposure and You, available from ARRL Publications Sales Department

Thank you for requesting the following information from the ARRL Technical Information Service (TIS) or the ARRL Automated Mail Server (info@arrl.org). ARRL HQ is glad to provide this information electronically free of charge as a service to League members and affiliated clubs.

For your convenience, you may reproduce this information, electronically or on paper, and distribute it to anyone who needs it, provided that you reproduce it in its entirety and do so free of charge.

Note! This information package provides introductory information only. Additional information on this subject and related topics can be found in back issues of *QST* and the following books:

The ARRL Handbook for Radio Amateurs
RF Exposure and You

A bibliography of articles on this topic is also available from ARRL HQ. To request this listing, send a short note with an SASE to the Technical Department Secretary.

If you have any questions concerning the reproduction or distribution of this material, please contact:

The American Radio Relay League, Inc
Technical Information Service
225 Main St.
Newington, CT 06111-1494
(860) 594-0278
(email: tis@arrl.org)

By Ed Hare, KA1CV

The FCC's New RF-Exposure Regulations

E very so often, an event gets the Amateur Radio community buzzing. On August 1, 1996, the FCC announced a significant rules change: Effective January 1, 1997, most radio services must comply with new requirements regulating human exposure to RF radiated fields. The new regulations include Amateur Radio; so, almost immediately, the telephones at ARRL Headquarters started ringing with members' questions. This overview accurately presents the best available information as *QST* goes to the printer. Sources for frequent updates appear under "Stay Tuned" at the end of this article.

BACKGROUND

In 1982, the IEEE developed the C95.1-1982 Standard that described appropriate limits for human exposure to RF energy.[1] Medical researchers, engineers and industry developed this Standard. Shortly, the FCC wrote a set of regulations that required radio services to comply with the limits set in the Standard.

While the FCC was developing those early regulations, ARRL requested that the Amateur Radio Service be categorically exempt from any *specific* requirements under the regulations. We urged the FCC to rely upon the demonstrated technical competence of amateur operators and self-education as sufficient tools to ensure continued Amateur Radio safety. The FCC agreed, and we were categorically exempt from any specific requirement to perform a station evaluation under the old RF-exposure regulations.

THE ARRL RF SAFETY COMMITTEE

To address what was then an emerging issue, in 1979 the ARRL Board of Directors formed the ARRL Bioeffects Committee. The ARRL Board has since reorganized this Committee as the ARRL RF Safety Committee. The committee consists of medical and research professionals. All of the current members hold Amateur Radio licenses.

Over the years, this committee has monitored developments in the medical and Standards communities and offered RF-safety input to the ARRL Board of Directors and

[1]Notes appear on page F.4.

Many hams fear that the new RF-safety regulations spell the end of Amateur Radio. In truth, the outlook is not so bad.

Headquarters staff. Based on information in the Standards and other scientific studies, the committee wrote (and updates) an extensive set of recommendations that appears in *The ARRL Handbook* and *The ARRL Antenna Book*.[2]

NEW STANDARDS

In 1991, IEEE published a new Standard, C.95.1-1991. (See the sidebar "How the IEEE C95.1 Standard Was Developed.") This Standard decreased the maximum recommended RF exposures and extended the frequency range covered by the original Standard. This set the stage for the rule changes that currently affect Amateur Radio.

ENTER THE FCC

On April 8, 1993, the FCC released a Notice of Proposed Rulemaking (ET Docket 93-62), announcing that it intended to develop a new set of regulations for all services, based on the C95.1-1991 Standard. ARRL filed comments asking that the Amateur Radio Service exemption continue, relying on the continued technical expertise and education of amateurs. The Amateur Radio Health Group filed comments requesting that Amateur Radio be included in the new regulations, citing some instances where amateur installations could exceed the exposure levels in the Standard and noting that not all hams have read the educational material available on the topic. The FCC took no further action until the US Congress added a mandate to the Telecommunications Act of 1996 for FCC to complete its work on revisions to the RF-exposure regulations.

It surprised ARRL when the FCC shortcut the process, going from a general proposal for new regulations to completed text in one

---

fell swoop. FCC announced the new regulations in the 96-326 Report and Order, "Guidelines for Evaluating the Environmental Effects of Radio-Frequency Radiation."[3]

THE REGULATIONS

First, let's look at the regulations as they stand at press time. (Also, see the sidebar, "ARRL Petitions the FCC for Change.") The most important change is that hams must now evaluate their stations for compliance with the FCC's RF-exposure regulations. (We were previously exempt from the evaluation, not the regulations.) Some hams think that these regulations apply *only* to hams. That's not true. The regulations have always applied to a wide range of services.

Most amateur stations already meet the exposure limits described in the regulations, especially considering things like duty cycle and antenna patterns. Most hams need only understand some new regulations and perform a "routine analysis" of their station operation.

The regulations cover RF *exposure*, not RF *emission*. The regulations limit our signal strength in areas where it affects people.

MAXIMUM PERMISSIBLE EXPOSURE (MPE)

The regulations have specific MPE requirements for radiated electric fields, magnetic fields and power density. (See Table 1.) MPEs are derived from the Specific Absorption Rate (SAR) at which tissue absorbs RF energy, usually expressed in watts per kilogram (W/kg). The FCC MPEs are not based strictly on IEEE C95.1, but rather on a hybrid between that Standard and one developed by the National Council on Radiation Protection and Measurements (NCRP),[4] a body commissioned to develop recommendations for federal agencies.

From a safe SAR, the Standards and regulations set MPEs that vary with frequency. The most stringent requirements are from 30 to 300 MHz because various human-body resonances fall in that frequency range.

MPEs assume continuous-duty and operation. The regulations, however, allow us to average the total power over 6 minutes for controlled environments and 30 minutes for uncontrolled environments. This average considers both the duty factor of the operating mode and the actual on and off times over the worst-case averaging period.

EXPOSURE "ENVIRONMENTS"

The regulations define two primary RF-exposure environments: "controlled/occupational" and "uncontrolled/general public." In a "controlled" RF environment people know that RF is present and can take steps to control their exposure. These are primarily occupational environments, but the FCC includes amateurs and their immediate households (families). This applies to areas where you control access. The limits for controlled environments are evaluated differently (less stringent) than those for uncontrolled environments.

"Uncontrolled" RF environments are those open to the general-public, where persons would normally be unaware of exposure to RF energy. This applies to all property near your station where you don't control public access: sidewalks, roads, neighboring homes and properties that might have some degree of public access.

The regulations require amateurs to evaluate their stations for both controlled and uncontrolled exposure areas.

CATEGORICAL EXCLUSIONS

All Amateur Radio stations must comply

How the IEEE C95.1 Standard Was Developed

I recently attended a one-day seminar conducted by the Chairperson of IEEE Standards Coordinating Committee 28, Non-Ionizing Radiation Hazards (SCC-28). This group has developed a number of IEEE Standards that relate to exposure to electromagnetic fields from 3 kHz to 300 GHz. This seminar educated engineers about the Standard and its development.

SCC-28 now has about 120 active members. About 200 more follow the Committee's work (including ARRL). SCC-28 is about 70% researchers, with others from various organizations and industry.

SCC-28 considers a large number of input sources and research papers. It evaluates these against scientific criteria. For example, they exclude papers that do not include measured RF field levels. The result included about 120 papers.

SCC-28 considered the topics and conclusions in these papers and combined them with the substantial collective knowledge of their learned membership. Finally, they reached a consensus that a standard for exposure could be set and did so.

An SAR (see the text of this article) of 4 W/kg determines the final Standard. This is the approximate level at which several animal species demonstrate temporary difficulty in performing complex tasks. (For example, a monkey trained to push a button six times to get a banana decided, when exposed to a 4-W/kg field, that he didn't want a banana. With removal of the field, he soon decided he was hungry, after all). The Committee deems these to be thermal effects. Human volunteers exposed to such fields usually asked, "Who turned on the sun?" They felt warm.

The Committee applied a safety factor of 10, setting an SAR of 0.4 W/kg for controlled/occupational exposure and an additional safety factor of 5 (SAR = 0.08 W/kg) for uncontrolled exposure. The MPEs in the Standard and regulations account for how much energy the human body absorbs over different frequency ranges.

Some have suggested that this whole topic is unfounded—there are no adverse effects of RF energy. Several ARRL committees and other technical experts advise us that these Standards are realistic and we should heed them. I serve on two US standards bodies, and have participated in others. I know how difficult it is to find common ground in a large group. Given that 120 members of SCC-28 agreed upon this Standard, it is almost certainly based on sound scientific principles.—*Ed Hare, KA1CV*

with the MPE limits, regardless of power, operating mode or station configuration. (Even Ed Hare's 10-mW station must comply.—*Ed.*) However, the FCC presumes that certain stations are safe without an evaluation. Those are:

• Amateur stations using a transmitter power of less than 50 W PEP at the transmitter output terminal.

• Mobile or portable stations using a transmitter with push-to-talk control.

PAPERWORK

Other than a short certification on Form 610 station applications, the regulations do not normally require hams to file proof of evaluation with the FCC. The Commission

recommends, however, that each amateur keep a record of the station evaluation procedure and its results, in case questions arise.

EXAMINATIONS

The regulations add five questions on the topic of RF exposure to each Amateur Radio examination for Novice, Technician and General class licenses. The Question Pool Committee (QPC) is addressing this in the normal cycle of changes to the question pools. The Novice and Technician pools were released on December 1, 1996. (ARRL has asked the FCC to extend the deadline for the General Class question pool to its normal cycle, December 1, 1997.)

This entire matter has very much been a

Table 1

Maximum Permissible Exposure (MPE) Limits

Frequency Range (MHz)	Controlled Exposure (6-Minute Average)			Uncontrolled Exposure (30-Minute Average)		
	Electric Field Strength (V/m)	Magnetic Field Strength (A/m)	Power Density (mW/cm²)	Electric Field Strength (V/m)	Magnetic Field Strength (A/m)	Power Density (mW/cm²)
0.3-3.0	614	1.63	(100)*			
3.0-30	1842/f	4.89/f	(900/f²)*			
0.3-1.34				614	1.63	(100)*
1.34-30				824/f	2.19/f	(180/f²)*
30-300	61.4	0.163	1.0	27.5	0.073	0.2
300-1500	—	—	f/300	—	—	f/1500
1,500-100,000	—	—	5	—	—	1.0

f = frequency, in MHz.
* = Plane-wave equivalent power. (This means the equivalent far-field strength that would have the E- or H-field component calculated or measured. It does not apply well in the near field of an antenna.)
— = Not specified.

ARRL Petitions the FCC for Change

No one, including ARRL, had an opportunity to comment on the specific regulations announced by the FCC. The regulations are significantly different from what the FCC proposed in the original Notice of Proposed Rulemaking. The FCC simply did not follow the "rules to make the rules." This lack of due process forms a significant part of several Petitions for Reconsideration.

There are petitions "on the plate" from industry and the amateur community. When the regulations were first announced, ARRL filed an emergency petition for relief from an implementation error that required question pools revision well before the effective date of the regulations.

Then our Laboratory staff, RF Safety Committee and outside experts pored over the 180+ page Report and Order (see note 3). We found many errors and flaws in the requirements as written.

The 50-W threshold for categorical exclusion is arbitrary: While the MPEs vary with frequency, the 50-W level does not. We ask that the 50-W level be increased at some frequencies, consistent with the MPEs. Some other services have exclusions when the antenna location is 10 meters from areas of exposure. At HF, 150 W to any antenna would be unconditionally safe when the antenna is 10 meters from areas of exposure—with a significant safety margin. We asked the FCC to add these criteria to the 50-W criterion already in the regulations.

We did not ask for any change to the 50-W criterion at VHF and higher, because some station and antenna configurations could result in fields that exceed the MPEs.

We considered higher limits, for HF, with a greater antenna separation. A safety margin similar to that for the 150-W scenario would require a rather great distance at some frequencies. We backed off this path because it might be misinterpreted. Local officials might assume that the worst-case distance for such high-power stations should apply to all amateur stations.

Part of the ARRL's petition for reconsideration asks the FCC to preempt local regulation of RF exposure. The congressional mandate to the FCC included the requirement to develop preemption of local regulation of RF exposure resulting from the operation of radios in the Personal Communications Services (of which we're not). In order to do so, they needed the federal RF-exposure regulations. The result is that the Amateur Radio Service bears the burden of these new regulations without the benefits of preemption.

As the FCC and amateur communities wrestled with understanding the requirements and rewriting Bulletin 65, it became apparent that neither the FCC nor the amateur community could meet the January 1, 1997, implementation date. If the FCC manages to complete Bulletin 65 by the target date of December 1, 1996, that would give amateurs only four weeks to obtain it, read it, understand it, perform the needed calculations and take steps to correct any problems. For example, if a ham wants to move a tower, it could require zoning approval and other paperwork. In some areas of the country, winter would prevent completion.

At their October meeting, the ARRL Board of Directors voted to ask the FCC to extend the implementation date by one year. The ARRL then joined the growing number of organizations and individuals seeking relief from the short deadlines for these regulations. At press time, there has been no decision on any of the petitions for reconsideration before the FCC (although this may have all been decided by the time you read this).—*Ed Hare, KA1CV*

moving target, with changes forthcoming from every direction. I commend all QPC members, including the ARRL/VEC, for their diligent work to meet the tight deadlines imposed by these regulations.

ROUTINE STATION EVALUATION

The regulations require amateur operators, whose stations are not categorically excluded, to perform a routine analysis of compliance with the MPE limits. The FCC is relying on the demonstrated technical skill of Amateur Radio operators to evaluate their own stations.

The FCC regulations do not require field-strength measurements. Measurements are one way to perform an analysis, but they're very tricky. With calibrated equipment and skilled measuring techniques, ±2 dB error is pretty good. In untrained hands, errors ex-

ceeding 10 dB are likely. A ham who elects to make measurements will need calibrated equipment (including probes) and knowledge of its use. Many factors can confound measurements in the near field.

Most evaluations will be comparisons against typical charts to be developed by the FCC, relatively straightforward calculations of worst-case scenarios or computer modeling of near-field signal strength. The FCC encourages flexibility in the analysis, and will accept any technically valid approach. Once an Amateur Radio operator determines that a station complies, operation may proceed. There's no need for FCC approval before operating.

FCC OFFICE OF ENGINEERING AND TECHNOLOGY "BULLETIN 65"

To help hams perform the routine evalu-

The FCC regulations do not require field-strength measurements.

ation, the FCC is revising an existing document: *Evaluating Compliance With FCC-Specified Guidelines for Human Exposure to Radio Frequency Radiation* (also known as "OET Bulletin 65.")

At press time, Bulletin 65 is not complete. The ARRL and others have been offering specific comments to the FCC, after reviewing the first draft. There has been considerable discussion about what the document should contain. So far, all parties agree on two points: The material should be easy to use, and there should be more than the three pages devoted to Amateur Radio in the draft copy! The ARRL has gathered a group of technically astute volunteers to help staff and the RF Safety Committee select the most useful course of action. When the document is complete, another article will discuss the details of Bulletin 65.

STAY TUNED ...

This article accurately presents the best available information as *QST* goes to the printer. (Every time we got to "where it's at"—it moved.) You can get frequent updates from *The ARRL Letter*, W1AW bulletins and our RF-Safety Resource page on the ARRLWeb site (look for the RF Safety News link on **http://www.arrl.org**) as new information develops. If the FCC grants our several Petitions for Reconsideration, we will have ample time to update ARRL publications and write additional *QST* articles to give you the specific information and tools you'll need to comply with the regulations.

Notes

[1] IEEE C95.1-1982 has been superseded by IEEE C95.1-1991. Copies are available from IEEE Sales Office, 445 Hoes Ln, PO Box 1331, Piscataway, NJ 08855-1331; tel 800-678-4333; fax 908-981-9667; e-mail **customer.service@ieee.org**; Web **http://stdsbbs.ieee.org/faqs/order.html**.

[2] ARRL publications are available from your local ARRL dealer or directly from ARRL. Mail orders to Pub Sales Dept, ARRL, 225 Main St, Newington, CT 06111-1494. You can call us toll-free at 888-277-5289; fax your order to 860-594-0303; or send e-mail to **pubsales@arrl.org**. Check out the full ARRL publications line on the World Wide Web at **http://www.arrl.org/catelog**.

[3] These are available electronically on the FCC's Office of Engineering and Technology Web page. See **http://www.fcc.gov/Bureaus/Engineering_Technology/Orders/fcc96326.txt**. Contact the FCC's Int'l Transcription Service 1270 Fairfield Rd, Gettysburg, PA 17325; tel 717-337-1433 for paper copies. Note: FCC documents may refer to ANSI/IEEE C95.1-1991 as C95.1-1992.

[4] NCRP Report No. 86, "Biological Effects and Exposure Criteria for Radio Frequency Electro-magnetic Fields," ISBN 0-913392-80-4. National Council on Radiation Protection and Measurements, 7910 Woodmont Ave, Bethesda, MD 20814; tel 301-657-2652, fax 301-907-8768, e-mail **ncrp@ncrp.com**; Web **http://www.ncrp.com/**

By Ed Hare, W1RFI

What's New About the FCC's New RF-Exposure Regulations?

The FCC has answered some of our questions about the RF-exposure rules. Changes to the regulations have a few pleasant surprises for Amateur Radio.

L ife seems to happen in spurts and jumps, and the FCC RF-exposure regulations have been no exception. In August of 1996, the FCC announced a new set of RF-exposure regulations. I described these in my January 1997 *QST* article, "The FCC's New RF-Exposure Regulations". The article discussed how the original standards were developed, the history of the rules, and included an explanation about what was, and was not, required of the Amateur Radio Service. It's must reading for anyone who wants to understand these rules.

The article you're reading now builds on that foundation, describing what is new, what rules changes and information the FCC has just announced, and what we can expect in the near future.

The January *QST* article didn't address one important issue: How to go about performing the evaluations. The FCC Office of Engineering and Technology offered to release information for all radio services to use—Bulletin 65—which would offer instructions on routine station evaluation.

Again, we waited.

On August 25, 1997, the wait was partly over—the FCC announced some helpful changes to the rules and released Bulletin 65, "Evaluating Compliance with FCC Guidelines for Human Exposure to Radiofrequency Electromagnetic Fields."

Things are not yet done—the bulletin and rule changes do not offer complete solutions for hams. The FCC is still preparing an amateur supplement to this bulletin, expected to contain more information and easy-to-use charts and tables specific to the Amateur Radio Service. Most hams will use the tables in the supplement to complete their evaluations.

WHAT HAS STAYED THE SAME?

Most of these rules have not changed from what was announced in 1996:
- The MPE levels are the same.
- Hams can still conduct their own evaluations without having to file paperwork with the FCC.

- Push-to-talk-operated portable and mobile stations are still exempt from the routine evaluation requirement.
- Amateur Radio examinations for some classes of license will contain additional questions about RF safety.

EFFECTIVE DATE—A MOVING TARGET

All services were originally given a transition period until January 1, 1997, to be in compliance. A number of petitions asked for more time. In response, the FCC extended the transition period for the Amateur Radio Service until January 1, 1998.

The FCC has added an additional transition period for existing installations, establishing September 1, 2000 as a "date certain" by which existing stations must be in full compliance with the rules. Starting January 1, 1998, if a licensee must file a 610 form with the FCC, such as for renewal or change of station location, the station must be in compliance when the 610 form is signed.

POWER LEVELS

The major change from the old rules is that the power level that triggers the need to do a station evaluation has been *increased* for the

Table 1

Power Thresholds for Routine Evaluation of Amateur Radio Stations

Band (Wavelength)	Transmitter Power (W)
160 m	500
75 m	500
80 m	500
40 m	500
30 m	425
20 m	225
17 m	125
15 m	100
12 m	75
10 m	50
6 - 1.23 m	50
70 cm	70
33 cm	150
23 cm	200
13 cm and up	250

Amateur Radio Service. Under the rules announced in 1996, all amateur stations using more than 50-W PEP were required to perform a routine evaluation. The ARRL asked that the 50 W be scaled by frequency to match the MPEs in the regulations. The FCC agreed.

Stations that use more power than the levels shown in Table 1 must be evaluated; those using less power must still comply with the exposure limits, but do not need to be evaluated because they are presumed to be in compliance. For the majority of amateurs, this change has virtually eliminated the need to perform station evaluations. Most HF transceivers are rated at 100-W PEP output; on 15 meters and below, stations using this power level need not be evaluated. Most VHF transceivers are rated at 50-W PEP or less; stations using this power level on VHF need not be evaluated. Statistically, most HF operators use "barefoot" rigs, typically 100-W PEP. While this change doesn't cover all barefoot HF operation, operators who wish to use 12 and 10 meters could either perform an evaluation for those two bands, or they could even reduce power to the levels in Table 1 and forgo the evaluation altogether.

GOOD NEWS FOR MULTIS AND REPEATER OWNERS

The new rules have a few provisions that will be helpful to stations located on sites shared with other transmitters. In the 1996 rules, stations at multitransmitter sites were jointly responsible for site compliance if their field exceeded 1% of the permitted MPEs. The new rules have been relaxed. They now exempt those stations whose exposure is less than 5% of that permitted. This actually covers a lot of small stations like amateur repeaters, although a station evaluation may be required to demonstrate that the exposure is below the 5% threshold. For VHF repeaters, though, this evaluation can be fairly straightforward, using the simple far-field formulas for field strength.

BULLETIN 65 OVERVIEW

Although the regulations are firm require

ments, the FCC intends that Bulletin 65 is advisory in nature. To quote directly from the bulletin:

> The bulletin offers guidelines and suggestions for evaluating compliance. *However, it is not intended to establish mandatory procedures, and other methods and procedures may be acceptable if based on sound engineering practice.*

This flexibility applies especially to the Amateur Radio Service; the FCC is relying on the technical ability of hams to select an appropriate method of analysis for the station evaluations.

Although the FCC is preparing an Amateur Radio supplement to Bulletin 65, the "core" bulletin does contain a section on Amateur Radio. Many hams will want to read it to help get a more complete picture of what is expected of operators in all radio services.

WHAT'S IN THE BULLETIN?

Bulletin 65 was written primarily for commercial radio stations, although the information can be used by any radio service. While hams can use this existing bulletin to complete their station evaluations, they need to be careful. It is easy to get lost in the complex formulas and explanations intended to be most helpful to other radio services. Most hams will find the pending amateur supplement a lot easier to use.

FIXING PROBLEMS

Most amateur stations are already in compliance with the MPE levels. A few hams may need to make some changes to their stations. Bulletin 65 offers guidance and flexibilty on what the FCC considers acceptable. Hams can adjust their power, mode, frequency, antenna location, antenna pointing or on-and-off times to bring their operation into compliance. For example, if you discovered that you were not in compliance after 25 minutes of operation when you pointed your antenna in a particular direction, you could either not point your antenna in that direction, or take a break for 5 minutes after 25 minutes of operation.

MEASUREMENTS

It is not likely that many hams will make actual measurements. Even so, Bulletin 65 discusses measurement techniques. Many hams will be surprised at the difficulty of making near-field measurements.

FORMULAS

Bulletin 65 describes how to use far-field formulas to obtain estimates of field strengths in the near field. *NEC4* modeling done by the ARRL shows that the formula applies well to antennas like dipoles and Yagis. We found, however, that it does *not* apply well to some antenna types such as small loops, so these formulas should be used with some caution.

The formulas apply to the field-strength

Table 2

Estimated Distances to meet RF power density guidelines in the main beam of a typical 3-element Yagi for the 28-MHz Amateur Radio band. Calculations include the EPA ground-reflection factor of 2.56.

Frequency: 28 MHz   Controlled limit: 1.15 mw/cm²

Antenna gain: 8 dBi   Uncontrolled limit: 0.23 mw/cm²

Transmitter power (watts)	Distance to controlled limited (feet)	Distance to uncontrolled limit (feet)
100	11	24.5
500	24.5	54.9
1000	34.7	77.6
1500	42.5	95.1

Table 3

Compliance Distances in Feet—Uncontrolled/General Public Environment.

10 meter three-element Yagi array Antenna Height = 40 feet

| Average Transmitter power (watts) | Height Above Ground | | |
	6 ft	12 ft	20 ft
100	0	0	0
500	0	0	28
1000	0	52	60
1500	54	76	93

Resources

See the ARRL RF-safety Web page at **http://www.arrl.org/news/rfsafety/** for the latest news and links to other Web sites.

An "RF-Exposure Information Package" is available from the ARRL Technical Department Secretary for $2 for members, $4 for non-members, postpaid.

The FCC has a number of pages devoted to the RF-exposure rules. Start with **http://www.fcc.gov/oet/dockets/et93-62/** and go from there.

levels *in the main beam of the antenna.* For this reason, they may result in an overly conservative estimate for many actual installations. The ARRL has supplied the FCC with data tables illustrating real antennas over real grounds to offer realistic compliance distances. The FCC will include some of these tables in the amateur supplement to Bulletin 65, along with some simple tables based on the worst-case formulas.

You don't need to resort to complicated formulas to do the worst-case analysis. Check the University of Texas Amateur Radio Club site at **http://www.cs.utexas.edu/users/kharker/rfsafety/.** You'll find a "form" that allows you to enter transmitter power, antenna gain and distance. After you enter the information, it calculates the field strength and tells you if you are in compliance. (If you're *not* in compliance, it tells you at what distance you would be in compliance.)

These simple calculations can be a good tool because if you pass "worst-case," you pass. If you use peak-envelope power in these estimates, this is truly a worst-case; the regulations

are specified in terms of *average* exposure, averaged over 30 minutes for uncontrolled exposure environments, 6 minutes for controlled environments. You also should use the ground-reflection options that are part of the formulas or programs if you want to ensure that you have a truly worst-case estimate.

WHAT IS COMING

This first part of Bulletin 65 and the rules changes have answered some of our questions. Many hams will find that they may not have to do station evaluations at all. Others can use the formulas to calculate their worst-case compliance. Some hams, though, will find it a lot easier to wait for the amateur supplement to use the simple lookup tables.

We can't say for certain what will appear in the supplement, but based on what we have learned from the FCC, we can offer an overview.

Most hams will probably use the worst-case lookup tables. Table 2 gives an example based on a 3-element Yagi at 28-MHz. If you pass with this table, you pass on 10 meters for this antenna. (Don't forget to use power averaged over the appropriate time period, not PEP for these calculations.)

For many antenna types, the antenna radiates more energy upward toward the ionosphere than it does downward toward people. The ARRL has developed a number of tables based on *NEC4* analysis of real antennas over real ground.

Refer to Table 3 for an example of a 10-meter, 3-element Yagi at 40 feet above ground. It shows how far you must be from the antenna to meet the requirements if the exposure point is at either (a) ground level, (b) first-story height of 12 feet, or (c) second-story height of 20 feet. Note that these distances are smaller than for those of the worst-case scenario. In several cases, the table takes some pretty wild jumps, as noted between 1000 W and 1500 W at the 6-foot compliance point level. This is due to the distribution of fields under the antenna; the field strength is actually less right under the antenna than it is some distance away.

The FCC is including these tables in the amateur supplement. They also suggest that hams can use various software approaches to calculating compliance. When this supplement is released, we'll follow up with another article telling you how to use it. An ARRL book on the RF-exposure regulations will be available near the end of the year.

By Ed Hare, W1RFI

FCC RF-Exposure Regulations—the Station Evaluation

Most hams can easily meet the requirements in the rules. The good news is that many amateur stations will not have to be evaluated at all!

It's been a long road, but Amateur Radio now has a clear light at the end of the tunnel leading us toward the implementation date of the RF exposure rules. The FCC has released "Supplement B," the Amateur Radio supplement to "OET Bulletin 65."[1] This answers our questions about the "routine environmental evaluation" required by the rules. The actual requirements are not nearly as onerous as they sound!

The rules were discussed in previous 1997 articles in *QST*. Reading those articles is a "must" to understanding this one.[2, 3]

The Sky Is Not Falling!

Most hams will not have difficulty meeting the requirements. In fact, most hams are already in compliance with the maximum permissible exposure (MPE) levels. Some fear, however, that they'll have to do difficult measurements, perform extensive calculations or file paperwork with the FCC. Wrong on all counts. The evaluation is often as easy as using tables to determine that your antenna is far enough away from people.

An Overview of the Rules

The rules set limits on the RF exposure levels people may be subjected to. The MPE limits vary with frequency. The MPE levels represent the amount of energy that can be present *where and when* people are being exposed. They do *not* limit the permitted radiated strength from a radio station and do *not* change the maximum power levels permitted to Amateur Radio operators. The actual MPE limits were explained in the January 1997 *QST* article.

The rules define two exposure environments, each with different MPE levels. The *uncontrolled* environment applies to areas where people would not normally know they are being exposed. This includes "public" areas such as your property line or a neigh-

boring apartment.

Controlled environments apply where people are aware of their exposure and have the ability and knowledge to control it. Greater MPE levels are permitted in controlled areas. A good rule of thumb is that the controlled exposure limit can be applied to those areas in which you can control access. An example of this is your fenced-in backyard. Your own household can also be a con-

trolled environment if your family or guests have been given instruction about RF exposure and safety. (You could show them the information on ARRL's Web page[4] or in the "ARRL RF Exposure" package.[5])

The rules also require that some amateur stations be evaluated to *verify* that they are in compliance with the MPE levels. It's this aspect of the rules that raises eyebrows among hams.

Who Needs to Do an Evaluation?

The good news is that most amateur stations *do not* need to be evaluated. The following classes of amateur stations are exempt from the evaluation requirement because their power levels or operating duty cycles are low enough that they are presumed to be in compliance with the MPE limits:

• Stations using the power levels at or below those shown in Table 1.

• Most mobile or portable stations (handheld).[6]

• Amateur repeaters using 500 W effective radiated power (ERP) or less, if they meet certain antenna-separation requirements.

The power levels in Table 1 are expressed in *PEP input to the antenna*, except for the repeater specification, which is in terms of ERP. To determine the PEP input to the antenna you will need to include the transmitter PEP output and any feed line losses. Hams whose power levels exceed these limits must perform an evaluation.

Who Can Do the Evaluation?

The FCC is relying on amateurs to perform their own station evaluations. Other than a simple statement on Form 610, the FCC does not require any paperwork from amateurs; once the evaluation is complete, the amateur can begin operation.

What is in Bulletin 65?

Let's take a look at what is found in Bulletin 65 and Supplement B. This article can't

---

Table 1

You must perform an RF environmental evaluation if the peak-envelope-power (PEP) input *to the antenna* exceeds these limits. (Use 500 W ERP for repeater stations.)

Band	Power (W)
160 meters	500
80	500
75	500
40	500
30	425
20	225
17	125
15	100
12	75
10	50
6	50
2	50
1.25	50
70 cm	70
33	150
23	200
13	250
SHF (all bands)	250
EHF (all bands)	250

Repeaters. Non-building-mounted antennas: If the distance between ground level and the lowest point of the antenna is less than 10 meters *and* the power is greater than 500 W ERP.

Building-mounted antennas: If the power exceeds 500 W ERP.

---

[1]Notes appear on page F.12.

Table 2
Operating Duty Factors by Mode

Mode	Duty Factor	Notes
Conversational SSB	20%	Note 1
Conversational SSB	40%	Note 2
Voice FM	100%	
FSK/RTTY	100%	
AFSK	100%	
Conversational CW	40%	
Carrier	100%	Note 3

Note 1: Includes voice characteristics and syllabic duty factor. No speech processing.
Note 2: Moderate speech processing employed.
Note 3: A full carrier is commonly used for tune-up purposes.

Table 3
Typical Antenna Gains in Free Space

Antenna	Gain dBi	Gain dBd
Quarter-wave ground plane or vertical	1.0	−1.1
Half-wavelength dipole	2.15	0.0
2-element Yagi array	6.0	3.9
3-element Yagi array	7.2	5.1
5-element Yagi array	9.4	7.3
8-element Yagi array	13.2	11.1
10-element Yagi array	14.8	12.7
17-element Yagi array	16.8	14.7

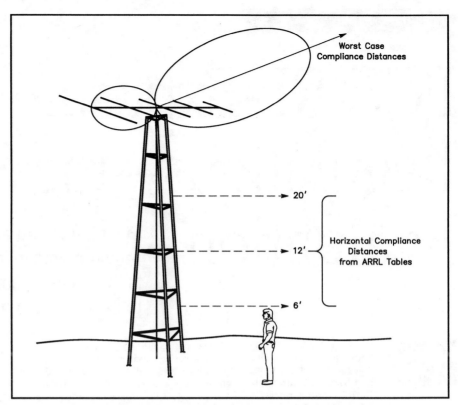

Figure 1—The power-density and field-strength formulas and tables give the compliance distance in the main beam of the antenna, at any angle, as the uppermost line shown on this drawing. If this same distance is applied to ground-level exposure, the estimate is generally conservative. The tables based on antenna modeling have calculated the horizontal compliance distances at ground level, and at first and second story exposure levels.

reprint the bulletin in its entirety, but it is available for download from the FCC. The most applicable parts will be reprinted in ARRL's upcoming book, *RF Exposure and You*. The bulletin outlines several ways that hams can evaluate their stations. However, hams may use any other technically appropriate methods. Many hams envision complicated measurements when they think about evaluating their stations. While precise, scientific measurements could be used, most hams will probably "pass" using one of the easier methods.

You can estimate compliance by using:

• Tables developed from power-density and field-strength formulas.
• Tables derived from antenna modeling.
• Antenna modeling software (*NEC*, *MININEC*, etc.)
• Power-density and field-strength formulas
• Software developed from power-density and field-strength formulas.
• Calibrated field-strength measurements.

Average Exposure

FCC rules define amateur power in PEP. (PEP is the average power of a single RF cycle at the peak of a modulation envelope.) The MPE limits, however, are based on *average exposure*, not peak exposure. This means that the total exposure for the averaging period must be below the limits. One way of factoring in average exposure could be to determine the average transmitter power.

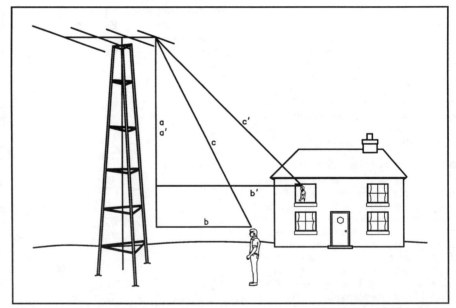

Figure 2— In calculating the actual worst-case horizontal compliance distances between the antenna and areas being evaluated, you must consider the antenna height, the height of the exposure and the horizontal distance between the antenna and the exposure point. This drawing illustrates exposures at ground and second-story levels. (Use the a′ and b′ for the second-story exposure.) From there, you can use the formula:

$$c = \sqrt{a^2 + b^2}$$

This distance can be used with the tables derived from the power-density formula. The ARRL tables (Tables 6 and 7) use distance "b."

To calculate average power, multiply PEP by the duty factor for the mode being used. The duty factors for various modes are shown in Table 2. Multiply that result by the percentage of time the transmitter could be in use during the appropriate averaging period—6 minutes for controlled exposure, 30 minutes for uncontrolled. A few examples are shown elsewhere in this article under the sidebar, "Step by Step."

Tables Developed from Formulas

Most amateurs will use the tables in Supplement B to estimate their compliance with the MPE levels. These tables show the distances people must be from any part of the antenna to avoid being exposed at levels exceeding the MPE limit. They have been calculated with a ground-reflection factor, which includes the ground gain of an antenna over typical ground. This allows hams to use manufacturer's antenna gain figures in dBi with confidence that the result represents a conservative real-world estimate. This model, although simplified, has been verified by the ARRL Laboratory staff using *NEC* antenna-modeling software with a number of antennas modeled over ground. These tables do not necessarily apply to all antenna types, though. *NEC* models of small HF loops, for example, give fields near the antenna that are much higher than the power-density formula predicts. A more accurate method was used to develop the small-loop table in the supplement.

The tables derived from formulas do have advantages: they generally offer conservative estimates and they are easy to use. If a ham "passes" using these conservative tables, the evaluation is complete. Doing a station evaluation can be just that simple!

Supplement B contains a number of these tables. Select the ones that best apply to your station, calculate your average power, then determine if your antenna is far enough away for each band and mode that you use. (One shortcut is to use the highest PEP you use on each band.) Figure 1 shows how this worst-case estimate applies to the main beam of the antenna. Figure 2 shows how to calculate the actual distance to the points you are evaluating.

Table 4

Estimated distances from transmitting antennas necessary to meet FCC power-density limits for Maximum Permissible Exposure (MPE) for either occupational/controlled exposures ("Con") or general-population/uncontrolled exposures ("Unc"). The estimates are based on typical amateur antennas and assuming a 100% duty cycle and typical ground reflection. (The figures shown in this table generally represent worst-case values, primarily in the main beam of the antenna.) The compliance distances apply to average exposure and average power, but can be used with PEP for a conservative estimate.

Frequency (MHz)	Gain (dBi)	100 W Con	100 W Unc	500 W Con	500 W Unc	1,000 W Con	1,000 W Unc	1,500 W Con	1,500 W Unc
2	0	0.5	0.7	1.0	1.6	1.5	2.2	1.8	2.7
	3	0.7	1.0	1.5	2.2	2.1	3.1	2.6	3.8
4	0	0.6	1.4	1.4	3.1	2.0	4.4	2.4	5.4
	3	0.9	2.0	2.0	4.4	2.8	6.2	3.4	7.6
7.3	0	1.1	2.5	2.5	5.7	3.6	8.1	4.4	9.9
	3	1.6	3.6	3.6	8.0	5.1	11.4	6.2	13.9
	6	2.3	5.1	5.1	11.4	7.2	16.1	8.8	19.7
10.15	0	1.6	3.5	3.5	7.9	5.0	11.2	6.1	13.7
	3	2.2	5.0	5.0	11.2	7.1	15.8	8.7	19.4
	6	3.2	7.1	7.1	15.8	10.0	22.4	12.2	27.4
14.35	0	2.2	5.0	5.0	11.2	7.1	15.8	8.7	19.4
	3	3.2	7.1	7.1	15.8	10.0	22.4	12.3	27.4
	6	4.5	10.0	10.0	22.3	14.1	31.6	17.3	38.7
	9	6.3	14.1	14.1	31.6	20.0	44.6	24.4	54.7
18.168	0	2.8	6.3	6.3	14.2	9.0	20.1	11.0	24.6
	3	4.0	9.0	9.0	20.0	12.7	28.3	15.5	34.7
	6	5.7	12.7	12.7	28.3	17.9	40.0	21.9	49.0
	9	8.0	17.9	17.9	40.0	25.3	56.5	31.0	69.2
21.45	0	3.3	7.5	7.5	16.7	10.6	23.7	13.0	29.0
	3	4.7	10.6	10.6	23.6	15.0	33.4	18.3	41.0
	6	6.7	14.9	14.9	33.4	21.1	47.2	25.9	57.9
	9	9.4	21.1	21.1	47.2	29.8	66.7	36.5	81.7
24.99	0	3.9	8.7	8.7	19.5	12.3	27.6	15.1	33.8
	3	5.5	12.3	12.3	27.5	17.4	39.0	21.3	47.7
	6	7.8	17.4	17.4	38.9	24.6	55.0	30.1	67.4
	9	11.0	24.6	24.6	55.0	34.8	77.7	42.6	95.2
29.7	0	4.6	10.4	10.4	23.2	14.7	32.8	18.0	40.1
	3	6.5	14.6	14.6	32.7	20.7	46.3	25.4	56.7
	6	9.2	20.7	20.7	46.2	29.3	65.4	35.8	80.1
	9	13.1	29.2	29.2	65.3	41.3	92.4	50.6	113.2

Frequency (MHz)	Gain (dBi)	50 W Con	50 W Unc	100 W Con	100 W Unc	500 W Con	500 W Unc	1,000 W Con	1,000 W Unc
50, 144, 222	0	3.3	7.4	4.7	10.5	10.5	23.4	14.8	33.1
	3	4.7	10.5	6.6	14.8	14.8	33.1	20.9	46.8
	6	6.6	14.8	9.3	20.9	20.9	46.7	29.5	66.1
	9	9.3	20.9	13.2	29.5	29.5	66.0	41.7	93.3
	12	13.2	29.5	18.6	41.7	41.7	93.2	59.0	131.8
	15	18.6	41.6	26.3	58.9	58.9	131.7	83.3	186.2
	20	33.1	74.0	46.8	104.7	104.7	234.1	148.1	331.1
420	0	2.8	6.3	4.0	8.8	8.8	19.8	12.5	28.0
	3	4.0	8.8	5.6	12.5	12.5	28.0	17.7	39.5
	6	5.6	12.5	7.9	17.7	17.7	39.5	25.0	55.8
	9	7.9	17.6	11.2	24.9	24.9	55.8	35.3	78.9
	12	11.1	24.9	15.8	35.2	35.2	78.8	49.8	111.4
	15	15.7	35.2	22.3	49.8	49.8	111.3	70.4	157.4
1240	0	1.6	3.6	2.3	5.2	5.2	11.5	7.3	16.3
	3	2.3	5.1	3.3	7.3	7.3	16.3	10.3	23.0
	6	3.2	7.3	4.6	10.3	10.3	23.0	14.5	32.5
	9	4.6	10.3	6.5	14.5	14.5	32.5	20.5	45.9
	12	6.5	14.5	9.2	20.5	20.5	45.8	29.0	64.8
	15	9.2	20.5	13.0	29.0	29.0	64.8	41.0	91.6

Repeaters

The power levels shown in Table 1 are in peak-envelope power (PEP) input *to the antenna*. FCC rules specify amateur power in PEP and most transmitters are rated in PEP. However, you must consider feed line losses to determine power to the antenna.

There is a little wrinkle in the rules when it comes to repeaters. The evaluation exemption for amateur repeater operation is determined by the *effective radiated power* (ERP) of the repeater. ERP is referenced to the gain of a half-wave dipole in free space (unlike equivalent isotropically radiated power, EIRP, which is referenced to an isotropic source). Supplement B describes how to calculate feed line losses and determine ERP for an amateur repeater.

All amateur repeaters using 500 W ERP or less generally do not need to be evaluated. Those that operate with more than 500 W ERP need to be evaluated if they have an antenna mounted on a building, or if any part of a nonbuilding-mounted antenna is less than 10 meters (32.8 feet) above ground. (This is another example that higher antennas generally create less field strength on the ground than lower antennas!)

Tables Based on Antenna Gain

Table 4 was prepared for the FCC Bulletin by the W5YI Group, working in cooperation with the ARRL. It shows the distances required to meet the power-density limits for different amateur bands, power and antenna gain, for occupational/controlled exposures ("con"), or for general population/uncontrolled exposures ("unc"). (All FCC tables give the all distances in meters; the tables in this article have been converted to feet.)

This table probably represents the *easiest* approach to doing a station evaluation. It can be conservatively applied to most antenna types. Select the appropriate band and "round up" antenna gain and power to match the table. The distances are the minimum separation that must be maintained between any part of the antenna and any area where people will be exposed. If the station passes, the evaluation is done. If not, a more precise evaluation can be done by using average power and exposure and the tables developed by ARRL, and discussed later in this article.

Tables for Specific Antenna Types

Supplement B also contains tables for specific antenna types. Table 5 is an example of those supplied by Wayne Overbeck, N6NB. It shows the estimated distances to meet RF power density guidelines for a typical 3-element "triband" Yagi HF antenna. These tables are also based on the power-density equations.

All of these simple tables are easy to use. They almost always give conservative estimates of compliance. They estimate the required distance one needs to be from the antenna *in the main beam of the antenna* (see Figure 1). In many cases, however, exposure *below* an antenna can be much less than that indicated by the far-field tables. If a station "passes" using the simple tables, this could be a moot point. Even so, some hams may find it useful to use other methods to demonstrate that the exposure from their station is much less than what the rules allow.

Like many tables, the ones shown in this article and Supplement B paint with a broad brush. They provide conservative answers to generalized conditions. If you need more precise evaluation methods, those are certainly available to you as well.

Tables Based on Antenna Modeling

A number of antenna modeling programs (see the sidebar, "Software") will give much more accurate estimates of field strength in the near field of an antenna. However, many hams do not have experience using them.

The ARRL Laboratory staff came up with a solution, but it involved considerable work on their part. To provide tables for specific antennas modeled at various heights over ground, they selected the *NEC4* software package. Using *NEC4* they modeled a number of antennas, heights and power levels and calculated the compliance distances at ground level, first story and second story exposure points. (My personal 75-MHz Pentium PC had to chew on some of these calculations for as long as four hours!)

The results were distilled into Table 6, showing the triband Yagi from Table 5, modeled 30 feet over average ground and operating on 10 meters. Table 7 shows the results with the antenna at 60 feet. Figure 1 shows how these tables relate to the areas being evaluated.

Tables 6 and 7 provide a more accurate estimate of actual exposure than the derived-from-formula tables. However, the antenna

---

Step by Step

Let's look at a hypothetical example of an amateur station and run through the evaluation steps. Assume that Al, N9AT, has the following station configuration:

• 80 meters, 100 W and 1000 W CW and SSB with a half wavelength dipole antenna 10 feet above ground. (I know this would be a terrible height for an 80-meter dipole, but bear with me!)

• 40 meters 100 W and 1000 W CW and SSB with a half wavelength dipole antenna 10 feet above ground. (Ditto my height comments above!)

• 10 meters, 100 W and 1500 W CW and SSB with a 3-element Yagi beam 30 feet above ground.

• 2 meters, 35 W FM, 100 W CW and SSB 60 feet above ground.

Al first looks at Table 1 to see which operation requires a station evaluation. In this case, his 100-W 80- and 40-meter operation and his 35-W 2-meter FM operation do not need to be evaluated. (Al intends to evaluate them anyway, just to learn more about the subject.)

He could calculate his average power for the remaining operation, but this may not be necessary. Al first tries his evaluation with PEP, using Table 3 in conjunction with Table 4. Rounding up to 3 dBi for the antenna gain, Table 4 estimates that on 80 meters at 1000 W his antenna needs to be located 2.8 feet from areas of controlled exposure and 6.2 feet from uncontrolled exposure. The antenna is located about 10 feet from the property line and is attached to the house with 5-feet of rope, so this band would be in compliance for operation at a 1000-W continuous carrier level.

On 40 meters at 1000 W, Al first rounds his dipole gain up to 3 dBi. Table 4 shows 5.1 feet for controlled exposure and 11.4 feet for uncontrolled exposure. On this band the end of his antenna is located 5 feet from the property line and tied to the house with a 4-foot rope. It doesn't quite pass with full power. Al has a few choices. He can relocate the antenna, reduce power, or calculate his average power and try again or use the antenna-specific table at the same height. In this case, he calculates his average power and determines that he is using 133 W average power on SSB and 266 W average power on CW. Rounding up, he selects 500 W in Table 4 and determines that his antenna needs to be 3.6 feet from controlled exposure and 8.0 feet from uncontrolled exposure. He passes for controlled exposure, but the antenna would be located 6.4 feet from a person standing on the property line, so the station may still not be in compliance. Al decides to move the antenna 10 feet from the property line sometime next week. In the meantime, he will reduce his power on 40 meters.

On 10 meters, he is using a 3-element Yagi 30 feet in the air. Rounding his gain up to 9 dBi, he determines that his antenna needs to be 50.6 feet from controlled exposure and 113.2 feet from uncontrolled exposure. The tower is located 40 feet from the house, and solving for the hypotenuse of the distance between his residence and the tower (his one-floor house has the top of the first floor 12 feet above ground), he calculates that the antenna is located 43.9 feet from areas of controlled exposure. This does not pass for full power, but passes easily when he calculates his average power. The tower is 50 feet from the property line, for a total distance of 55.5 feet from ground level exposure *on the property line*. This does not pass for uncontrolled exposure. Al doesn't give up, though, he goes to Table 6 and determines that at ground level, the *NEC* model shows that the compliance distance needs to be 57.1 feet from the center of the antenna at 1500 W average power. He clearly cannot do 30 minutes of tuneup if his neighbor is on the property line. At 500 W average power, however, Al notes that his antenna could be built on the property line and ground-level exposure would be below the limits. He has passed and does not need to make any changes to his station except to limit his tuneup time.

On 2 meters, his antenna has 8 dBi of gain. Rounding up to 9 dBi, he determines that at 100 W his antenna needs to be 13.2 feet from controlled exposure and 29.5 feet from uncontrolled. This antenna is at the top of his 45 foot tower, so he can run continuous power on 2 meters. Al gathers all of the papers containing these calculations (along with his notes) and files them with his station records. Within 20 minutes he has completed his station evaluation!

Table 5

Estimated distances (in feet) to meet RF power density guidelines in the main beam of a typical three-element "triband" (20-15-10 meter) Yagi antenna assuming surface (ground) reflection. Distances are shown for controlled (con) and uncontrolled (unc) environments.

	14 MHz, 6.5 dBi		21 MHz, 7 dBi		28 MHz, 8 dBi	
	con	unc	con	unc	con	unc
100	4.7	10.4	7.4	16.5	11.0	24.6
500	10.4	23.1	16.5	36.8	24.6	54.9
1000	14.7	32.7	23.3	51.9	34.8	77.7
1500	17.9	40.1	28.5	63.6	42.6	95.1

Table 6

Estimated distances (in feet) to meet RF power density guidelines with a typical three-element "triband" (20-15-10 meter) Yagi antenna operating on a frequency of 29.7 MHz at a height of 30 feet above ground. Distances are shown for controlled (con) and uncontrolled (unc) environments.

Height above ground where exposure occurs (feet)

Average Power (W)	6.0		12.0		20.0		30.0	
	con	unc	con	unc	con	unc	con	unc
50	0	0	0	0	0	0	8.6	13.2
100	0	0	0	0	0	0	9.9	18.1
150	0	0	0	0	0	15.5	11.2	21.7
250	0	0	0	0	0	24.6	13.2	26.9
500	0	0	0	47.0	0	49.2	18.1	36.1
750	0	35.2	0	59.1	15.1	70.9	21.4	45.0
1000	0	46.0	0	68.0	21.0	83.0	24.0	60.1
1500	0	57.1	0	79.1	27.9	100.1	28.9	103.0

Table 7

Estimated distances (in feet) to meet RF power density guidelines with a typical three-element "triband" (20-15-10 meter) Yagi antenna operating on a frequency of 29.7 MHz at a height of *60* feet above ground. Distances are shown for controlled (con) and uncontrolled (unc) environments.

Note: The 68.9-foot distance shown in the 12.0-foot column is not a typo! *NEC4* modeling indicates that at this distance the E field just exceeds the MPE at 1500 W continuous power.

Height above ground where exposure occurs (feet)

Average Power (W)	6.0		12.0		20.0		60.0	
	con	unc	con	unc	con	unc	con	unc
50	0	0	0	0	0	0	8.6	13.2
100	0	0	0	0	0	0	9.9	18.1
150	0	0	0	0	0	0	11.2	22.0
250	0	0	0	0	0	0	13.2	27.9
500	0	0	0	0	0	0	18.1	42.0
750	0	0	0	0	0	0	22.0	51.9
1000	0	0	0	0	0	0	25.0	57.8
1500	0	0	0	68.9*	0	0	30.9	66.0

*NOTE: This is not a typographical error.

and its height must match the table to be applicable. (If the antenna is located higher than the heights in these tables, in general, the exposure should be less than the predicted values.) The ARRL offered a number of these tables to the FCC for inclusion in Supplement B. The remainder will be made available in the ARRL's new book, *RF Exposure and You.* Supplement B features a number of these antennas at heights of both 30 feet *and* 60 feet, helping us demonstrate that "higher is better!"

Antenna Modeling

The software used to create our tables can model virtually any antenna system. Hams sometimes use some exotic antennas and it is not practical to create a table for each one. Some hams may want to evaluate the effect of multiple antennas or other conductors in proximity to their antennas to have a more accurate answer than can be derived from any other calculation method.

Many of these modeling programs are based on *NEC* or *MININEC* versions. They can give good results *if the model is accurate,* which usually means modeling all nearby conductors. These programs can calculate the near-field electric and magnetic-field strengths. See the "Software" sidebar.

Measurements

The subject of making field-strength measurements has been treated in Bulletin 65. In general, however, most hams will not have the accurate calibrated equipment to do actual field-strength measurements. The relative field-strength meters in common use by hams do not have the necessary accuracy, especially when dealing with fields of different frequencies. Even in the most qualified hands, repeatability of more than a few dB is difficult to achieve.

Multi-Transmitter Environments

Some amateur stations use multiple transmitters, such as an HF DX or contest station that also accesses a VHF PacketCluster. Other stations might be located at sites also occupied by transmitters in other radio services. Two or more transmitters could be operating at the same time, each adding to the exposure level. In these cases, the operators must take steps to ensure that the *total* exposure does not exceed the MPE level.

In most cases, multi-transmitter amateur station operators are responsible for the total exposure. In cases where an amateur station location is shared with other transmitters, all operators may be jointly responsible for the site compliance.

It is relatively easy to calculate total exposure at multitransmitter sites. For any point being evaluated, determine what percentage of the permitted MPE will actually be contributed by each transmitter. Then, add up the percentages for any transmitters that could be in operation simultaneously. If the total percentage exceeds 100%, the site is not in compliance. For example, if a 2-meter transmitter creates exposure at 40% of what is permitted on that frequency, and a simultaneous transmission is occurring by a 1.5-GHz commercial transmitter at the same at 70% of the limit, the total is 110% and the *site* is out of compliance, even though each transmitter is being operated below its own limit.

The FCC has determined that any transmitter that operates at an exposure level greater than 5% of the power density *permitted to its own operation* is jointly responsible *with all the other operators within its exposure area* who are also exceeding 5% for site compliance. In those areas where the exposure from the transmitter is less than 5% of the MPE level for the repeater, the operator is not jointly responsible.

In addition, however, those stations that are not required to be evaluated generally are *presumed* not to be responsible for site compliance. The FCC can require any operator to conduct an evaluation if they believe that there could be a problem.

The antenna tables elsewhere in this article cannot be used to determine actual power-density levels. The field-strength formulas in this article and in Bulletin 65 or various antenna-modeling programs can be used instead.

In some cases, amateurs may not be able to obtain full information about the other transmitters on the site. If you find yourself in this situation, you should attempt to secure information from the site owner. If that isn't available, make the best estimates possible of other transmitter powers and antenna gains on the site to determine compliance.

Mobile Stations

As described in the FCC rules, there is no specific requirement that mobile and portable (hand-held) devices used under Part 97 (Amateur Radio) be evaluated. Bulletin 65 explained that this applies to amateur mobile operation using push-to-talk operation. This is because of the low power, low operating duty cycles usually employed and the expected shielding of the vehicle occupants by the vehicle body. Most Amateur Radio mo-

Software

The calculations used to create the far-field tables have been written into BASIC by Wayne Overbeck,N6NB, and made available for download from the Web at **ftp://members.aol.com/cqvhf/97issues/rfsafety.bas**. This software has also been written into a Web-page calculator by Ken Harker, KM5FA. It can be accessed at **http://www.utexas.edu/students/utarc**.

Brian Beezley, K6STI, has made a scaled-down version of his *Antenna Optimizer* software available. Download *NF.ZIP* at from the Web at **http://oak.oakland.edu:8080/pub/hamradio/arrl/bbs/programs/**. These programs are based on *MININEC* and will generally give the same results as you can obtain from using Tables 6 or 7. Contact Brian Beezley, K6STI, 3532 Linda Vista Drive, San Marcos, CA 92069; tel 760-599-4962, e-mail **k6sti@n2.net**.

Roy Lewellan, W7EL, offers *ELNEC* and *EZNEC* antenna modeling software. *ELNEC* is based on *MININEC*, but does not have near-field capability. *EZNEC* is based on *NEC2* and can be used to predict the near field strength. This software is available from W7EL Software, PO Box 6658, Beaverton, OR 97007; tel 503-646-2885; fax 503-671-9046; e-mail **w7el@teleport.com**; **ftp://ftp.teleport.com/vendors/w7el/**.

NEC2 and documentation is available from the "NEC Home—Unofficial" at **http://www.dec.tis.net/~richesop/nec/index.html**. Beware, however, that *NEC* is *not* a user-friendly program. These are used best in the hands of experienced antenna modelers.

bile or portable stations that meet these general criteria do not need to be evaluated.

If You Don't Need to do an Evaluation...

If the regulations do not specifically require you to perform an evaluation, there could be a number of reasons to do one anyway. If nothing else, doing an evaluation now would be good practice for the day when you upgrade your station (by adding an amplifier or antenna, for instance) in such a way that makes an evaluation necessary. More importantly, the results of your evaluation will certainly demonstrate to yourself, and possibly your neighbors, that your station is operating well within FCC guidelines and is no cause for concern. Finally, if you have an antenna that is located close to people, you may be operating in excess of the MPEs. It's a good idea to evaluate and be on the safe side, just in case.

Correcting Problems

The vast majority of stations will pass their evaluations handily. But some stations

whose antennas are close to areas of exposure may not meet the MPE limits. The FCC gives amateurs considerable flexibility in correcting problems. You can choose another frequency band where the MPE limits are higher, or another operating mode that results in a lower average power. You could also adjust your operating power to produce less exposure, or reduce the percentage of time the station is on the air during the averaging period. Those with directional antennas can simply ensure that they are not pointing toward areas that would be out of compliance *if people are present in the areas.*

The FCC Worksheet

As this article was going to press, the FCC sent a draft of an evaluation worksheet included in Supplement B. This optional worksheet has instructions on how to include the various factors necessary to do a station evaluation and provides a handy way to maintain a record of the evaluation. It runs step by step through the procedures outlined in this article, using the methods outlined in Supplement B. The worksheet describes the methods to calculate power to the antenna using feed line losses, and how to calculate ERP using both feed line losses and antenna gain. This is another example of how the FCC has made the evaluation process as easy as possible for the Amateur Radio Service.

Conclusion

The biggest impact of the new rules is that many hams will have to learn something new. Learning new things is what attracted most hams to the avocation anyway, so this is not a major problem. The FCC has worked closely with the amateur community to fine tune both the rules and the FCC's own materials. (See "Exam Info" on page 93 for information concerning changes to Form 610.) All that remains is for us to put all of this information to good use in our own stations!

Notes

[1] The FCC Office of Engineering and Technology Web page has all the FCC information on RF exposure. See **http://www.fcc.gov/oet/rfsafety/**. Paper copies of OET Bulletin 65 are available from the National Technical Information Service, Springfield, VA 22161; tel 800-553-6847; e-mail **orders@ntis.fedworld.gov**; **http://www.ntis.gov**

[2] Ed Hare, KA1CV, "The FCC's New RF-Exposure Regulations," *QST*, Jan 1997, p 47.

[3] Ed Hare, W1RFI, "What's New About the FCC's New RF-Exposure Regulations," *QST* Oct 1997, p 51.

[4] A copy of all applicable *QST* articles, and links to all FCC material and RF-exposure rules can be found on the ARRL Web at **http://www.arrl.org/news/rfsafety/**.

[5] The ARRL "RF Exposure Package," is available for $2 for ARRL members, $4 for non-members, from the ARRL Technical Department Secretary, 225 Main St, Newington, CT 06111, tel 860-594-0278, e-mail **tis@arrl.org**.

[6] The FCC defines "portable" devices as those generally operated with the antenna located within 20 centimeters of the body. This is different from taking a base station "portable" to a remote location.

Index

FEEDBACK

Please use this form to give us your comments on this book and what you'd like to see in future editions, or e-mail us at **pubsfdbk@arrl.org** (publications feedback). If you use e-mail, please include your name, call, e-mail address and the book title, edition and printing in the body of your message. Also indicate whether or not you are an ARRL member.

Where did you purchase this book?
□ From ARRL directly   □ From an ARRL dealer

Is there a dealer who carries ARRL publications within:
□ 5 miles   □ 15 miles   □ 30 miles   of your location?   □ Not sure.

License class:
□ Novice   □ Technician   □ Technician Plus   □ General   □ Advanced   □ Amateur Extra

Name _____

ARRL member? □ Yes □ No

_____ Call Sign _____

Daytime Phone (   ) _____ Age _____

Address _____ E-mail address _____

City, State/Province, ZIP/Postal Code _____

If licensed, how long? _____

Other hobbies_____

_____

Occupation _____

From _____

_____

_____

EDITOR, RF EXPOSURE AND YOU
AMERICAN RADIO RELAY LEAGUE
225 MAIN STREET
NEWINGTON CT 06111-1494

— — — — — — — — — — — — — — — — — — — please fold and tape — — — — — — — — — — — — — — — — — —